Industrial and Laboratory Pyrolyses

Industrial and Laboratory Pyrolyses

Lyle F. Albright, EDITOR
Purdue University

Billy L. Crynes, EDITOR
Oklahoma State University

Symposiums sponsored by the Division of Industrial and Engineering Chemistry at the First Chemical Congress of the North American Continent, Mexico City, Mexico, Dec. 1–2, 1975, and by the Division of Petroleum Chemistry at the 169th Meeting of the American Chemical Society, Philadelphia, Pa., April 7–8, 1975.

ACS SYMPOSIUM SERIES 32

AMERICAN CHEMICAL SOCIETY
WASHINGTON, D. C. 1976

Library of Congress CIP Data

Industrial and laboratory pyrolyses

(ACS symposium series; 32, ISSN 0097-6156)

Includes bibliographical references and index.

1. Pyrolysis—Congresses.

I. Albright, Lyle F., 1921- II. Crynes, Billy L., 1938- . III. American Chemical Society. Division of Industrial and Engineering Chemistry. IV. Chemical Congress of the North American Continent, 1st, Mexico City, 1975. V. American Chemical Society. Division of Petroleum Chemistry. VI. American Chemical Society. VII. Series: American Chemical Society. ACS symposium series; 32.

TP156.P915 660.2'842 76-28733
ISBN 0-8412-0337-7 ACSMC8 32 1-510 (1976)

PRINTED IN THE UNITED STATES OF AMERICA

ACS Symposium Series

Robert F. Gould, *Editor*

FOREWORD

The ACS SYMPOSIUM SERIES was founded in 1974 to provide a medium for publishing symposia quickly in book form. The format of the SERIES parallels that of its predecessor, ADVANCES IN CHEMISTRY SERIES, except that in order to save time the papers are not typeset but are reproduced as they are submitted by the authors in camera-ready form. As a further means of saving time, the papers are not edited or reviewed except by the symposium chairman, who becomes editor of the book. Papers published in the ACS SYMPOSIUM SERIES are original contributions not published elsewhere in whole or major part and include reports of research as well as reviews since symposia may embrace both types of presentation.

CONTENTS

PREFACE

The 28 papers in this book discuss and present new information on all important phases of pyrolyses and thermal cracking operations varying from theory and chemistry to commercial plants. All except one of these papers were presented at two symposiums during 1975. All 18 papers presented at the symposium "Thermal Cracking Operations" held at Mexico City, December 1 and 2, 1975 as part of the First Chemical Congress of the North American Continent are included in this book. Numerous speakers and attendees were from other continents, including Europe and Asia. This symposium, chaired by the editors of this book, was sponsored by the Division of Industrial and Engineering Chemistry of the American Chemical Society. Nine papers in this book were presented at the symposium "New Approaches to Fuels and Chemicals by Pyrolysis," held in Philadelphia, April 7 and 8, 1975. This symposium was chaired by H. G. Davis of Union Carbide Corp. and J. C. Hill of Monsanto Co., and was under the sponsorship of the Division of Petroleum Chemistry of the American Chemical Society. In addition, a paper that was submitted too late to be presented at the Mexico City symposium is also included. Each of these papers has been carefully reviewed by the editors and sometimes by others in preparation for presentation and publication in this book.

Several developments within the last few years have emphasized the need for obtaining a better understanding of the chemistry of pyrolysis and of the thermal cracking processes themselves. In the past, light paraffins (and especially ethane and propane) were the major feedstocks used for production of ethylene and, to a lesser extent, propylene. Limited availability of these light paraffins has necessitated consideration of and increased use of heavier petroleum-based stocks including naphthas, gas oils, and even crude oil. With increased use of heavy feedstocks, more propylene, dienes, and aromatics are being produced by pyrolysis. Yet, the understanding of the chemistry of pyrolysis especially of heavier hydrocarbons is relatively limited. Furthermore, increased attention is being given to the conservation of energy during pyrolysis. This is especially important as plant sizes reach the scale of more than one billion pounds of ethylene per year. Consideration is also being given to the use of non-petroleum feedstocks including coal and oil shales; and pyrolysis reactions likely will be of importance when these latter feedstocks are used for production of olefins and dienes.

The editors of this book concluded that a symposium to review the present state of knowledge would be quite timely. To do this, they felt that chemists, chemical engineers, and other scientific personnel should be brought together from both the industrial and academic establishments. The Mexico City symposium resulted with an obvious interaction between workers of various disciplines and employment.

We also included in this book several papers from the earlier Philadelphia symposium. Most of the results of one paper (Chapter 16) were presented at the Philadelphia symposium, but parts were given at the Mexico City meeting.

The first group of papers, Chapters 1–12, emphasizes primarily the chemistry and mechanism of pyrolyses of various light hydrocarbons—methane through C_6's. Clarification of the key reactions and the mechanism of gas-phase free-radical steps have been reported. One can conclude that significant progress is being made in developing mechanistic models for pyrolysis of many hydrocarbons.

The second broad grouping of papers (Chapters 13–17) describes useful design modifications for commercial plants. Surface reactions resulting in both the formation of coke and carbon oxides, and the destruction of olefins and other desired products are described. In quartz or Vycor glass reactors, however, such surface reactions are relatively unimportant and often have not been considered. Yet such reactions can be most significant in metal reactors when steam is used as a diluent of the entering feedstock. Data obtained in metal reactors are of value to the designer of plant equipment.

The third grouping of papers (Chapters 18–24) describes both pilot plants and commercial units for the pyrolyses of various hydrocarbons. Methods of collecting reliable data, correlating the data, and eventually designing commercial units are all discussed.

The fourth group of papers (Chapters 25–28) describes pyrolyses of various compounds that are present in coal-derived liquids, oil shale, waste products, and miscellaneous organic materials. This particular type of pyrolysis will certainly increase in importance in the future, and the results of this section indicate that a good foundation is being established.

In conclusion, we would like to record our thanks to all who made possible the symposiums and this subsequent publication. First, our appreciation goes to the Divisions of Industrial and Engineering Chemistry and of Petroleum Chemistry for sponsorship, and equally important, to the authors, without whom it could never have existed, and finally, to those who have worked behind the scenes, particularly our two secretaries, Mrs. Doris Lidester (West Lafayette) and Mrs. Wanda Dexter (Stillwater) who have been burdened with considerably more than their normal duties as a result.

LYLE F. ALBRIGHT
West Lafayette, Ind.

B. L. CRYNES
Stillwater, Okla.

1

Mechanism of the Thermal Decomposition of Methane

C.-J. CHEN and M. H. BACK

Chemistry Department, University of Ottawa, Ottawa, Canada K1N 6N5

R. A. BACK

Chemitsry Division, National Research Council, Ottawa, Canada

Methane is not used as a feedstock in thermal cracking operations because the decomposition temperature is too high and yields of useful products too low. Nevertheless the elementary reactions occurring in the decomposition of methane, particularly the secondary reactions involving ethane, ethylene, acetylene and coke formation, will be important in many commercial thermal cracking systems. Although methane is the simplest hydrocarbon the mechanism of its decomposition still poses some unresolved questions (1-4). Because the C-H bonds of methane are considerably stronger than the C-C bonds of the products, secondary and tertiary reactions become important at very early stages of the reaction and have usually obscured the details of the initial process. In static and flow systems, at temperatures between 800 and 1100°K, induction periods and autocatalysis have been observed and the kinetic parameters are not well defined (5-8). The role of carbon as a catalyst and its mechanism of formation are not well understood (9,10). In shock-tube experiments the pyrolysis has been effectively limited to the very early stages and it has been possible to measure the initial rate of dissociation (11-13). Experiments with mixtures of CH_4 and CD_4 suggest that under these conditions dissociation occurs primarily into CH_3 + H, (14) although the alternative dissociation into CH_2 + H_2 has also been postulated (15,16). The pressure-dependence of the initial dissociation reaction has not been established.

The present work was undertaken to resolve some of these uncertainties, particularly those of the low-temperature pyrolysis, by a careful study of the static pyrolysis in its very early stages. It will be seen that the time course of the reaction can be resolved into three stages, all within a total conversion of methane of less than 2%. The first stage may be described as the primary reaction; ethane and hydrogen are the only products in this region. In the second stage the rate of formation of ethane gradually falls off towards a plateau value while ethylene is observed as a secondary product. In the third stage the tertiary products acetylene and propylene appear and at about the same time a sharp increase in the rate of formation of ethane and other products is observed. In the first two stages it will be shown that a very simple homogeneous mechanism can satisfactorily describe the results. The third stage which apparently involves autocatalysis, is a more complex process. In most previous work the autocatalysis has been attributed to carbon formation, either as nuclei in the gas phase or on the surface of the vessel (6,17, 18). In the present paper the possibility that homogeneous reactions of acetylene and propylene may explain the autocatalysis will be discussed.

Experimental

Materials: Research grade methane, from Matheson, Canada Ltd., was treated with Ridox reagent (Fisher Scientific), which reduced the oxygen impurity to less than one ppm. The ethane impurity was reduced to less than 0.1 ppm by a gas chromatographic separation. Methane-d_4 was obtained from Merck, Sharp and Dohme, Canada, and also treated with Ridox reagent. Other gases used for calibration were Research grade from Matheson.

Apparatus: The reaction vessel was a quartz cylinder, 478 cc, used as a static reactor. Nitrogen gas was passed continuously through a jacketed space surrounding the vessel, to prevent diffusion of oxygen through the walls. The reactor was enclosed in an Autoclave High Temperature furnace which maintained the temperature to within ± 0.75°. The temperature gradient along the length of the vessel was less than one degree. After each experiment oxygen was admitted to the reactor to burn off any accumulation of carbon.

Analysis: The products hydrogen, ethane, ethylene, acetylene, and propylene were analyzed by gas chromatography. The non-condensable gases were separated on a 3 m molecular seive column at room temperature. The bulk of the unreacted methane was separated from the higher molecular weight products with a short silica gel column held at -78°C. The remainder of the methane and the products were collected and analyzed on a 1.5 m silica gel column maintained at 30°C. Hydrogen, HD and D_2 were analyzed on a 2 m column of alumina at -196°C. Deuterated methanes were analyzed by mass spectrometer.

Results and Discussion

Figure 1 shows a typical yield-time plot for the main products of the decomposition. Hydrogen and ethane are seen to be the only primary products; ethylene is clearly secondary, while acetylene and propylene are tertiary products appearing still later in the course of the reaction. The three stages of the decomposition are evident; the initial production of ethane and hydrogen; the fall off in the rate of ethane formation, approaching a plateau, with ethylene yield rising sharply, and finally the

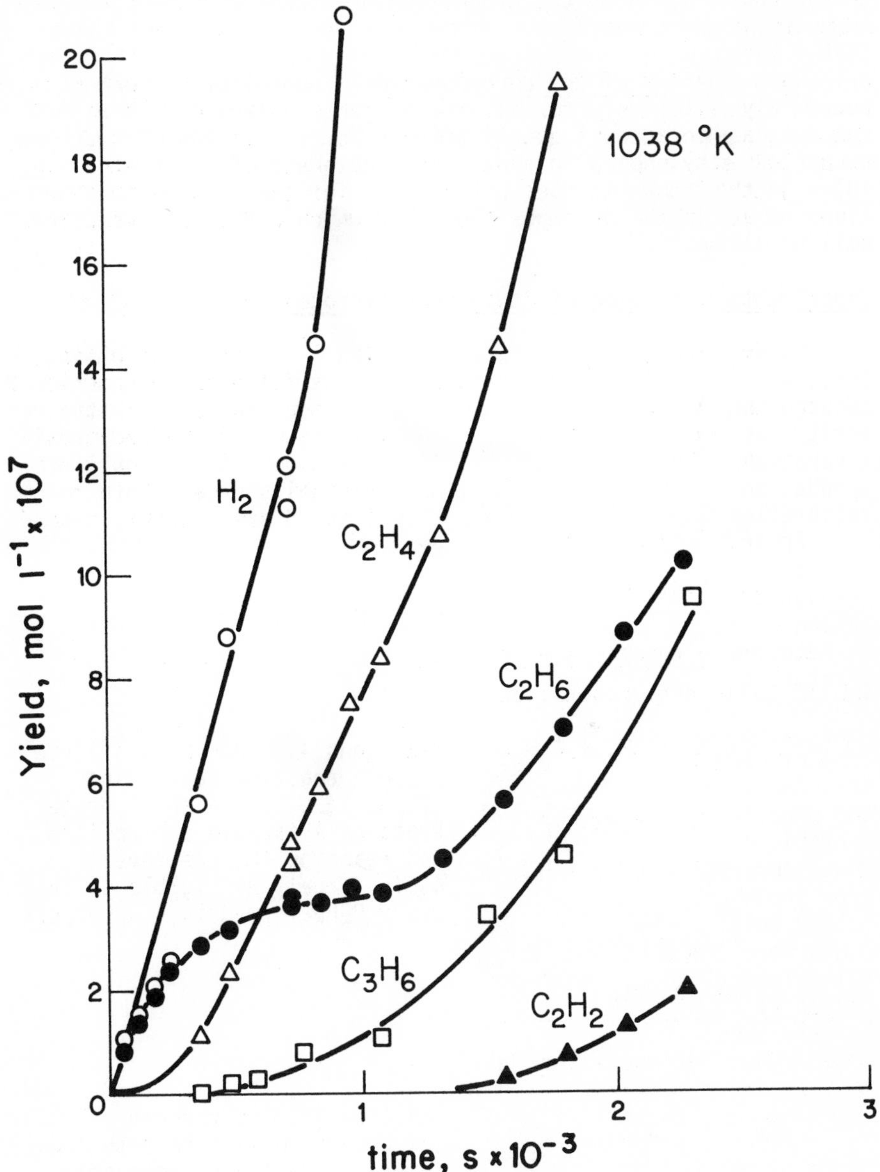

Figure 1. Yield of the principal products as a function of time

marked acceleration in the production of ethane and the other products at longer times.

The mechanism of the decomposition The following mechanism accounts qualitatively for all the products observed. It will also be seen to account quantitatively for the yields of hydrogen, ethane and ethylene up to about the attainment of the plateau in ethane production. It does not account for the subsequent sharp acceleration in the decomposition, for which additional reactions must be invoked.

Primary formation of ethane and hydrogen

(1) $CH_4 \rightarrow CH_3 + H$

(2) $H + CH_4 \rightarrow CH_3 + H_2$

(3) $2\ CH_3 \rightarrow C_2H_6$

$2\ CH_4 \rightarrow C_2H_6 + H_2$

(net reaction)

Reaction 1 is rate controlling, and always followed by reaction 2 which is much faster. In the initial stages of the decomposition reaction 1 is the only primary radical source. Both reaction 1 and 3 are pressure dependent.

Secondary reactions of ethane

Radical chain dehydrogenation

(4) $CH_3 + C_2H_6 \rightarrow CH_4 + C_2H_5$

(5) $C_2H_5 \rightarrow C_2H_4 + H$

(2) $H + CH_4 \rightarrow CH_3 + H_2$

$C_2H_6 \rightarrow C_2H_4 + H_2$

(net reaction)

Reactions 4, 5 and 2 constitute a chain reaction converting ethane to ethylene and hydrogen. Reaction 4 is rate controlling and reaction 4R, the reverse of 4, is largely negligible so that every C_2H_5 formed in 4 decomposes.

Unimolecular decomposition

(3R) $C_2H_6 \rightarrow 2\ CH_3$

Reaction 3R, the reverse of 3, can become an important secondary source of radicals under some conditions.

Secondary reactions of ethylene

Radical chain dehydrogenation

(6) $CH_3 + C_2H_4 \rightarrow CH_4 + C_2H_3$

(7) $C_2H_3 \rightarrow C_2H_2 + H$

(2) $H + CH_4 \rightarrow CH_3 + H_2$

$C_2H_4 \rightarrow C_2H_2 + H_2$

(net reaction)

Reaction sequences 6, 7 and 2, and 8, 9 and 2 constitute two parallel chain decompositions of ethylene, propagated by methyl radicals, and initiated by abstraction and addition respectively. Reaction 6 is rate controlling in the first sequence, and its reverse, 6R, is probably negligible. In the addition sequence on the other hand, reaction 8R is probably much faster than reaction 9, making the latter rate controlling.

Radical chain methylation

(8) $CH_3 + C_2H_4 \rightleftarrows n\text{-}C_3H_7$

(9) $n\text{-}C_3H_7 \rightarrow C_3H_6 + H$

(2) $H + CH_4 \rightarrow CH_3 + H_2$

$C_2H_4 + CH_4 \rightarrow C_3H_6 + H_2$

(net reaction)

Secondary reactions of acetylene

Radical chain dehydrogenation

(10) $CH_3 + C_2H_2 \rightleftarrows CH_4 + C_2H$

(11) $C_2H \rightarrow C_2 + H$

This dehydrogenation sequence does not occur because reaction 11 is much too endothermic ($\sim$ 139 kcal/mole) to be of any importance at the present temperatures; this cannot therefore be a source of carbon in the methane decomposition.

Radical chain methylation

(12) $CH_3 + C_2H_2 \rightarrow CH_3CHCH$

(13) $CH_3CHCH \rightarrow CH_3CCH + H$

(2) $H + CH_4 \rightarrow CH_3 + H_2$

$C_2H_2 + CH_4 \rightarrow CH_3CCH + H_2$

(net reaction)

As with ethylene, the reverse of the addition reaction, 12R, is probably much faster than reaction 13, so that the latter is rate controlling.

Secondary reactions of propylene

Radical chain dehydrogenation

(14) $CH_3 + C_3H_6 \rightarrow CH_4 + C_3H_5$

(15) $C_3H_5 \rightarrow CH_2CCH_2 + H$

(2) $H + CH_4 \rightarrow CH_3 + H_2$

$C_3H_6 \rightarrow CH_2CCH_2 + H_2$

(net reaction)

Both addition and abstraction sequences can occur with propylene. Abstraction appears to predominate, as yields of allene were much higher than those of butene.

Radical chain methylation

(16) $CH_3 + C_3H_6 \rightarrow CH_3CH_2CHCH_3$

(17) $CH_3CH_2CHCH_3 \rightarrow CH_3CH_2CH{=}CH_2 + H$

(2) $H + CH_4 \rightarrow CH_3 + H_2$

$C_3H_6 + CH_4 \rightarrow CH_3CH_2CH{=}CH_2 + H_2$

(net reaction)

Unimolecular decomposition

(18) $C_3H_6 \rightarrow H + C_3H_5$ or $CH_3 + C_2H_3$

This can become a significant source of radicals when the propylene concentration becomes high enough.

The mechanism outlined above is in essence very simple, and depends on the following premises:

(1) The CH_3 radical is the only radical present in significant concentrations.

(2) Hydrocarbon products disappear rapidly solely through reaction with methyl radicals in chain sequences with no net consumption of radicals. A minor exception is dissociation reactions 3R and 18.

(3) Other hydrocarbon radicals dissociate so rapidly that they take part in no other reactions. An exception is C_2H which instead is maintained at a negligibly low concentration by back reaction with methane (10R).

(4) Hydrogen atoms always react by reaction 2 to form hydrogen.

These premises are made tenable by the large excess of methane always present, so that even at the highest conversions attained products were always less than 3% of the methane, and by the relatively high temperature, which renders the higher radicals very shortlived with respect to decomposition.

The several chain reaction sequences propagated by methyl radicals follow a common pattern, with initiation by abstraction or addition, and propagation by radical decomposition, but there are notable differences. Thus with ethane, only abstraction is possible, but with ethylene both abstraction and addition sequences occur; with acetylene the abstraction sequence is blocked by the stability of the C_2H radical so that only addition is important, while with propylene, abstraction of the allylic hydrogen is apparently much faster than addition, and the former predominates. Finally, it should be noted that initially there is no radical-chain decomposition of methane; further, that the chain sequences initiated by abstraction do not consume methane, and only with the addition sequences, beginning with ethylene, is methane consumed; even then it is a limited chain, limited by the amount of olefin present.

The rate of the initial dissociation of methane Measurements of the initial rate of dissociation have been described in detail elsewhere (19) and will be only briefly summarized here. It is clear from the proposed mechanism that the rate of reaction 1 may be equated to the initial rate of formation of ethane or hydrogen, or as reactions 4 and 5 become important, to the sum of the rates of formation of ethane and ethylene. These several measures of reaction 1 gave essentially the same rates; the most accurate and reliable was found to be the initial rate of formation of ethane extrapolated to zero time by computer fitting, and values of k_1 obtained in this way are shown in fig. 2 as a function of methane pressure at four temperatures. From isotopic experiments with CH_4-CD_4 mixtures, and from thermochemical arguments, it was concluded that the initial dissociation was in fact reaction 1 and not the alternative dissociation to $CH_2 + H_2$. It was also concluded that no chain decomposition of methane was occurring so that initial product yields were indeed simply a measure of the primary dissociation. Finally, experiments in the packed reaction vessel showed virtually no surface effects in the initial stages of the decomposition. It was therefore concluded that the values of k_1 shown in fig. 2 correspond to the homogeneous unimolecular dissociation of methane according to reaction 1, and the variation of k_1 with methane pressure was attributed to the pressure dependence of this unimolecular dissociation. The points in fig. 2 are experimental data; the curves are calculated from RRKM theory (20) and show a good fit over the range of pressure and temperature of the experiments. Limiting Arrhenius parameters and other details of the RRKM calculations are summarized in table I.

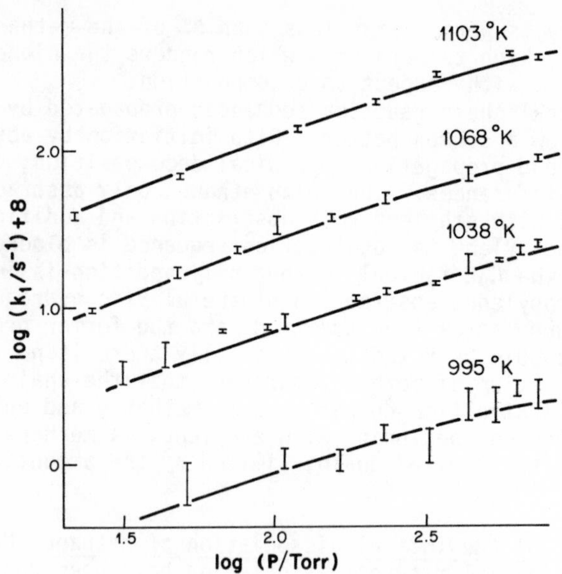

Figure 2. Variation of k_1 *with pressure. Solid line: calculated; I: experimental.*

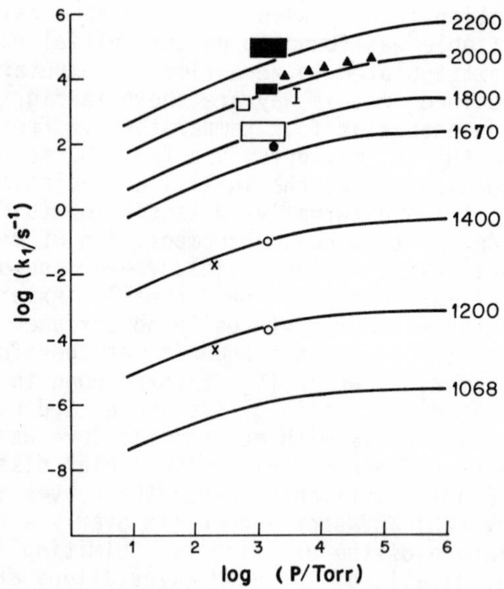

Figure 3. Variation of k_1 *with pressure. Solid line, calculated;* ○, (11); *x*, (21); ●, (15); □, (16); *I*, (12); ■, (33); ▲, (13).

TABLE I

Properties of the Activated Complex

The methyl group in the complex was assumed to be planar with the vibrational frequencies of the methyl radical. The dissociating C-H bond was extended at right angles to the plane of the methyl radical with a length of 3.00 Å. The doubly degenerate bending frequency of the dissociating bond was reduced to 280 cm^{-1}.

Moments of Inertia: $I_a = 5.84 \times 10^{-40}$ gm cm^2

$I_b = I_c = 17.0 \times 10^{-40}$ gm cm^2

Collision diameter: 3.8 Å

E_o = 103.0 Kcal/mole

Collision efficiency, $\lambda = 1$

Statistical factor, $L^{\ddagger} = 4$

Limiting Arrhenius Parameters $A^{\infty} = 2.8 \times 10^{16}$ S^{-1}

E^{∞} = 107.6 Kcal/mole

Finally, in fig. 3 the present results are compared with other experimental measurements of k_1; again, the points are the measured values, while the curves are calculated from RRKM theory using the parameters in table I obtained in the present work.

It is evident from fig. 3 that calculated values of k_1 based on the present measurements are in good agreement with the shock-tube data except at the highest temperature. Shock-tube experiments are expected to measure the initial rate of dissociation without complications due to secondary reactions, even though conversions are usually much higher than in the present work; this is because the higher temperatures employed lead to much higher CH_3 concentrations, thus favoring radical combination (reaction 3) over reactions of CH_3 with the products, and effectively limiting the decomposition to its initial stage. Data from one flow system (21) in which secondary reactions were minimized by reaction with a carbon probe are also included in fig. 3 and are in good agreement with our data. The present study shows clearly that the dissociation of methane remains considerably pressure-dependent even at the highest pressures used in the shock-tube experiments. Only those of Hartig, Troe and Wagner (13) reported such a dependence; in most other studies it was assumed that the reaction was in its high-pressure limit; this is clearly not so.

Data from the many other static pyrolyses in the literature

are not included in fig. 3, as it appears that in every case conversions were much too high and the reaction was well into its acceleration stage. Many of these studies depended on measurements of pressure change, and it can be seen from the mechanism that no change is expected until ethylene begins to appear; it seems doubtful whether any useful kinetic information about the primary reaction can be extracted from pressure measurements.

Secondary reactions of ethane Initially, reactions 1, 2 and 3 describe the course of the decomposition, but as ethane accumulates it begins to disappear in the dehydrogenation sequence, reactions 4, 5 and 2, and should reach a steady-state concentration if no other processes intervene. At the steady state, $R_3 = \alpha\ k_4\ [CH_3][C_2H_6]_{ss}$, where R_3 is the rate of formation of ethane in reaction 3, and $\alpha = k_5/(k_5 + k_{4R}[CH_4])$, the fraction of ethyl radicals formed in reaction 4 which decompose via reaction 5; if reaction 4R is negligible, $\alpha = 1$. The concentration of methyl radicals will be given by $[CH_3] = ((R_3 + k_{3R}[C_2H_6]_{ss})/k_3)^{\frac{1}{2}}$. Note that while reaction 3R, the reverse of reaction 3, must be taken into account in calculating the radical concentration and in fact becomes the dominant source of radicals, it does not enter directly into the steady-state equation for ethane because it does not lead to a net consumption of methyl radicals, since these ultimately can only recombine to reform ethane in the present system.

Experimentally, ethane is observed to reach a more or less well defined steady-state plateau before the reaction begins to accelerate (fig. 1), and from such plateau values, $\alpha\ k_4$ has been calculated from the steady-state equation. R_3 was obtained from the initial rate of formation of ethane. Values of k_{3R} were derived from RRKM calculations based on model I described by Lin and Laidler (22) for the ethane decomposition. Values of k_3 were then calculated from k_{3R} via the equilibrium constant, which was in turn calculated using values of $\Delta S_{1000} = 40.73$ cal/mole degree and $\Delta H_{1000} = 91.05$ Kcal/mole (23). Values of $\alpha\ k_4$ and the data from which they were derived are shown in table II. No variation of $\alpha\ k_4$ with methane pressure was observed, from which it was concluded that reaction 4R was negligible under these experimental conditions, i.e., $\alpha = 1$. Average values of k_4 are shown in an Arrhenius plot in fig. 4 together with other estimates of k_4 taken from the literature. In recent years there has been growing evidence that Arrhenius plots for some simple hydrogen abstraction reactions show marked upward curvature in the region 700-1500 K so that rate constants measured at these temperatures are considerably higher than predicted from the simple Arrhenius equation measured at lower temperatures. This behaviour has been discussed by Clark and Dove (24) who have shown that the observed increase in activation energy for reactions such as 2 and 4 may be accounted for within the framework of activated complex theory. The present results provide confirmation of a "high" value of k_4 around 1000°K first reported by Pacey and Purnell (25) from a

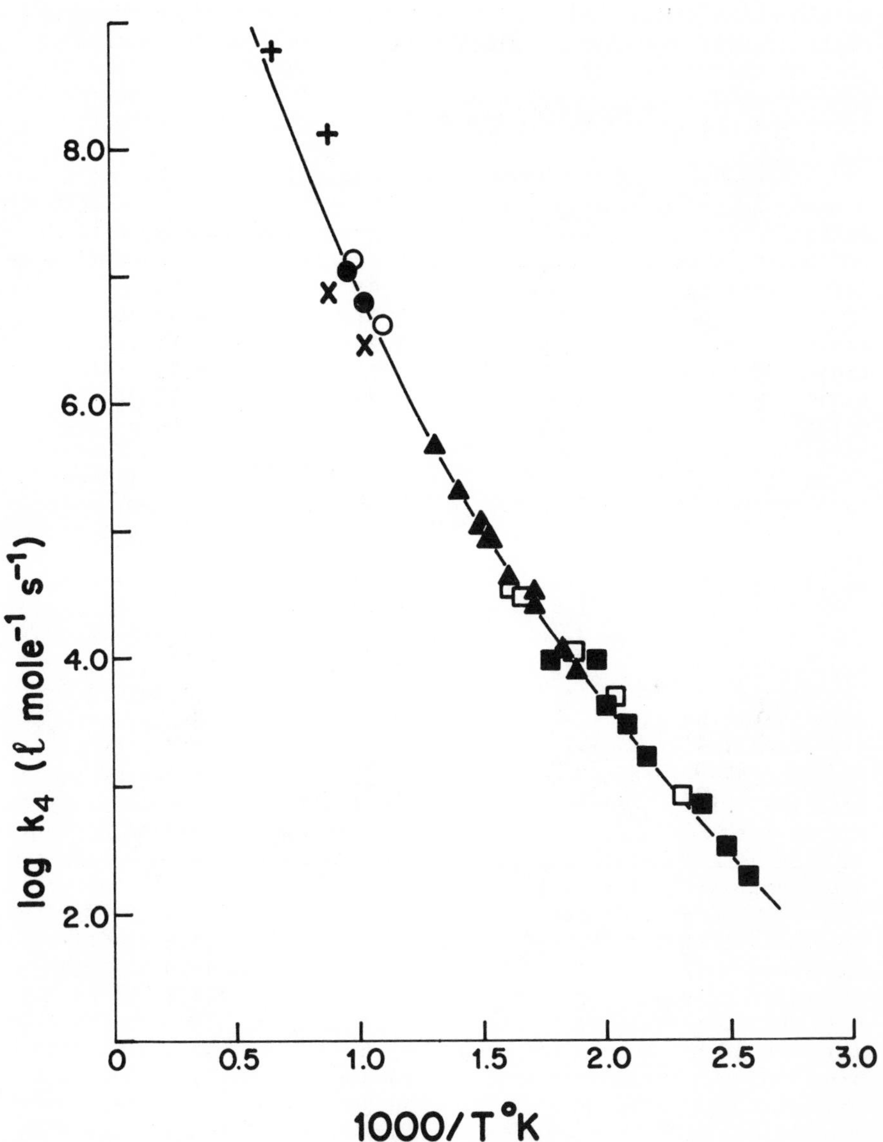

Figure 4. Arrhennius plot of k. +, (26); ▲, (27); □, (28); ■, (29); ○, (25); *x*, (30); ●, *this work.*

TABLE II

Values of k_4

$T^oK = 995 \quad K_3 = 2.78 \times 10^{13} \frac{\ell}{mole}$

PCH_4 Torr	$[C_2H_6]_{ss}$ $\frac{mole}{\ell}\times10^8$	R_3 $\frac{mole}{\ell sec}\times10^{10}$	k_{3R} $sec^{-1}\times10^3$	k_4 $\frac{\ell}{mole\ sec.}\times 10^{-6}$
51	2.09	.044	0.84	6.85
108	3.98	.152	1.01	8.62
440	23.03	1.33	1.32	5.31
540	24.08	1.79	1.36	6.46
640	33.50	2.29	1.39	5.13
741	28.26	2.83	1.42	7.60
			average	6±1

$T^oK = 1038 \quad K_3 = 4.31 \times 10^{12} \frac{\ell}{mole}$

PCH_4 Torr	$[C_2H_6]_{ss}$ $\frac{mole}{\ell}\times10^8$	R_3 $\frac{mole}{\ell sec}\times10^{10}$	k_{3R} $sec^{-1}\times10^3$	k_4 $\frac{\ell}{mole\ sec.}\times 10^{-6}$
31.9	1.675	.166	4.64	14.5
108	5.49	1.31	6.66	18.2
188	10.99	3.15	7.60	15.1
236	15.08	4.50	7.98	13.6
338	21.99	7.79	8.57	13.2
441	37.69	11.6	9.00	9.90
543	31.41	15.8	9.32	15.0
642	46.06	20.2	9.57	10.1
741	50.25	24.8	9.77	10.1
			average	13±3

$T^oK = 1068 \quad K_3 = 1.28 \times 10^{12} \frac{\ell}{mole}$

PCH_4 Torr	$[C_2H_6]_{ss}$ $\frac{mole}{\ell}\times10^8$	R_3 $\frac{mole}{\ell sec}\times10^{10}$	k_{3R} $sec^{-1}\times10^3$	k_4 $\frac{\ell}{mole\ sec.}\times 10^{-6}$
25.2	2.09	0.422	12.7	14.6
48.6	4.18	1.324	15.9	15.9
75.5	7.12	2.79	18.2	15.1
104	8.38	4.74	19.9	19.5
154	15.3	8.99	22.0	15.1
233	20.9	17.2	24.3	17.6
339	36.9	30.6	26.2	13.5
440	44.0	45.3	27.6	15.1
600	67.0	71.6	29.1	13.7
742	73.3	97.4	30.1	14.6
			average	15±2

study of the pyrolysis of butane, and support the suggestion that the activation energy is temperature dependent.

Secondary reactions of ethylene, acetylene and propylene As ethylene accumulates, it will begin to disappear in the dehydrogenation and methylation sequences reactions (6, 7 and 2, and 8, 9 and 2) yielding acetylene and propylene respectively. The rate of formation of propylene was typically several times that of acetylene, and because the reverse of reaction 8 is probably 5 to 10 times faster than reaction 9 at 1000 K (3), the addition reaction, 8, must be 20 to 50 times faster than the abstraction reaction, 6 to account for the observed product ratio.

Acetylene and propylene in turn will disappear through the chain reactions of methyl radicals as outlined in the mechanism. It is noteworthy that the dehydrogenation sequence probably does not occur with acetylene because the C_2H radical is too stable to decompose, making this an unlikely route to carbon formation. Our present analysis does not distinguish clearly between methyl acetylene and allene, and thus between reactions 12 and 14, but formation of a C_3H_4 product is observed as acetylene and propylene begin to accumulate. Little or no butene is observed, suggesting that abstraction from propylene, leading to dehydrogenation, is more important than addition. The unimolecular dissociation of propylene, reaction 18, is probably also a significant loss process (3), and possibly an important secondary source of radicals at higher conversions.

Autocatalysis, carbon formation and surface effects The third or autocatalytic stage in the methane pyrolysis, in which the yield of ethane begins to rise sharply again after the steady-state plateau (fig. 1) is not predicted or explained by the reaction mechanism postulated above. The autocatalysis is most evident in the yield of ethane, but almost certainly affects the other products as well, although it is less obvious because their yields are already rising sharply. Autocatalysis has frequently been reported in the decomposition of methane, and under various conditions of pressure, temperature, conversion or surface, may have a variety of causes. It is most commonly associated with the formation of carbon, and attributed to reactions occuring at a carbon surface.

The effect of carbon deposits in the present system was investigated in several ways. Ordinarily, oxygen was admitted to the reaction vessel after each experiment to oxidize any carbon that might be present, and the vessel was thoroughly pumped out; this procedure yielded highly reproducible results over long periods of time. In some experiments a carbon deposit was first laid down in the vessel by pyrolyzing methane for a few hours; subsequent pyrolyses in the presence of this carbon surface showed faster than normal and somewhat erratic rates of reaction. In further series of experiments begun with the usual clean surface, methane and volatile products were removed at intervals, but any carbon deposit was retained; a new sample of methane was then admitted and the reaction continued. If the accumulation of a non-volatile carbon deposit were responsible for the autocatalysis, this should not be affected by periodic removal of volatile products, and autocatalysis should set in at the usual total time of reaction. This was not observed; experiments with products removed at intervals showed no acceleration in rate even though carried on far past the usual time for autocatalysis. Even experiments in which each interval was well into the autocatalytic stage showed no increase in rate over many intervals. It can only be concluded that carbon deposits play no part in autocatalysis in the present system.

Surface effects were also investigated by experiments in a reaction vessel filled with quartz tubes which increased the surface/volume ratio by a factor of 10. Initial rates of ethane formation were quite unaffected, and the early stages of the decomposition were undoubtedly homogeneous. Formation of secondary products and of ethane, in the autocatalytic stage showed some surface enhancement, suggesting a small surface component in the secondary reactions, probably less than 10% of the whole. This relative lack of sensitivity to surface/volume ratio again indicates that surface deposits of carbon are not very important in the reaction mechanism.

Two plausible alternatives may be suggested to account for the observed autocatalysis. The first of these is that carbon is indeed responsible, but a fine smoke-like suspension of carbon particles rather than a surface deposit (6). These would have an enormous surface area and therefore be much more effective than a surface deposit of carbon; the particles would also be swept out of the reaction vessel with the methane and volatile products, so that no cumulative effects would be observed in the "interval" experiments. Experiments are in progress to test this hypothesis by direct observation of carbon particles by light scattering.

The second alternative is that homogeneous reactions can cause the autocatalysis. A new source of radicals must become important in the autocatalysis region, and the rate of this new initiation process must rise very sharply with time. Simple computer simulation of the reaction system shows which type of reactions might have the right properties. Disproportionation reactions between olefins and methane such as $C_2H_4 + CH_4 \rightarrow C_2H_5 + CH_3$ are kinetically wrong, as they cannot cause a sharp enough rise in the reaction rate, as well as being too slow. Bimolecular reactions of olefins or acetylene such as $2\ C_2H_2 \rightarrow C_2H + C_2H_3$ are more promising in as much as they yield autocatalysis curves of the proper shape, but again the reaction rates are probably too slow. A bimolecular reaction of acetylene yielding a diradical which can initiate radical chains would also have the right kinetic properties. This type of reaction would have a lower activation energy than the bimolecular reaction giving two radicals (31) and could be fast enough to give the observed yields of products. Diradical formation has been postulated in the pyrolysis of acetylene itself (32) and seems to offer the most plausible mechanism for homogeneous autocatalysis in the present system, but these suggestions must be regarded as speculative.

Acknowledgment

The authors thank the National Research Council of Canada for financial support of this work.

Literature Cited

1. Steacie, E.W.R., Atomic and Free Radical Reactions, Reinhold (1954).
2. Khan, M.S. and Crynes, B.L., Ind. and Eng. Chem. 62, 54 (1970).
3. Benson, S.W. and O'Neal, H.E., Kinetic Data on Gas Phase Unimolecular Reactions, NSRDS - NBS 21 (1970).
4. Laidler, K.J. and Loucks, L.F., The Decomposition and Isomerization of Hydrocarbons. Comprehensive Chemical Kinetics, Vol. 5, Ed. C.H. Bamford and C.F.H. Tipper, Elsevier Publishing Co. New York (1972).
5. Schneider, I.A. and Murgulescu, I.G., Z. Phys. Chem. (Leipzig) 237, 81 (1968); 223, 231 (1963); 218, 338 (1961).
6. Palmer, H.B., Lahaye, J. and Hou, K.C., J. Phys. Chem. 72, 348 (1968).
7. Eisenberg, B. and Bliss, H., Chem. Eng. Prog. Sym. Series 63, 3 (1967).
8. Tardieu de Maleissye, J. et Delbourgo, R., Bull. Soc. Chim. France, 60 (1970); 1108 (1969).
9. Lieberman, M.L. and Noles, G.T., Carbon 12, 689 (1974).
10. Odochian, L. and Schneider, I.A., Rev. Roum. de Chimie 15, 1647 (1970); 16, 1467 (1971); 17, 441 (1972).
11. Skinner, G.B. and Ruehrwein, R.A., J. Phys. Chem. 63, 1736 (1959).
12. Napier, D.H. and Subrahmanyam, N., J. Appl. Chem. Biotechnol. 22, 303 (1972).
13. Hartig, R., Troe, J. and Wagner, H.G., 13th Comb. Sym. 147 (1970).
14. Yano, T., Bull. Chem. Soc. Japan, 46, 1619 (1973).
15. Kevorkian, V., Heath, C.E. and Boudart, M., J. Phys. Chem. 64, 964 (1960).
16. Kozlov, G.I. and Knorre, V.G., Combustion and Flame, 6, 253 (1960).
17. Shantarovich, P.S. and Pavlov, B.V., Int. Chem. Eng. 2, 415 (1962).
18. Makarov, K.I. and Pechik, V.K., Carbon 12, 391 (1974).

19. Chen, C.J., Back, M.H. and Back, R.A., Can. J. Chem. 53, 3580 (1975).
20. Robinson, P.J., Holbrook, K.A., Unimolecular Reactions, Wiley, New York (1972).
21. Palmer, H.B. and Hirt, T.J., J. Phys. Chem. 67, 709 (1963).
22. Lin, M.C. and Laidler, K.J., Trans. Faraday Soc. 64, 79 (1968).
23. JANAF Thermochemical Tables, Second Ed. NSRDS-NBS 37 (1970).
24. Clark, T.C. and Dove, J.E., Can. J. Chem. 51, 2147 (1973).
25. Pacey, P.D. and Purnell, J.H., J. Chem. Soc. Faraday Trans. I, 68, 1462 (1972).
26. Clark, T.C., Izod, T.P.J. and Kistiakowsky, G.B., J. Chem. Phys. 54, 1295 (1971).
27. McNesby, J.R., J. Am. Chem. Soc. 64, 1671 (1960).
28. Wijnen, M.H.J., J. Chem. Phys. 23, 1357 (1955).
29. Trotman-Dickenson, A.F., Birchard, J.R. and Steacie, E.W.R., J. Chem. Phys. 19, 163 (1951).
30. Yampolskii, Y.P. and Rybin, V.M., Reaction Kin. and Cat. Letters 1, 321 (1974).
31. Quick, L.M., Knecht, D.A. and Back, M.H., Int. J. Chem. Kin. 4, 61 (1972).
32. Stehling, F.C., Frazee, J.D. and Anderson, R.C., 8th Combustion Symposium, Cal. Inst. Techn. (1960) p. 775.
33. Glick, H.S., 7th Combustion Symposium, 98 (1959).

2

Acceleration, Inhibition, and Stoichiometric Orientation, by Hydrogenated Additives, of the Thermal Cracking of Alkanes at ca. 500°C

M. NICLAUSE, F. BARONNET, G. SCACCHI, J. MULLER, and J. Y. JEZEQUEL

Institut National Polytechnique de Lorraine et Université de Nancy I, Equipe de Recherche n° 136 Associée au C.N.R.S., E.N.S.I.C., 1, rue Grandville, 54042 NANCY Cedex, France

The main purpose of this article is to present a survey of the results, which have been obtained in our laboratories and by other researchers, on the pyrolysis of light alkanes from C_2 to C_5, around 500°C, in the presence of three hydrogenated additives : hydrogen chloride, HCl, hydrogen bromide, HBr, hydrogen sulphide, H_2S ; the thermal decompositions have been performed at relatively small extents of reaction.

The investigations which have been carried out with ethane and neopentane (2,2-dimethylpropane) and those performed with the mixtures n C_4H_{10}-H_2S have already been described in some notes and articles, published or to be published shortly. The first results of the experiments carried out with isobutane and propane have not yet been published.

From an experimental point of view, the reactions have been performed in a conventional static reaction vessel (at constant volume close to 250 cm^3) made of PYREX glass, unpacked (ratio surface/volume approximately equal to 0.9 cm^{-1}). Reaction products have been analyzed by gas chromatography after expansion and quenching.

Alkanes which decompose according to one major primary stoichiometric equation

Let us first consider the case of two alkanes which decompose according to one major primary stoichiometric equation, neopentane and ethane. Neopentane decomposes into isobutene and methane, according to (1) :

$$\text{(A)} \quad \text{neo } C_5H_{12} = \text{i } C_4H_8 + CH_4$$

We have observed that this pyrolysis is strongly accelerated

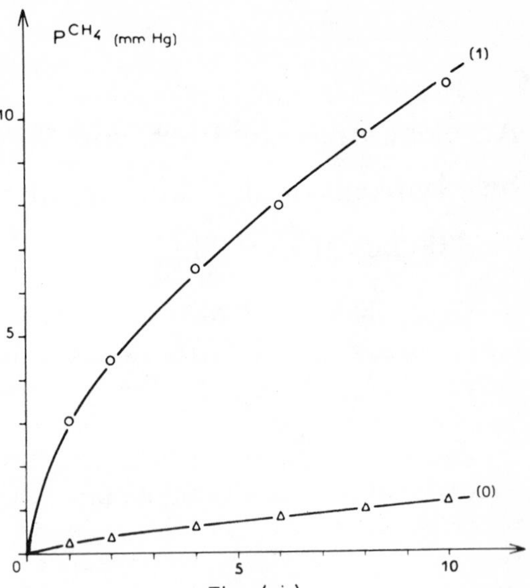

Figure 1. Acceleration, by HCl, of methane formation in the pyrolysis of neopentane. $P_o^{neo\text{-}C_5H_{12}} = 100$ *mm Hg; T = 480°C.* △ *pure neopentane; curve (0).* ○ *neopentane in the presence of HCl (60 mmHg); curve (1).*

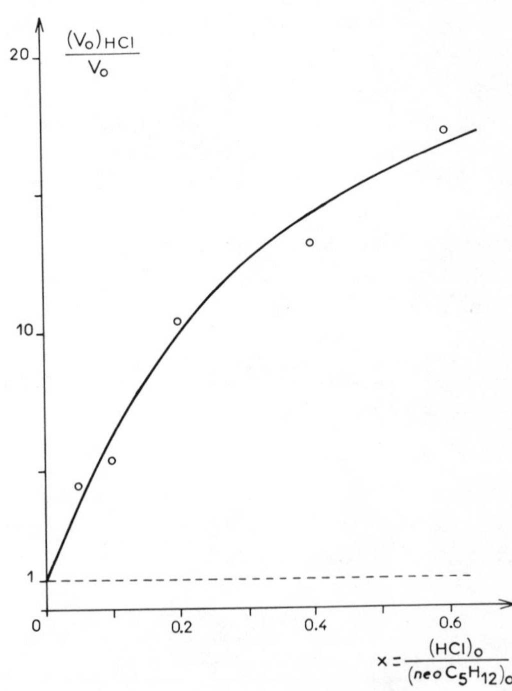

Figure 2. Influence of initial concentration of HCl on the initial rate of formation of CH_4 (or i-C_4H_8) in the pyrolysis of neopentane. $P_o^{neo\text{-}C_5H_{12}} = 100$ *mmHg; T = 480°C.*

by addition of HCl (2) (3), all other conditions being the same, as shown in Figure 1. In Figure 2, we have plotted the ratio of the initial reaction rates, measured by methane formation, in the presence, $(V_o)_{HCl}$, and in the absence, V_o, of hydrogen chloride, as a function of the ratio of the initial concentrations of additive and neopentane.

It is worth noting that this accelerating effect has been already pointed out in a note published in 1964 (4) by ANDERSON and BENSON.

When additive is HBr, we have also observed a strong accelerating effect upon the formation of methane (or isobutene) as shown in Figure 3 (2). Other experiments using H_2S as an additive lead to an effect close to that of HBr (5, 6).

Ethane is the second example of a light alkane which decomposes according to one major primary stoichiometric equation which is (7) :

$$\text{(B)} \qquad C_2H_6 = C_2H_4 + H_2$$

By using the same additives, we have observed the following effects.

Ethylene (or hydrogen) formation is strongly inhibited by addition of hydrogen bromide (8, 9, 10) as shown in Figure 4. In Figure 5, we have plotted the ratio of the initial rates in the presence, $(V_o)_{HBr}$, or in the absence, V_o, of HBr, as a function of the ratio of the initial concentrations of additive and ethane.

It has also been shown that hydrogen sulphide has a strong inhibiting influence upon the pyrolysis of ethane (6, 11). This effect of hydrogen sulphide upon the formation of the major products of ethane pyrolysis has also been observed by Mc LEAN and Mc KENNEY in 1970 (12).

On the contrary, hydrogen chloride has no noticeable effect, neither accelerating, nor inhibiting, upon this reaction (8, 9, 10). The ratio of the initial rates, in the presence or in the absence of HCl, remains close to 1 (see Figure 6).

In the different cases which have been investigated, we have not observed any noticeable consumption of hydracid inside the reaction vessel.

These results can be relatively easily interpreted by taking into account a theoretical general mechanism of acceleration or inhibition of the pyrolysis of an organic substance, μH, by an hydrogenated compound, YH, which was published in 1966 (13) and the main kinetic consequences of which have been recently summarized (14).

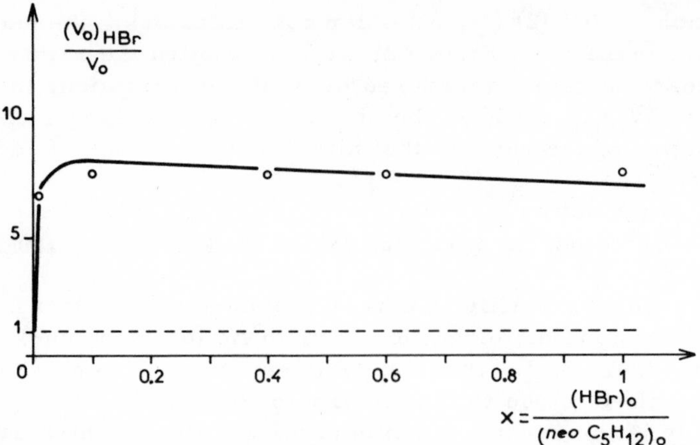

Figure 3. Influence of initial concentration of HBr on the initial rate of formation of CH_4 (or i-C_4H_8) in the pyrolysis of neopentane. $P_o^{neo\text{-}C_5H_{12}} = 100$ mm Hg; T = 480°C.

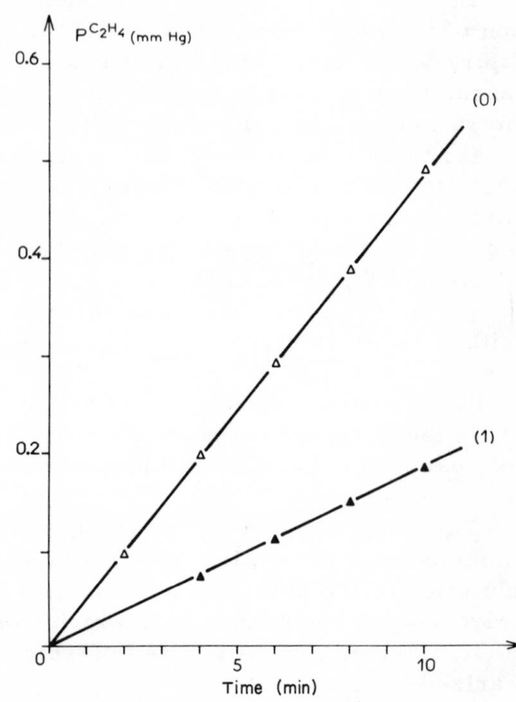

Figure 4. Inhibition, by HBr, of ethylene formation in the pyrolysis of ethane. $P_o^{C_2H_6} = 50$ mmHg; T = 540°C. △ pure ethane; curve (0). ▲ ethane in the presence of HBr (1.15 mm Hg); curve (1).

At small extents of reaction, the mechanism of the chain radical pyrolysis of a pure alkane into two major primary products (one stoichiometry) can be outlined as follows, by using the symbolism "β, μ" (15, 16).

Initiation :	μH	$\xrightarrow{k_1}$	free radicals	(1)
Propagation	$\mu\cdot$	$\xrightarrow{k_2}$	$m + \beta.$	(2)
	$\beta\cdot + \mu H$	$\xrightarrow{k_3}$	$\beta H + \mu.$	(3)
Termination	$\beta\cdot + \beta\cdot$	$\xrightarrow{k_{\beta\beta}}$	products	($\beta\beta$)
	$\beta\cdot + \mu\cdot$	$\xrightarrow{k_{\beta\mu}}$	products	($\beta\mu$)
	$\mu\cdot + \mu\cdot$	$\xrightarrow{k_{\mu\mu}}$	products	($\mu\mu$)

In this scheme, μH represents the organic compound submitted to pyrolysis, $\mu\cdot$ stands for the free radical chain carrier which decomposes according to a monomolecular reaction (propagation process 2) and $\beta.$ stands for the free radical (or atom) chain carrier which reacts with μH by a bimolecular process, abstracting an H atom and recreating the free radical $\mu\cdot$ (propagation process 3). Lastly, m and βH are the major primary products of μH pyrolysis :

$$\mu H = m + \beta H$$

If the chains are long, then :

$$\frac{(\mu.)}{(\beta.)} = \frac{k_3\,(\mu H)_o}{k_2}$$

Therefore, the relative values of $k_3(\mu H)_o$ and k_2 govern the relative concentrations of the chain carriers $\mu.$ and $\beta.$ and, consequently, the relative magnitudes of the three possible processes of chain termination, $\mu\mu$, $\beta\mu$ and $\beta\beta$.

In the case of the pyrolysis of neopentane (1), one has :

$$\frac{(\mu.)}{(\beta.)} = \frac{(\text{neo } C_5H_{11}\cdot)}{(CH_3\cdot)} \ll 1$$

and the principal chain termination process belongs to the $\beta\beta$ type :

$$CH_3\cdot + CH_3\cdot \longrightarrow C_2H_6$$

In the case of the pyrolysis of ethane (7), one has :

$$\frac{(\mu.)}{(\beta.)} = \frac{(C_2H_5\cdot)}{(H.)} \gg 1\,.$$

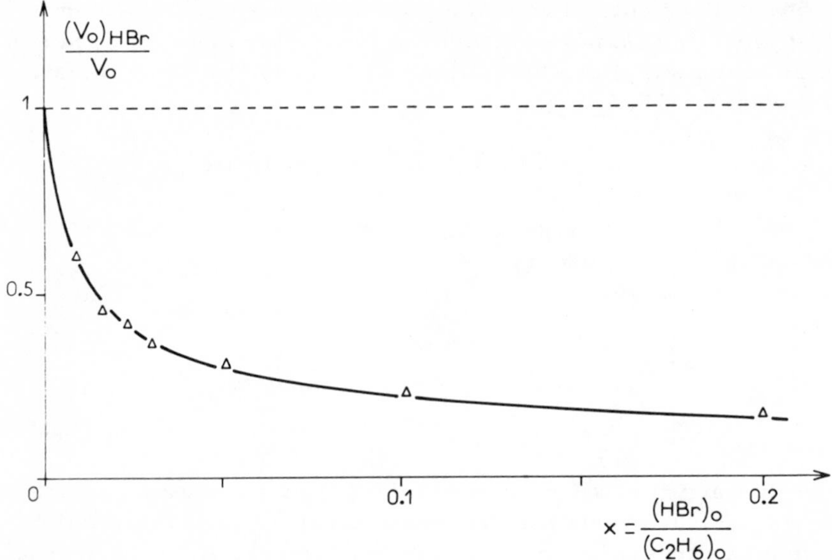

Figure 5. Influence of initial concentration of HBr upon the initial rate of formation of C_2H_4 (or H_2) in the pyrolysis of ethane. $P_o^{C_2H_6}$ *= 50 mmHg;* T = *560°C.*

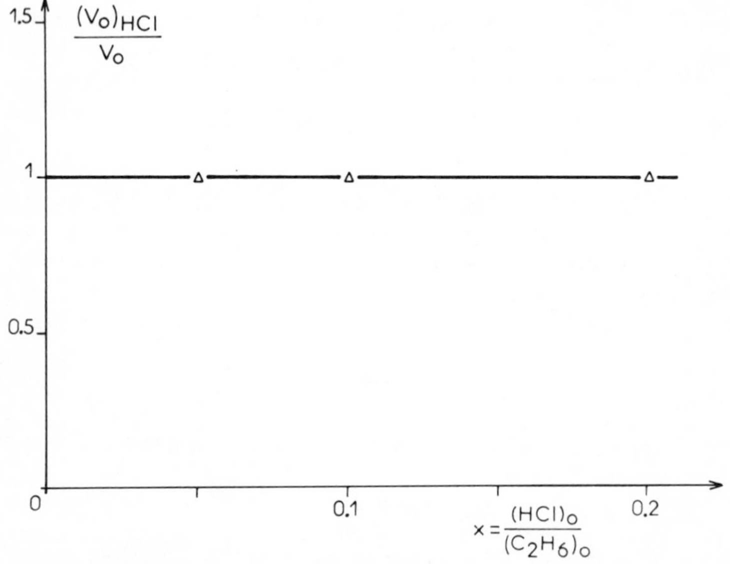

Figure 6. Influence of initial concentration of HCl on the initial rate of formation of C_2H_4 (or H_2) in the pyrolysis of ethane. $P_o^{C_2H_6}$ *= 50 mmHg;* T *= 540°C.*

and the principal chain termination process belongs to the μμ type:

$$C_2H_5\cdot + C_2H_5\cdot \longrightarrow \begin{cases} n\,C_4H_{10} \\ C_2H_4 + C_2H_6 \end{cases}$$

In the presence of an additive YH also including a "mobile" H atom (like HCl, HBr or H_2S), the reaction mechanism can be written, if we assume that YH does not increase the chain initiation rate :

Initiation : $\mu H \xrightarrow{k_1}$ free radicals (1)

Propagation I

$$\mu\cdot \xrightarrow{k_2} m + \beta\cdot \quad (2)$$

$$\beta\cdot + \mu H \xrightarrow{k_3} \beta H + \mu\cdot \quad (3)$$

Propagation I'

$$\mu\cdot \xrightarrow{k_2} m + \beta\cdot \quad (2)$$

$$\beta\cdot + YH \xrightarrow{k_{4\beta}} \beta H + Y\cdot \quad (4\beta)$$

$$Y\cdot + \mu H \underset{k_{4\mu}}{\overset{k_5}{\rightleftharpoons}} YH + \mu\cdot \quad (5)\ \&\ (4\mu)$$

Termination : processes ββ, βμ, μμ, βY, μY, and YY of combination or disproportionation of free radicals β., μ. and Y.

This scheme is essentially based on the assumption that the free radicals β. and μ., chain carriers in the decomposition of pure μH, can react with YH to give a new free radical (or atom) Y., thermally stable, but capable of reacting with μH to recreate YH.

In the absence of YH, the free radicals β. and μ. can propagate only the chain decomposition of μH according to I. But in the proposed mechanism, the assumption is made that, in the presence of YH, another type I' of chain decomposition of μH can propagate as well, involving the new radical (or atom) Y., but giving the same stoichiometry as I :

$$\mu H = m + \beta H$$

Lastly, the proposed mechanism implies that YH tends to make new chain termination processes appear.

Let us note that, in this reaction scheme, there is no consumption of YH in the chain propagation ; a slight consumption of YH may only arise from some new termination processes.

The theoretical study of the influence of additive YH upon the initial rate of pyrolysis of μH is very complex in the general case. It becomes comparatively more simple when one considers one or the other of the two following limiting cases which are a priori possible (13, 14) :

$$\text{First limiting case :} \quad k_3 (\mu H)_o \ll k_2$$

to which belongs, as shown before, the thermal decomposition of neopentane.

This first limiting case implies that the radical process 3 is definitely more difficult than radical process 2. It follows that, in the pyrolysis of pure μH, process 3 is the rate determining step of chain propagation ; one has $(\mu .) \ll (\beta .)$ and the chain termination chiefly involves process ββ.

It can be shown that YH has the two following effects on the pyrolysis of μH :

- an inhibiting effect, due to the fact that YH makes new chain termination processes appear,

- an accelerating effect, due to the fact that a new propagation type I' in μH chain decomposition appears through YH, since the additive YH makes a new reaction path appear (propagation processes 4β and 5) starting from the same initial state (β . and μH) and arriving at the same final state (βH and μ.), like the difficult step 3 ; therefore, YH catalyzes the propagation of μH chain decomposition.

It can be shown (13, 14) that the overall result of the two opposed kinetic influences of YH is an acceleration of μH pyrolysis on condition that, as a first approximation, the free radical (or atom) Y. is more reactive than the free radical (or atom) β. when abstracting an H atom from μH : $k_5 > k_3$.

Our experimental results, obtained with the mixtures neo C_5H_{12}-HCl can be explained by the well known fact [see tables of rate constants of bimolecular reactions (17)] that chlorine free atom Cl. is much more reactive than a methyl free radical when abstracting an H atom from an alkane molecule, like neopentane ($k_5 > k_3$). Likewise, in the case of HBr, bromine free atom Br. is more reactive than CH_3., but definitely less than chlorine free atom Cl., in the same type of elementary step.

Results which have been obtained with H_2S suggest that a free radical HS. is more reactive than a methyl free radical when abstracting an hydrogen atom from a neopentane molecule. A detailed investigation of HCl and HBr influence upon neopentane pyrolysis is to be published shortly (2).

On the contrary, the overall effect of the two opposed kinetic influences of YH, again in this first limiting case $k_3(\mu H)_o << k_2$, is an inhibition when $k_5 < k_3$, which means that the free radical (or atom) Y. is less reactive than the free radical (or atom) β. when abstracting an H atom from a neopentane molecule.

The inhibition of neopentane pyrolysis by various alkenes, including propene (5, 18), and the inhibition and self-inhibition of this pyrolysis by isobutene (5, 19, 20) have been interpreted on this basis.

Therefore, we have been able to find the kinetic mechanism accounting for the self-inhibition of neopentane pyrolysis ; we have determined a mathematical model, derived from this mechanism, which describes the time course of this reaction and appears to be in good agreement with our experimental results (21).

Second limiting case : $k_2 << k_3 (\mu H)_o$

This second limiting case supposes that, if we consider the two radical processes 2 and 3, process 2 is definitely more difficult and, as a consequence, is the rate determining step in the chain propagation ; therefore, in the pyrolysis of pure μ H, one has : (β.) <<(μ.) and process μμ mainly accounts for the chain termination. The thermal decomposition of ethane belongs to this limiting case, as seen before.

It can be shown that (13, 14), in such a case, YH can have only an inhibiting effect upon the pyrolysis of μH, due to the fact that YH makes new termination processes μH and YY appear (13) ; in this second limiting case, the additive YH cannot have an accelerating effect upon the pyrolysis of μH, as the overall rate of chain propagation in the presence of YH remains limited by the relatively difficult process 2, which is common to the two propagation sequences I and I'.

Our experimental results obtained when pyrolyzing ethane in the presence of an hydrogenated additive are consistent with the relative values of the bond dissociation energies. HBr (bond dissociation energy ≃ 88 kcal/mole) and H_2S (bond dissociation energy of HS-H ≃ 90 kcal/mole) strongly inhibit ethane pyrolysis ; on the other hand, HCl has no noticeable effect upon the thermal decomposition of ethane, in agreement with the ascribed value of the bond dissociation energy of HCl (103 kcal/mole) higher than that of C_2H_5-H (98 kcal/mole) (10).

Our proposed mechanism has proved to be successful in the case of alkanes [or other organic compounds, like acetaldehyde (22)] which decompose according to one major primary stoichiometric equation. But most alkanes decompose according to several major stoichiometric equations; let us try to investigate the case of light alkanes like isobutane or propane which decompose according to two major primary stoichiometric equations.

Alkanes which decompose according to several major primary stoichiometric equations

Example of isobutane. It is well known (23, 24) that the pyrolysis of isobutane can be described by two major primary stoichiometric equations :

$$(A') \qquad i\,C_4H_{10} = C_3H_6 + CH_4$$

$$(B') \qquad i\,C_4H_{10} = i\,C_4H_8 + H_2$$

the relative magnitudes of which are approximately 30 % for (A') and 70 % for (B') around 500°C and 100 mm Hg.

These analytical results are interpreted by a chain radical mechanism ; at zero time, the chains are propagated by the processes :

$$A' \left\{ \begin{array}{ll} i\,C_4H_9\cdot\ (p) \xrightarrow{k_{2a}} C_3H_6 + CH_3\cdot & (2a) \\ CH_3\cdot + i\,C_4H_{10} \xrightarrow{k_{3a}} CH_4 + i\,C_4H_9\cdot\ (p \text{ or } t) & (3a) \end{array} \right.$$

$$B' \left\{ \begin{array}{ll} i\,C_4H_9\cdot\ (t) \xrightarrow{k_{2b}} i\,C_4H_8 + H\cdot & (2b) \\ H\cdot + i\,C_4H_{10} \xrightarrow{k_{3b}} H_2 + i\,C_4H_9\cdot\ (p \text{ or } t) & (3b) \end{array} \right.$$

where there are two isomeric forms of the isobutyl free radical:

primary isobutyl : $(CH_3)_2CH\text{-}\dot{C}H_2$;

tertiary isobutyl : $(CH_3)_2\dot{C}\text{-}CH_3$

It has also been shown (8) that it is adequate to write a second order isomerization reaction of the tertiary and primary radicals.

$$i\,C_4H_9\cdot\ (t) + i\,C_4H_{10} \rightleftarrows i\,C_4H_{10} + i\,C_4H_9\cdot\ (p)$$

(A') is a reaction of loss of methane, propagated by a methyl free radical and which therefore resembles the reaction of neopentane

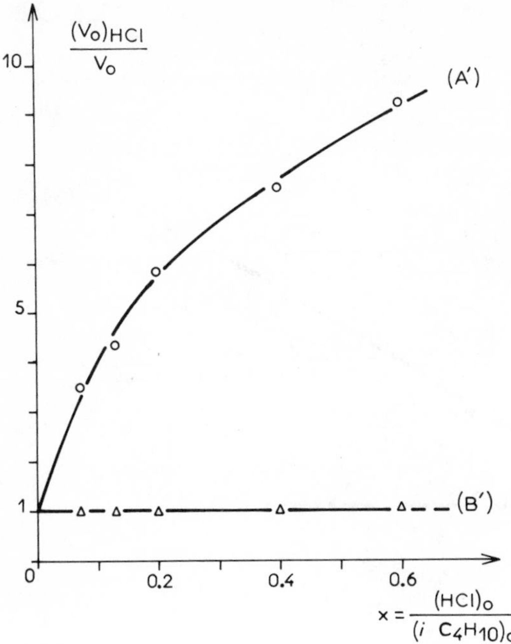

Figure 7. Influence of initial concentration of HCl on the initial rates of formation of: CH_4 (or C_3H_6), curve (A′); and i-C_4H_8 (or H_2), curve (B′) in the pyrolysis of isobutane. $P_o^{i\text{-}C_4H_{10}} = 50$ *mmHg;* $T = 480°C$.

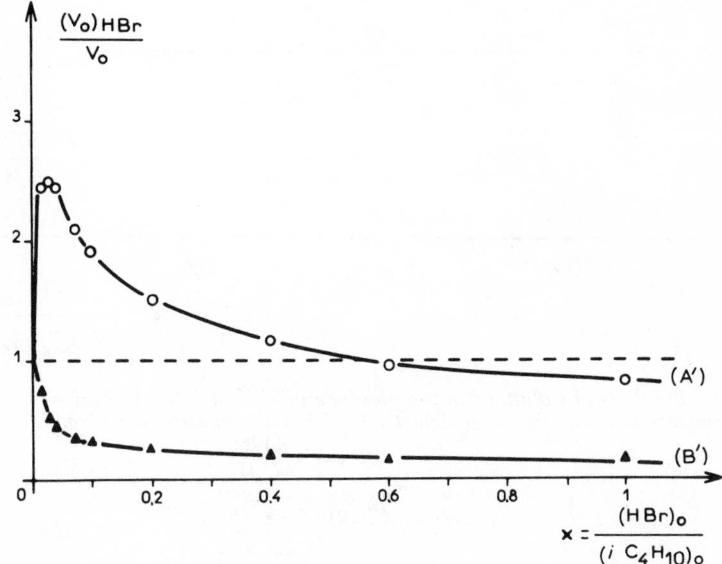

Figure 8. Influence of initial concentration of HBr on the initial rates of formation of: CH_4 (or C_3H_6), curve (A′), and i-C_4H_8 (or H_2), curve (B′) in the pyrolysis of isobutane. $P_o^{i\text{-}C_4H_{10}} = 50$ *mmHg;* $T = 480°C$.

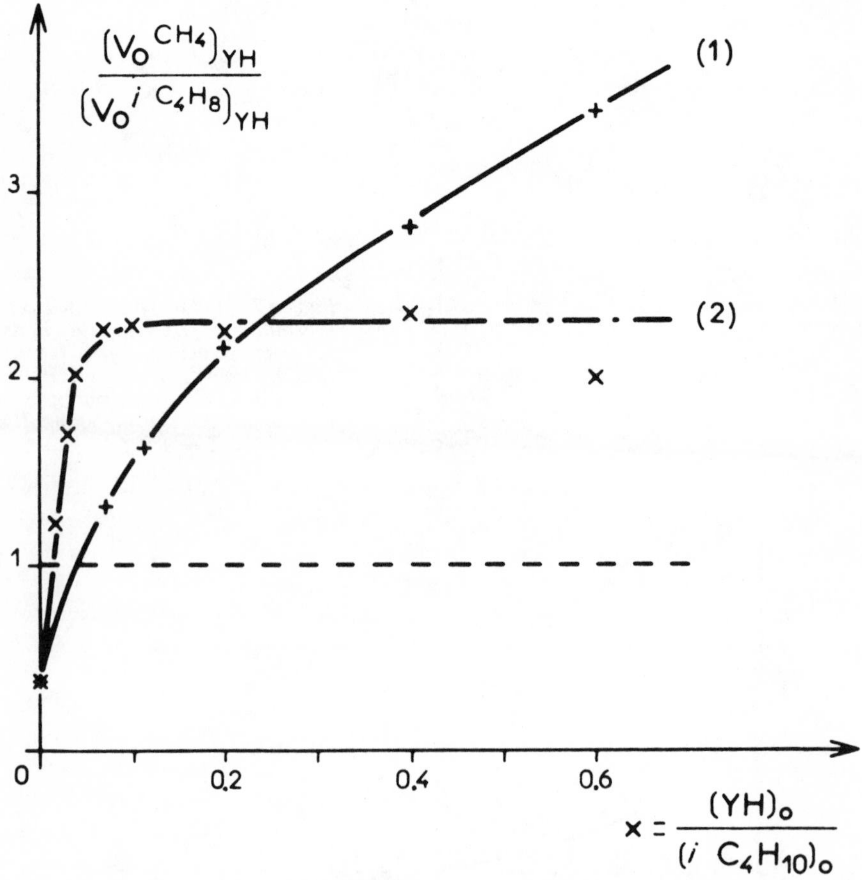

Figure 9. Pyrolysis of isobutane in the presence of HCl or HBr. Variations of the relative magnitude of the two stoichiometries vs. initial concentration of additive YH.

$$\frac{(V_o^{CH_4})_{YH}}{(V_o^{i\text{-}C_4H_8})_{YH}} \; vs. \quad x = \frac{(YH)_o}{(i\text{-}C_4H_{10})_o}$$

$P^{i\text{-}C_4H_{10}} = 50$ *mmHg*; T = 480°C.

(1) YH ≡ HCl; (2) YH ≡ HBr.

thermal decomposition (A).
(B') is a reaction of dehydrogenation, propagated by a free hydrogen atom and which therefore resembles the reaction of ethane pyrolysis (B).

What is then the effect of additives HCl, HBr and H_2S upon the pyrolysis of isobutane ?

Systematic experiments performed at an initial partial pressure of isobutane of 50 mm Hg and at 480°C, in the presence of each of the three above-mentioned hydracids led to the following results :

- HCl strongly accelerates the formation of methane and propene [reaction (A')]. Isobutene and hydrogen formations [reaction (B')] remain unmodified (see Figure 7).
- HBr definitely inhibits the formation of hydrogen and isobutene (B'). Formation of methane and propene (A') is accelerated at low concentration of HBr and, as one continues to increase HBr concentration, the accelerating effect reaches a maximum, then decreases and is even transformed into an inhibition (see Figure 8).
- H_2S definitely inhibits the dehydrogenation reaction (B') and slightly accelerates the reaction of loss of methane (A').

It is important to observe that, in each case, the stoichiometry (A') (into C_3H_6 + CH_4) becomes and remains predominant compared with the stoichiometry (B') (into i C_4H_8 + H_2), as shown in Figure 9, in the cases where additives are HCl (curve 1) and HBr (curve 2), by plotting the ratio of the initial rates of loss of methane (A') and dehydrogenation (B') as a function of the ratios of initial concentrations.

Measurements by infra-red spectrophotometry have not shown a noticeable consumption of the added hydracid inside the reaction vessel.

It can be shown (8) that the previous observations can be qualitatively interpreted by assuming that additive YH (HCl, HBr or H_2S) makes new processes appear. These new processes are similar to those previously written in the simple cases of neopentane and ethane (see first section), especially the following propagation processes :

$$CH_3\cdot + YH \xrightarrow{k_{4\beta a}} CH_4 + Y\cdot \qquad (4\ \beta\ a)$$

$$H\cdot + YH \xrightarrow{k_{4\beta b}} H_2 + Y\cdot \qquad (4\ \beta b)$$

$$Y. + i\,C_4H_{10} \underset{k_{4\mu a}}{\overset{k_{5a}}{\rightleftarrows}} YH + i\,C_4H_9.\ (p) \quad (5a)\ \&\ (4\mu a)$$

$$Y. + i\,C_4H_{10} \underset{k_{4\mu b}}{\overset{k_{5b}}{\rightleftarrows}} YH + i\,C_4H_9.\ (t) \quad (5b)\ \&\ (4\mu b)$$

The ratio $R = \dfrac{(i\,C_4H_9.\,(p))}{(i\,C_4H_9.\,(t))}$ of the concentrations of the primary and tertiary forms of the isobutyl free radical is therefore a function of the concentration of additive YH, and the ratio of the initial rates of reactions (A') and (B') is given by the relationship :

$$\frac{V_o^{A'}}{V_o^{B'}} \simeq \frac{k_{2a}}{k_{2b}}\,R$$

We have seen before that addition of YH (HCl, HBr or H_2S) increases the ratio $V_o^{A'}/V_o^{B'}$. This shows that ratio R increases too by addition of YH, as a consequence of the conversion of free radicals i $C_4H_9.$ (t) into free radicals i $C_4H_9.$ (p) by processes 4 μ b and 5a.

Example of propane. The pyrolysis of propane (25, 26), like that of isobutane, can be described by two major primary stoichiometric equations.

$$(A'') \qquad C_3H_8 = C_2H_4 + CH_4$$

$$(B'') \qquad C_3H_8 = C_3H_6 + H_2$$

This reaction has been accounted for by a chain radical mechanism ; the chains are propagated, at zero time, by the processes :

$$A'' \left\{ \begin{array}{lll} C_3H_7.\ (p) & \xrightarrow{k_{2a}} & C_2H_4 + CH_3. \quad (2a) \\ CH_3. + C_3H_8 & \xrightarrow{k_{3a}} & CH_4 + C_3H_7.\ (p\ or\ s) \quad (3a) \end{array} \right.$$

$$B'' \left\{ \begin{array}{lll} C_3H_7.\ (s) & \xrightarrow{k_{2b}} & C_3H_6 + H. \quad (2b) \\ H. + C_3H_8 & \xrightarrow{k_{3b}} & H_2 + C_3H_7.\ (p\ or\ s) \quad (3b) \end{array} \right.$$

There are two isomeric forms of the propyl free radical :

$$\text{primary } C_3H_7.\ (p)\ :\ CH_3 - CH_2 - \dot{C}H_2$$

$$\text{secondary } C_3H_7.\ (s)\ :\ CH_3 - \dot{C}H - CH_3$$

with a bimolecular process of isomerization (26) :

$$C_3H_7\cdot \text{ (s)} + C_3H_8 \rightleftharpoons C_3H_8 + C_3H_7\cdot \text{ (p)}$$

We have performed a few experiments to find out the influence of HCl, HBr and H_2S upon the pyrolysis of propane. In Figures 10 and 11, we have represented the formation curves of methane [representative of reaction (A")] and propene [representative of reaction (B")] when pyrolyzing 100 mm Hg of propane at 490°C, in the absence of additive (curves 0) and in the presence of 5 mm Hg of hydrogen chloride of bromide (curves 1).

Figure 10 shows that HCl strongly accelerates reaction (A''), whereas it has only a very slight accelerating effect on reaction (B'').

The effect of HBr (see Figure 11) is a strong acceleration of reaction (A'') and, on the contrary, an inhibition of reaction (B''). Results similar to those obtained with HBr have been obtained when additive is H_2S.

It can be noticed that, as in the case of isobutane, addition of HCl, HBr or H_2S increases the relative magnitude of the reaction of loss of methane compared with the reaction of dehydrogenation. Such an effect is due to a conversion of free radicals $C_3H_7\cdot$ (s) into free radicals $C_3H_7\cdot$ (p).

Example of n-butane. Normal butane decomposes according to 3 major primary stoichiometries (27, 28, 29).

$$(A_1''') \quad n\ C_4H_{10} = C_3H_6 + CH_4$$
$$(A_2''') \quad n\ C_4H_{10} = C_2H_4 + C_2H_6$$
$$(B''') \quad n\ C_4H_{10} = 2\ C_2H_4 + H_2$$

It has been shown (29, 30) that, around 500°C, H_2S accelerates reactions (A_1''') and (A_2''') ; on the contrary, H_2S inhibits reaction (B''') as were inhibited reactions (B) (B') and (B'') of dehydrogenation seen before.

It seems to us that mechanisms close to those previously described are relevant to interpret these experimental facts.

Conclusion

In our opinion, these investigations on the pyrolyses of light alkanes in the presence of three hydrogenated additives HCl, HBr and H_2S are interesting from two points of view : fundamental and practical.

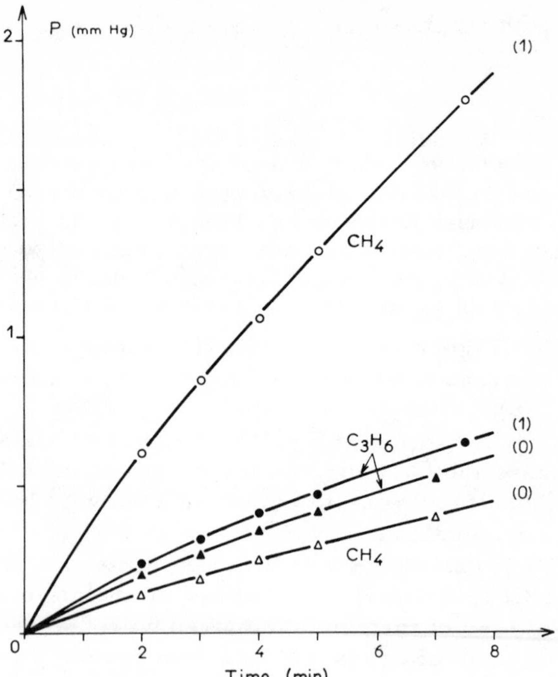

Figure 10. Influence of HCl upon the formation of CH_4 (or C_2H_4) and C_3H_6 (or H_2) in the pyrolysis of propane. $P_o^{C_3H_8}$ = 100 mmHg; T = 490°C. Pure propane, curves (0); propane in the presence of HCl (5 mm Hg), curves (1).

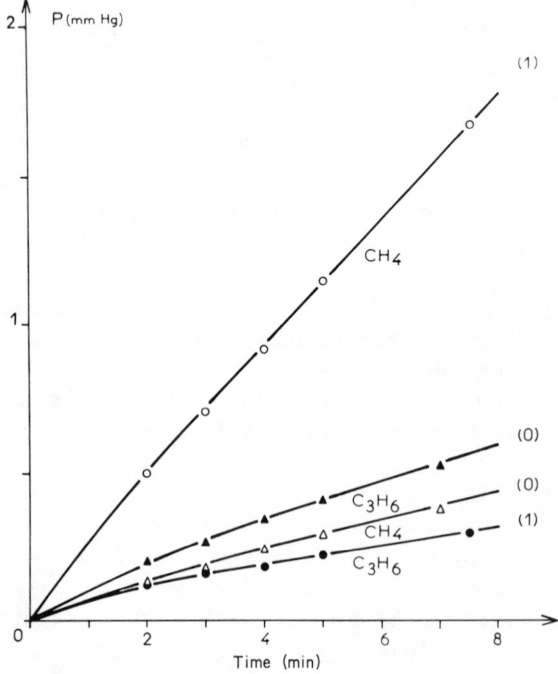

Figure 11. Influence of HBr on the formation of CH_4 (or C_2H_4) and C_3H_6 (or H_2) in the pyrolysis of propane. $P_o^{C_3H_8}$ = 100 mmHg; T = 490°C. Pure propane, curves (0); propane in the presence of HBr (5 mm Hg), curves (1).

From a fundamental point of view, the reactions studied in the presence of these additives have permitted to point out without ambiguity the elementary processes which are the rate determining steps of the pyrolysis. Such a method can be a useful means to determine the details of a reaction mechanism. We think that it might be useful when investigating the mechanisms of olefin thermal decompositions which are usually more complex than those of alkane pyrolyses. Investigations on olefin pyrolyses in the presence of H_2S have been performed at Nancy (31) and in the U. S. A. (32).

Moreover, the study of the mechanisms of reactions alkane-hydracid has indirectly permitted a check of the estimates of a number of rate constants of elementary steps (2, 8, 10).

More generally, the concept of chain co-reactions with mutual interactions, which is the basis of the mechanisms previously proposed to account for the effect of hydrogenated additives upon the pyrolyses of alkanes, has also proved to be useful in two other cases :

- to interpret the effect of trace amounts of oxygen on the pyrolyses of alkanes around 500 °C (33) ;
- to investigate the reactions between alkanes and compounds with at least one "mobile" atom other than an hydrogen atom, for instance CCl_4. The mixtures which have been studied are neo C_5H_{12} - CCl_4 and i C_4H_{10} - CCl_4 (34).

From a practical point of view, the use of hydrogenated additives enables to modify the relative yields of products in the thermal decomposition of light alkanes. This might be useful in industrial cracking reactions, especially in the case of countries which, like the USA, produce an important part of their ethylene by cracking of ethane and propane. In the case of propane, hydrogenated additives are able to reduce the yield of propene and to increase that of ethylene in the thermal decomposition products. The aim of this survey is mainly to point out the role of hydrogenated additives for this purpose. It seems to us that investigations based on this principle would be worth developing.

Abstract

H_2S, HBr and HCl accelerate the main primary reaction of neopentane pyrolysis : neo C_5H_{12} = CH_4 + i C_4H_8 (A).

On the contrary, that of ethane : C_2H_6 = H_2 + C_2H_4 (B) is inhibited by H_2S and HBr (HCl has practically no effect).

In the case of isobutane, which decomposes according to two main primary stoichiometric equations :

- one of loss of methane : i $C_4H_{10} = CH_4 + C_3H_6$ (A')
- the other of dehydrogenation : i $C_4H_{10} = H_2$ + i C_4H_8 (B'),

reaction (A') is accelerated by H_2S, HBr and HCl, whereas reaction (B') is inhibited by H_2S and HBr (HCl has practically no effect).

Therefore, the above-mentioned hydrogenated additives permit an orientation of the thermal cracking of isobutane. Similar effects are observed with propane and n-butane.

A mechanistic interpretation of these effects is proposed and some possible applications of these observations are considered.

Literature cited

(1) ENGEL J., COMBE A., LETORT M. and NICLAUSE M., C.R. Acad. Sc. (1957), 244, 453 ; BARONNET F., DZIERZYNSKI M., MARTIN R. and NICLAUSE M., C.R. Acad. Sc. (1968), 267 C, 937 ; BARONNET F., DZIERZYNSKI M., CÔME G.M., MARTIN R. and NICLAUSE M., Communication at the First International Symposium "Gas Kinetics", Szeged (Hungary), July 8-11, 1969, Preprints p. 73 ; Intern. J. Chem. Kinetics (1971), 3, 197 ; HALSTEAD M.P., KONAR R.S., LEATHARD D.A., MARSHALL R.M. and PURNELL J.H., Proc. Roy. Soc. (1969), A 310, 525.

(2) MULLER J., BARONNET F., DZIERZYNSKI M. and NICLAUSE M., to be published.

(3) MULLER J., BARONNET F. and NICLAUSE M., C.R. Acad. Sc. (1972), 274 C, 1772 ; Communication at the 3rd Int. Symposium on Gas Kinetics, Free University of Brussels, Brussels (Belgium), August 27-31, 1973.

(4) ANDERSON K.H. and BENSON S.W., J. Chem. Phys. (1964), 40, 3747.

(5) BARONNET F., Thèse de Doctorat ès Sciences Physiques, Nancy (1970).

(6) SCACCHI G., BARONNET F., MARTIN R. and NICLAUSE M., J. Chim. Phys. (1968), 65, 1671.

(7) PURNELL J.H. and QUINN C.P., Proc. Roy. Soc. (1962), A 270, 267 ; QUINN C.P., Proc. Roy. Soc. (1963), A 275, 190 ; LIN M.C. and BACK M.H., Canad. J. Chem. (1966), 44, 505, 2357 and 2369 ; TRENWITH A.B., Trans. Faraday Soc. (1966), 62, 1538 ; (1967), 63, 2452 ; SCACCHI G., MARTIN R. and NICLAUSE M., Bull. Soc. Chim. (1971), 731.

(8) MULLER J., Thèse de Doctorat ès Sciences Physiques, Nancy (1973).

(9) MULLER J., BARONNET F., SCACCHI G. and NICLAUSE M., C.R. Acad. Sc. (1971), 272 C, 271 ; MULLER J., SCACCHI G. and NICLAUSE M., Communication at the 3rd Int. Symposium on Gas Kinetics, Free University of Brussels, Brussels (Belgium), August 27-31, 1973.

(10) MULLER J., SCACCHI G., DZIERZYNSKI M. and NICLAUSE M., to be published.

(11) SCACCHI G., Thèse de Doctorat ès Sciences Physiques, Nancy (1969) ; SCACCHI G., DZIERZYNSKI M., MARTIN R. and NICLAUSE M., Communication at the First Int. Symposium on Gas Kinetics, Szeged (Hungary), July 8-11, 1969, Preprints p. 236 ; Intern. J. Chem. Kinetics (1970), 2, 115.

(12) Mc LEAN P.R. and Mc KENNEY D.J., Canad. J. Chem. (1970), 48, 1782.

(13) NICLAUSE M., MARTIN R., BARONNET F. and SCACCHI G., Rev. Inst. Fr. Pétrole (1966), 21, 1724.

(14) NICLAUSE M., BARONNET F., SCACCHI G. and MULLER J., to be published.

(15) GOLDFINGER P., LETORT M. and NICLAUSE M., "Volume commémoratif Victor Henri", 283, Desoer, Liège, 1947-1948.

(16) NICLAUSE M., Rev. Inst. Fr. Pétrole (1954), 9, 327 and 419.

(17) TROTMAN-DICKENSON A.F. and MILNE G.S., "Tables of Bimolecular Gas Reactions", NSRDS-NBS 9, Washington 1967; KONDRATIEV V.N., "Rate Constants of Gas Phase Reactions", Academy of Sciences of the U.S.S.R., Moscow 1970.

(18) BARONNET F., MARTIN R. and NICLAUSE M., J. Chim. Phys. (1972), 69, 236.

(19) BARONNET F., MARTIN R. and NICLAUSE M., C.R. Acad Sc. (1969), 268 C, 1744.

(20) BARONNET F., CÔME G.M. and NICLAUSE M., J. Chim. Phys. (1974), 71, 1214.

(21) CÔME G.M. and BARONNET F., C.R. Acad. Sc. (1969) 268 C, 1917.

(22) RICHARD C., Thèse de Doctorat ès Sciences Physiques, Nancy (1972) ; RICHARD C., MARTIN R. and NICLAUSE M., J. Chim. Phys. (1973), 70, 1151 ; Reaction Kinetics and Catalysis Letters (1974), 1, 37.

(23) STEACIE E.W.R. and PUDDINGTON J.E., Canad. J. Research (1938), B 16, 260 ; STEPUKHOVITCH A.D., KOSYREVA R.V. and PETROSYAN V.I., Russ. J. Phys. Chem. (1965), 35, 653 ; KONAR R.S., PURNELL J.H. and

QUINN C.P., Trans. Faraday Soc. (1968), 64, 1319.
(24) FUSY J., Thèse de Doctorat ès Sciences Physiques, Nancy (1966); FUSY J., MARTIN R., DZIERZYNSKI M. and NICLAUSE M., Bull. Soc. Chim. Fr. (1966), 3783.
(25) STEACIE E.W.R. and PUDDINGTON J.E., Canad. J. Research (1938), B 16, 411 ; MARTIN R., DZIERZYNSKI M. and NICLAUSE M., J. Chim. Phys. (1964), 60, 286 ; LEATHARD D.A. and PURNELL J.H., Proc. Roy. Soc. (1968), A 305, 517.
(26) JEZEQUEL J.Y., BARONNET F. and NICLAUSE M., Communication at the 4th International Symposium on Gas Kinetics, Heriot-Watt University, Edinburgh (U.K.), August 25-30, 1975.
(27) PURNELL J.H. and QUINN C.P., Proc. Roy. Soc. (1962), A 270, 267.
(28) LARGE J.F., MARTIN R. and NICLAUSE M., Bull. Soc. Chim. Fr. (1972), 961.
(29) LARGE J.F., Thèse de Doctorat ès Sciences Physiques, Nancy (1970).
(30) LARGE J.F., MARTIN R. and NICLAUSE M., C.R. Acad. Sc. (1972), 274 C, 322.
(31) GOUSTY Y. and MARTIN R., Reaction Kinetics and Catalysis Letters (1974), 1, 189.
(32) HUTCHINGS D.A., FRECH K.J. and HOPPSTOCK F., Communication at the 169th ACS National Meeting, Philadelphia (Pa), U.S.A., April 7-11, 1975.
(33) MARTIN R., NICLAUSE M. and SCACCHI G., Communication at the Symposium on Thermal Cracking Operations, 1st Chemical Congress of the North American Continent, Mexico-City (Mexico), Nov. 30 -Dec. 5, 1975.
(34) PHILARDEAU Y., Thèse de Doctorat ès Sciences Physiques, Nancy (1970) ; PHILARDEAU Y., MARTIN R. and NICLAUSE M., C.R. Acad. Sc. (1971), 273 C, 1302 ; PHILARDEAU Y., DZIERZYNSKI M., MARTIN R. and NICLAUSE M., Bull. Soc. Chim. Belges (1972), 81, 7 ; PHILARDEAU Y., BARONNET F., MARTIN R. and NICLAUSE M., Communication at the 2nd Int. Symposium on Gas Kinetics, Swansea (U.K.), July 5-9, 1971.

3

Influence of Small Quantities of Oxygen on the Thermal Cracking of Alkanes at ca. 500°C

RENÉ MARTIN, MICHEL NICLAUSE, and GÉRARD SCACCHI

Institut National Polytechnique et Université de Nancy I, Equipe de Recherche N° 136 Associée au C.N.R.S., 1, rue Grandville, 54042 NANCY, Cedex, France

After the discovery of the promoting effect of oxygen traces on the thermal decomposition of ethanal (1), it was reported that small quantities of oxygen also accelerate the pyrolyses of n-butane (2) (3) and other alkanes (3), at least in unpacked pyrex vessels ($s/v \simeq 1\ cm^{-1}$). But, in these studies, the role of the walls was not examined, nor was the influence of oxygen on the distribution of the products of hydrocarbon pyrolysis investigated.

Therefore, we made a systematic study, by gas chromatography, of the influence of small amounts of oxygen on alkane pyrolyses, in static systems and different types of pyrex reaction vessels. The first observations, referring to propane, have already been published (4) (5).

This paper aims at summarizing our researches on the pyrolyses, at about 500°C, of four other alkanes -ethane (6a), isobutane (6b), n-butane (6c) and isopentane (6d)- in the presence of small oxygen concentrations (0.01 up to several %) and at initial pressures between 10 and 100 mm Hg. In an attempt to reach the chemical and kinetic features of the reactions without interference arising from the products, these reactions were studied at low percentage of conversion. Some preliminary results obtained in these investigations have already been succinctly presented (5) (7) (8) (9).

Experimental Results

1-Chemical Aspects. The primary thermal decomposition of pure ethane can be essentially represented by a unique stoichiometry :

$$\text{(A)} \qquad C_2H_6 = C_2H_4 + H_2$$

When ethane is pyrolyzed in the presence of small oxygen quantities, it primarily produces more ethylene than hydrogen ; this is due to the oxidation stoichiometry :

- Table I -

Relative importances (%) of the primary stoichiometries of *isobutane decomposition* and *oxidation* at low extent of reaction

T = 490°C ; $P_o^{i\ C_4H_{10}}$ = 50 mm Hg ; PbO-coated pyrex vessel (s/v ≃ 2.2 cm^{-1})

	$(O_2)_o$ %	0	0.2	1	4.3
decomposition	i C_4H_{10} = i C_4H_8 + H_2	71.5 %	63.5 %	12.5 %	3.5 %
	i C_4H_{10} = C_3H_6 + CH_4	28.5 %	34.5 %	35.5 %	35.5 %
oxidation	i C_4H_{10} + O_2 (or 1/2 O_2) = i C_4H_8 + H_2O_2 (or H_2O)	—	2 %	52 %	61 %

(B) $C_2H_6 + O_2$ (or $\frac{1}{2}O_2$) $= C_2H_4 + H_2O_2$ (or H_2O)

which then adds to (A).

For instance, in a PbO-coated pyrex vessel (s/v $\simeq$ 10.5 cm^{-1}) at about 540°C, with 50 mm Hg initial pressure of ethane and 0.5 % initial concentration of oxygen, the importance of (A) as compared with (A) + (B) is 42 %.

In the case of isobutane, n-butane or isopentane, several stoichiometries are necessary to account for the primary decomposition of each of these alkanes (see Tables I, II and III).

Here again, in the presence of small amounts of oxygen and as a consequence of the oxidation, we observe the formation of hydrogen peroxide (or water) and extra yields of the olefins which already appear in the decomposition reactions of the alkanes by loss of H_2, i.e. : isobutene in isobutane pyrolysis, ethylene and n-butenes in n-butane pyrolysis, ethylene, propene and isopentenes in isopentane pyrolysis (see Tables I, II and III).

It has already been pointed out that, above approximately 400°C, the oxidation of several alkanes yields olefins and only small quantities of oxygenated organic compounds (10). In connection with the problem of olefin production through oxidative cracking of alkanes, let us note that two patents have been taken out (11).

J.E. BLAKEMORE et al. (12) recently studied the pyrolysis of n-butane in the presence of trace quantities of oxygen, using a gold microreactor operated under plug-flow conditions, at 530-600°C and atmospheric pressure. For a given initial concentration of n-butane, increasingly larger fractions of 1- and 2-butenes, 1,3-butadiene and water appeared in the products as the oxygen level in the n-butane was increased. Except for the formation of 1,3-butadiene, which we do not find around 500°C at low extents of reaction, the results of BLAKEMORE et al. are in agreement with our own work (see Table II).

2-Kinetic Aspects. Let us first consider the case of ethane. In a pyrex vessel with uncoated walls, 0.5 % oxygen accelerates the production of hydrogen (and ethylene) ; large at the very beginning of the reaction, the accelerating effect decreases, then quickly stops (see Figure 1).

By contrast, in a vessel of the same surface to volume ratio but with PbO-coated walls, small amounts of oxygen strongly inhibit the production of hydrogen (and ethylene) ; this inhibiting effect lasts a long time (see Figure 2).

In the presence of oxygen, ethylene yield is higher than hydrogen because ethylene arises from both decomposition (A) and oxidation (B) of ethane, whereas hydrogen is representative of this alkane decomposition (A) (see § 1).

Oxygen is swiftly consumed when it accelerates ethane pyrolysis. On the contrary, it is slowly consumed when it inhibits this reaction (see Figure 3).

- Table II -

Relative importances (%) of the primary stoichiometries of n-butane *decomposition* and *oxidation* at low extent of reaction

T = 494°C ; $P_o^{n\,C_4H_{10}}$ = 50 mm Hg ; PbO-coated pyrex vessel (s/v ≃ 5 cm^{-1})

	$(O_2)_o$ %	0	0.01	0.07	0.13
decomposition	n C_4H_{10} = C_3H_6 + CH_4	71 %	68.7 %	47.5 %	37 %
	n C_4H_{10} = C_2H_4 + C_2H_6	21 %	10.7 %	1.5 %	≃ 0 %
	n C_4H_{10} = 2 C_2H_4 + H_2	8 %	4.7 %	≃0 %	≃ 0 %
	n C_4H_{10} = n C_4H_8 + H_2	< 1 %	≃0 %	≃0 %	≃ 0 %
oxidation	n C_4H_{10} + O_2 (or 1/2 O_2) = 2 C_2H_4 + H_2O_2 (or H_2O)	—	10.7 %	21 %	19.5 %
	n C_4H_{10} + O_2 (or 1/2 O_2) = n C_4H_8 + H_2O_2 (or H_2O)	—	5.2 %	30 %	43.5 %

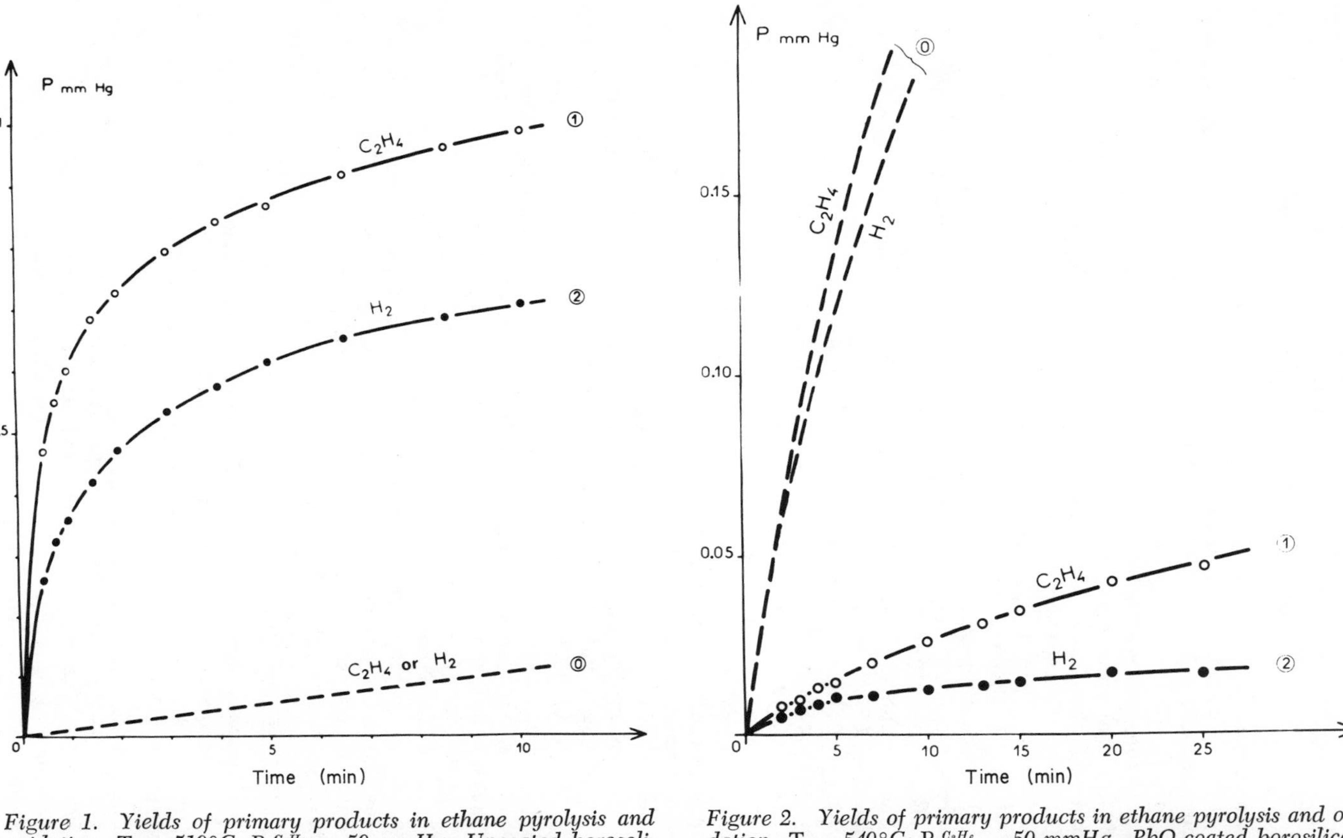

Figure 1. Yields of primary products in ethane pyrolysis and oxidation. $T = 510°C$; $P_0^{C_2H_6} = 50$ *mmHg. Uncoated borosilicate vessel* ($s/v \simeq 10.5$ cm^{-1}). *Curve (0): without oxygen. Curves (1) and (2):* $(O_2)_0 \simeq 0.5\%$.

Figure 2. Yields of primary products in ethane pyrolysis and oxidation. $T = 540°C$; $P_0^{C_2H_6} = 50$ *mmHg. PbO-coated borosilicate vessel* ($s/v \simeq 10.5$ cm^{-1}). *Curve (0): without oxygen. Curves (1) and (2):* $(O_2)_0 \simeq 0.45\%$.

We made similar observations for the pyrolyses of isobutane, n-butane and of isopentane in the presence of oxygen traces.

The case of isopentane, for instance, is illustrated on Figures 4 and 5.

In Figure 4 the pressure of pyrolyzed isopentane (computed from results obtained in chromatographic analyses of the decomposition products of this hydrocarbon) is plotted as a function of time, oxidized isopentane being not taken into account (see Table III). Curve (0) represents the pyrolysis of pure isopentane in either of the four vessels which have been used. Curves (1), (2), (3) and (4) show that about 1% oxygen can have a large effect, accelerating or inhibiting, according to the extent and the nature of the walls, upon isopentane pyrolysis.

Figure 5 shows that oxygen is swiftly consumed in Vessels (1) and (2) where isopentane decomposition is initially accelerated; by sharp contrast, oxygen is slowly consumed in Vessels (3) and (4), where the alkane decomposition is inhibited.

These observations are similar to those we previously made on propane (4) (5).

In short, according to experimental conditions (especially to vessel wall conditions), oxygen accelerates or, on the contrary, inhibits the pyrolysis of several alkanes (ethane, propane, isobutane, n-butane, isopentane). In the presence of oxygen, an increase of the vessel wall surface or a PbO (or KCl) wall coating causes a strong diminution of the alkane decomposition rate* and oxygen consumption rate. When accelerating, the oxygen effect, first decreases, then very quickly vanishes, whereas it lasts a long time, when inhibiting. Correlatively, oxygen is consumed quickly in the first case and slowly in the second one.

J. E. BLAKEMORE et al. (12) find that the initial rate of n-butane decomposition in a gold reactor of relatively high s/v ratio (32 in^{-1}) is depressed by the presence of small quantities of oxygen at 535 and 595°C. This result is similar to our own observations in pyrex vessels, at about 500°C (6c).

B. L. CRYNES and L. F. ALBRIGHT (14) have noted that the course of the propane decomposition at 700°C is different after a stainless steel reactor has been treated with oxygen. The product composition and propane conversion change significantly, especially in the initial stages of the run.

Interpretation and Discussion

The thermal decomposition of a pure alkane (μH) is an essentially homogeneous chain radical reaction.

Let us recall, for instance, the mechanism of the homogeneous pyrolysis of ethane, around 500°C and a low extent of reaction (15):

* On the contrary, the rates of pyrolyses of pure alkanes are nearly independent (case of n-butane and of isopentane) or slightly dependent (case of ethane and of isobutane) on wall extent and coating of the reaction vessel (13).

- Table III -

Relative importances (%) of the primary stoichiometries of *isopentane decomposition* and *oxidation* at low extent of reaction

T = 480°C ; $P_o^{i\,C_5H_{12}}$ = 25 mm Hg ; PbO-coated pyrex vessel (s/v ≃ 0.8 cm^{-1})

	$(O_2)_o$ %	0	1
decomposition	i C_5H_{12} = C_4H_8 + CH_4	73.5 %	70 %
	i C_5H_{12} = C_3H_6 + C_2H_4 + H_2	14 %	7.5 % (both rows)
	i C_5H_{12} = i C_5H_{10} + H_2	0.5 %	
	i C_5H_{12} = C_3H_6 + C_2H_6	10 %	5.5 %
	i C_5H_{12} = C_2H_4 + C_3H_8	2 %	1 %
oxidation	i C_5H_{12} + O_2 (or 1/2 O_2) = C_3H_6 + C_2H_4 + H_2O_2 (or H_2O)	—	16 % (both rows)
	i C_5H_{12} + O_2 (or 1/2 O_2) = i C_5H_{10} + H_2O_2 (or H_2O)	—	

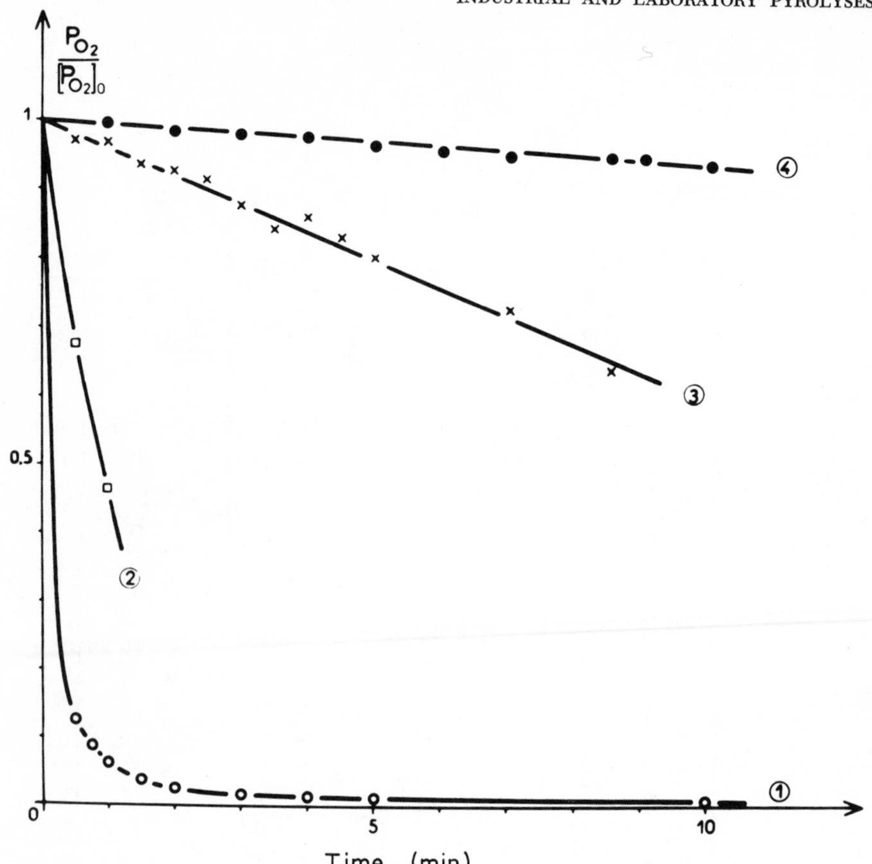

Figure 3. Oxygen consumption vs. time in pyrolysis and oxidation of ethane ($P_o^{C_2H_6}$ = 50 mmHg) at 540°C

	Pyrex vessels		$(O_2)_o$	ethane decomposition
	s/v(cm^{-1})	coatings		
(1)*	10.5	uncoated	0.5 %	strong *acceleration*
(2)	0.8	PbO (light)	0.4 %	weak *acceleration*
(3)	0.8	PbO (heavy)	0.45%	weak *inhibition*
(4)	10.5	PbO (heavy)	8.8 %	strong *inhibition*

* 510°C

Initiation :

$$C_2H_6 \longrightarrow 2\ CH_3\cdot \qquad (1)$$

$$CH_3\cdot + C_2H_6 \longrightarrow CH_4 + C_2H_5\cdot \qquad (x)$$

Propagation

$$\left\{ \begin{array}{ll} C_2H_5\cdot \longrightarrow C_2H_4 + H\cdot & (2) \\ H\cdot + C_2H_6 \longrightarrow H_2 + C_2H_5\cdot & (3) \end{array} \right.$$

Termination :

$$2\ C_2H_5\cdot \longrightarrow n\text{-}C_4H_{10} \quad (\text{or } C_2H_4 + C_2H_6) \qquad (\mu\mu)$$

When the alkane (μH) is pyrolyzed <u>in the presence of oxygen</u>, <u>new</u> processes can appear.

1 - Chain carriers of the decomposition can also become oxidized and yield oxygenated free radicals (particularly $HO_2\cdot$ or $.OH$), which then <u>propagate</u> the alkane chain <u>oxidation</u> (into olefin + H_2O_2 or H_2O). This reaction <u>competes</u> with the alkane chain decomposition.

For example, chain carriers of ethane decomposition, $C_2H_5\cdot$ and H., can be oxidized as indicated in the following scheme :

$$\left\{ \begin{array}{lll} C_2H_5\cdot + O_2 \longrightarrow C_2H_4 + HO_2\cdot & & (4\mu) \\ \text{or} \left[\begin{array}{l} C_2H_5\cdot \longrightarrow C_2H_4 + H\cdot \\ H\cdot + O_2\ (+M) \longrightarrow HO_2\cdot\ (+M) \end{array} \right. & & \begin{array}{l} (2) \\ (4\beta) \end{array} \\ HO_2\cdot + C_2H_6 \longrightarrow H_2O_2 + C_2H_5\cdot & & (5) \end{array} \right.$$

and/or

$$\left\{ \begin{array}{lll} C_2H_5\cdot \longrightarrow C_2H_4 + H\cdot & & (2) \\ H\cdot + O_2 \longrightarrow .OH + O: & & (4\beta') \\ O: + C_2H_6 \longrightarrow .OH + C_2H_5\cdot & & (4) \\ .OH + C_2H_6 \longrightarrow H_2O + C_2H_5\cdot & & (5') \end{array} \right.$$

This scheme accounts for extra yield of C_2H_4 and H_2O_2-H_2O formation.

2 - Besides, oxygen may cause a new <u>initiation</u> process :

$$\mu H + O_2 \longrightarrow \mu\cdot + HO_2\cdot \qquad (1')$$

which adds to the quasi monomolecular process :

$$\mu H \longrightarrow \text{free radicals} \qquad (1)$$

Then, oxygen would have a first <u>accelerating</u> influence on the decomposition of the alkane (μH).

Moreover, a <u>branching</u> process like $4\beta'$ also leads to an <u>accelerating</u> effect of oxygen.

At last, as soon as oxidation processes such as 5 have produced traces of hydrogen peroxide, the <u>degenerate branching</u> process:

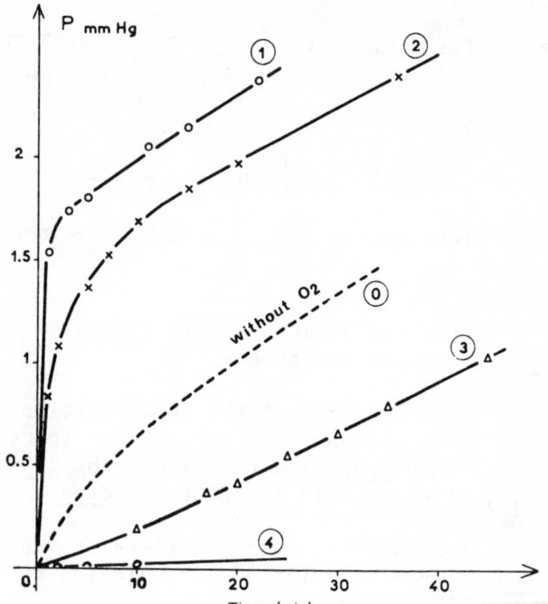

Figure 4. Pyrolysis of isopentane ($P_0^{i\text{-}C_5H_{12}}$ = 25 mmHg), at 480°C, in presence of oxygen [$(O_2)_0 \simeq 1.1\%$] and in various pyrex vessels: (1) s/v ≃ 0.8 cm⁻¹, uncoated; (2) s/v ≃ 0.8 cm⁻¹, PbO coated; (3) s/v ≃ 2 cm⁻¹, PbO coated; (4) s/v ≃ 11 cm⁻¹, PbO coated. Decomposed isopentane vs. time.

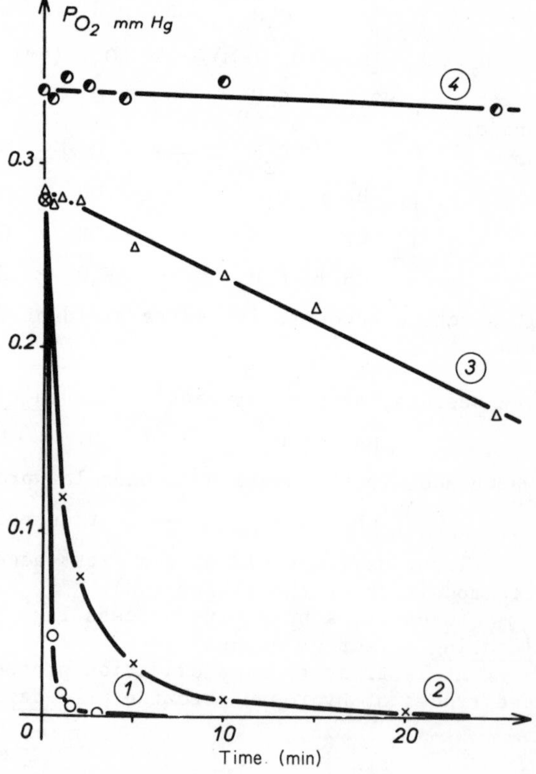

Figure 5. Pyrolysis of isopentane ($P_0^{i\text{-}C_5H_{12}}$ = 25 mmHg), at 480°C, in presence of oxygen [$(O_2)_0 \simeq 1.1\%$] and in various pyrex vessels: (1) s/v ≃ 0.8 cm⁻¹, uncoated; (2) s/v ≃ 0.8 cm⁻¹, PbO coated; (3) s/v ≃ 11 cm⁻¹, PbO coated; (4) s/v ≃ 11 cm⁻¹, PbO coated; Oxygen consumption vs. time.

$$H_2O_2 \ (+M) \longrightarrow 2\ .OH\ (+M) \qquad (1'')$$

increases the rate of the free radical formation ; so, H_2O_2 has a self-accelerating effect, which could appear very early in the course of the reaction.

3 - But the presence of new (oxygenated) free radicals in the system, tends to make new chain termination processes appear ; these processes are heterogeneous, some of them at least, for it has been shown (16) that, unlike alkyl radicals, oxygenated radicals (particularly HO_2.) are easily destroyed on solid surfaces, especially when they are covered with PbO (or KCl) coatings. Under these circumstances, oxygen would have also an inhibiting effect on the alkane decomposition ; this effect would be all the more important as vessel walls are larger or covered with PbO (or KCl) coatings.

Therefore we are led to ascribe two opposed kinetic effects to oxygen, the global result of which (acceleration or inhibition) depends upon experimental conditions. The mechanism thus put forward allows us to account qualitatively for the influences of small oxygen concentrations and vessel walls on the formation rate of major primary products and for the inhibiting effect of the walls on oxygen consumption (6) (7).

J.E. TAYLOR and D.M. KULICH (17) using a wall-less reactor at temperatures above 600°C, also observed an accelerating effect of oxygen on ethylene production in the reaction of a mixture of 1 % ethane and 0.4 % oxygen in nitrogen. But this accelerating effect would be preceded, at very low extent of reaction (less than 0.1 % at about 630°C), with an inhibiting effect of oxygen. After these authors, at the very beginning of the reaction, a relatively unreactive free-radical, HO_2., is formed which is gradually converted to a more reactive species .OH* ; as this conversion becomes significant, the reaction is accelerated. However, let us note that TAYLOR and KULICH did not make hydrogen analyses : hydrogen yields are representative of ethane decomposition whereas ethylene arises from both decomposition and oxidation of this alkane.

Let us recall that several stoichiometries are necessary to represent the primary decomposition of isobutane, n-butane and isopentane. The presence of oxygen makes oxidation reactions appear. If we only consider the decomposition reactions (the oxidation ones being set apart), we see that the addition of oxygen modifies the relative weights of the decomposition stoichiometries :

---

* through processes :

$$HO_2. + C_2H_6 \longrightarrow H_2O_2 + C_2H_5.$$

$$2\ HO_2. \longrightarrow H_2O_2 + O_2$$

$$H_2O_2\ (+M) \longrightarrow 2\ .OH\ (+M)$$

particularly, it favours the reaction of "demethanation" of the alkane (see Tables I, II and III).

This perturbation can be explained. Indeed, when the pyrolysis of an alkane (μH) involves several stoichiometries, the reaction chain includes several kinds of decomposition sequences such as :

$$\begin{cases} \mu\cdot \longrightarrow m + \beta\cdot & (2) \\ \beta\cdot + \mu H \longrightarrow \beta H + \mu\cdot & (3) \end{cases}$$

There are then several kinds of free radicals β. (β_1., β_2., ...) and several isomeric forms of the free radical μ.. Now, the chain reactions of decomposition and oxidation of the alkane are competing. Thus, if different chain carriers of the decomposition are oxidized with different efficiencies -and this is very likely- oxygen addition will disturb the relative weights of the decomposition stoichiometries of the hydrocarbon. Since the presence of oxygen favours the relative weight of the "demethanation" stoichiometry :

$$C_n H_{2n+2} = C_{n-1} H_{2(n-1)} + CH_4$$

which is propagated by the free radical CH_3., it seems that this methyl radical is less easily oxidized than other chain carriers of the alkane decomposition.

Conclusion

The pyrolysis of four alkanes (ethane, isobutane, n-butane and isopentane) was studied in the presence of small quantities of oxygen, at low extent of reaction.

The organic products of primary oxidation are mainly olefins.

The kinetic effect of oxygen on the thermal decomposition of these alkanes is twofold : accelerating and inhibiting. The global result of these two opposed kinetic effects depends on experimental conditions, especially on wall conditions. Acceleration or inhibition of the pyrolysis of these paraffins can be important on particular conditions. When accelerating, oxygen is quickly consumed, whereas it slowly disappears when inhibiting.

The interpretation of these results is based on the facts that, in the presence of oxygen, chain reactions of decomposition and of oxidation are competitive and that it appears new homogeneous processes of free-radical production (which result in an accelerating effect of O_2) and new termination processes at least are partly heterogeneous (which result in an inhibiting effect).

At last, in the cases of isobutane, n-butane and isopentane, the presence of oxygen increases the relative weight of the "demethanation" reaction (with regard to the other decomposition reactions of the alkane). This indicates that the free-radical CH_3. becomes less easily oxidized than other chain carriers of the alkane decomposition.

Summary

Results obtained in an investigation of the pyrolyses of four alkanes (ethane, n-butane, isobutane and isopentane) in the presence of trace amounts of oxygen, at low extent of reaction and around 500°C, are reported. The organic products of the primary oxidation are mainly olefins. According to experimental conditions (particularly to wall conditions of the reaction vessel), oxygen accelerates or inhibits the alkane decomposition. Walls inhibit the oxygen consumption. These experimental facts are interpreted and compared with recent results in literature.

Literature Cited

(1) LETORT M., Compt. Rend. Acad. Sc., Paris, (1933) 197, 1042 ; (1935) 200, 312 ; J. de Chimie phys., (1937) 34, 428.

(2) RICE F.O. and POLLY O.L., Trans. Faraday Soc., (1939) 35, 850.
MAIZUS Z.K., MARKOVICH V.G. and NEIMAN M.B., Doklady Akad. Nauk. S.S.S.R., (1949) 66, 1121.
APPLEBY W.G., AVERY W.H., MEERBOTT W.K. and SARTOR A.F., J. Am. Chem. Soc., (1953) 75, 1809.
SKARTCHENKO V.K. and KOUZMITCHEV S.P., Kinetika i Kataliz, (1970) 11, 845.

(3) ENGEL J., COMBE A., LETORT M. and NICLAUSE M., Bull. Soc., Chim. Fr., (1954) 1202 ; Rev. Inst. Fr. Pétrole, (1957) 12, 627.

(4) MARTIN R., NICLAUSE M. and DZIERZYNSKI M., Compt. Rend. Acad. Sc., Paris, (1962) 254, 1786 ; J. de Chimie phys., (1964) 61, 286, 790 and 802.

(5) NICLAUSE M., MARTIN R., COMBES A. and DZIERZYNSKI M., Canad. J. Chem., (1965) 43, 1120.

(6) See last parts of the following theses maintained at Nancy :
a) SCACCHI G. (1969) ; b) FUSY J. (1966) ;
c) LARGE J.F. (1970) ; d) COMBES A. (1965).

(7) NICLAUSE M., MARTIN R., COMBES A., FUSY J. and DZIERZYNSKI M., Communication n° 244 (Section n° 20) at the 36th International Congress of Industrial Chemistry, Brussels (Sept. 1966) ; Ind. Chim. Belge, (1967) 32, special issue, vol. II, 674.

(8) SCACCHI G., MARTIN R. and NICLAUSE M., Compt. Rend. Acad. Sc., Paris, (1972) 275 C, 1347.

(9) SCACCHI G., LARGE J.F., MARTIN R. and NICLAUSE M., Communication P 12 at the Symposium on the Mechanisms of Hydrocarbon Reactions, Siófok - Hungary (June 1973).

(10) See for ex. : SATTERFIELD C.N. and coll., Ind. Eng. Chem., (1954) 46, 1001 ; J. Phys. Chem., (1955) 59, 283 ; 5th Symposium (International) on Combustion, Reinhold Publ. Corp., New York (1955), p. 511.
KNOX J.H. and WELLS C.J., Trans. Faraday Soc., (1963) 59, 2786 and 2801.

SAMPSON R.J., J. Chem. Soc., (1963) 5095 ; communication at the Symposium "Oxidation in Organic Chemistry", Manchester (March 1965).

BENSON S.W., Advan. Chem. Ser., (1968) 76, 145.

BALDWIN R.R., HOPKINS D.E. and WALKER R.W., Trans. Farad. Soc., (1970) 66, 189.

(11) Patents of Imperial Chemical Industries Ltd :

(a) SAMPSON R.J., Brit. 976, 966 (Cl C 07 c),Dec. 2nd 1964, Appl. Oct. 13rd 1960.

(b) SAMPSON R.J. and TAYLOR J.S., Brit. 945, 448 (Cl C 07 c), Jan. 2nd 1964, Appl. Jan. 4th 1962.

(12) BLAKEMORE J.E., BARKER J.R. and CORCORAN W.H., Ind. Eng. Chem. Fundam., (1973) 12, 147.

(13) See, for ex. : FUSY J., SCACCHI G., MARTIN R., COMBES A. and NICLAUSE M., C.R. Acad. Sc. Paris, (1965) 261, 2223.

(14) CRYNES B.L. and ALBRIGHT L.F., I. and E.C. Proc. Des. and Dev., (1969) 8, 25.

(15) See, for ex. : SCACCHI G., MARTIN R. and NICLAUSE M., Bull. Soc. Chim. France, (1971) 731.

(16) See, for ex. : LEWIS B. and von ELBE G., Combustion, Flames and Explosions of Gases, Acad. Press, New York, 2nd edition (1961).

SHARMA R.K. and BARDWELL J., Combustion and Flame, (1965) 9, 106.

(17) TAYLOR J.E. and KULICH D.M., Int. J. Chem. Kin., (1973) 5, 455.

4

Mechanistic Studies of Methane Pyrolysis at Low Pressures

K. D. WILLIAMSON and H. G. DAVIS
Union Carbide Corp., South Charleston, W.Va. 25303

Introduction

In recent years, most workers have adopted an abridged version of the Rice-Herzfeld mechanism(1-5) for ethane pyrolysis

$$C_2H_6 \longrightarrow 2\ CH_3\cdot \quad \text{(Initiation)} \quad 1)$$

$$CH_3\cdot + C_2H_6 \longrightarrow CH_4 + C_2H_5\cdot \quad 2)$$

$$C_2H_5\cdot \longrightarrow C_2H_4 + H\cdot \quad \text{(Propagation)} \quad 3)$$

$$H\cdot + C_2H_5 \longrightarrow H_2 + C_2H_5\cdot$$

$$2\ C_2H_5\cdot \longrightarrow C_4H_{10} \quad \text{(Termination)} \quad 5a)$$

$$\text{or } C_2H_4 + C_2H_6 \quad 5b)$$

Other possible termination reactions, such as (6), (7), or (8) are of minor importance because the

$$CH_3\cdot + C_2H_5\cdot \longrightarrow C_3H_8 \quad 6)$$

$$H\cdot + C_2H_5\cdot \longrightarrow C_2H_6 \quad 7a)$$

$$\text{or } C_2H_4 + H_2 \quad 7b)$$

$$2\ CH_3 \longrightarrow C_2H_6 \quad 8)$$

concentrations of hydrogen atom and methyl radical are small compared to the concentration of ethyl radical, at common temperatures and pressures. Termination reactions for reactions (9) and (10) are ruled out also because they require three-body collisions

$$2\,H\cdot \longrightarrow H_2 \qquad 9)$$

$$\text{and } H\cdot + CH_3\cdot \longrightarrow CH_4 \qquad 10)$$

to stabilize the newly-formed molecules, and are improbable on that account.

The over-all reaction becomes more complicated than indicated by reactions (1)-(5) before much pyrolysis has occurred. Data on the rates of the primary reactions (obtained photochemically and otherwise) show that reactions (3R) and (4R), the reverse of (3) and (4), must become significant very quickly (2). These reactions offer the simplest explanation of the self-inhibition of the over-all reaction which develops when only a few per cent decomposition of ethane has occurred. Moreover, the primary product, ethylene, reacts with the free radicals of the system, principally ethyl, because of its relative abundance, to form molecules in the C_3-C_4 range, and eventually molecules of even higher molecular weight. Additionally, the butane which accumulates as a chain-termination product, pyrolyzes approximately four times as fast as its parent ethane--making only slight stoichiometric changes, but fast obscuring the mechanistic clues associated with minor products.

Obviously, as pyrolysis of the ethane proceeds, the over-all picture becomes ever more complicated. Any present hope of understanding the mechanism must be based on satisfactory knowledge of the initial phenomena. Ideally, data would be taken at essentially zero decomposition of the ethane. However, this is extremely difficult experimentally, especially at the higher temperatures of commercial interest. The necessary practical compromise is to take data at a few per cent decomposition, where product compositions, reaction times, etc., are finite--then extrapolate or correct back to zero composition. That is the course followed here.

The primary purpose of the present work was to check some phenomena which follow from the projected mechanism. Two principal effects to be looked for were: (1) An increase in the specific rate of the over-all pyrolysis as the partial pressure of ethane is lowered, and (2) an indication that ethyl combination was not the exclusive reaction for chain termination at very low pressure. These phenomena are deduced from a steady-state treatment of reactions (1)-(5), with the back reactions (3R) and (4R) included. This treatment gives for the over-all rate of ethane pyrolysis*:

$$-\frac{d\ \ln [C_2H_6]}{dt} = \frac{(k_1/k_5)^{1/2}\ k_3(1-\frac{Q}{K})\ [C_2H_6]^{-1/2}}{1+\frac{k_4}{k_{3R}}\frac{[C_2H_4]}{[C_2H_6]}} \qquad \text{(A)}$$

where $Q = \frac{[H_2]\ [C_2H_4]}{[C_2H_6]}$ at a given time, and K is the value of Q at equilibrium. The left-hand side of (A) is in the form of a first-order rate constant. However, the right-hand side, containing variable concentration terms, cannot be constant. Thus, if the proposed mechanism is valid, the over-all reaction cannot be truly first-order--although it may appear approximately first-order over limited ranges of reaction. At "zero" decomposition, (A) takes the simplified form,

$$-\frac{d\ \ln [C_2H_6]}{dt} = \frac{C}{[C_2H_6]^{1/2}}$$

where $$C = \left(\frac{k_1}{k_5}\right)^{1/2} k_3$$

Thus, at very low conversions, the apparent first-order constant for ethane pyrolysis should be inversely proportional to the half-power of the ethane concentration.

Other expressions arising from the steady-state treatment are:

$$[CH_3\cdot] = 2\ k_2/k_1$$

---

*This expression neglects the approximately 1 to 3% of the ethane that forms methane or butane in the initiation and termination steps at low decompositions (somewhat more at higher decompositions).

$$[C_2H_5\cdot] = \left(2\, k_1/k_5\right)^{1/2} [C_2H_6]^{1/2}$$

$$[H\cdot] = \frac{\left\{k_3 + k_{4R}\right\}[H_2] \left(\frac{2\, k_1}{k_5}\right)^{1/2} [C_2H_6]^{1/2}}{k_4 [C_2H_6] + k_{3R} [C_2H_4]}$$

and at low decomposition of ethane,

$$[H\cdot]_o \doteq \left(\frac{2\, k_1}{k_5}\right)^{1/2} \left(\frac{k_3}{k_4}\right) [C_2H_6]^{-1/2} .$$

Thus, as $[C_2H_6]$ decreases, $[CH_3\cdot]$ is invariant, $[C_2H_5\cdot]$ decreases and $[H\cdot]$ increases.

Previous work (2) indicated that at one atmosphere pressure and 775°C,

$$[C_2H_5\cdot] \doteq 10\, [CH_3\cdot] \doteq 400\, [H\cdot]$$

so that ethyl is much the predominant radical, and ethyl + ethyl much the most favored radical recombination reaction. However, lowering the ethane concentration greatly alters these ratios, as shown below:

Radical	Pressure (atm.) at 1000K $C_2H_6 = 1$	$C_2H_6 = 0.1$	$C_2H_6 = 0.01$
$C_2H_5\cdot$	1.0×10^{-5}	3.2×10^{-6}	1.0×10^{-6}
CH_3	9.7×10^{-7}	9.7×10^{-7}	9.7×10^{-7}
$H\cdot$	2.4×10^{-8}	7.6×10^{-8}	2.4×10^{-7}

Considering the uncertainties with which these radical concentrations are known, it is reasonable to say that the steady-state concentrations of the three radicals during ethane pyrolysis at 0.01 atmosphere are in the same range. Evidence of this equalization would be increased ratio of propane/butane in the product, or in finding a deficit of chain termination between hydrogen atom and ethyl occurred, e.g.

$$H\cdot + C_2H_5\cdot \longrightarrow C_2H_6$$

$$\text{or } H_2 + C_2H_4$$

since the products of these chain terminations could not be distinguished from other major amounts of these species in the reaction.

Eventually, of course, if the chain termination steps change from (5) to (6), etc., the over-all rate expression and the expressions for radical concentrations will change. For chain termination dominated by (7), $CH_3\cdot$ remains the same as previously, but

$$[H\cdot]_o = \left(\frac{k_1k_3}{k_4k_7}\right)^{1/2} \quad \text{(not affected by } [C_2H_6]\text{)}$$

and

$$\frac{d \ln [C_2H_6]}{dt} = \left(\frac{k_1k_3k_4}{k_7}\right)^{1/2} \quad \text{(a true first-order expression).}$$

Thus, a transition from ethyl-ethyl termination to ethyl-hydrogen atom termination would exhibit a lining-out of the over-all decomposition rate at some maximum level.

These various points will be checked against the experimental data reported here.

Experimental Work

The experimental work reported here was performed in different quartz flow reactors, having surface-volume ratios from 1 to 20. Different reactors were used for two reasons--first, to demonstrate that results were relatively insensitive to the surface-volume ratio, and other features relating to a particular reactor; second, to permit maintenance of proper flow ratios, for desired reaction times, without appreciable pressure drop. Flow reactors were required because of the relatively high temperatures used, and the short reaction times required for low conversions.

Virtually all longitudinal temperature gradient was removed from two annular reactors used (B and C) by mounting them within fluidized beds, then making final adjustment on segment heaters along the reactors. This type of reactor is illustrated in Figure 1. Gradients were minimized along the 1" tubular reactor (A) by mounting it within a massive copper block. Reactors were seasoned by passing ethane through them for several hours prior to a run. At the relatively high temperatures used here, surface reactions should be less important than for some reported in the literature. However, as discussed

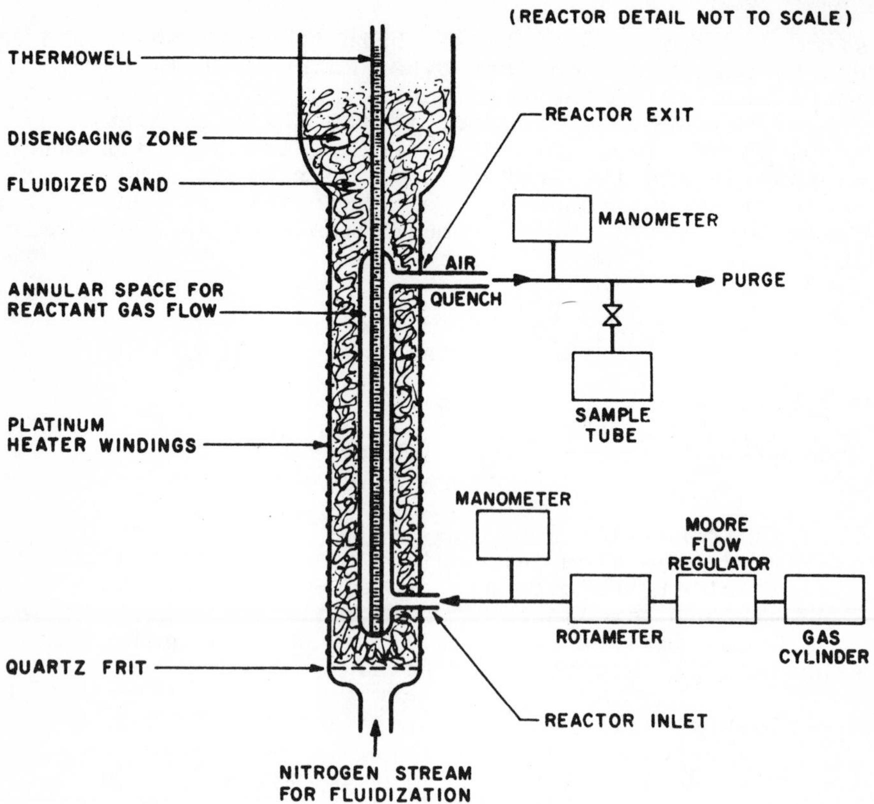

Figure 1. Schematic of apparatus for pyrolysis of hydrocarbons

later, the possibility of surface reactions must be considered at the lower partial pressures of ethane explored in the present work.

Gas flowing out of the reactor was immediately quenched by a small quartz water condenser just beyond the reaction zone. Beyond the condenser was a sampling point, for collecting samples for gas chromatography. All components in the analysis were checked quantitatively against standard samples. In addition, some analyses were performed by mass spectrometer, as a check against the GC. Complete atom balances were made on each run. These atom balances provided the most quantitative basis for calculating the expansion of gas due to reaction, a factor required for calculating accurate residence times.

TABLE I

PYROLYSIS OF ETHANE AT ATMOSPHERIC PRESSURE

Temp., °C	673	698	723	744	768
Reaction Time (seconds)	2.52	2.37	2.17	1.88	1.66
Products	Moles/100 Moles of Ethane Fed				
H_2	4.9	11.5	21.1	31.9	45.6
CH_4	0.088	0.33	0.91	2.06	4.71
C_2H_2		0.004	0.007	0.033	0.107
C_2H_4	4.9	11.6	21.4	31.9	44.0
C_2H_6	95.0	88.1	77.8	66.2	50.3
C_3H_6	0.005	0.033	0.14	0.35	0.66
C_3H_8			0.003	0.020	0.051
C_4H_6		0.013	0.060	0.149	0.390
1-C_4H_8	0.003	0.019	0.043	0.054	0.072
2-C_4H_8		0.004	0.005	0.010	0.038
C_4H_{10}	0.028	0.074	0.128	0.149	0.144
Chains initiated* / Chains terminated	1.24	1.32	1.09	1.33	1.20

*Estimated number of chains initiated divided by estimated number of chains ended by ethyl combination and disproportionation. See text.

Scope of Data. Experiments covered the temperature range 625-775°C, absolute pressures from 0.13-1.67 atm., and steam dilutions up to 100:1 molar. Reaction times were from 0.5-2.5 seconds, depending on other conditions, to give a range of conversions from 0.2-75%. For present purposes, the data at low conversions are most pertinent.

Sample data for one set of runs is shown in Table I. Characteristics of the various sets of data used for this paper are given in Table II.

Discussion

The first prediction of the model to be checked is variation of the initial rate of disappearance of ethane with pressure. As mentioned earlier, the data were taken at appreciable decomposition of ethane,

TABLE II

DATA USED IN THIS STUDY

PYROLYSIS OF ETHANE AT CONDITIONS SHOWN

Reactor	Pressure (atm.)	Steam Dilution	Nominal Reaction Time (sec.)
A	1	0:1	2.5
A	1	0.64:1	2.5
A	1	8:1	2.5
A	1.67	0:1	2.5
A	1.67	0.77:1	2.5
A	1	9:1	1.0
A	1	100:1	1.0
A	1	100:1	2.5
B	1	0:1	0.5
C	0.13	0:1	0.5

A = Quartz reactor 1" I.D., 90 cm long in copper block, 147 ml vol.

B = Quartz annular reactor 1 mm annulus, 60 cm long in fluid bed, 38 ml. vol.

C = Quartz annular reactor 1 mm annulus, 60 cm long in fluid bed, 16 ml vol.

and must be corrected back to the values at "zero" decomposition for valid comparison. The method of achieving this will be described.

From expression (A) the ratio of the rate constant at time zero to that at time t is derived as

$$R = \left(\frac{1 + \frac{k_4}{k_{3R}} \frac{[C_2H_4]_t}{[C_2H_6]_t}}{1 - Q_t/K} \right) \left(\frac{[C_2H_6]_t}{[C_2H_6]_o} \right)^{1/2}$$

where the values with subscript t are those existing at time t. R can be set up as a function of the fraction of ethane decomposed, with one adjustable parameter, k_4/k_{3R}. This parameter can be evaluated by matching R with real data for ethane decomposition at constant temperature and variable time. This has been done for several sets of data, not presented here, to give $k_4/k_{3R} = 2$, a reasonable value in terms of photochemical data (2). The function R is then used to generate an integral function k/k_o, where $\underline{k}$ is the virtual rate constant, which has been measured, and $\underline{k_o}$ is the rate constant at zero time. Figure 2 gives a plot of this function. If, for example, a rate constant is calculated for a run with 30% decomposition of ethane, then that rate constant is divided by 0.75 to obtain $\underline{k_o}$. Analysis shows that k/k_o should be nearly linear until equilibrium is closely approached. Near equilibrium, the value of k/k_o drops sharply. The departure from linearity, occurring at different decompositions at different temperatures is not shown here. For the relatively low decompositions considered here, it was not important.

Rate constants, so adjusted, are shown as Arrhenius plots in Figure 3. The somewhat different slopes of the various lines probably reflect some error in the correction technique for runs at very high decomposition, and errors in measuring decomposition for runs at very low decomposition. Another point of interest is comparison of data from different reactors. Rate constants measured in the annular reactor (B) seem to run about 25% higher than those in the reactor of one-inch diameter (A). This possibly measures a difference in rate of heat transfer leading to errors in opposite directions in the estimate of absolute temperature level. Despite these relatively small discrepancies, the data for the different reactors over a wide span of decomposition demonstrate the general validity of the experimental work. Closer comparisons can be drawn, of course, between sets of runs done in the same reactor than between sets done in different reactors.

The rate constants for runs at one atmosphere, but 9:1 dilution, are much higher than their counterparts for undiluted ethane at one atmosphere. Furthermore, rate constants for 100:1 dilution are still higher. The comparison between rate constants at these three conditions at 725°, and predictions from the model are shown below. Shown also is k_o for another set at 1.67 atmosphere pressure (no dilution), which is not included on the Arrhenius plot.

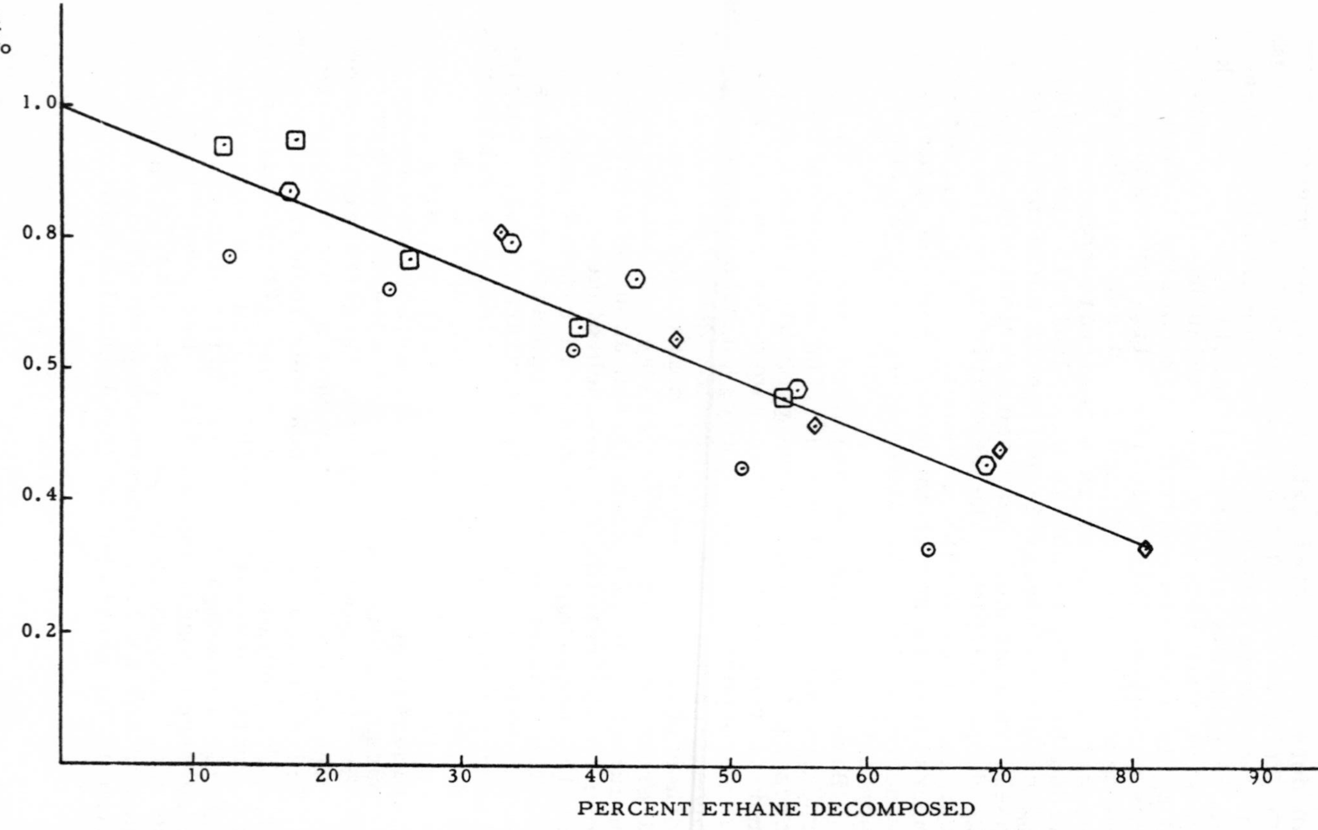

Figure 2. "Universal" curve for correcting ethane decomposition rate constants to zero time. ○ 800°; □ 850°; 875°; ◇ 900°.

k_o (rel.) at Partial Pressure Indicated	Relative Rate Constants for Ethane Pyrolysis	
	Measured	Predicted
1.67	0.81	0.76
1.	1.0	1.
0.1	1.9	3.2
0.01	4.1	10.

In contrast to the above, a real anomaly exists with the data for 100 mm. pressure, with no steam dilution. The k's for this set actually fall below

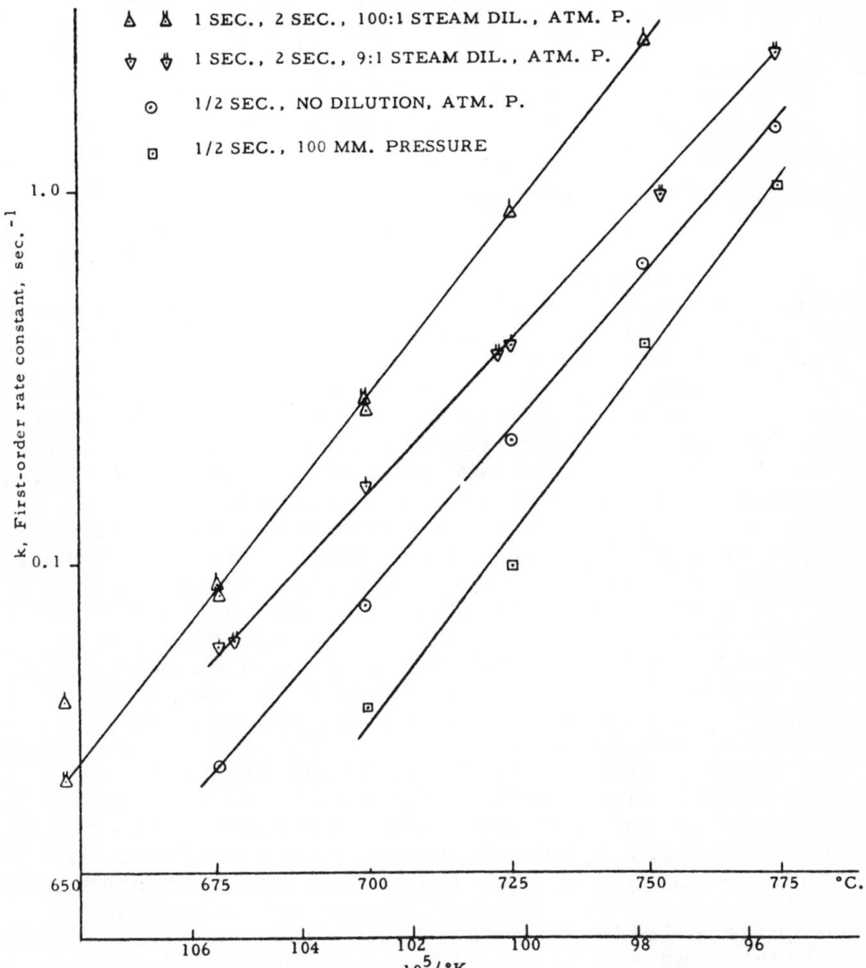

Figure 3. Ethane decomposition rate constants for zero time. Comparison at different partial pressures of ethane.

TABLE III

RATE CONSTANT FOR $C_2H_6 \longrightarrow 2\ CH_3\cdot$ AT 675°C

Conditions

Reactor	Pressure (atm.)	Dilution	Residence Time (sec.)	k_1 (corr) $\times 10^4 (sec^{-1})$
A	1	0:1	2.52	1.65
A	1	8:1	2.60	1.59
A	1.67	0:1	2.48	1.63
A	1.67	0.77:1	2.46	1.77
A	1	9:1	1.01	1.91
A	1	100:1	1.01	2.42
A	1	100:1	1.98	2.47
B	1	0:1	0.57	3.25
C	0.13	0:1	0.59	1.61

(See Table I for identification of reactors.)

their counterparts at one atmosphere. The explanation for this exception seems to lie in the pressure sensitivity of one or more of the rate constants within the cluster $C = (k_1/k_5)^{1/2}k_3$. A pressure effect on k_5 would be in the wrong direction and is unlikely in any case, so k_5 can not be the cause of the low k_o. Previous reports (3,4a,6) have indicated that k_1, for the cleavage of ethane into two methyls, becomes pressure-sensitive at about 100 mm., the present pressure. However, data shown later in this paper indicate no appreciable drop in k_1 at 100 mm., as compared with its value at higher pressure. Probably the principal cause for the abnormally low ethane decomposition at 100 mm. is the pressure sensitivity of k_3. This has also been documented previously (2-4) though it is hard to quantify. Present evidence suggests that in the rate determining cluster $(k_1/k_5)^{1/2}\ k_3\ [C_2H_6]^{-1/2}$, lowering the total pressure reduces k_3 proportionately more than it increases $[C_2H_6]^{-1/2}$, thus leads to a net decrease in the over-all rate. Evidently when the total pressure is maintained at one atmosphere by substituting steam for ethane, the dominant change is in $[C_2H_6]^{-1/2}$.. However, water molecules may not be quite so effective in transferring energy to ethane molecules as are other ethane molecules. This may account for the k_o's in steam dilution runs being somewhat less than predicted.

Constancy of Initiation Reaction. Despite the variation in the over-all rate constant at a particular temperature of nearly an order of magnitude, depending on the pressure and dilution, the rate constant for the initiation reaction is remarkably constant. This is illustrated by Table III, which gives rate constants k_1 for 675°C. Relatively minor corrections for secondary methane were necessary at this low temperature--the ethane decomposition never

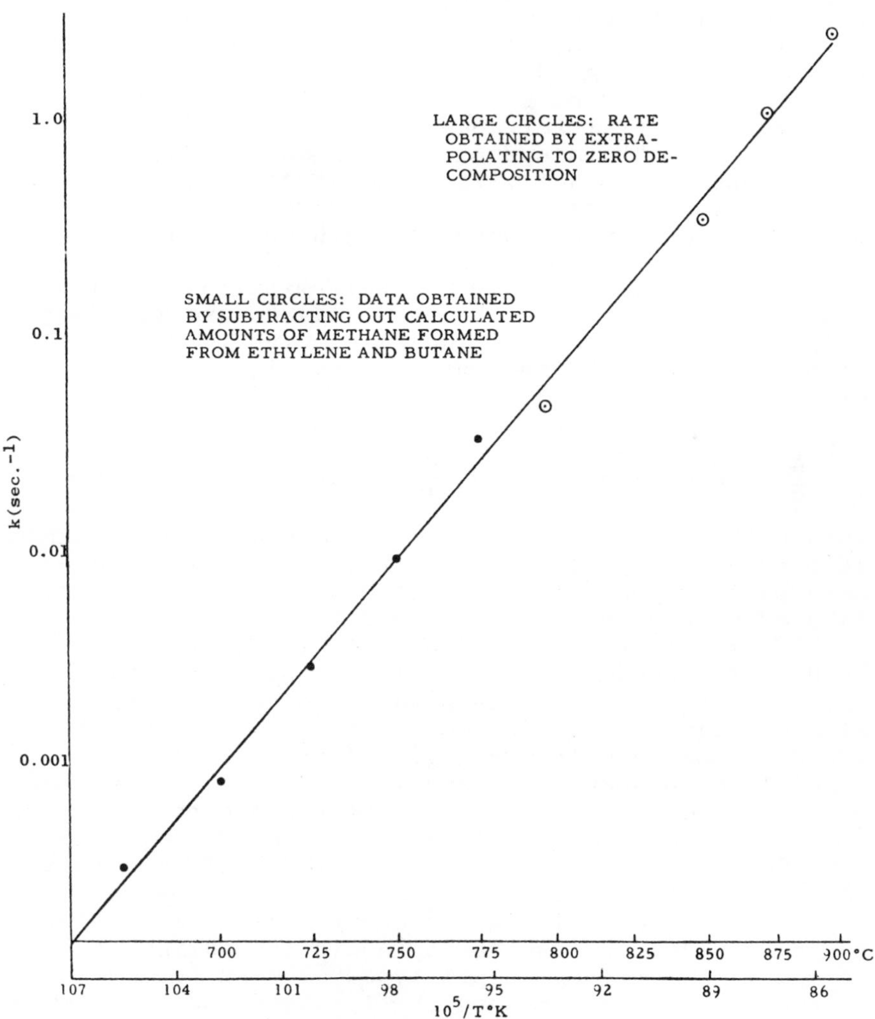

Figure 4. Rate constant for $C_2H_6 \rightarrow 2\,CH_3$. Variation with temperature.

exceeding 15%. From the lowest value to the highest, there is about a factor of two difference in k_1. It would appear that k_1 is slightly higher for the runs at 100:1 dilution than for other runs in the same reactor. The only other high value of k_1 is for one atmosphere, no steam dilution, in an annular reactor.

The low value of k_1 are in especially good agreement with the value calculated from the equation given by Lin and Back (4a), derived from experiments in the range 550-620°C: 1.50×10^{-4} $sec.^{-1}$.

It is evident that the major part of the dependency of the over-all rate constant on pressure is not due to the pressure sensitivity of k_1, in the present range of study.

The behavior of k_1 at atmospheric pressure through a temperature range of over 200°C is shown in Figure 4. The slope of the Arrhenius plot, corresponds to an activation energy of 87 kcal./mole, consistent with the accepted strength at these temperatures for the C-C bond strength in ethane.

Chain Initiation and Chain Termination. Via reactions (1)-(5), all chains initiated are manifested by production of methane and all chains terminated by production of butane (or alternately, ethylene + ethane). Then,

$$\frac{\text{Chains initiated}}{\text{Chains terminated}} = \frac{CH_4}{2\{C_4H_{10} + \text{contrib. from (5b)}\}} = 1.$$

Though the contribution from (5b) cannot be determined in pyrolysis experiments, the ratio k_{5b}/k_{5a} has been determined repeatedly as about 0.15 (7). That factor is used in present calculations.

This picture is modified slightly by the small amount of propane, evidently formed by reaction (6). Each propane formed in this way denotes the stopping of one chain. Thus, the moles of propane should be added to the denominator. (The formation of a molecule of propane consumes two radicals, of course. However, the methyl really is intercepted before it reacts via reaction (2) to start a chain.)

Secondary reactions almost immediately add to the amount of methane recovered and subtract from the butane recovered. These secondary reactions must be corrected for, in considering chain initiation and termination. Two principal types of reaction are significant in this respect: (1) secondary reactions involving the product, ethylene, and (2) decomposition of butane.

The ethylene reactions are presumed to take a form such as (8)

$$C_2H_5\cdot + C_2H_4 \longrightarrow CH_3CH_2CH_2CH_2\cdot \xrightarrow{+C_2H_4}$$

$$CH_3\dot{C}HCH_2CH_2CH_2CH_3$$

$$\downarrow$$

$$CH_3CH{=}CH_2 + \cdot CH_2CH_2CH_3$$

$$\downarrow$$

$$C_2H_4 + CH_3\cdot$$

In this sequence, there is no net gain or loss of radicals--thus no chains should be counted as being initiated or terminated. However, the methane generated here will count as a chain started unless it is subtracted out. By the scheme outlined here, the moles of excess methane equal the moles of propylene. This will not be strictly true. For example, formation of 3-hexyl in the isomerization step would lead to other products; also, small amounts of propylene can be formed by other routes such as loss of H· from propyl or decomposition of butene or butane. Moreover, as mentioned later, the propylene itself decomposes, and its loss must be calculated.

The loss of butane is estimated from a knowledge of the relative rates of decomposition of admixed butane and ethane. Over a wide range of conditions, the rate constant for decomposition of butane is about four times that for ethane (11). If the butane is initially generated from the ethane, then, at first, the rate of butane generation will greatly exceed the rate of its decomposition. Since the ethane concentration will not change much during this period, the net accumulation of butane will be nearly linear, as shown by Figure 5. The amount of butane which does disappear from time 0 to time t can be approximated from the average concentration of butane during this period. If

$$k_b \doteq -\frac{1}{t}\ln\frac{C_{tb}}{C_{ob}} \doteq 4\ k_e \doteq -\frac{4}{t}\ln\frac{C_{te}}{C_{oe}}$$

where the subscripts b and e stand for butane and ethane, respectively, then

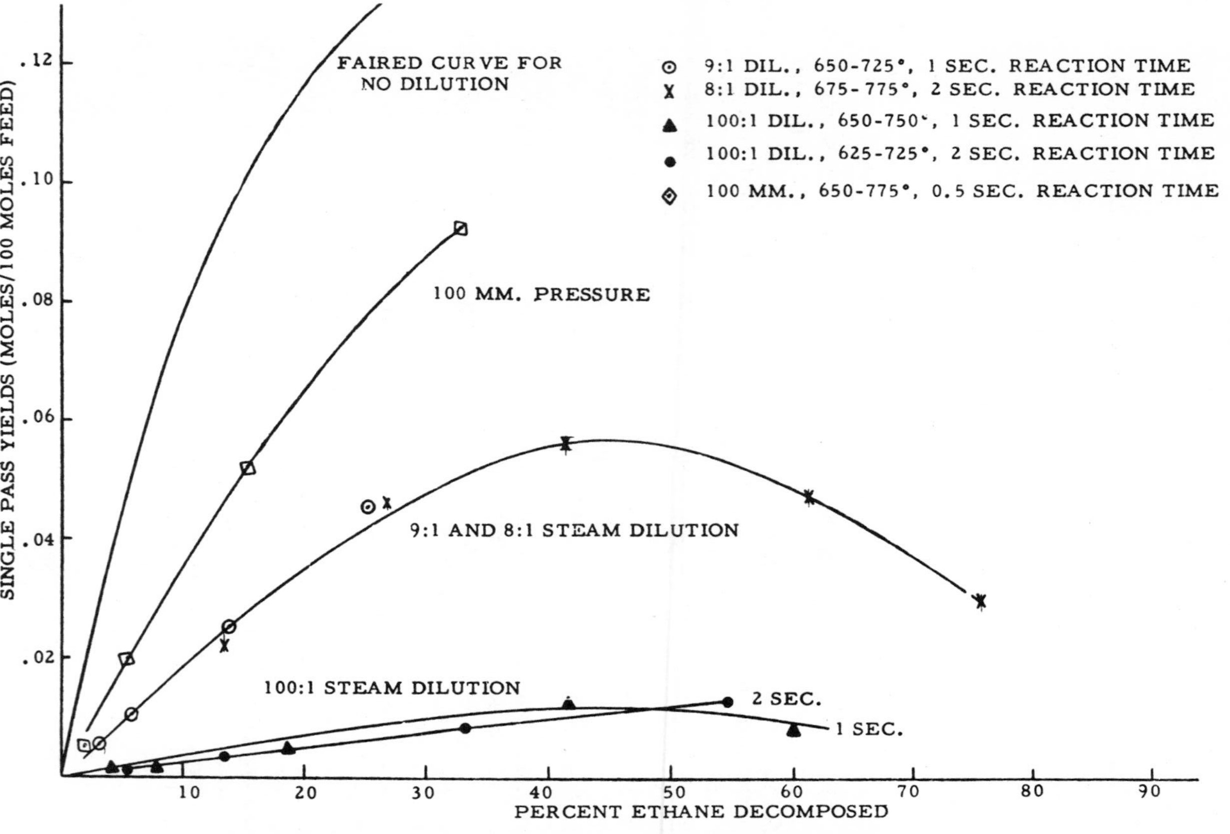

Figure 5. n-*Butane yields in ethane pyrolysis. Effects of dilution (1 atm total pressure).*

$$\frac{C_{tb}}{C_{ob}} = \left(\frac{C_{te}}{C_{oe}}\right)^4$$

and from the reaction of ethane that has decomposed, the fraction of the average butane concentration that has decomposed can be calculated. This scheme has not been used to correct for butane deficiency at levels of decomposition of ethane greater than 15%. At this level, the concentration of butane in the product must be multiplied by a factor of 1.24.

The decomposition of the butane is important, not only because of its own disappearance, but also because its decomposition generates additional secondary methane. The butane, like the methane, will decompose principally by reaction with free radicals in the system (9),

$$R\cdot + C_4H_{10} \longrightarrow RH + \begin{cases} \xrightarrow{\text{either}} CH_3CH_2CH_2CH_2\cdot \longrightarrow C_2H_5\cdot + C_2H_4 \\ \xrightarrow{\text{or}} CH_3CH_2\dot{C}HCH_3 \longrightarrow CH_3\cdot + C_3H_6 \end{cases}$$

Again, this set of secondary reactions does not result in net formation or termination of chains. However, the extra methane so formed, must be subtracted from total methane, in calculating the number of chains initiated. As with the secondary reactions involving ethylene, a propylene tends to be formed for each secondary methane. Insofar as this remains unreacted, the propylene can be used to calculate the secondary methane, from whichever source.

Previous work has shown that propylene, in excess ethane, decomposes much faster than does neat propylene (2). As a first approximation, the specific rate of propylene disappearance in excess ethane can be taken as twice the specific rate of ethane disappearance. If the average value for propylene concentration is taken as half the final value, then the total propylene formed during the reaction can be calculated by dividing the propylene recovered by the fraction ethane undecomposed. This corrected value for propylene should be approximately equal to the secondary methane for as high as 15% decomposition of ethane. It has been used for this purpose, and within this range, with the present data.

Results of applying these guidelines to the calculation of chains initiated and chains terminated

TABLE IV

CALCULATED RATIOS OF CHAIN INITIATION TO CHAIN TERMINATION

Pressure (atm.)	Dilution (steam)	Temperature °C	Residence Time (sec.)	Calculated Chain Initiation/Chain Termination
1	0	673	2.52	1.18
1	0	698	2.37	1.49
1	0.64:1	687	2.28	1.29
1	0.64:1	698	2.15	1.37
1	8:1	678	2.60	1.39
1.67	0	675	2.48	1.09
1.67	0	701	2.21	1.28
1.67	0.77:1	676	2.46	1.38
1.67	0.77:1	699	2.32	1.55
1	9:1	650	1.06	1.40
1	9:1	675	1.01	1.47
1	100:1	650	1.06	6.32
1	100:1	675	1.01	11.30
1	100:1	650	2.25	10.50
1	100:1	675	1.98	9.70
1	0	675	0.57	3.10
1	0	700	0.55	1.40
1	0	725	0.51	1.30
0.132	0	650	0.63	5.00
0.132	0	675	0.59	8.30
0.132	0	700	0.51	5.20
0.132	0	725	0.58	3.40

are shown in Table IV. These results show a rough balance between observed initiations and terminations for the experiments at one atmosphere undiluted ethane, or up to 9:1 dilution with steam. At 100:1 steam dilution, only about 10% of the chain terminations can be accounted for; at 100 mm. pressure of undiluted ethane, only about 20% of the chain termination is evident.

The cause for this apparent lack of terminators cannot be unambiguously identified from present results. The mechanism indeed predicts that this situation will occur, with chains being terminated more and more by reaction of H· and C_2H_5·, as the partial pressure of ethane is decreased. However, there is no evidence of any intermediate stage, where reaction (6), forming propane, assumes greater importance. If we assume that termination is by H· + C_2H_5· (e.g., reaction 7) and use the approximate radical concentrations estimated earlier, we can calculate an order of magnitude value for the reaction rate. At 1048K the second order reaction rate constant needs to be roughly $10^{14} cm^3 mol^{-1} sec^{-1}$, or nearly two orders of magnitude larger than has been estimated by Camilleri, Marshall and Purnell (12). Obviously, the homogeneous reaction is more important than has been suggested, or termination is by some other mechanism, such as a surface reaction.

Certainly, the possibility of termination of chains by surface reactions cannot be disregarded. Several authors have discussed this possibility for ethane pyrolysis at lower temperatures (10). For example, chemisorption of hydrogen atom on a wall site could be followed by reaction by this site with an ethyl radical, forming ethane, which is later desorbed. More work will have to be done with reactors with different kinds of surfaces, with diffusion paths of different length, and with different surface/volume ratios, before even the qualitative aspects of the possible surface reactions are understood.

Conclusions

With modern instrumentation and analytical techniques, much more can be done to confirm or reject mechanisms for gas pyrolysis than was possible when theories of chain reactions were first propounded. The mechanism for ethane pyrolysis devised by Rice and Herzfeld was based on limited data on rates and product formation at relatively high decompositions of ethane, where both rates and products were strongly affected by secondary reactions. The present work and other recent studies look at the over-all reaction under circumstances relatively free of secondary effects, with rational bases for correcting for such incipient effects. Even so, detailed experimental data pose new questions, some of which are not readily soluble by direct experimentation on the system involved.

The uncertainties and ambiguities discussed in this report might well discourage any attempt to apply a fundamental approach to industrial cracking. However, some of the problems encountered here, though interesting, would be absent or attenuated in industrial cracking furnaces. Pressure dependency of rate constants and chain termination by surface reactions should be quite unimportant there. The number of fundamental free-radical reactions that must be considered in high-conversion cracking of mixed feeds is, of course, enormous. This problem becomes less insurmountable as more rate data for the fundamental reactions becomes available, and computer technology progresses. An intermediate approach, applying fundamental concepts to simplified models, has recently been reported as an effective means of handling complex cracking systems (11).

Abstract

At partial pressures near one atmosphere, ethane decomposes by a simple Rice-Herzfeld mechanism, with combination or disproportionation of ethyl radicals as the predominant chain-ending step. However, at a total pressure of 100 mm., or at a partial pressure of 0.01 atm. another chain-ending step predominates. Unlike butane formed from ethyl, the products of this step cannot be distinguished analytically from the major products of the reaction chain. It is therefore believed to involve reaction of H• and C_2H_5•, either homogeneously or at the reactor wall. Quantitative rate and yield data are given, as are methods of correction for secondary reactions and of extrapolation to zero reaction time.

Literature Cited

(1) Rice, F. O., and Herzfeld, K. F., J. Amer. Chem Soc. 56,284(1934).
(2) Davis, H. G., and Williamson, K. D., Proc. 5th World Petrol. Congr., Vol. IV, p. 37(1960).
(3) Quinn, C. P., Proc. Roy. Soc. London A275, 190(1963).

(4) Lin, M. C., and Back, M. H., (a) Can. J. Chem. 44, 505(1966); (b) ibid. 44, 2357(1966); (c) ibid 44, 2369(1966).
(5) Leathard, D. A., and Purnell, J. H., Rev. Phys. Chem. 1970, p. 202, 206, 209-15.
(6) Trenwith, A. B., (A) Trans. Faraday Soc. 59, 2543(1966); (b) ibid 60, 2452(1967).
(7) Terry, J. O., and Futrell, J. H., Can. J. Chem. 45, 2327(1967).
(8) Quinn, C. P., Trans. Faraday Soc. 59, 2543(1963).
(9) Pacey, P. D., and Purnell, J. H., Int. J. Chem. Kinetics 4, 657(1972).
(10) Marshall, R. M., and Quinn, C. P., Trans. Faraday Soc. 61, 2671(1965).
(11) Davis, H. G., and Farrell, T. J., Ind. Engr. Chem. Process Des. and Develop. 12, 171(1973).
(12) Camilleri, P., Marshall, R. M., and Purnell, J. H., J. C. S. Faraday I, 70, 1434(1974).

5

Oxidative Pyrolyses of Selected Hydrocarbons Using the Wall-less Reactor

JAY E. TAYLOR and DONALD M. KULICH

Department of Chemistry, Kent State University, Kent, Ohio

The homogeneous high temperature pyrolysis of hydrocarbons normally gives similar products in either the presence or absence of limited amounts of oxygen. The effect of oxygen is often to increase the rate of reaction and to vary product ratios. It is the purpose of this paper to examine the oxidative pyrolyses of a series of hydrocarbons under surface free conditions. Achievement of a totally surface free environment has been made possible by the development of the wall-less reactor(1, 2).

Since the rates of oxidative pyrolyses may be strongly affected by surfaces(2,3,4,5), a study of the absolute effects of surface is also included. Surface effects are evaluated by comparing reactions done in a homogeneous environment to those with a surface inserted into the reactor.

Experimental

In principle the wall-less reactor encompasses a flowing stream of hydrocarbon injected into a protective cylinder of inert flowing gas. There is no contact of the hydrocarbon with the containing surface after attainment of the reaction temperature since both the injection port and the sampling tube are kept well below the reaction temperature. Plug flow is maintained in the nitrogen stream permitting excellent reproducibility of the flow pattern and of the kinetic data. The gross uncertainties due to variable surface effects are entirely eliminated as compared to a conventional static or flow reactor. The details of the reactor have been previously described(1,2).

The major disadvantage of the present design of the wall-less reactor is the diffusion of the injected center stream into the surrounding nitrogen. Accordingly, if more than one reactant is involved or if the reaction is higher order, the reproducibility is equally good, but calculations of the rate parameters are made much more complicated. In this paper calculations of the kinetic constants are made only from the nonoxidative rate data.

Agreement of data from the wall-less reactor with conventional reactors is excellent when valid comparisons can be made. Neopentane which exhibits significant surface effects has the same kinetic constants (within experimental error) when pyrolyzed either in the wall-less(1,6) or shock tube reactor(7). Oxygen-free ethane exhibits only a very small surface effect(8), and there is excellent agreement between the wall-less and conventional static reactors as is noted in this paper.

For the oxygen work the percent of oxygen indicated is that added to the nitrogen stream and is not indicative of the oxygen to hydrocarbon ratio. At 0.4% oxygen in the nitrogen stream the average oxygen to hydrocarbon ratio is about 1:1 with smaller ratios at the center and inlet of the hydrocarbon stream and higher values at the outer portions of the stream.

The hydrocarbons are from commercial sources and are purified by removal of oxygen and unwanted hydrocarbons in a purification train prior to introduction into the reactor.

The reactants and products were analyzed by standard gas chromatographic techniques. In all cases the added surface is 1/16" oxidized stainless steel rods set parallel to the direction of gas flow.

Results

Ethane. The data on ethane have been previously described (2) and are reviewed in this paper for comparative purposes. At very low conversions the homogeneous rate of pyrolysis is less with oxygen present than in its absence. A reversal occurs above 0.13% conversion after which oxygen is seen to increase the rate of reaction. The main product is ethylene; in addition, 1% methane is formed in the absence of oxygen and 2% methane with oxygen present. Since there is no indication of a direct homogeneous reaction of elemental oxygen with ethane, the initiation steps appear to be the same regardless of the presence or absence of oxygen.

There was no detectable effect of surface on ethane alone, but with oxygen present a surface inhibition was noted at the higher conversions, and a promoting effect was seen at very low conversions.

For pure ethane the first order rate constant is

$$k = 10^{14.8} e^{-71800/RT} \sec^{-1}$$

The above values are equal within experimental error to those given by Laidler and Wojciechowski(8).

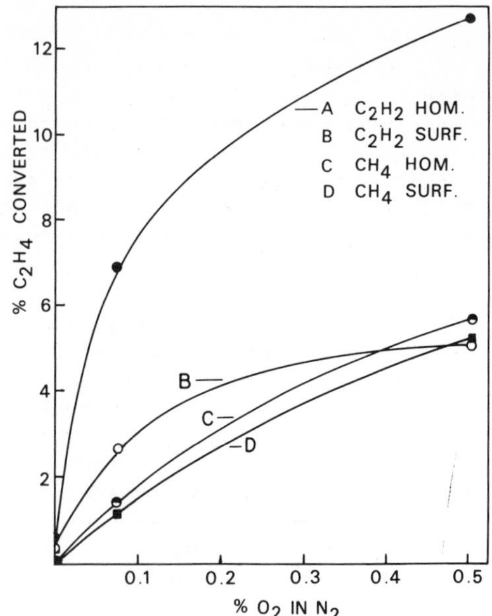

Figure 1. Effect of added oxygen on the pyrolysis of ethylene. Reaction time: homogeneous, 0.19 sec; stainless steel surface, 0.18 sec. Surface-to-volume ratio, 0.7 cm^{-1}. 0.4% O_2 is roughly 1:1 O_2-to-hydrocarbon ratio.

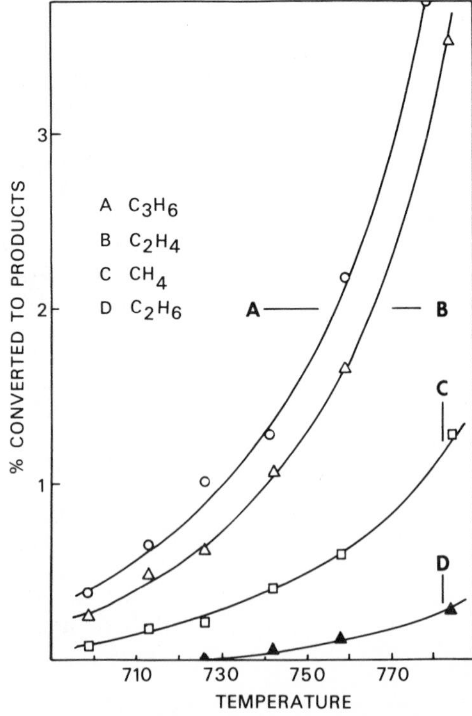

Figure 2. Percent propane converted to products. Homogeneous. Reaction time $\simeq$ 0.22 sec. Variation in temperature in °C is used to attain varying product conversions.

Ethylene. Good yields of acetylene are obtained from ethylene at modest conversions either in the presence or absence of oxygen. The only other identifiable product is methane which amounts to about 10% of the acetylene. The oxygen does, however, increase the rate of reaction appreciably. This is shown in Figure 1 where the effects of increasing concentrations of oxygen are recorded. Partial oxidative loss to nonidentified products (presumably carbon monoxide) does occur at the higher oxygen concentrations. Upon adding 0.08% oxygen to the nitrogen, 98-99% of the original ethylene was accounted for; with 0.26% oxygen, the indicated recovery is 87-90%. As previously noted, 0.26% oxygen would represent an estimated average ratio of 1:1.5 of oxygen to hydrocarbon.

Upon placing an oxidized stainless steel surface in the reactor, oxygen absent, the rate decreases and the activation energy is increased which makes ethylene unique with respect to the other hydrocarbons of this paper. Due to the lability of the triple bond to surface effects (see section on propylene), there is a large uncertainty in the surface data for ethylene since carbon deposition could markedly alter the calculated rate constants.

The kinetic rate constants for these reactions are:

$$\text{Homogeneous:} \quad k = 10^{11.0} e^{-65400/RT} \text{sec}^{-1}$$

$$\text{At } S/V = 1.4 \text{ cm}^{-1}: \quad k \simeq 10^{11.3} e^{-67500/RT} \text{sec}^{-1}$$

$$S/V = \text{surface/volume}$$

With oxygen present a significant surface effect is seen. As oxygen is increased, the formation of acetylene does not increase proportionally to the other products. See Figure 1.

The data and activation energies for the oxygen-free reaction are in keeping with the mechanism proposed by Benson and Haugen(9).

Propane. Propane pyrolyzes to form propylene, ethylene and methane with propylene predominating in the absence of surface. See Figure 2. Similar data taken in the presence of a surface are seen in Figure 3. Comparison of these two sets of graphs indicates that upon adding a surface the propylene decreases and ethylene increases along with a slight increase in rate. The kinetic rate constants are:

$$\text{Homogeneous:} \quad k = 10^{12.1} e^{-60400/RT} \text{sec}^{-1}$$

$$\text{At } S/V = 0.7 \text{ cm}^{-1}: \quad k = 10^{12.3} e^{-60200/RT} \text{sec}^{-1}$$

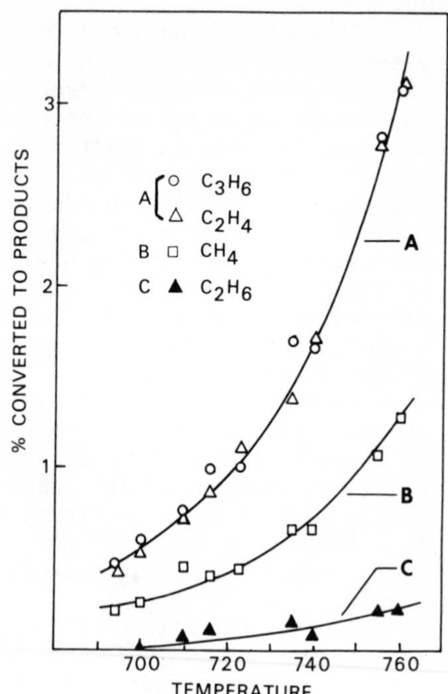

Figure 3. Percent propane converted to products. Conversions varied with temperature in °C; oxidized stainless steel surface; surface-to-volume ratio, 0.7 cm^{-1}; reaction time $\simeq$ 0.22 sec.

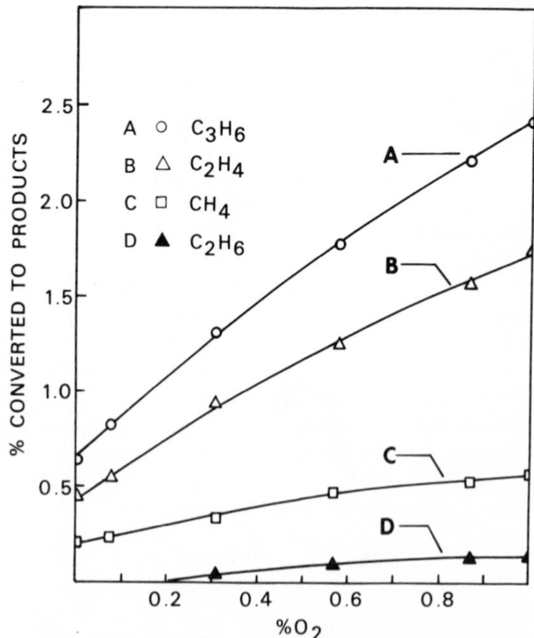

Figure 4. Percent propane converted to products with oxygen present. Homogeneous; temperature, 718°C; time, 0.227 sec. 0.4% O_2 is roughly 1:1 O_2-to-hydrocarbon ratio.

The overall effect of oxygen on propane is to increase the rate of pyrolysis with no significant changes in the product ratios from the homogeneous reaction. See Figure 4.

However, at very low conversions the oxygen effect is reversed (similar to ethane) and oxygen becomes an inhibitor. This reversal is demonstrated in Figure 5. In taking the data for this graph the device of varying temperature is used to vary the extent of the conversion to products at a given reaction time. In effect Figure 5 is a plot of temperature versus percent products.

In contrast to ethane the presence of surface increases the rate of oxidative pyrolysis. The data are given in Table I.

Table I

Effect of an Oxidized Stainless Steel Surface on the Propane-Oxygen Reaction[1]

Conditions	Temp (^oC)	k, sec^{-1}	% C_3H_8 to products			
			CH_4	C_2H_6	C_2H_4	C_3H_6
homogeneous	745	0.28	0.82	0.12	2.2	2.8
stainless steel[2]	745	0.42	1.13	0.21	3.2	4.1
homogeneous	785	0.93	2.44	0.68	7.2	7.8
stainless steel[2]	785	1.51	3.68	1.12	11.9	9.9

[1]Oxygen added = 0.4% in the nitrogen blanket gas stream. Rate constants were calculated assuming a first order reaction.

[2]Oxidized stainless steel surface with S/V = 0.7 cm^{-1}.

Propylene. Good yields of a combination of allene and methylacetylene are formed upon pyrolyzing propylene in the absence of surface. With a surface present the yields of these compounds are decreased to the extent that only trace amounts are detectable. It appears that the decomposition of these compounds is surface promoted although this has not been clearly established. There also appears to be an overall increase in rate upon adding surface to the reactor. The homogeneous activation energy is high but has not been carefully determined.

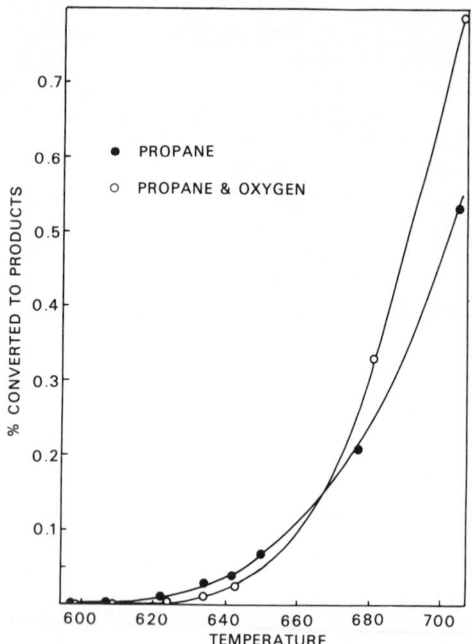

Figure 5. Inversion of rate of product formation with varying conversion as determined by temperature in °C. Homogeneous; reaction time ≃ 0.22 sec

The addition of oxygen at very low conversions results in increased allene formation. In the absence of oxygen the product analysis is 0.07% methane, 1.4% ethylene and 1.5% allene. With 0.5% oxygen in the nitrogen stream the products are 0.08% methane, 1.8% ethylene and 2.9% allene. Very small amounts of methylacetylene and ethane are included with the allene and ethylene, respectively.

Isobutane. The predominant products are isobutylene, propylene, ethylene, methane and small amounts of ethane. See Figure 6. In the absence of surface conversion to isobutylene predominates, but with added surface the propylene is increased and the isobutylene is decreased. The extent of this trend is seen upon comparing Figure 6 with Figure 7. Surface also has an influence upon the rate and activation energy as seen in the following constants:

$$\text{Homogeneous:} \quad k = 10^{11.7} e^{-56300/RT} \sec^{-1}$$

$$\text{At } S/V = 0.7 \text{ cm}^{-1}: \quad k = 10^{10.7} e^{-50700/RT} \sec^{-1}$$

Upon adding oxygen the rate of pyrolysis is increased though not as extensively as with ethane and propane. See Figure 8. Oxygen also effects a change in product ratios since more propylene is formed at the expense of isobutylene.

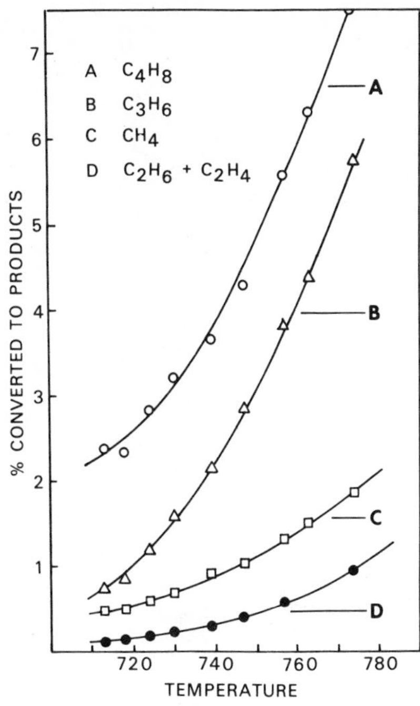

Figure 6. Percent isobutane converted to products. Conversions varied with temperature in °C. Homogeneous; reaction time ≃ 0.22 sec.

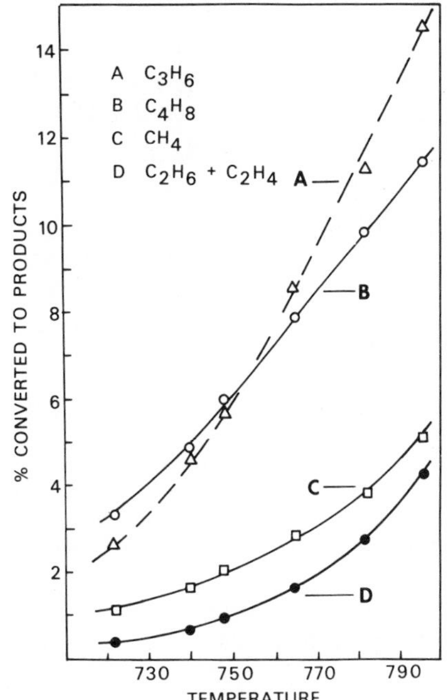

Figure 7. Percent isobutane converted to products. Conversions varied with temperature in °C; oxidized stainless steel surface present; surface-to-volume ratio, 0.7 cm⁻¹; reaction time ≃ 0.22 sec.

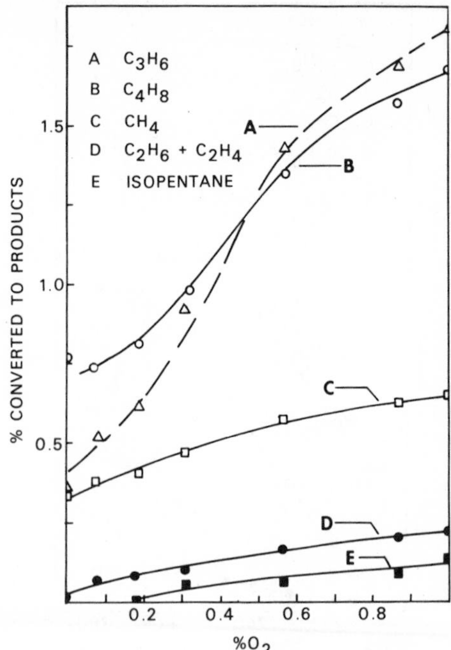

Figure 8. Percent isobutane converted to products. Homogeneous; temperature, 701°C; reaction time ≃ 0.227 sec. 0.4% O_2 is roughly 1:1 O_2-to-hydrocarbon ratio.

Conclusions

The thermal decomposition of pure ethane is a much studied reaction, and the mechanism is well defined. For purposes of comparison with the oxygen reactions, the ethane pyrolysis is redescribed below.

Initiation: $$C_2H_6 \rightarrow 2CH_3\cdot \qquad [1]$$

$$CH_3\cdot + C_2H_6 \rightarrow CH_4 + C_2H_5\cdot \qquad [2]$$

Propagation: $$C_2H_5\cdot \rightarrow C_2H_4 + H\cdot \qquad [3]$$

$$H\cdot + C_2H_6 \rightarrow C_2H_5 + H_2 \qquad [4]$$

followed by termination steps.

Since it was noted with ethane(2) that the direct homogeneous reaction of oxygen with the hydrocarbon does not appear to be an important reaction, the initiation steps, [1] and [2], are the same in either the presence or absence of oxygen. Oxygen becomes important only during the propagation steps.

Ethane with oxygen is as follows.

Propagation:

a. Period of decreased rate (compared to [3] and [4]):

$$C_2H_5\cdot + O_2 \rightarrow C_2H_4 + HO_2\cdot \text{ (preferred)} \qquad [5]$$

$$\text{or} \begin{pmatrix} C_2H_5\cdot \rightarrow H\cdot + C_2H_4 \\ M + H\cdot + O_2 \rightarrow HO_2\cdot + M \end{pmatrix} \qquad [6]$$

$$HO_2\cdot + C_2H_6 \rightarrow H_2O_2 + C_2H_5\cdot \qquad [7]$$

b. Period of increased rate (compared to [3] and [4]):

$$M + H_2O_2 \rightarrow 2HO\cdot + M \qquad [8]$$

$$HO\cdot + C_2H_6 \rightarrow H_2O + C_2H_5\cdot \qquad [9]$$

followed by the termination steps.

A computer simulation treatment(10) gives excellent support to the above set of equations.

Propane exhibits a similar mechanism but with added complexity. As with ethane the oxygen-propane reaction does not take place at a lower temperature than the oxygen-free pyrolysis of propane; therefore, the initiation steps are concluded to be the same in either case.

Initiation:

$$C_3H_8 \rightarrow C_2H_5\cdot + CH_3\cdot \qquad [10]$$

$$CH_3\cdot(\text{or } C_2H_5\cdot) + C_3H_8 \rightarrow C_3H_7\cdot + CH_4(\text{or } C_2H_6) \qquad [11]$$

Propagation:

a. Period of decreased rate:

$$C_3H_7\cdot + O_2 \rightarrow C_2H_4 + HO_2\cdot \qquad [12]$$

$$\text{or} \begin{pmatrix} C_3H_7\cdot \rightarrow H\cdot + C_3H_6 \\ M + H\cdot + O_2 \rightarrow HO_2\cdot + M \end{pmatrix} \text{preferred} \qquad [13]$$

$$HO_2\cdot + C_3H_8 \rightarrow H_2O_2 + C_3H_7\cdot \qquad [14]$$

b. Period of increased rate:

$$M + H_2O_2 \rightarrow 2HO\cdot + M \qquad [8]$$

$$HO\cdot + C_3H_8 \rightarrow C_3H_7\cdot + H_2O \qquad [15]$$

c. Ethylene formation:

$$C_3H_7\cdot \rightarrow C_2H_4 + CH_3\cdot \qquad [16]$$

followed by the appropriate termination steps.

The preference of [5] over [6] in the case of ethane oxidation and the reversed preference of [13] over [12] with propane are based on calculations of relative rates as pertains to these reactions.

The period of decreased rate for ethane with oxygen, equations [3] to [6] has been discussed at length(2). The oxygen inhibition of propane pyrolysis, as seen in the 0-0.15% conversion section of Figure 5 and in equations [12]-[14] is very similar to the observations for ethane. In the presence of oxygen the more reactive hydrogen atom is replaced by the less reactive hydroperoxy radical. See [12] and [13]. As the hydrogen peroxide concentration increases by [14], the formation of very reactive hydroxy radicals increases by [8], and the rate becomes greater than the oxygen free reaction by virtue of equation [15].

The two alkenes, ethylene and propylene, have been studied in a preliminary manner only. It is assumed that the oxygen promotion is due also to the formation of hydroxy radicals by a process similar to the ethane and propane mechanisms.

In comparing Figures 4 and 8 with the corresponding graph for ethane(2), it is seen that the homogeneous oxygen effect decreases with the complexity of the alkane; i.e., ethane>propane>isobutane. In contrast the tendency of the C-C bond to break increases with the size of the alkane, and this is due to the lability of the corresponding radical formed by hydrogen abstraction. Although negligible amounts of methane are formed from the ethyl radical, the propyl radical readily forms ethylene by reaction [16]. The isobutyl radical is even more easily decomposed. Thus, the radicals lost by cracking are unavailable for oxidation, and the complementary nature of these effects are thereby explained.

The effect of surface (oxygen absent) shows a similar trend in that isobutane is the most affected and ethane the least. With oxygen present the surface effects are more intense but show a less clearly defined pattern of reactivity as seen in comparing the ethane and propane data. Surface effects on the olefins are even more complex due to decomposition of the acetylenes and/or allenes. Continued studies on the absolute effects of surfaces are planned for the near future.

Abstract.

The oxidative pyrolyses of ethane, propane, isobutane, ethylene and propylene have been studied under various conditions using a wall-less reactor. A comparison of the alkanes indicates that: (1) the promoting effect of oxygen (homogeneous) lessens with increased molecular weight, (2) surface effects using stainless steel increase with increased molecular weight, and (3) with added oxygen there are no clear trends in surface effects. The rate of homogeneous pyrolysis of the alkenes is also promoted by oxygen, but heterogeneous effects upon the acetylenic or allenic products complicate the study of surface effects. The mechanism of interaction of oxygen with propane is described. As with ethane, hydroperoxy radicals are first generated causing an induction period, then the formation of hydroxyl radicals accelerates the reaction.

Literature Cited

(1) Taylor, J. E., Hutchings, D. A., and Frech, K. J., J. Amer. Chem. Soc. (1969) 91, 2215.
(2) Taylor, J. E. and Kulich, D. M., Int. J. Chem. Kinet. (1973) 5, 455.
(3) Martin, R., Dzierzynski, M., and Niclause, M., J. Chim. Phys. (1964) 61, 790.
(4) Baldwin, R. R., Hopkins, D. E., and Walker, R. W., Trans. Faraday Soc. (1970) 66, 189.
(5) Scacchi, G., Martin, R., Niclause, M., C. R. Acad. Sci. (1972) 275, 1347.
(6) Hutchings, D. A., Taylor, J. E., and Frech, K. J., J. Phys. Chem. (1969) 73, 3167.
(7) Tsang, W., J. Chem. Phys., (1966) 44, 4283.
(8) Laidler, K. J., and Wojciechowski, B. W., Proc. Roy. Soc. (London) (1961) A260, 91.
(9) Benson, S. W. and Haugen, G. R., J. Phys. Chem. (1967) 71, 1735.
(10) Kulich, D. M. and Taylor, J. E., Inter. J. Chem. Kinet, (1975) 7, 895.

6

Kinetics and Mechanism of Hydrogenolysis of Propylene in the Presence of Deuterium

YOSHINOBU YOKOMORI, HIROMICHI ARAI, and HIROO TOMINAGA

Faculty of Engineering, University of Tokyo, Hongo 7-3-1, Bunkyo-ku, Tokyo, Japan

Introduction

In our previous study, (1) the introduction of large amounts of hydrogen into a hydrocarbon pyrolysis system was found to enhance the rate of pyrolysis and to result in increased yields of ethylene. The role of hydrogen was discussed in terms of the reaction kinetics and mechanism where hydrogenolysis of higher olefins into ethylene has an important role. In this connection, hydrogenolysis of 1-butene (2) and isobutene (3) had been investigated to demonstrate the kinetics of consecutive demethylation of the olefins into ethylene.

On the other hand, a kinetic investigation had been reported on thermal hydrogenolytic demethylation of propylene by Amano, A. and Uchiyama, M. (4) The reaction can be expressed by the stoichiometry,

$$C_3H_6 + H_2 = C_2H_4 + CH_4 ,$$

and is sufficiently clean for rate measurements. The observed rate was described by

$$d[C_2H_4]/dt = k[C_3H_6][H_2]^{1/2}, \text{ mol/l, sec}$$

$$k = 10^{12.3} \exp(-55,900/RT).$$

The values of overall three-halves order and Arrhenius kinetic parameters were interpreted in terms of a free radical chain mechanism, which was subsequently supported by Benson, S.W. and Shaw, R. (5)

The free radical chain mechanism, which also could be successfully applied to the hydrogenolytic demethylation of toluene, (6,7,8) is, to our understanding, essentially as follows.

$$C_3H_6 \underset{b}{\overset{a}{\rightleftharpoons}} CH_2{=}CHCH_2\cdot + H\cdot$$

$$H\cdot + C_3H_6 \underset{d}{\overset{c}{\rightleftharpoons}} H_2 + CH_2{=}CHCH_2\cdot$$

$$H\cdot + C_3H_6 \xrightarrow{e} C_2H_4 + \cdot CH_3$$

$$\cdot CH_3 + H_2 \xrightarrow{f} CH_4 + H\cdot$$

$$\cdot CH_3 + C_3H_6 \underset{h}{\overset{g}{\rightleftharpoons}} CH_4 + CH_2{=}CHCH_2\cdot$$

Reaction (e) was postulated by Amano (7) as a process involving a hot n-propyl radical. Benson and Shaw (5) suggested that reaction (e) was a stepwise one consisting of addition of hydrogen atom to propylene followed by elimination of methyl radical, where the former step being rate controlling. Most interesting is the behavior of the hot n-propyl radical, namely its decomposition into ethylene and a methyl radical or into propylene and a hydrogen atom, or stabilization to give propane after hydrogen abstraction. Furthermore, the terminal addition of a hydrogen atom to propylene to give a hot sec-propyl radical and its behavior are also noteworthy.

It was attempted in this study, therefore, to have a more detailed investigation of propylene pyrolysis in the presence of hydrogen. Furthermore propylene was subjected to pyrolysis in the presence of deuterium. Kinetic isotope effect was measured and deuterium distributions in the pyrolysis products were analyzed. The experimental results were interpreted in line with the reaction mechanism above mentioned, and discussed in the light of RRKM theory which quantitatively deals with the behavior of chemical activation system such as the decomposition rates of hot alkyl radicals.

Experimental

Materials. Cylinder propylene, 99.7 mol% purity with principal impurities being methane and ethane, was used without further purification. Cylinder hydrogen, 99.99 mol% was used after purification over Deoxo catalyst and zeolite 13Y. Cylinder deuterium, 99.5 mol% purity with principal impurities being H_2 and HD and nitrogen, was used after purification over Pd/Al_2O_3 and 13Y. Cylinder nitrogen, 99.99 mol% was used without further purification.

Apparatus and Procedure. A conventional flow system was used. Hydrogen (or deuterium) and propylene were separately supplied at controlled and constant

Table I. Pyrolysis of Propylene in the Presence of Hydrogen

Temp. (°C)	Methane (mol%)	Ethane (mol%)	Ethylene (mol%)	Reaction time (sec)	$\frac{H_2}{C_3H_6}$	k_{obs} ($l^{1/2}/mol^{1/2}.sec$)
800	57.6	4.8	55.6	0.92	9.5	8.36
800	27.5	1.0	27.8	0.45	10.2	7.17
775	36.1	2.0	33.6	1.00	10.0	4.05
775	14.9	0.4	15.0	0.46	10.2	3.42
750	22.5	—	19.8	1.02	10.0	2.19
750	11.6	—	9.6	0.53	13.0	1.84
725	19.2	1.0	18.9	2.20	11.1	0.904
725	8.1	0.2	8.1	1.00	10.7	0.802
700	7.5	0.2	7.9	1.98	10.0	0.391
700	4.4	0.1	4.6	1.13	9.8	0.393
700	1.9	0.1	1.9	0.51	10.5	0.355
675	3.7	0.1	3.6	2.14	10.6	0.158
675	1.6	0.03	1.7	1.08	10.1	0.147

flow rates, mixed at desired mole ratios, and then introduced into a reactor. The reactor was consisted of two quartz tubes (6 mm outer diameter of the smaller tube and 11 mm inner diameter of the bigger tube); their length was 100 mm. The reactants flowed through the annular space of 2.5 mm width. The reactor was fitted in an stainless-steel block heated by an electric furnace. Temperature profiles of the reactor were measured during each run by a movable chromelalumel thermocouple along the central axis of the vessel. The average effective temperature and reactor volume were determined by usual method. (9) The product gas was passed through a trap (immersed in an ice-methanol bath) in which condensation products collected. The reaction was studied under conditions varied over the following range; temperature, 675 ∿ 800°C; contact time, 0.5 ∿ 3.8 sec; hydrogen or deuterium/propylene mole ratio, up to approximately 10.

Analysis. Gaseous products, consisting of hydrocarbons from methane up to C_3's were analyzed by gas chromatography using a Porapak Q column (2 m long) at 40°C with nitrogen as carrier gas.

Distribution of deuterated products were determined by gas chromatographic-mass spectrography using a Porapak Q column (1 m long) at 40 ∿ 90°C with helium as carrier gas. Pattern coefficients of ethylene-d_0, -d_1 and propylene-d_0 were obtained from mass spectral data of API Research Project 44. The fragmentation mode of propylene-d_1 was assumed to be the same with propylene-d_0. It was assumed that both d_0 and d_1 compounds had the same sensitivity.

Results

The experimental results with hydrogen and deuterium are summarized in Tables I and II. The overall rates of ethylene formation were described by 1.5 order rate equation as follows:

$$d[C_2H_4]/dt = k_{obs}[C_3H_6][H_2 \text{ or } D_2]^{1/2}$$

Figure 1 shows Arrhenius plots of the 1.5 order rate constants, which are respectively given by,

$$(k_{obs})_{H_2} = 10^{14.0} \exp(-64{,}400/RT),\ l^{1/2}/mol^{1/2}\ sec$$

$$(k_{obs})_{D_2} = 10^{15.1} \exp(-70{,}600/RT),\ l^{1/2}/mol^{1/2}\ sec$$

Table II. Pyrolysis of Propylene in the Presence of Deuterium

Temp. (°C)	Methane (d_0+ d_1) (mol%)	Ethane (mol%)	Ethylene (d_0+d_1+d_2) (mol%)	Reaction time (sec)	$\frac{D_2}{C_3H_6}$	k_{obs} ($l^{1/2}/mol^{1/2}\cdot sec$)
800	59.1	3.0	53.6	1.47	9.7	5.24
800	42.7	1.6	42.3	1.12	9.7	5.40
775	34.1	1.4	26.9	1.28	9.8	2.21
775	28.0	0.7	23.6	1.02	12.3	2.56
750	9.4	0.1	9.4	0.83	10.8	1.14
725	9.3	0.2	9.1	2.11	9.3	0.433
725	3.7	0.1	3.9	1.02	9.7	0.373
700	5.8	0.2	5.9	3.00	10.5	0.189
700	3.6	0.1	3.6	2.12	9.7	0.163
675	3.2	0.1	3.1	3.76	11.2	0.077
675	2.5	0.1	2.5	3.00	10.3	0.078

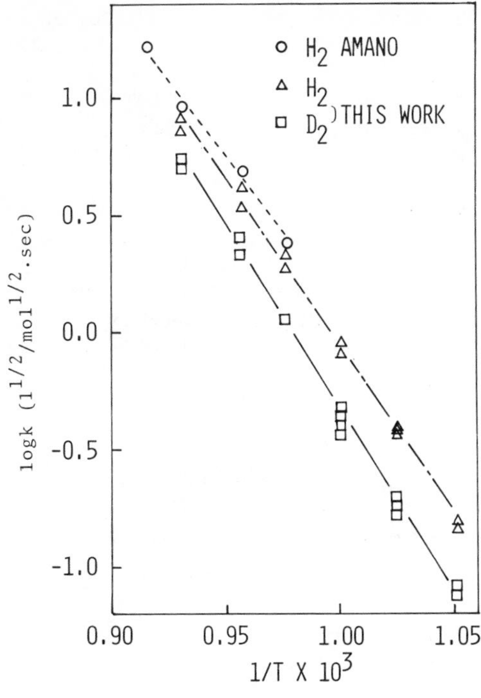

Figure 1. Arrhenius plots of propylene hydrogenolysis rate constants

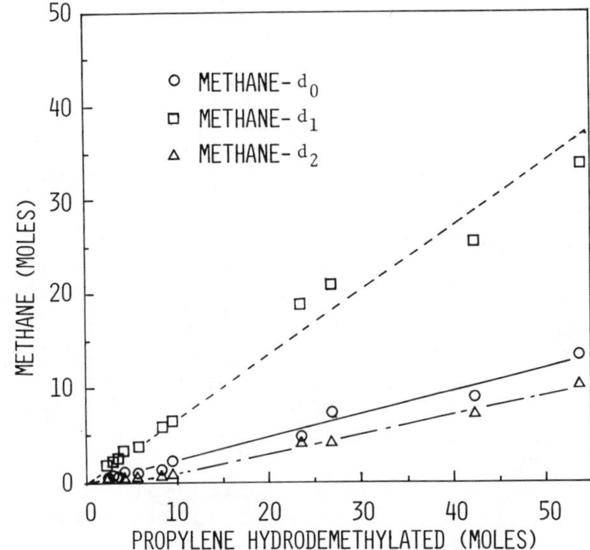

Figure 2. Deuterium distribution in methane (675° ~ 800°C). D_2/C_3H_6 = ca. 10.

The rate constant obtained in the presence of hydrogen is in fairly good agreement with that obtained by Amano. (4)

The overall kinetic isotope effect is defined as the ratio of k values for hydrogen and deuterium, and the values were found to vary as $(k_{obs})_{H_2}/(k_{obs})_{D_2}$ = 1.45 ∿ 2.13 in the temperature range studied. Table III shows distributions of deuterated products. The expression "fraction" indicates a mole composition of the respective $d_0 \sim d_i$ compounds. The expression "yield" indicates the moles of products obtained from 100 mol of propylene. Figures 2, 3, and 4 show the changes in deuterium distributions as a function of propylene hydrodemethylation.

Discussion

Kinetic Isotope Effect. The steady-state assumption was applied to the free radical chain mechanism, (a) ∿ (h), giving rise to the overall rate equation as follows.

$$d[C_2H_4]/dt = k_e[H\cdot][C_3H_6] = k_e\left(\frac{k_a\ k_d}{k_b\ k_c}\right)^{1/2}[C_3H_6][H_2]^{1/2}$$

The overall rate constant can be separated into two terms; k_e and $(k_a\ k_d/k_b\ k_c)^{1/2}$. The isotope effects are examined for both terms below:

[1] k_e term

Equation (e) is separated into two steps, that is, addition and decomposition steps.

$$H\cdot + C_3H_6 \xrightarrow{e-1} (\cdot CH_2\underset{|}{C}HCH_3)^* \xrightarrow{e-2} CH_2{=}CH_2 + \cdot CH_3$$

(with H attached to the central carbon)

(a) Addition step (e-1)

The C-D bond formed is stronger than the C-H bond by about 1 ∿ 2 kcal/mol, so that $(k_{e-1})_H/(k_{e-1})_D$ term is expected to be smaller than unity, where the symbol $(k_{e-1})_D$ represents the rate constant for addition of deuterium atom, etc. Based on the kinetic isotope effect at 25°C reported by Daby et al., (10) $(k_{e-1})_H/(k_{e-1})_D$ at 1000°K is estimated at 0.91.

(b) Decomposition step (e-2)

This is an unimolecular decomposition process of a hot n-propyl radical. This rate constant k_{e-2} can be estimated by RRKM theory. Details of the calculation are given in the appendix.

$$(k_{e-2})_H = 3.39 \times 10^9,\ sec^{-1}$$

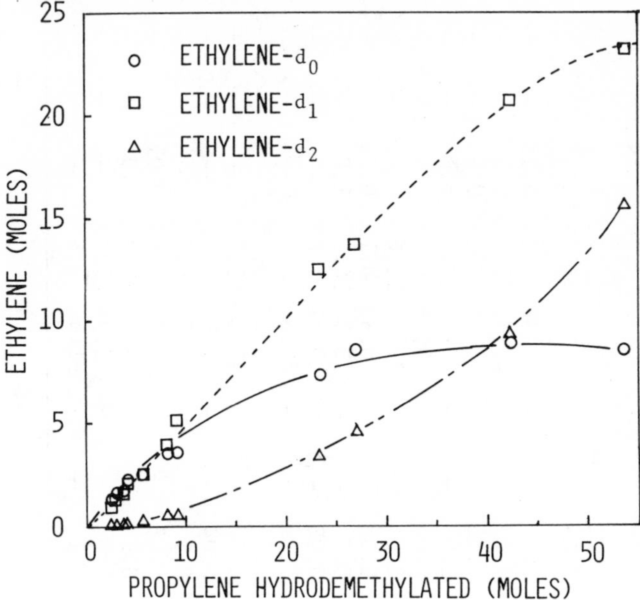

Figure 3. Deuterium distribution in ethylene (675° ~ 800°C). D_2/C_3H_6 = ca. 10.

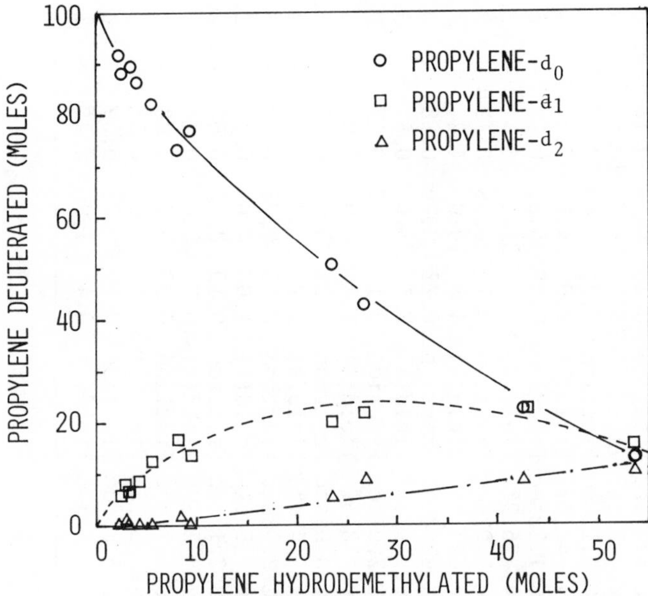

Figure 4. Deuterium distribution in propylene (675° ~ 800°C). D_2/C_3H_6 = ca. 10.

Table III. Deuterium Distribution in the Pyrolysis Products of Propylene

		Methane				Ethylene					Propylene				
		d_0	d_1	d_2	d_3	d_0	d_1	d_2	d_3	d_4	d_0	d_1	d_2	d_3	d_4
800°C 1.47sec	Fraction	0.23	0.57	0.18	0.02	0.16	0.43	0.30	0.10	0.01	0.31	0.36	0.24	0.06	0.03
conv. 53.6%	Yield(%)	13.6	33.7	10.6	1.2	8.6	23.1	16.1	5.4	0.5	13.4	15.6	10.4	2.6	1.3
800°C 1.12sec	Fraction	0.21	0.60	0.17	0.02	0.21	0.49	0.22	0.06	0.02	0.39	0.39	0.16	0.06	—
conv. 42.3%	Yield(%)	9.0	25.6	7.3	0.9	8.9	20.7	9.3	2.5	0.8	22.5	22.5	9.2	3.5	—
775°C 1.28sec	Fraction	0.25	0.62	0.13	—	0.32	0.51	0.17	—	—	0.58	0.29	0.13	—	—
conv. 26.9%	Yield(%)	8.5	21.1	4.4	—	8.6	13.7	4.6	—	—	42.4	21.2	9.5	—	—
775°C 1.02sec	Fraction	0.17	0.68	0.15	—	0.32	0.53	0.15	—	—	0.66	0.26	0.08	—	—
conv. 23.6%	Yield(%)	4.8	19.0	4.2	—	7.5	12.5	3.5	—	—	50.5	19.9	6.1	—	—
750°C 0.84sec	Fraction	0.22	0.70	0.08	—	0.39	0.55	0.06	—	—	0.85	0.15	—	—	—
conv. 9.4%	Yield(%)	2.1	6.6	0.8	—	3.7	5.2	0.6	—	—	77.0	13.6	—	—	—
725°C 2.01sec	Fraction	0.17	0.72	0.11	—	0.45	0.47	0.08	—	—	0.80	0.18	0.02	—	—
conv. 8.3%	Yield(%)	1.4	6.0	0.9	—	3.7	3.9	0.7	—	—	73.3	16.5	1.8	—	—
725°C 0.99sec	Fraction	0.24	0.67	0.09	—	0.49	0.46	0.05	—	—	0.91	0.09	—	—	—
conv. 4.1%	Yield(%)	1.2	3.2	0.4	—	2.4	2.2	0.2	—	—	86.6	8.6	—	—	—
700°C 2.98sec	Fraction	0.18	0.72	0.10	—	0.47	0.47	0.06	—	—	0.87	0.13	—	—	—
conv. 5.6%	Yield(%)	1.0	4.0	0.6	—	2.6	2.6	0.3	—	—	82.2	12.3	—	—	—
700°C 2.12sec	Fraction	0.19	0.74	0.07	—	0.50	0.46	0.04	—	—	0.93	0.07	—	—	—
conv. 3.59%	Yield(%)	0.7	2.6	0.2	—	1.8	1.7	0.1	—	—	89.7	6.7	—	—	—
675°C 3.76sec	Fraction	0.20	0.72	0.08	—	0.54	0.43	0.03	—	—	0.91	0.08	0.01	—	—
conv. 3.12%	Yield(%)	0.6	2.3	0.3	—	1.7	1.3	0.1	—	—	88.2	7.8	1.0	—	—
675°C 3.00sec	Fraction	0.16	0.75	0.09	—	0.57	0.40	0.03	—	—	0.94	0.06	—	—	—
conv. 2.54%	Yield(%)	0.4	1.9	0.2	—	1.4	1.0	0.1	—	—	91.6	5.8	—	—	—

$$(k_{e-2})_D = 3.48 \times 10^9, \ sec^{-1}$$

$$(k_{e-2})_H/(k_{e-2})_D = 0.97$$

This computation shows that there is no appreciable isotope effect in the decomposition step.

These considerations indicate that k_e term, $(k_e)_H/(k_e)_D = 0.9$, is unable to account for our experimental results, that is $(k_{obs})_{H2}/(k_{obs})_{D2} = 1.45 \sim 2.13$. The implication is that the second term should be much larger than unity.

[2] $(k_a\ k_d/k_b\ k_c)^{1/2}$ term

The $(k_a\ k_d/k_b\ k_c)^{1/2}$ term is correlated with hydrogen atom concentration as follows.

$$[H\cdot] = \left(\frac{k_a\ k_d}{k_b\ k_c}\right)^{1/2} [H_2]^{1/2}$$

In this equation, k_a and k_b should remain nearly constant in the presence of hydrogen or deuterium, and thus these terms do not contribute to kinetic isotope effect. Small differences in k_c can be estimated from the Evans-Polanyi-Semenov rule. (11)

	Ea kcal/mol	log A 1/mol sec
$H\cdot + C_3H_6 \longrightarrow H_2 + CH_2{=}CHCH_2\cdot$	5.0	10.1
$D\cdot + C_3H_6 \longrightarrow HD + CH_2{=}CHCH_2\cdot$	4.8	10.1

$$\left(\frac{(k_c)_D}{(k_c)_H}\right)^{1/2} = \left\{\exp(0.2 \times 10^3/RT)\right\}^{1/2}$$

On the other hand for reaction (d), the allyl radical is much less reactive, so that zero point energy difference between H-H and D-D bonds possibly contributes to the isotope effect to a large extent. If it is assumed that the difference, 2 kcal/mol, is fully reflected, the isotope effect is calculated as follows.

$$\left(\frac{(k_d)_H}{(k_d)_D}\right)^{1/2} = \left\{\exp(2.0 \times 10^3/RT)\right\}^{1/2}$$

In summary,

$$\frac{(k_{obs})_{H2}}{(k_{obs})_{D2}} = \frac{(k_e)_H}{(k_e)_D} \times \left(\frac{(k_d)_H\ (k_c)_D}{(k_d)_D\ (k_c)_H}\right)^{1/2}$$

$$= \exp(0.9 \times 10^3/RT).$$

The kinetic isotope effect can be calculated from this equation. The calculated and observed values are shown as follows.

Overall Kinetic Isotope Effect	Temperature (°C) 800	750	700
Observed	1.45	1.68	1.96
Calculated	1.53	1.56	1.60

The relatively good agreement between the calculated and observed values seemingly supports the validity of the proposed reaction mechanism.

Deuterium Distribution of Products.

Methane. Figure 2 indicates that methane-d_1 is produced in largest concentrations of all methanes produced when deuterium is mixed with the propylene, and methane-d_0 is the second largest. The methyl radical, produced by hydrogenolysis of propylene or by reaction (e), is converted either to methane-d_1 by reaction (f') or to methane-d_0 by reaction (g)

$$\cdot CH_3 + D_2 \longrightarrow CH_3D + D\cdot$$

Experimental values of $[CH_3D]/[CH_4] = 3$ suggests that $k_{f'}/k_g$ equals to 0.3. This seems reasonable from a kinetic standpoint. Formation of methane-d_2 in the advanced stage of hydrogenolysis is remarkable. This probably comes from $CH_2D\cdot$ which is produced by hydrogenolysis of deuterated propylene given by the reactions (d') and (i'-1 & -2),

$$D\cdot + CH_2{=}CH{-}CH_3 \xrightarrow{i'-1} (\underset{D}{CH_2}{-}CH{-}CH_3)^*$$

$$\xrightarrow{i'-2} CH_2D{-}CH{=}CH_2 + H\cdot$$

$$\xrightarrow{i'-3} CHD{=}CH{-}CH_3 + H\cdot$$

This explanation is apparently supported by the observation that $[CH_2D_2]/[CH_3D] = [C_3H_5D]/[C_3H_6]$

* Analysis by Professor K. Kuratani and T. Yano of mono-deuterated propylenes by use of microwave spectroscopy has revealed that they are mainly $CH_2D{-}CH{=}CH_2$. (Private communication)

Ethylene. It should be noted in Figure 3 that both ethylene-d_1 and -d_0 are the primary products in the initial stage of hydrogenolysis. Formation of ethylene-d_1 is naturally expected by the presence of reaction (e'),

$$CH_2{=}CH{-}CH_3 + D\cdot \xrightarrow{e'} CH_2{=}CHD + \cdot CH_3$$

The formation of ethylene-d_0 is rather unexpectedly large, but the previously mentioned reaction (i') can explain the production of H· in the reaction system where the hydrogen molecule is initially absent, and the attack of H· on propylene may produce ethylene-d_0. The decrease in ethylene-d_0 yield in the advanced stage is probably due to the H-D exchange of propylene and ethylene. The gradual increase in ethylene-d_2 yield is partly accounted by the hydrogenolysis, with D·, of propylene-d_1 produced by reaction (i'-3).

Propylene. Figure 3 shows the successive deuteration of propylene along with demethylation. The rate of deuteration of propylene is roughly two times larger than that of hydrogenolytic demethylation. Furthermore, analysis of propylene-d_1 and of ethylene-d_1 in the initial stage, or below ca. 10% conversion of propylene, gives the rate ratio of deuteration (mostly via terminal addition of D· to propylene) to demethylation (via internal addition) as approximately 4.5. In view of the high temperature employed in this study, this ratio is not incompatible with the observation at room temperature that the ratio of terminal addition to internal addition is more than 10. (12)

Conclusions

The kinetic isotope effect observed on the overall rates of hydrogenolytic demethylation of propylene in the presence of deuterium was successfully interpreted in terms of a free radical chain mechanism. Little differences were inferred to exist between the rates of addition of H· or D· to propylene and also between those of unimolecular decomposition of the produced hot n-propyl radicals, and thus the kinetic isotope effect was ascribed mainly to the difference between the steady state concentrations of [H·] in the presence of hydrogen and the concentrations of [D·] + [H·] in the presence of deuterium. In more detail, conversion of rather inactive allyl radical by metathesis with deuterium into an active D· is relatively slow. This was concluded to be the main cause of the observed kinetic isotope effect, which agrees well with the calculated values based on the above free radical chain mechanism. Deuterium distribution in the reaction products also supports the mechanism proposed. In summary, the most important role of hydrogen molecule in pyrolysis of propylene is to convert the allyl radical, which otherwise easily polymerizes, into an active hydrogen atom that promotes the key reaction, namely hydrogenolytic demethylation of propylene.

Acknowledgement. We wish to thank professors Akira Amano and Osamu Horie for their helpful discussions on chemical activation system. We are also grateful to professor Tadao Yoshida for his suggestion on RRKM program. We are favoured to have the assistance of Mr. Shozo Yamamiya who contributed his experimental skill.

Summary

The kinetics and mechanism of hydrogenolysis of propylene to give ethylene and methane were investigated in a flow reactor made of quartz, at temperatures between 675 and 800°C. The molar ratio of deuterium or hydrogen to propylene employed was up to approx. 10. The main objective of this investigation is to elucidate the role played by hydrogen in the pyrolysis of hydrocarbons when hydrogen is employed to enhance the rates of hydrocarbon decomposition and to promote ethylene formation. The overall rates of ethylene formation were described by 1.5 order rate equations as follows:

$$d[C_2H_4]/dt = k_{obs}\ [C_3H_6][H_2 \text{ or } D_2]^{1/2}, \text{ mol/l sec},$$

where the rate constants are respectively given by,

$$(k_{obs})_{H_2} = 10^{14.0} \exp(-64{,}400/RT)$$

$$(k_{obs})_{D_2} = 10^{15.1} \exp(-70{,}600/RT).$$

The kinetic isotope effect observed on the overall rates, ranging from 2.0 at 700°C to 1.5 at 800°C, was interpreted in terms of a free radical chain mechanism where a unimolecular decomposition of a hot n-propyl radical, produced by addition of H· or D· to propylene, plays a key role. The difference in the observed overall rates was mainly ascribed to that in steady state concentrations of H· and D· which are produced, in part, through the reaction between allyl radical and H_2 or D_2. No appreciable difference between the decomposition rates of C_3H_7 and C_3H_6D was infered by RRKM theory.

Literature Cited

(1) Kunugi, T., Tominaga, H., and Abiko, S., Proceedings of the 7th World Petroleum Congress (Mexico City) (1967).
(2) Kunugi, T., Tominaga, H., Abiko, S., Uehara, K., and Ohno, T., Kogyo Kagaku Zasshi, (1967) 70, 1477.
(3) Tominaga, H., Arai, H., Moghul, K.A., Takahashi, N., and Kunugi, T., ibid., (1971) 74, 371
(4) Amano, A. and Uchiyama, M., J. Phys. Chem., (1963) 67, 1242.
(5) Benson, S.W. and Shaw, R., J. Chem. Phys., (1967) 47, 4052.
(6) Matsui, H., Amano, A., and Tokuhisa, H., Bull. Japan Petrol. Inst., (1959) 1, 67.
(7) Amano, A., Tominaga, H., and Tokuhisa, H., ibid., (1965) 7, 59.
(8) Tominaga, H., Arai, H., Kunugi, T., Amano, A., Uchiyama, M., and Sato, Y., Bull. Chem. Soc. Japan, (1970) 43, 3658.
(9) Hougen, O.A. and Watson, K.M., "Chemical Process Principles". p.884, John Wiely & Sons, Inc., New York, (1950).
(10) Daby, E.E., Niki, H., and Weinstock, B., J. Phys. Chem., (1971) 75, 1601.
(11) Semenov, N.N., "Some Problems in Chemical Kinetics and Reactivity", Pergamon Press and Princeton University, (1958).
(12) Falconer, W.E., Rabinovitch, B.S., and Cvetanovic, R.J., J. Chem. Phys., (1963) 39, 40.
(13) Rabinovitch, B.S. and Setser, D.W., Adv. Photochem., (1964) 3, 1.

Appendix

RRKM Calculation on the Unimolecular Decomposition of Hot n- and sec-Propyl Radicals. The basic concept of the computation is given elsewhere. (12) The thermochemical data were obtained from the literature. (13) Vibration models and the other parameters were mostly adopted from the work by Falconer, W.E. et al., (12) but with some modifications. The calculated rate constants shown below are at 998°K and atmospheric pressure.

Elementary Reactions			Rate Constant, sec^{-1}
$CH_2{=}CHCH_3 \xrightarrow{D\cdot} (CH_2D\dot{C}HCH_3)^*$	k_1	$CH_2{=}CHCH_3 + D\cdot$	0.400×10^9
	k_2	$CH_2DCH{=}CH_2 + H\cdot$	1.88×10^9
	k_3	$CHD{=}CHCH_3 + H\cdot$	0.757×10^9
$CH_2{=}CHCH_3 \xrightarrow{D\cdot} (\dot{C}H_2CHDCH_3)^*$	k_4	$CH_2{=}CHCH_3 + D\cdot$	1.18×10^9
	k_5	$CH_2{=}CDCH_3 + H\cdot$	1.89×10^9
	k_6	$CH_2{=}CHD + \cdot CH_3$	$3.48 \times 10^9 = (k_{e-2})_D$
$CH_2{=}CHCH_3 \xrightarrow{H\cdot} (CH_3\dot{C}HCH_3)^*$	k_7	$CH_2{=}CHCH_3 + H\cdot$	1.95×10^9
$CH_2{=}CHCH_3 \xrightarrow{H\cdot} (\dot{C}H_2CH_2CH_3)^*$	k_8	$CH_2{=}CH_2 + \cdot CH_3$	$3.39 \times 10^9 = (k_{e-2})_H$

7

Kinetics of the Pyrolysis of Propane-Propylene Mixtures

T. HARAGUCHI, F. NAKASHIO, and W. SAKAI

Department of Organic Synthesis, Kyushu University, Fukuoka, Japan

It is well-known that propylene acts in the capture of free radicals in the same manner as nitric oxide and propylene inhibits the rate of reaction, which proceeds through the chain mechanism. Therefore, in the progress of a reaction such as the pyrolysis of propane in which one of the main products is propylene, the inhibition effect to the reaction will be observed. In order to analyze the kinetics of the inhibited reaction, it will be necessary to investigate how much propylene influences the rate of pyrolysis of other hydrocarbons. There are, however, a few quantitative works about this effect. Stubbs & Hinshelwood (9) and Laidler & Wojciechowski (7) have researched the thermal decomposition of propane, sufficiently inhibited by propylene, and were able to discuss the reaction mechanisms of the thermal decomposition of normal paraffin-hydrocarbons. Kershenbaum and Martin (5) have carried out experiments on pyrolysis of propane with small amounts of propylene in the feed and diluted with nitrogen, and they noted the effect of propylene, if any, on the pyrolysis.

Whereas, the investigations on the influence of other compounds on the rate of decomposition of propylene itself are quite scare. Szwarc (10), Kunugi et al (6) and Robinson & Weger (8) have only reported that the rate of decomposition of propylene had a tendency to be accelerated by the addition of small amounts of diallyl, 1-butene and propane in the feed, respectively.

In the present work, propane-propylene mixtures with various ratios were pyrolyzed at temperatures near 900°C and at an atmospheric pressure in an annular flow reactor. Hydrogen was used as a diluent. Under these experimental conditions, it was rather difficult to maintain an uniform temperature throughout the reactor since the reaction rate was high, and consequently, thermal effects due to the heat reaction were significant. In this work, therefore, experimental data at the initial stage of decomposition were analyzed using the effective temperature method to obtain kinetic rate parameters, activation energy and frequency factor, for propane and propylene decompositions. From the relations between

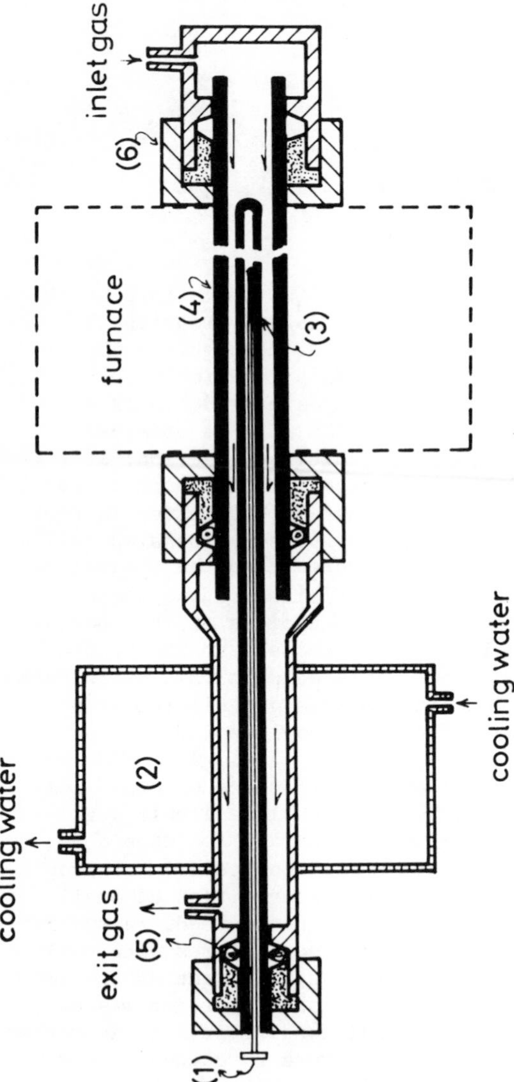

(1): Thermocouple (Pt–Pt·Rh), (2): Water jacket

(3): Inner tube (o.d. 6mm , i.d. 4mm , l. 70 cm)

(4): Outer tube (o.d. 12mm , i.d. 8mm , l. 100cm)

(5): asbestine seal

Figure 1. Reactor details

the propane-propylene ratio in the feed and each decomposition rate, the inhibition effect of propylene and the acceleration effect of propane were obtained.

Furthermore, the concentration profiles of products in the pyrolysis of propane were calculated. In this calculation, a reaction model was postulated which involved only stable chemical species detected in the exit gas of the pyrolysis of propane. The rate equations, which were formulated on the basis of the reaction model taking account of these inhibition and acceleration effects and the axial temperature distribution of the reactor in each run, were solved.

Experimental

Figure 1 shows details of the reactor. After the propane, propylene and hydrogen flows were separately adjusted by needle valves to desired rates and were metered by capillary tubes, these gases were then mixed. The resulting gas mixture was passed through a drying tube and then transfered to the reactor. The exit gas of the reactor was cooled and its flow rate was measured using a wet gas meter. The inlet and exit gases of the reactor were analyzed with a gas chromatograph.

The reactor used was an annular one which consisted of two concentric tubes made of alumina. The inner diameter of the outside tube was 8mm and the outer diameter of the inner tube was 6mm. The catalytic activity of the tube surface was considered negligible because soot formed on the surface. A large section of the reactor (60cm) was placed inside of a high capacity electric furnace, the temperature of which was controlled within $\pm$ 1°C. The axial temperature distribution of the reactor was measured by inserting a 70cm long Pt-Pt·Rh thermocouple into the reactor. The thermocouple was moved at 5cm intervals along the axis of the reactor to obtain the temperature profile. The cross-sectional area of the reactor used in the present study was so small that the temperature measurements thus obtained were assumed to give the gas temperatures within the reactor.

Commercially available hydrogen, nitrogen, propane and propylene were used without further purification. The purities of the propane and propylene were higher than 99.5%; the former was obtained from the Maruzen Petroleum Company and the latter from the Phillips Petroleum Company. The purities of the hydrogen and nitrogen were higher than 99.8%.

Analysis of reactant and product gases were performed with a two-stage gas chromatograph which was equipped with an activated alumina-2% squalene column (3m) and with a molecular sieve 5A column (4m). Helium was used as a carrier gas with a flow rate of 30cc/min.

Prior to the experimental measurements, the maximum reactor temperature, T_{max}, was set approximately at the desired value. The entire experimental apparatus was filled with nitrogen and then this was replaced by the mixture of hydrogen and hydrocarbons. The flow rates of these three kinds of gases and the maximum temperature were set at the specified values. After the experimental conditions had attained a steady state, we measured the concentrations of inlet and exit gases, the gas flow rate, the temperature profiles of the reactor, and the pressure drop through the reactor.

The experimental conditions used in this study were as follows: maximum reactor temperatures, T_{max}, 820-940°C; inlet concentrations of hydrocarbons $[(x_{C_3H_8})_i + (x_{C_3H_6})_i]$ 0.07-0.2; propane-propylene ratio in feed ($\gamma_{C_3H_8}$) 0 - 1, where $\gamma_{C_3H_8}$ can be calculated from Equation (1)

$$\gamma_{C_3H_8} = 1 - \gamma_{C_3H_6} = \frac{(x_{C_3H_8})_i}{\{(x_{C_3H_8})_i + (x_{C_3H_6})_i\}} \quad (1)$$

total feed rates at reactor inlet, F_i, 4.0×10^{-3} - 1.1×10^{-2} moles/sec; and apparent contact time, τ_e, 0.006 - 0.7 sec, where τ_e can be calculated from Equation (2).

$$\tau_e = \frac{S\ L}{F_i} \left(\frac{\bar{P}}{R\ T_{max}}\right) ^{*} \quad (2)$$

Theoretical

When a kinetic study is undertaken, the most desirable experimental condition is a constant temperature throughout the reaction vessel. However, it is impossible to keep the reactor under such an isothermal condition when the fast reactions such as the pyrolysis of hydrocarbons proceed at high temperatures or with considerable heats of reaction. Towell & Martin (11) developed a useful method to analyze the kinetic data accounting for nonuniform temperature distributions, but this method is troublesome in the case of processing of many experimental data. In the present work, a simple and convenient method was devised to determine Arrhenius parameters from experimental results under nonisothermal conditions by the least squares estimation. The outline of the method will be described below.

When an irreversible, first-order gaseous reaction

$$A \rightarrow B$$

takes place in a plug flow reactor in which the temperature varies considerably in the axial direction, the basic equation to obtain the decomposition rate of component A can be expressed as

$$-\frac{F_i dx_A}{S\ dl} = \alpha \exp\{-E/RT(l)\}\left[\frac{x_A\ P(l)}{R\ T(l)}\right] \quad (3)$$

Using Equation (2) which defines apparent contact time, τ_e, and Equation (4),

*Nomenclature listing at end.

$$\frac{S\ \bar{F}}{T_{max}} = \lambda \tag{4}$$

Equation (3) becomes

$$-\frac{dx_A}{dl} = \{\tau_e/(\lambda L)\}\ \alpha k'(l)\ x_A \tag{5}$$

where

$$k'(l) = \{P(l)/T(l)\}\ \exp\{-E/RT(l)\} \tag{6}$$

If the profiles of temperature, T(l), and pressure, P(l), in the reactor can be measured, Equation (5) is integrated along the length of the reactor from 0 to L and from concentration $x_{A,i}$ to $x_{A,o}$, and we obtain

$$\ln\ (x_{A,i}/x_{A,o}) = (\tau_e/\lambda)\ (\alpha/L)\int_0^L k'(l)dl$$

$$= (\tau_e/\lambda)\ \bar{k}_o \tag{7}$$

where

$$\bar{k}_o = (\alpha/L)\int_0^L k'(l)\ dl \tag{8}$$

In Equation (7) only two unknowns, activation energy E and frequency factor α, exist. Substituting the next relations into Equation (7),

$$\bar{k} = (L/\bar{P})(\lambda/\tau_e)\ \ln(x_{A,i}/x_{A,o}) \tag{9}$$

$$\beta = \int_0^L \{P(l)/\bar{P}\}\{1/T(l)\}\ \exp\{-E/RT(l)\}dl \tag{10}$$

Equation (7) is rewritten to

$$\bar{k} = \alpha \times \beta \tag{11}$$

and

$$\log\alpha = \log\bar{k} - \log\beta \tag{12}$$

Towell & Martin (11) assumed the following straight line for Equation (12)

$$\log\alpha = \log\bar{k} + B\frac{E}{R\ T(l)} \tag{13}$$

and proposed the graphical method of plotting the relation between the assumed E values and the corresponding calculated log α values for each run. In this method, true Arrhenius parameters, α and E, are theoretically determined from the intersection of E vs log α plots for two runs, but actually this intersection for many experiments does not indicate one point clearly because of experimental errors. By differentiating Equation (12) with respect to E to examine the slope of E vs log α plots, we obtain

$$\frac{d(\log \alpha)}{dE} = \frac{d(-\log \beta)}{dE} = \frac{1}{R} \cdot \frac{1}{T_{eff}} \tag{14}$$

where

$$\frac{1}{T_{eff}} = \frac{\int_0^L \{1/T(l)\}^2 \{P(l)/\overline{P}\} \exp\{-E/RT(l)\} dl}{\int_0^L \{1/T(l)\} \{P(l)/\overline{P}\} \exp\{-E/RT(l)\} dl} \tag{15}$$

The temperature, T_{eff}, defined by Equation (15) is named here as "effective temperature" since it is a weighted average temperature in consideration of the distribution of the reaction rate constant along the reactor. Therefore, it is found that the slope of the E vs log α curve or the E vs $-\log \beta$ curve at some E is proportional to the reciprocal of the effective temperature for the E value. In our previous (1) paper, it was shown that in the range of $E>30$kcal/mole, the slope of the E vs $-\log \beta$ curve was almost constant and was dependent only upon the maximum reactor temperature, T_{max}. Consequently, it is found that the T_{eff} is closer to the T_{max} value. In the range of $E<30$kcal/mole, the relationship between E and $-\log \beta$ can be considered to be linear within a very short range of E. In these cases the relationship between E and $-\log \beta$ is expressed as

$$-\log \beta = \frac{1}{R} \cdot \frac{1}{T_{eff}} E - \log \beta_0 \tag{16}$$

where β_0 is an extrapolated value of the tangent of the E vs $-\log \beta$ curve to E=0 and varies with the difference of temperature profile in the reactor. Substituting Equation (16) in Equation (12), we obtain

$$\log \overline{k} = -\frac{E}{R} \cdot \frac{1}{T_{eff}} + \log \alpha\beta_0 \tag{17}$$

As the primary step, the first approximate value of the activation energy, $E^{(1)}$, is obtained from the slope of $1/T_{max}$ vs log $\overline{k}$ plot using a series of runs which was made at different maximum reactor temperatures, T_{max}. Substituting $\beta^{(1)}$ and $T_{eff}^{(1)}$ corresponding to $E^{(1)}$ in Equation (16), log β_0 is expressed as

$$\log \beta_0 = \frac{E^{(1)}}{R} \cdot \frac{1}{T^{(1)}_{eff}} + \log \beta^{(1)} \tag{18}$$

Also substituting this relation in Equation (17),

$$\log \left(\frac{\bar{k}}{\beta^{(1)}}\right) = - \frac{E}{R} \cdot \frac{1}{T_{eff}} + \frac{E^{(1)}}{R} \cdot \frac{1}{T^{(1)}_{eff}} + \log \alpha \tag{19}$$

is obtained. If $E^{(1)}$ is not too different from E, $1/T_{eff} = 1/T^{(1)}_{eff}$ is valid and then Equation (19) is rewritten as

$$\log\left(\frac{\bar{k}}{\beta^{(1)}}\right) = - \frac{\Delta E}{R} \cdot \frac{1}{T^{(1)}_{eff}} + \log \alpha \tag{20}$$

where

$$\Delta E = E - E^{(1)} \tag{21}$$

Here ΔE and α are obtained from the relation between $1/T^{(1)}_{eff}$ and $\log (\bar{k}/\beta^{(1)})$ for the series of runs by the least squares estimation and also the second approximate value $E^{(2)}$ is obtained with the next equation.

$$E^{(2)} = E^{(1)} + \Delta E \tag{22}$$

This procedure is repeated until ΔE becomes nearly zero.

Although this method seems to be apparently troublesome because of repetition of least squares estimation, iteration cycle is from two to five. Therefore, it is possible to determine the true activation energy and frequency factor simply and conveniently from many experimental results by using the digital computer. Furthermore, this method can be applied to the analysis of complex reactions $A \rightarrow B \rightarrow C$, $A \rightarrow D$ by using a series of suitable (2) experimental data obtained under the nonisothermal conditions. This method was named "effective temperature method."

Results

Propane and propylene were first pyrolyzed separately to determine general trends in the product distribution. The typical product distributions of the pyrolysis of propane and propylene mixtures are given in Table 1. The product distributions when the major reactant was propane or propylene were similar to those when propane and propylene were separately pyrolyzed, respectively. From the material balances for carbon and hydrogen, it was

Table 1. Typical results in pyrolysis of propane–propylene mixtures

Run	T_{max}	$F_{in} \cdot 10^3$	τ_e	Composition (mol%)											$Z_{C_3H_8}$	$Z_{C_3H_6}$
				Inlet gas			Exit gas									
No.	(°C)	(mol/sec)	(sec)	C_3H_8	C_3H_6	H_2	H_2	CH_4	C_2H_6	C_2H_4	C_3H_8	C_3H_6	C_2H_2	C_4	(%)	(%)
522	899	4.44	0.0199	0.220	10.1	89.7	89.1	0.579	0.0162	0.535	0.207	9.53	0.0147	0.0165	6.89	6.61*
502	902	4.33	0.0205	0.508	10.1	89.5	88.4	1.00	0.0130	0.829	0.454	9.26	0.0119	0.00907	11.1	8.67*
507	901	4.32	0.0205	1.02	9.47	89.5	88.4	1.14	0.0351	0.984	0.889	8.52	0.0330	0.0236	12.7	10.2*
512	900	4.44	0.0200	2.01	8.23	89.8	88.3	1.60	0.0511	1.35	1.64	7.00	0.0233	0.0217	18.6	16.9
516	900	4.41	0.0201	3.21	6.75	90.0	88.2	1.94	0.0683	1.65	2.50	5.57	0.0200	0.0207	22.0	20.4
527	900	7.05	0.0127	5.00	5.21	89.8	88.0	1.75	0.0595	1.61	3.97	4.51	0.0473	0.0926	21.7	21.0
531	900	7.16	0.0125	7.36	2.35	90.3	87.8	2.42	0.0949	2.06	5.11	2.37	0.0528	0.0697	31.0	25.2
535	900	11.1	0.00817	9.01	0.955	90.0	87.6	2.32	0.0802	1.99	6.23	1.66	0.0392	0.0841	30.8	13.9

* Correction was neglected

determined that the formation of carbon and liquid components were almost negligible.

Here, the conversion of propylene in the copyrolysis of propane and propylene could not be calculated straightforwardly since propylene is one of the major products of propane decomposition. Therefore, the conversion of propylene was evaluated by the same way as reported by Robinson et al (8). From the preliminary pyrolysis of propane, it was found that the relationship between propane conversion, $z_{C_3H_8}$, and its selectivity to propylene, $y_{C_3H_6}$, was approximately expressed by the formula given below.

$$Y_{C_3H_8} = 0.45 - 0.20\ z_{C_3H_8} \tag{23}$$

If this equation is also valid in the pyrolysis of propane-propylene mixtures, the amount of propylene available for the copyrolysis is defined as the sum of the propylene fed plus propylene produced from propane evaluated from Equation (23). However, the selectivity was small and could be negligible in the case of low propane to propylene ratio in feed ($\gamma_{C_3H_8} < 0.2$). Methane yield from propane and propylene was larger than that of ethylene as shown in Table 1. This result will be caused by hydrogenation and hydrogenolysis of each component.

Discussion

Determination of Rate Constant in the Pyrolysis of Propane-Propylene Mixtures

Using the experimental results with low conversions and the special experimental technique by Towell & Martin (11), the order of propane or propylene decomposition was obtained in the copyrolysis system. It was revealed that the rate equations for the component in question were a function of only one concentration variable as follows:

$$-r_{C_3H_8} = k_{C_3H_8}(\gamma_{C_3H_6}) \times C_{C_3H_8} \tag{24}$$

$$-r_{C_3H_6} = k_{C_3H_6}(\gamma_{C_3H_8}) \times C_{C_3H_6} \tag{25}$$

where $k_{C_3H_8}(\gamma_{C_3H_6})$ and $k_{C_3H_6}(\gamma_{C_3H_8})$ are the rate constants of propane and propylene decomposition, respectively, which are the functions of the ratio of both components.

To clarify the relationship between these rate constants in Equations (24) and (25) and the propane-propylene ratios, the decomposition rate constant of propane or propylene was determined from the experimental data with various ratios of both components

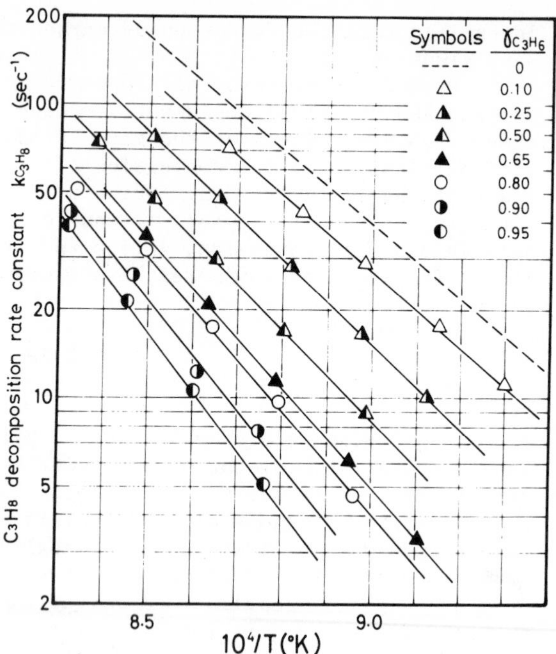

Figure 2. Arrhenius plots for propane decomposition rate in pyrolysis of propane–propylene mixtures

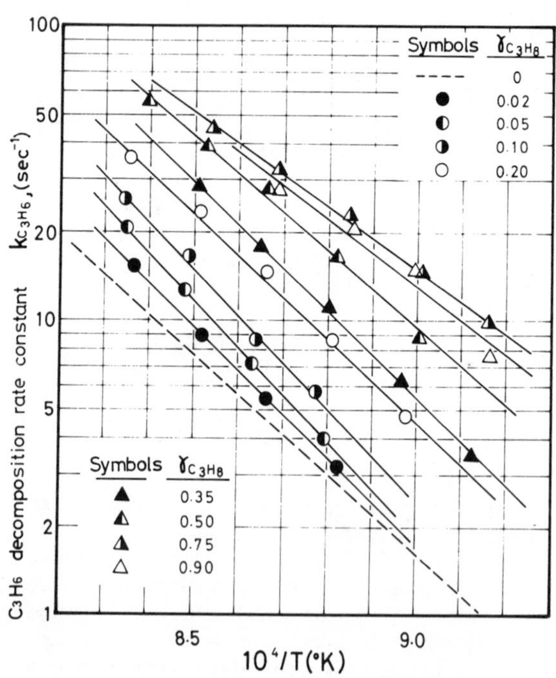

Figure 3. Arrhenius plots for propylene decomposition rate in pyrolysis of propane–propylene mixtures

in the feed. As the reactor was operated under nonisothermal conditions in this study, the effective temperature method detailed above was used to estimate the Arrhenius parameters of the decompositions. The experimental results which were obtained under the conditions of different maximum reactor temperature, T_{max}, and a constant propane-propylene ratio, $\gamma_{C_3H_8}$ or $\gamma_{C_3H_6}$, were analyzed with this method.

The rate constants obtained are plotted as a function of the reciprocal of the effective temperature in Figure 2 for propane and in Figure 3 for propylene and the Arrhenius parameters in Figures 2 and 3 are given in Tables 2 and 3, respectively. From Figure 2 and Table 2 it was found that the propane decomposition rate was decreased with increasing propylene ratio in the feed and the activation energy had an increasing tendency. These results show clearly the inhibition effects of propylene on the kinetics of propane decomposition. On the contrary, Figure 3 and Table 3 suggest that the propylene decomposition rate was increased with propane ratio in feed. The acceleration effect of propane on propylene decomposition might be explained by the fact that an additional source of chain carrier species allows the chain propagation steps to proceed independent of the propylene initiation step.

Figure 4 shows the relationships between the rate constant for propane or propylene decomposition at 900°C and propane-propylene ratio in the feed. Stubbs et al (9) reported that the

Table 2. Arrhenius kinetic parameters of propane decomposition rates in pyrolysis of propane–propylene mixtures

Symbols in Fig.2	$\alpha_{C_3H_8}$ (sec^{-1})	$E_{C_3H_8}$ (kcal/mol)
-----	1.76×10^{14}	64.5
△	1.29×10^{13}	59.4
◭	1.36×10^{14}	65.9
◮	5.73×10^{14}	70.4
▲	7.01×10^{15}	77.0
○	6.17×10^{15}	77.2
◑	5.21×10^{18}	93.7
◐	1.01×10^{20}	101

Table 3. Arrhenius kinetic parameters of propylene decomposition rates in pyrolysis of propane–propylene mixtures

Symbols in Fig.3	$\alpha_{C_3H_6}$ (sec^{-1})	$E_{C_3H_6}$ (kcal/mol)
------	1.27×10^{12}	60.7
●	4.09×10^{13}	68.0
◐	4.71×10^{14}	73.3
◑	4.09×10^{14}	72.3
○	2.94×10^{13}	65.2
▲	9.24×10^{13}	67.3
◮	6.87×10^{12}	60.4
◭	8.23×10^{10}	49.6
△	3.17×10^{11}	52.8

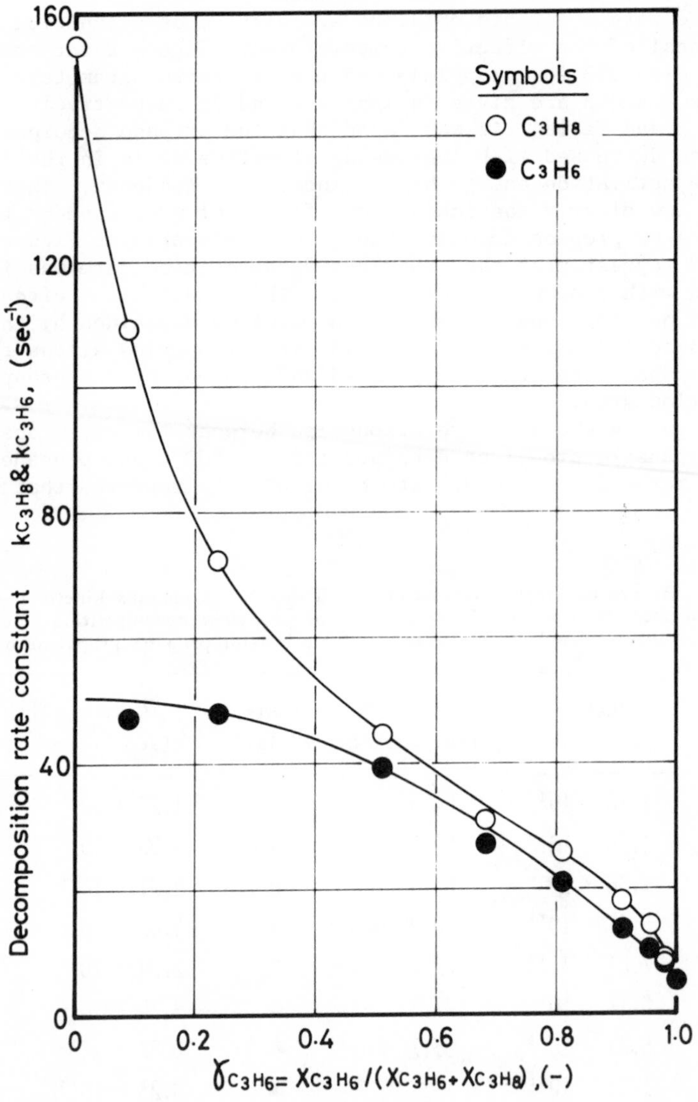

Figure 4. Relationships between propane or propylene decomposition rate constant in copyrolysis at $T_{eff} = 900°C$ and propane–propylene ratio in feed

Table 4. Reaction model for pyrolysis of propane

Reaction			α (sec^{-1})	E (kcal/mol)
$C_3H_8 \rightarrow C_2H_4 + CH_4$	(i)	K_1	1.14×10^{14}	64.7
$C_3H_8 \rightarrow C_3H_6 + H_2$	(ii)	K_2	1.00×10^{14}	65.2
$C_3H_8 + H_2 \rightarrow C_2H_6 + CH_4$	(iii)	K_3	1.43×10^{12}	62.2
$C_3H_8 + 2H_2 \rightarrow 3CH_4$	(iv)	K_4	1.04×10^{12}	62.2
$C_3H_6 + H_2 \rightarrow C_2H_4 + CH_4$	(v)	K_5	1.12×10^{12}	60.5
$C_3H_6 + 3H_2 \rightarrow 3CH_4$	(vi)	K_6	1.67×10^{11}	62.9
$C_2H_4 + H_2 \rightleftarrows C_2H_6$	(vii)	K_7	1.98×10^{7}	38.5
		K_{-7}	$2.24\ 10^{14}$	73.0
$C_2H_4 \rightleftarrows C_2H_2 + H_2$	(viii)	K_8	3.94×10^{16}	93.5
		K_{-8}	4.54×10^{9}	49.4
$C_2H_6 + H_2 \rightarrow 2CH_4$	(ix)	K_9	4.50×10^{14}	77.2

rate for the decomposition of paraffin hydrocarbons was lowered to a limiting value at temperatures near 500°C by the addition of a considerable amount of propylene. But, Figure 4 shows that the rate constant for propane decomposition falls off very rapidly as the ratio of propylene in the feed is increased. This result was explained by the fact that a large number of hydrogen atoms are produced under the conditions and these atoms promote the chain propagation steps.

Evaluation of Product Distribution in the Pyrolysis of Propane

An attempt was made to calculate the product distributions of propane pyrolysis on the basis of the reaction model considering both inhibition and acceleration effects observed in the pyrolysis of propane-propylene mixtures. The reaction model for propane pyrolysis used in this work is shown in Table 4. The rate constants given in Table 4 were measured in our previous works (3, 4). At the initial stage of propane pyrolysis, the formation of the primary products such as methane, ethylene and propylene is predominant but as the reaction proceeds, the consecutive decomposition of each product is also remarkable. Therefore, a reaction model was postulated which consisted of major stoichiometric reactions for propane {(i) - (iv)}, for propylene {(v),(vi)} and for ethylene {(vii)-(ix)}. Methane was scarcely decomposed under the experimental conditions.

The basic equations to calculate the concentration of each species in the exit gas were written as follows:

$$-\frac{dx_{C_3H_8}}{dl} = \{\tau_e/(\lambda L)\}\{\alpha_1 k'_1(l) + \alpha_2 K'_2(l) + \alpha_3 k'_3(l) + \alpha_4 k'_4(l)\}x_{C_3H_8} \quad (26)$$

$$\frac{dx_{C_3H_6}}{dl} = \{\tau_e/(\lambda L)\}[\alpha_2 k'_2(l)x_{C_3H_8} - \{\alpha_5 k'_5(l) + \alpha_6 k'_6(l)\}x_{C_3H_6}] \quad (27)$$

$$\frac{dx_{C_2H_6}}{dl} = \{\tau_e/(\lambda L)\}[\alpha_3 k'_3(l)x_{C_3H_8} + \alpha_7 k'_7(l)x_{C_2H_4} - \alpha_{-7}k'_{-7}(l) + \alpha_9 k'_9(l)\}x_{C_2H_6}] \quad (28)$$

$$\frac{dx_{C_2H_4}}{dl} = \{\tau_e/(\lambda L)\}[\alpha_1 k'_1(l)x_{C_3H_8} + \alpha_5 k'_5(l)x_{C_3H_6} + \alpha_{-7}k'_{-7}(l)x_{C_2H_6} + \alpha_{-8}k'_{-8}(l)x_{C_2H_2} - \{\alpha_7 k'_7(l) + \alpha_8 k'_8(l)\}x_{C_2H_4}] \quad (29)$$

$$\frac{dx_{C_2H_2}}{dl} = \{\tau_e/(\lambda L)\}\{\alpha_8 k_8(l)\, x_{C_2H_4} - \alpha_{-8}k'_{-8}(l)x_{C_2H_2}\} \quad (30)$$

$$\frac{dx_{CH_4}}{dl} = \{\tau_e/(\lambda L)\}[\{\alpha_1 k'_1(l) + \alpha_3 k'_3(l) + 3\alpha_4 k'_4(l)\}x_{C_3H_8} + \{\alpha_5 k'_5(l) + 3\alpha_6 k'_6(l)\}x_{C_3H_6} + 2\alpha_9 k'_9(l)x_{C_2H_6}] \quad (31)$$

$$\frac{dx_{H_2}}{dl} = \{\tau_e/(\lambda L)\}\ [\{\alpha_2 k'_2(1) - \alpha_3 k'_3(1) - 2\alpha_4 k'_4(1)\}x_{C_3H_8} - \{\alpha_5 k'_5(1) + 3\alpha_6 k'_6(1)\}x_{C_3H_6} + \{\alpha_{-7} k'_{-7}(1) - \alpha_9 k'_9(1)\}x_{C_2H_6} - \{\alpha_7 k'_7(1) - \alpha_8 k'_8(1)\}x_{C_2H_4} - \alpha_{-8} k'_{-8}(1) x_{C_2H_2} \quad (32)$$

where

$$k'_j(1) = \{P(1)/T(1)\}\exp\{-E_j/RT(1)\} \quad (6)'$$

$$(j=1,2,---9)$$

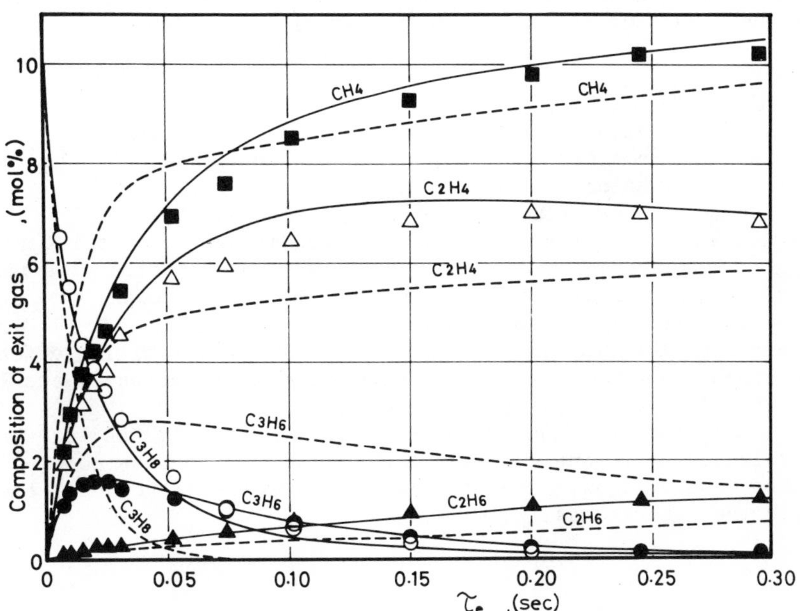

Figure 5. Comparison between experimental product distributions and calculated values for propane pyrolysis, $T_{max} = 900°C$, $(C_{C_3H_8})_i = 10\%$

A Runge-Kutta method was applied to obtain numerical solutions of these simultaneous equations. For the temperature distribution, T(l), in this numerical calculation, an experimentally observed temperature was used. For the pressure distribution, P(l), it was assumed that, due to the pressure drop in the reactor, the pressure decreases linearly along the axis of the reactor.

The dotted lines of Figure 5 show the calculated results using only the rate constants given in Table 4 and the dots are the observed concentrations for each component. It can be seen from this figure that the calculated values deviate from the observed values as the amount of propylene produced increases. Consequently, the numerical calculations were tried again using the interactions between propane and propylene in addition to the rate constants in Table 4. In this calculation, it was assumed that the ratios of rate constants for the stoichiometric reactions of propane decomposition ($k_1 : k_2 : k_3 : k_4$) and of propylene decomposition ($k_5 : k_6$) are independent of the propane-propylene ratio. The calculated concentrations of each component at the reactor exit are shown with the solid line of Figure 5. It was found that the calculated values showed good agreement with observed values up to the region of higher conversion.

Conclusions

In this study a method was developed for determining Arrhenius parameters from nonisothermal kinetic data by the least squares estimation.

Using the experimental results of the pyrolysis of propane-propylene mixtures under the conditions of temperatures near 900°C, atmospheric pressure and hydrogen dilution, the relation between the decomposition rate constant of propane or propylene and the ratio of both reactants was obtained. It was found from the results that propylene had an inhibition effect on propane decomposition, and conversely, propane had an acceleration effect on propylene decomposition.

The concentration of each species in the exit gas for the pyrolysis of propane was calculated with the reaction model and these interactions. The calculated values showed good agreement with the experimental results.

Abstract

Propane-propylene mixtures were pyrolyzed in a flow reactor near 900°C and atmospheric pressure with a large amount of hydrogen dilution. Under these experimental conditions, the temperature distributions in the reactor axis were highly nonuniform. To obtain Arrhenius kinetic parameters of propane and propylene decompositions in this system, the experimental results with low conversions were analyzed with the effective temperature method. From the relation between propane-propylene ratio in the feed and these decomposition rate constants at 900°C, it was shown that the propane decomposition was inhibited by the addition of propylene and conversely the propylene decomposition was accelerated by the addition of propane. By using the reaction model in which these interactions were considered, the product distributions of pyrolysis of propane were estimated. The calculated values showed good agreements with the experimental results.

Nomenclature

A	reactant
B	product
C	concentration [moles/cc]
E	activation energy [kcal/mole]
F_i	total feed rate [moles/sec]
k	rate constant [sec^{-1}]
$\bar{k}$	apparent rate constant defined by eq. (9) [sec^{-1} $°K^{-1}$]
k_o	value defined by eq. (8) [mmHg/PK sec]
k'	value defined by eq. (6) [mmHg/°K]
L	total reactor length [cm]
l	distance from reactor inlet [cm]
P	pressure [mmHg]
$\bar{P}$	average pressure [mmGh]
R	gas constant [cal °K/mole or cc mmHg/mole °K]
r	reaction rate [moles/cc sec]
S	reactor cross-sectional area [cm^2]
T	temperature [K°]
T_{eff}	effective temperature [°K]
T_{max}	maximum temperature [°K]
x	mole fraction
y	selectivity
z	conversion
α	frequency factor [sec^{-1}]
β	integral value defined by eq. (10)
γ	ratio of propane or propylene in feed defined by eq. (1)
λ	value defined by eq. (4) [cm^2mmHg/°K]
τ_e	apparent contact time defined by eq. (2) [sec]
(Subscripts)	
i,o	indicate reactor inlet and exit, respectively

Literature Cited

1) Haraguchi, T., F. Nakashio, and W. Sakai. Kagaku Kōgaku, 33, 786 (1969).
2) Haraguchi, T., F. Nakashio, and W. Sakai. Op Cit., 35, 761 (1971).
3) Haraguchi, T., F. Nakashio, and W. Sakai. Op Cit., 36, 1223 (1972).
4) Haraguchi, T., F. Nakashio, and W. Sakai. Op Cit., 37, 837 (1973).

5) Kershenbaum, L. J., and J. J. Martin. A.I.Ch.E. Journal, 13, 148 (1967).
6) Kunugi, T., H. Tominaga, S. Abiko, and A. Namatame. J. Japan Petrol. Inst., 9, 890 (1966).
7) Laidler, K. J., and B. W. Wojciechowski. Proc. Roy. Soc. (London). A259, 257 (1960).
8) Robinson, K. K., and E. Weger. Ind. Eng. Chem., Fundamentals, 10, 198 (1971).
9) Stubbs, F. J., and F. R. S. Sir Cyril Hishelwood. Proc. Roy. Soc. (London). A200, 458 (1950).
10) Szwarc, M. J. Chem. Phys., 17, 284 (1949).
11) Towell, G. D., and J. J. Martin. A.I.Ch.E. Journal, 7, 693 (1961).

8

Pyrolysis of 1-Butene and *Cis*-2-Butene

DALE R. POWERS

Corning Glass Works, Corning, N.Y. 14830

WILLIAM H. CORCORAN

Department of Chemical Engineering, California Institute of Technology, Pasadena, Calif. 91125

The pyrolysis of olefins has not been as well studied as for paraffin pyrolysis. The lesser interest in olefins can be attributed to both more complex product mixtures and lower rates of pyrolysis for the olefins.

Investigation of pyrolysis of the butenes was chosen here as a complement to previous work in this laboratory on pyrolysis and partial oxidation of n-butane (1)(2)(3). Previous investigations of butene pyrolysis typically have employed static systems and/or high conversions (4)(5)(6)(7)(8)(9).

In this investigation, a micro-reactor was used which consisted of 12 inches of 0.125-inch-I.D. gold tubing. The gold surface is thought to have a minimal effect on pyrolysis reactions (1)(3). Flows of butenes were diluted with argon so that the mole fraction of the hydrocarbon fraction could be varied from 1 to 90 mole percent. Atmospheric pressure and residence times of 4 seconds were used.

Product analyses were made on an F & M Model 5750 gas chromatograph using a 33-ft column of 80/100 Chromosorb P containing 22 percent adiponitrile. The column was operated at room temperature and separated all products from methane to pentenes in about 32 minutes.

The rate of decomposition of 1-butene was found to be much faster than that of cis-2-butene. Pyrolysis of 1-butene was studied in the temperature range of 530 to 605°C, whereas the temperatures used for the pyrolysis of cis-2-butene were 575 to 620°C. As can be seen in Figure 1 and Tables I and II, the product distribution for these two pyrolyses were very different. Cracking is the main course of the reaction for 1-butene, but isomerization is the main result of cis-2-butene pyrolysis.

The overall pyrolysis for both 1-butene and cis-2-butene was found to be first-order. In spite of that, the formation of some individual products deviated significantly from first order. Order of formation of a product is here defined as the slope of the logarithm of the rate of its formation as a function of the logarithm of the initial concentration of the butene being

Table I

A comparison of experimental results and predictions by mechanism of the molar percentage of each product in the pyrolysis of 1-butene

Temperature (°C)	560		560		575	
Concentration 1-Butene (moles/liter)	0.001		0.008		0.001	
Experimental or Mechanism	Exp	Mech	Exp	Mech	Exp	Mech
Methane	27.3	25.2	26.8	19.7	26.9	25.6
Ethane	3.5	5.1	5.0	5.3	3.2	4.6
Ethylene	11.3	12.0	12.2	6.7	12.0	12.8
Propylene	21.9	24.6	21.8	18.9	23.5	24.8
T-2-Butene	5.2	2.4	11.0	13.8	5.5	1.9
C-2-Butene	5.0	2.4	8.2	13.9	4.8	2.0
1,3-Butadiene	21.2	27.0	9.2	20.4	19.1	27.1
1-Pentene	2.6	1.6	3.3	1.7	2.8	1.7
T-2-Pentene	2.2	1.3	2.7	0.7	2.0	1.4

Table II

A comparison of the experimental results and predictions by mechanism of the molar percentages of each product in the pyrolysis of cis-2-butene

Temperature (°C)	575		575		590	
Concentration Cis-2-Butene (moles/liter)	0.001		0.008		0.001	
Experimental or Mechanism	Exp	Mech	Exp	Mech	Exp	Mech
Methane	3.7	4.2	2.7	3.0	5.18	4.5
Ethane	0.03	0.00	0.07	0.001	0.07	0.00
Ethylene	0.28	0.001	0.13	0.001	0.42	0.001
Propylene	2.7	4.2	2.0	3.0	3.9	4.6
1-Butene	1.4	0.7	4.1	3.8	1.2	0.6
T-2-Butene	85.9	86.6	84.4	87.2	82.5	85.7
1,3-Butadiene	6.1	4.2	6.6	3.0	6.7	4.5

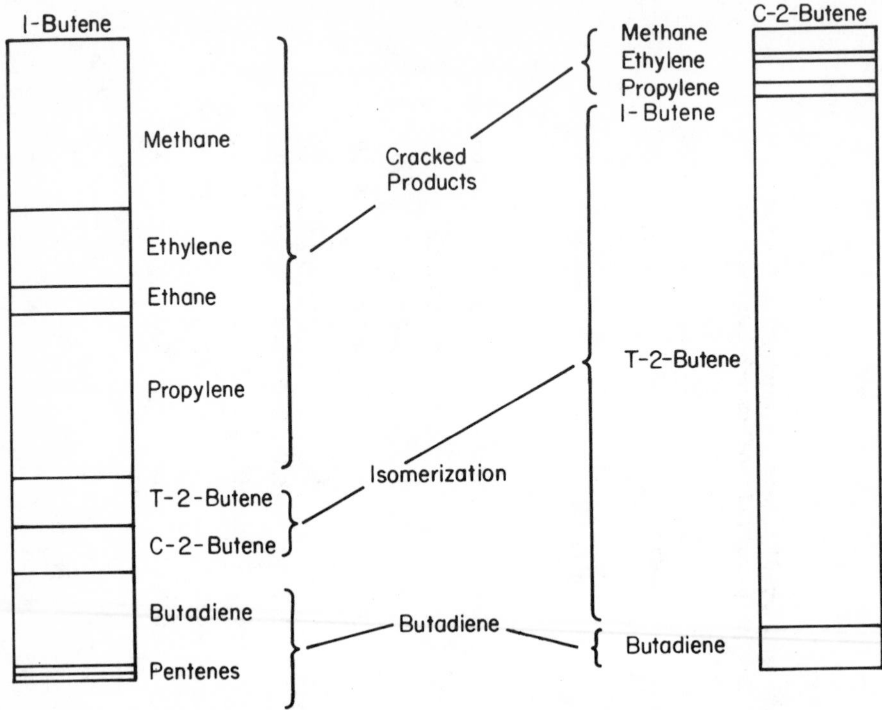

Figure 1. Comparison of product distributions for the pyrolyses of 1-butene and cis-2-*butene*

pyrolyzed. For the pyrolysis of 1-butene, the order of formation of butadiene was 0.7 whereas it was 1.5 for trans-2-butene. Similarly in the case of the pyrolysis of cis-2-butene, the orders for ethylene and 1-butene were 0.7 and 1.3, respectively.

In the pyrolysis of 1-butene, the overall activation energy was 76 kcal/mole, and in the case of cis-2-butene it was 66 kcal/mole. Activation energies for the formation of each of the individual products for these pyrolyses are given in Tables III and IV along with the orders of formation.

The mechanism of butene pyrolysis is obviously more complex than the mechanism operative in butane pyrolysis. The complexity of butene pyrolysis is emphasized by the larger number of products and the larger number of possible free radicals.

An attempt was made to formulate a free-radical mechanism for the pyrolyses of 1-butene and cis-2-butene. The relative importance of 76 elementary steps was investigated. Estimates were made for the individual rate constants for these reactions. In general, the literature (2)(7)(10)(11)(12)(13)(14)(15)was consulted for estimates of these constants. These rate constants are listed in Table V.

Table III

Comparison of the experimental results with the predictions by the mechanism of the orders of reaction and activation energies for the pyrolysis of 1-butene

Experimental or Mechanism	Order of Reaction with respect to 1-Butene at 575°C		Activation Energy at 0.008 moles/liter (kcal/mole)	
	Exp	Mech	Exp	Mech
Methane	0.91	0.87	76	68
Ethane	1.17	1.04	72	66
Ethylene	0.97	0.70	76	71
Propylene	0.91	0.86	78	68
T-2-Butene	1.50	1.83	77	45
C-2-Butene	1.26	1.83	76	47
1,3-Butadiene	0.71	0.86	81	68
Overall	1.01	1.02	76	62

For a complete set of rate constants, the prediction of the mechanism was made by a computer. Steady-state concentrations of free radicals were obtained numerically by use of a Newton-Raphson iteration after obtaining an initial estimate. These steady-state concentrations of free radicals were used to determine the rate of formation of each product. Integration of these rates of formation over the residence time was performed to give the predicted concentration of products in the effluent. From these predicted effluent concentrations, product ratios, orders of reaction, and activation energies could be calculated.

Predictions from the mechanism could be used to compare with any of the experimental results. Likewise, it was possible to determine what alterations in the predictions would be caused by a particular change in one or more rate constants.

Tables VI and VII give the most important elementary steps for the pyrolyses of 1-butene and cis-2-butene, respectively. These tables also give estimates of the frequency of occurrence of

Table IV

Comparison of the experimental results with the predictions by the mechanism of the orders of reaction and activation energies for the pyrolysis of cis-2-butene

Experimental or Mechanism	Order of Reaction with respect to Cis-2-Butene at 590°C		Activation Energy at 0.001 moles/liter (kcal/mole)	
	Exp	Mech	Exp	Mech
Methane	0.78	0.90	72	75
Ethane	1.28	1.96	83	118
Ethylene	0.72	1.28	76	145
Propylene	0.78	0.90	77	75
1-Butene	1.30	1.88	59	54
T-2-Butene	0.92	1.02	66	64
1,3-Butadiene	0.90	0.90	61	75
Overall	0.90	1.02	66	65

the various elementary steps. Frequency of occurrence of an elementary step is defined as the frequency at which that particular step occurs relative to the rate of decomposition of the molecules being pyrolyzed.

An important feature of butene pyrolysis is the existence of free radicals with resonant stabilization energy. The allyl and the 2-butenyl (methylallyl) radicals are stabilized by the fact that they have delocalized electrons. Added stability of these free radicals is about 10 to 13 kcal/mole (16), which significantly increases the rate of their production and decreases the rate of their consumption.

There are five initiation steps included in the mechanism. Step 1 is for the initiation of cis-2-butene pyrolysis, step 2 for trans-2-butene, and steps 3 to 5 for 1-butene. Step 3 is known to be the principle initiation step for 1-butene (17). Step 4 is expected to have an activation energy of 12.5 kcal/mole higher than step 3, and step 5 is 20 kcal/mole higher than step 3. Because of their higher activation energies, steps 4 and 5 are expected to be at least orders of magnitude less important than step 3. Likewise, initiation steps involving 1-propenyl radicals

Table V

Proposed Mechanism for the Pyrolysis of Butenes

	ELEMENTARY STEP	LOG (PRE-EXPONENTIAL)	ACTIVATION ENERGY
1	C-2-Butene ⟶ 2-Butenyl + H	16.00	87.00
2	T-2-Butene ⟶ 2-Butenyl + H	16.00	87.00
3	1-Butene ⟶ Allyl + Methyl	16.00	74.50
4	1-Butene ⟶ 2-Butenyl + H	16.00	87.00
5	1-Butene ⟶ Vinyl + Ethyl	16.00	94.00
6	H + C-2-Butene ⟶ 2-Butenyl + Hydrogen	9.00	13.00
7	H + T-2-Butene ⟶ 2-Butenyl + Hydrogen	9.00	13.00
8	H + 1-Butene ⟶ 2-Butenyl + Hydrogen	9.00	13.00
9	Methyl + C-2-Butene ⟶ 2-Butenyl + Methane	11.00	16.00
10	Methyl + T-2-Butene ⟶ 2-Butenyl + Methane	11.00	16.00
11	Methyl + 1-Butene ⟶ 2-Butenyl + Methane	11.00	16.00
12	Vinyl + 1-Butene ⟶ 2-Butenyl + Ethylene	11.00	13.00
13	Ethyl + C-2-Butene ⟶ 2-Butenyl + Ethane	10.30	17.00
14	Ethyl + T-2-Butene ⟶ 2-Butenyl + Ethane	10.30	17.00
15	Ethyl + 1-Butene ⟶ 2-Butenyl + Ethane	10.30	17.00
16	2-Propyl + C-2-Butene ⟶ 2-Butenyl + Propane	10.00	17.00
17	2-Propyl + 1-Butene ⟶ 2-Butenyl + Propane	10.00	17.00
18	Allyl + C-2-Butene ⟶ 2-Butenyl + Propylene	9.00	20.00
19	Allyl + T-2-Butene ⟶ 2-Butenyl + Propylene	9.00	20.00
20	Allyl + 1-Butene ⟶ 2-Butenyl + Propylene	9.00	20.00
21	2-Butenyl + C-2-Butene ⟶ T-2-Butene + 2-Butenyl	8.30	20.00
22	2-Butenyl + T-2-Butene ⟶ 1-Butene + 2-Butenyl	8.30	20.00
23	2-Butenyl + T-2-Butene ⟶ C-2-Butene + 2-Butenyl	7.90	20.00
24	2-Butenyl + T-2-Butene ⟶ 1-Butene + 2-Butenyl	7.90	20.00
25	2-Butenyl + 1-Butene ⟶ C-2-Butene + 2-Butenyl	7.90	20.00
26	2-Butenyl + 1-Butene ⟶ T-2-Butene + 2-Butenyl	7.90	20.00
27	H + C-2-Butene ⟶ 2-Butyl	9.95	2.00

Table V (cont.)

	ELEMENTARY STEP							LOG (PRE-EXPONENTIAL)	ACTIVATION ENERGY
28	H	+	T-2-Butene	——>	2-Butyl			9.95	2.00
29	H	+	1-Butene	——>	1-Butyl			9.90	2.00
30	H	+	1-Butene	——>	2-Butyl			9.90	0.50
31	H	+	Propylene	——>	2-Propyl			9.48	0.50
32	H	+	Propylene	——>	1-Propyl			9.48	2.00
33	H	+	Ethylene	——>	Ethyl			10.00	0.50
34	H	+	Butadiene	——>	2-Butenyl			10.70	0.50
35	Methyl	+	Propylene	——>	2-Butenyl			7.00	2.00
36	2-Butenyl			——>	Butadiene	+	H	11.76	42.00
37	1-Butyl			——>	Ethyl	+	Ethylene	13.30	24.00
38	1-Butyl			——>	H	+	1-Butene	14.41	38.00
39	2-Butyl			——>	Propylene	+	Methyl	13.30	26.00
40	2-Butyl			——>	1-Butene	+	H	13.90	37.50
41	2-Butyl			——>	C-2-Butene	+	H	13.90	37.50
42	2-Butyl			——>	T-2-Butene	+	H	14.00	37.50
43	2-Propyl			——>	Propylene	+	H	14.41	37.80
44	1-Propyl			——>	Ethylene	+	Methyl	13.30	28.00
45	1-Propyl			——>	Propylene	+	H	14.41	37.80
46	Ethyl			——>	Ethylene	+	H	13.23	40.50
47	C-2-Butene			——>	T-2-Butene			15.50	70.00
48	T-2-Butene			——>	C-2-Butene			16.20	70.65
49	Methyl	+	Methyl	——>	Ethane			9.30	0.0
50	Methyl	+	Ethyl	——>	Propane			9.30	0.0
51	Methyl	+	Vinyl	——>	Propylene			9.00	0.0
52	Methyl	+	2-Propyl	——>	I-Butane			9.00	0.0
53	Methyl	+	1-Propyl	——>	N-Butane			9.00	0.0
54	Methyl	+	2-Butenyl	——>	1-Butene			9.00	0.0

Table V (cont.)

	ELEMENTARY STEP					LOG (PRE-EXPONENTIAL)	ACTIVATION ENERGY
55	Methyl	+	2-Butenyl	——>	2-Pentene	9.00	0.0
56	Methyl	+	H	——>	Methane	7.00	0.0
57	Ethyl	+	Ethyl	——>	N-Butane	9.20	0.0
58	Ethyl	+	Ethyl	——>	Ethane + Ethylene	8.40	0.0
59	Vinyl	+	Ethyl	——>	1-Butene	8.00	0.0
60	Ethyl	+	Allyl	——>	1-Pentene	9.00	0.0
61	Ethyl	+	2-Butenyl	——>	3-Hexene	8.50	0.0
62	Ethyl	+	H	——>	Ethane	10.00	0.0
63	Vinyl	+	Allyl	——>	1,3-Pentadiene	9.00	0.0
64	Allyl	+	1-Propyl	——>	1-Hexene	9.00	0.0
65	Allyl	+	2-Propyl	——>	4-Methyl-2-Pentene	9.00	0.0
66	Allyl	+	Allyl	——>	1,5-Hexadiene	7.00	0.0
67	Allyl	+	2-Butenyl	——>	1,5-Heptadiene	7.00	0.0
68	Methyl	+	Allyl	——>	1-Butene	9.00	0.0
69	Vinyl	+	2-Butenyl	——>	1,5-Hexadiene	9.00	0.0
70	2-Butenyl	+	1-Propyl	——>	2-Heptene	7.00	0.0
71	2-Butenyl	+	2-Propyl	——>	5-Methyl-2-Hexene	7.00	0.0
72	2-Butenyl	+	2-Butenyl	——>	2,6-Octadiene	5.50	0.0
73	H	+	2-Butenyl	——>	1-Butene	10.00	0.0
74	H	+	2-Butenyl	——>	C-2-Butene	10.00	0.0
75	H	+	2-Butenyl	——>	T-2-Butene	10.00	0.0
76	H	+	H	——>	Hydrogen	7.00	0.0

Pre-exponentials have units of liter/mole-sec or sec^{-1} given as logarithms to the base 10.

Activation energies are in kcal/mole.

Table VI

Most Important Elementary Steps for 1-Butene Pyrolysis

Frequency of occurrence of the elementary steps in the pyrolysis of 1-butene are given in the columns on the right. (The sum of the steps that consume 1-butene has been normalized to be 1.00).

	ELEMENTARY STEP	Temperature:	560°	560°	575° C
		Initial 1-Butene Concentration:	0.001	0.008	0.001 M
3	1-Butene ⟶	Allyl + Methyl	0.064	0.061	0.071
11	Methyl + 1-Butene ⟶	Methane + 2-Butenyl	0.35	0.24	0.36
15	Ethyl + 1-Butene ⟶	Ethane + 2-Butenyl	0.071	0.065	0.067
20	Allyl + 1-Butene ⟶	Propylene + 2-Butenyl	0.035	0.042	0.036
25	2-Butenyl + 1-Butene ⟶	C-2-Butene + 2-Butenyl	0.032	0.171	0.026
26	2-Butenyl + 1-Butene ⟶	T-2-Butene + 2-Butenyl	0.032	0.171	0.026
29	H + 1-Butene ⟶	1-Butyl	0.126	0.077	0.132
30	H + 1-Butene ⟶	2-Butyl	0.31	0.19	0.32
36	2-Butenyl ⟶	Butadiene + H	0.40	0.26	0.41
37	1-Butyl ⟶	Ethyl + Ethane	0.126	0.077	0.132
39	2-Butyl ⟶	Methyl + Propylene	0.31	0.19	0.32
55	Methyl + 2-Butenyl ⟶	2-Pentene	0.018	0.009	0.021
72	2-Butenyl + 2-Butenyl ⟶	Products	0.007	0.027	0.006

are expected to be of little importance and thus have not been included.

The propagation reactions in the pyrolyses of butenes is of an interesting form. Propagation occurs by four sequential phases.

The first of these four phases is a hydrogen-abstraction reaction. Using the pyrolysis of cis-2-butene as an illustration, step 9 makes up the first phase. In this step, a hydrogen atom is abstracted from a cis-2-butene molecule by a methyl radical to form a 2-butenyl radical. The decomposition of the 2-butenyl radical in step 36 makes up the second phase. Decomposition of the 2-butenyl radical produces a butadiene molecule and a hydrogen atom.

The third phase of the propagation is the addition of a hydrogen atom to a cis-2-butene molecule to form a 2 butyl radical in step 27. The fourth phase is the decomposition of the 2-butyl radical in step 39. Decomposition of the 2-butyl radical produces a methyl radical used in the first phase of the propagation. Thus the four phases are:

1. hydrogen-atom abstraction
2. a 2-butenyl radical decomposition
3. a hydrogen atom addition
4. a butyl radical decomposition

The propagation steps in the pyrolysis of 1-butene also have these cited four phases. These phases are somewhat more complex in the case of 1-butene because of the larger number of hydrogen-atom-abstracting radicals present. Along with the methyl radical also found in the cis-2-butene case, ethyl and allyl radicals act as hydrogen-atom-abstracting radicals. A further complication is that the hydrogen atoms may add to 1-butene to form either 1-butyl or 2-butyl radicals. Literature values for the ratios of rates of these reactions predict the product ratios (12)(15). The ratio of the rate of formation of 1-butyl and 2-butyl radicals determines the relative amounts of methane and propylene as opposed to the amounts of ethane and ethylene.

The main termination reactions are thought to be the recombination of 2-butenyl radicals and cross-combinations of methyl and 2-butenyl radicals.

Isomerization of cis-2-butene to trans-2-butene is thought to occur primarily by a molecular pathway. Steps 47 and 48 represent this isomerization and its reverse.

The mechanism models fairly well the experimental results as shown in Tables I - IV. The product ratios are predicted quite well except for the minor products. The greatest weakness is the mechanism's ability to predict the amount of ethane and ethylene produced in the pyrolysis of cis-2-butene. Although these are very minor products amounting to less than 0.5 percent on a molar basis, it appears that they are formed by other pathways.

One other possible pathway would involve the isomerization

Table VII

Most Important Elementary Steps for Cis-2-Butene Pyrolysis

Frequency of occurrence of the elementary steps in the pyrolysis of cis-2-butene are given in the columns on the right. (The sum of the steps that consume cis-2-butene has been normalized to be 1.00).

	ELEMENTARY STEP							575°	575°	590° C
							Initial Cis-2-Butene Concentration:	0.001	0.008	0.001 M
1	C-2-Butene			⟶	2-Butenyl	+	H	0.00012	0.00012	0.00015
9	Methyl	+	C-2-Butene	⟶	2-Butenyl	+	Methane	0.045	0.031	0.049
21	2-Butenyl	+	C-2-Butene	⟶	T-2-Butene	+	2-Butenyl	0.007	0.040	0.006
22	2-Butenyl	+	C-2-Butene	⟶	1-Butene	+	2-Butenyl	0.007	0.040	0.006
27	H	+	C-2-Butene	⟶	2-Butyl			0.046	0.031	0.050
36	2-Butenyl			⟶	Butadiene	+	H	0.045	0.031	0.050
39	2-Butyl			⟶	Propylene	+	Methyl	0.045	0.031	0.050
47	C-2-Butene			⟶	T-2-Butene			0.93	0.88	0.97
48	T-2-Butene			⟶	C-2-Butene			0.016	0.016	0.035
55	Methyl	+	2-Butenyl	⟶	2-Pentene			0.00010	0.00005	0.00013
72	2-Butenyl	+	2-Butenyl	⟶	Products			0.00003	0.00011	0.00003

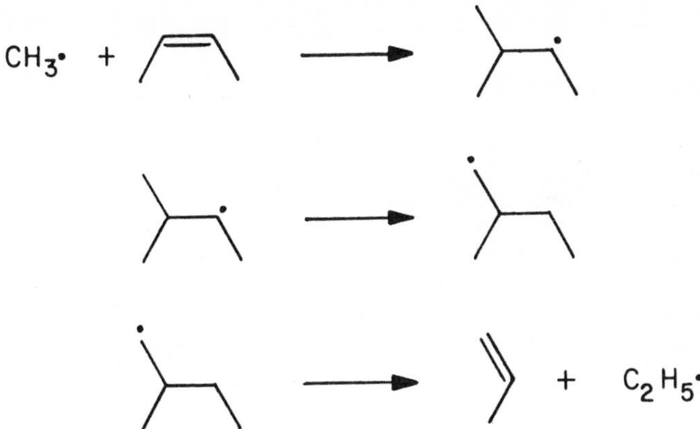

Figure 2. Other possible elementary steps occurring in the pyrolysis of cis-2-*butene*

of the 2-butyl radical. The amount of isomerization of 2-butyl to 1-butyl need only be 5 percent to produce the small amount of ethane and ethylene formed in the reaction. Another possible pathway (4) would involve the addition of a methyl radical to a cis-2-butene molecule, as shown in Figure 2.

The net result is the formation of a propylene molecule and an ethyl radical. The ethyl radical could either abstract a hydrogen atom to form ethane or eliminate a hydrogen atom to form ethylene. This pathway would have the unfortunate tendency of increasing the amount of propylene produced while decreasing the amount of methane formed.

Another weakness of the mechanism is that the orders of formation are not predicted well for the isomerization of cis-2-butene to 1-butene, and 1-butene to cis- and trans-2-butene. There may exist other pathways. The most obvious possibility would be molecular pathways.

Abstract. Pyrolysis of 1-butene and cis-2-butene was studied in a gold, tubular reactor over the temperature range of 530 to 620°C and at a total pressure of 1 atmosphere. Argon was used as a diluent. Mole fractions of the butenes ranged from 0.06 to 0.9 mole percent. Reaction models were established and reasonable agreement with experiment observed.

Literature Cited

(1) Blakemore, J. E., Barker, J. R. and Corcoran, W. H., Ind. Eng. Chem., Fundam. (1973) 12, 147.

(2) Blakemore, J. E. and Corcoran, W. H., Ind. Eng. Chem., Process Des. Develop. (1969) 8, 207.

(3) Powers, D. R. and Corcoran, W. H., Ind. Eng. Chem., Fundam. (1974) 13, No. 4, 351.

(4) Bryce, W. A. and Kebarle, P., Trans. Faraday Soc. (1958) 54, 1660.
(5) Bryce, W. A. and Ruzicka, D. J., Can. J. Chem. (1960) 38, 835.
(6) Danby, C. J., Spall, B. C., Stubbs, F. J. and Hinshelwood, C., Proc. Roy. Soc. A (1955) 228, 448.
(7) Mellotte, H. and Delbourgo, R., Bull. Soc. Chim. (1970) 10 3473.
(8) Molera, M. J. and Stubbs, F. J., J. Chem. Soc. (1952) 381.
(9) Jeffers, P. and Bauer, S. H., Int. J. Chem. Kinet. (1974) 6, 763.
(10) Cvetanovic, R. J. and Doyle, L. C., J. Chem. Phy. (1969) 50, 4705.
(11) Daby, E. E., Niki, H. and Weinstock, B., J. Phys. Chem. (1971) 75, 1601.
(12) Falconer, W. E. and Sunder, W. A., Int. J. Chem. Kinet. (1972) 3, 395.
(13) Grotewald, J. and Kerr, J. A., J. Chem. Soc. (1963) 4337.
(14) James, D. G. L. and Kambanis, S. M., Trans. Faraday Soc. (1969) 65, 2081.
(15) Knox, J. H. and Dalglish, D. G., Int. J. Chem. Kinet. (1969) 1, 69.
(16) Benson, S. W., "Thermochemical Kinetics," John Wiley and Sons, Inc., New York, 1968.
(17) Lossing, F. P., Ingold, K. U. and Henderson, I. H. S., J. Chem. Phy. (1954) 22, 621.

9

Pyrolysis of 2, 2-Dimethylpropane in a Continuous Flow Stirred Tank Reactor

J. A. RONDEAU and G. M. CÔME

ENSIC, Institut National Polytechnique de Lorraine, Laboratoire de Cinétique Appliquée, 1, rue Grandville, F-54042 NANCY Cedex, France

J. F. LARGE

Université de Technologie, Départment de Génie Chemique, B.P. 233, F-60206 COMPIEGNE, France

Fundamental work on the kinetics of hydrocarbon cracking is fully justified by strong industrial interest (1 to 5). Up to now kinetic studies of complex radical reactions in the gaseous phase have been conducted mainly in batch reactors. Nevertheless, the use of a continuous flow stirred tank reactor (CFSTR) seems very promising because it gives a direct measurement of reaction rates (6) and also it is well adapted to the study of fast reactions.

The performance of an actual continuous reactor with complex reactions depends on the following factors :

- the kinetic laws
- the residence time distribution characterizing the "macro-mixing" (7)
- the segregation characterizing the degree of mixedness (8).

In view of the interaction between these kinetic and hydrodynamic phenomena, the residence time distribution (RTD) never gives sufficient information for the calculation of the reactor yield except in the case of first order reactions (9 to 11). In spite of such difficulties, the hypothesis of maximum mixedness might be valuable for gaseous systems, when the reactor is sufficiently stirred, and might allow accurate determination of reaction rates.

We have studied the pyrolysis of 2-2 dimethyl-propane (neopentane) in the gaseous phase at about 500° C and at small extents of reaction. The kinetic parameters obtained from these investigations in a continuous stirred reactor have been compared to the values previously determined from batch experiments (12 to 19).

Experimental Design and Procedure

It is difficult to design a good CFSTR for two main reasons :

- the actual reactor must have a RTD very close to the

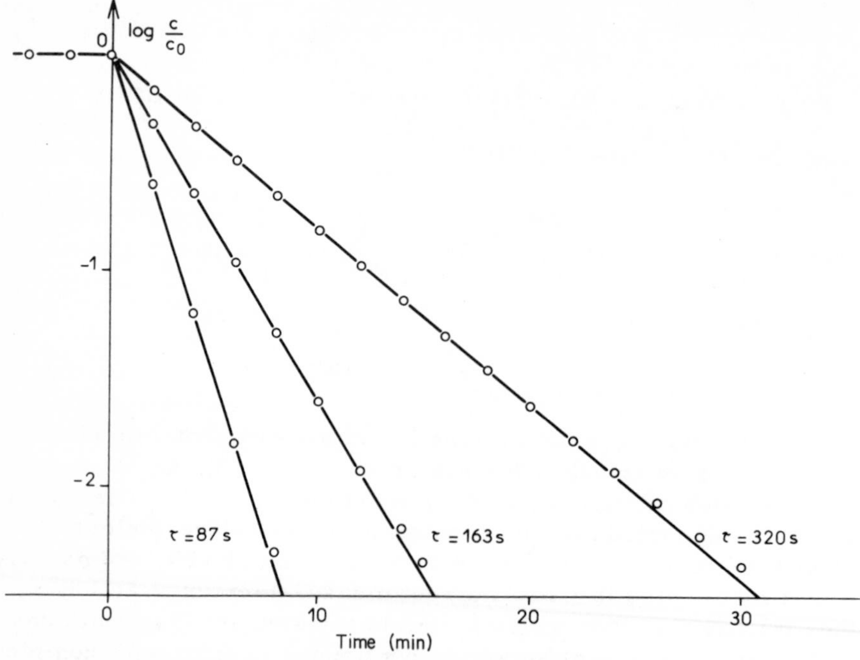

Figure 1. Residence time distribution. Downstream signal as a response to an upstream step input signal of methane:
$C = C_o\,[1 - U(t)]$, *with* $U(t) = 0$ *if* $t < 0$, *and* $U(t) = 1$ *if* $t \geqslant 0$.

inlet
fixed helix
well for thermocouple
semi-capillary tubes (I.D. 1.5 mm)
to pressure transducer
outlet

Figure 2. Reactor: spherical vessel in borosilicate glass with a reflexing baffle (fixed helix) between inlet and outlet; volume, 150 cm^3

theoretical one

- it is difficult to control the gas concentrations inside the reactor at low flow rates because of flow fluctuations.

Reactor design. Several spherical PYREX vessels, with a volume between 100 and 200 cm^3 were used. The gases were introduced into the reactor through capillary tubes (internal diameter : 1 mm) and they were mixed with a helix. The RTD of the reactors were measured, under operating conditions, by three following tracer methods :

- tracing methane was introduced via an inert gas feed (concentration ratio : 1/100) and the response signal to a step change in the methane concentration was determined by measurements at given time intervals. The flame ionization detector of the chromatograph used for the gas analysis had a high sensitivity and this enabled a precise determination of the "tail" of the signal to be made (Figure 1).

- the response signal to a pulse concentration of hydrogen (DIRAC delta function) added to the ingoing stream was continuously recorded by an ultra micro bead CARLE detector system. This method was very convenient to detect any short-circuits or dead spaces of the experimental reactor and its deviation from the theoretical CFSTR.

- a signal of ozone with an arbitrary but reproducible shape was super imposed on a flow of oxygen. Thus, the transfer function of the reactor was obtained from a convolution analysis of the entrance and exit concentration signals (20).

In addition to the RTD experiments, the exact structure of the gaseous jets was obtained by means of a chemiluminescent reaction of phosphorous vapor (21). It was observed that the gas mixture was approximately homogeneous.

All these experiments have indicated that good macro-mixing of gases can be obtained without mechanical stirring. Nevertheless, it was found essential to position a reflecting baffle between the gas flow inlet and the oulet. For example, in the present study, the blade of a fixed helix was positioned in front of the inlet (Figure 2).

In the following, we shall assume that the maximum mixedness of the gases is produced essentially by natural diffusion. Furthermore, at low extent of reaction (less than 5 %) the molecular state of mixing has little influence on the kinetics of the reaction (22).

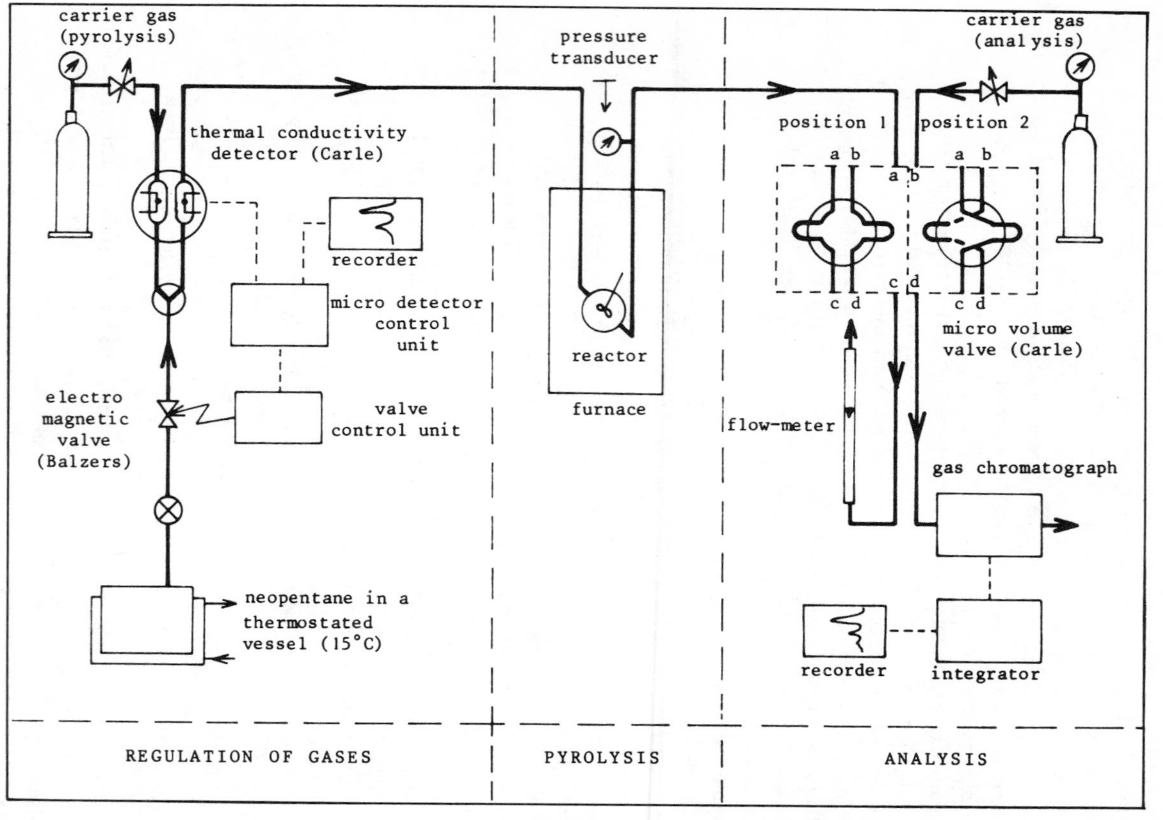

Figure 3. Schematic of experimental apparatus

Experimental procedure (Figure 3). Neopentane (MATHESON N 30 with less than 1/1000 of n-butane as the main impurity) was added continuously to an inert gas feed (nitrogen N48 with less than 2 vpm or known but small amounts of oxygen). The neopentane flow rate, and thus its concentration, was regulated with an electromagnetic valve (R. M. E. -BALZERS A. G.) and monitored by a thermistor system (CARLE) composed of two separate detecting cells (one reference cell for the carrier gas, the other for the mixture). Such an assembly of two cells in series has been proposed by KAISER (23). The same technics could provide a mixture of nitrogen, neopentane and isobutene, with or without oxygen, when the valve and regulating systems were doubled. The gaseous mixture then flowed through the reaction vessel situated inside an isothermal oven. The analysis of the outgoing gas stream was performed with a gas chromatograph equipped with a flame ionizing detector. The 6. 5 m long column was packed with fire brick (60-80 mesh) impregnated with squalane. The internal diameter of the column was 4 mm and it was operated at 37° C.

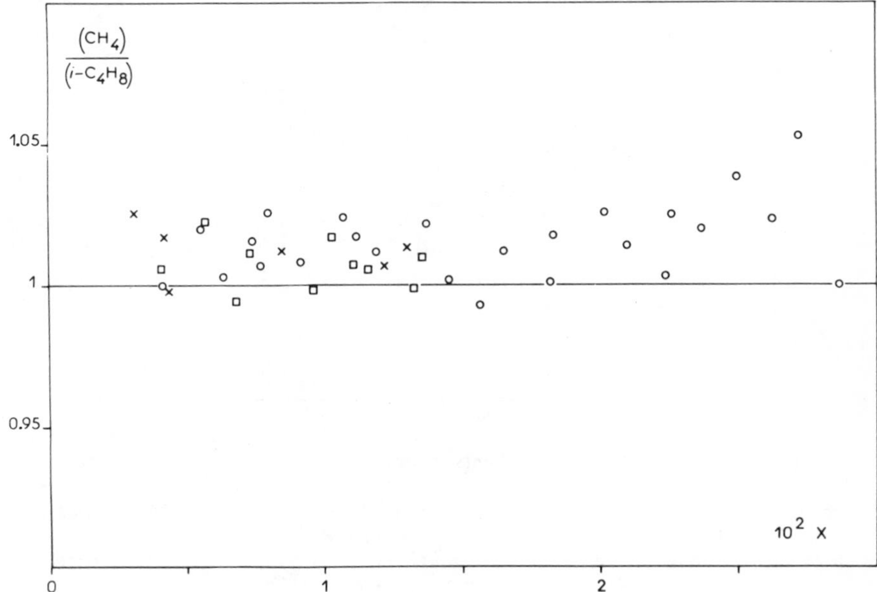

Figure 4. Stoichiometric equation: neo-C_5H_{12} = CH_4 + i-C_4H_8. ○ T = 794°K, *uncoated walls;* □ T = 775°K, *uncoated walls;* × T = 775°K, *PbO coated walls.*

Pyrolysis of Neopentane at no Extent of the Reaction

According to previous studies (14, 15, 18), the pyrolysis of neopentane μH would imply the main primary stoichiometric equation :

$$\text{neo-}C_5H_{12} = CH_4 + \text{i-}C_4H_8 \ .$$

The experiments carried out in our continuous flow reactors at 502° C and 521° C, with neopentane concentrations between 9×10^{-6} and 9×10^{-8} mole/cm^3 resulted in the same concentrations of methane and isobutene :

$$(CH_4) = (\text{i-}C_4H_8)$$

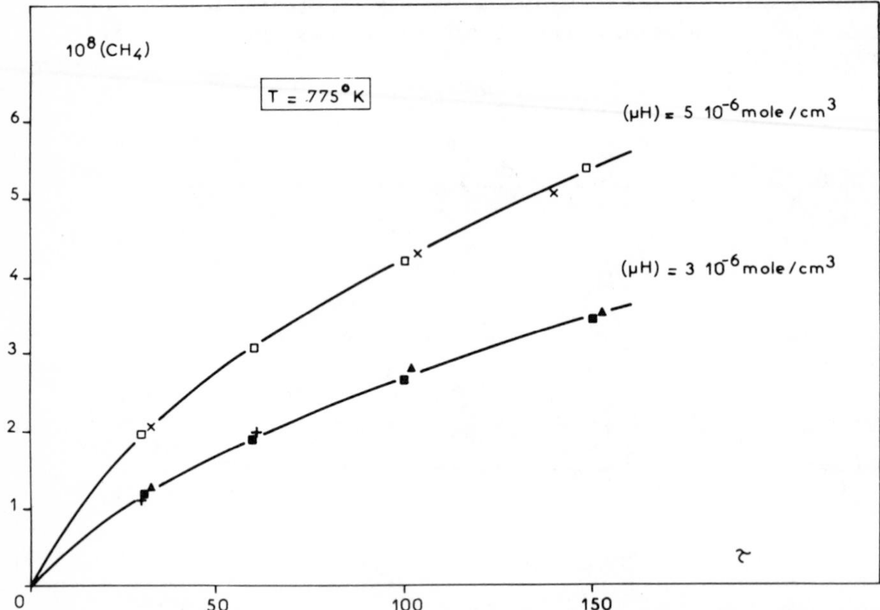

Figure 5. Influence of very small amounts of oxygen on pyrolysis of neopentane; T = *775°K.*

□ *(neo-C_5H_{12}) = 5 × 10^{-6} mol/cm^3;* ■ *(neo-C_5H_{12}) = 3 × 10^{-6} mol/cm^3. Carrier gas: nitrogen 99.998% of purity containing less than 2 vpm O_2.*

× *(neo-C_5H_{12}) = 5 × 10^{-6} mol/cm^3;* ▲ *(neo-C_5H_{12}) = 3 × 10^{-6} mol/cm^3. Carrier gas: nitrogen, containing less than 5 vpm O_2.*

+ *(neo-C_5H_{12}) = 3 × 10^{-6} mol/cm^3. Carrier gas: nitrogen purified by using a trap containing manganous oxide for the removal of O_2*

—— *theoretical model*

Only small amounts of methane in concentrations greater than isobutene were observed (Figure 4) and this agreed with the findings of previous investigations (15). The stoichiometric equation still remains valid when the walls of the vessels are PbO coated (Figure 4). Unwanted oxygen was traped (25) but a few experiments were carried out in the presence of very small amounts of oxygen (up to 5 vpm) and with different grades of commercial nitrogen. The experimental results (Figure 5) show that traces of oxygen have no noticable influence on the kinetics of the reaction.

The following long chain and homogeneous radical mechanism accounts for the global kinetic features at no extent of reaction (14, 15, 18) :

Initiation :

$$\text{neo-}C_5H_{12} \longrightarrow (CH_3)_3C\cdot + CH_3\cdot \qquad (1)$$

$$(CH_3)_3C\cdot \longrightarrow \text{i-}C_4H_8 + H\cdot \qquad (x)$$

$$H\cdot + \text{neo-}C_5H_{12} \longrightarrow H_2 + \text{neo-}C_5H_{11}\cdot \qquad (y)$$

Propagation (I) :

$$\text{neo-}C_5H_{11}\cdot \longrightarrow \text{i-}C_4H_8 + CH_3\cdot \qquad (2)$$

$$CH_3\cdot + \text{neo-}C_5H_{12} \longrightarrow CH_4 + \text{neo-}C_5H_{11}\cdot \qquad (3)$$

Termination :

$$CH_3\cdot + CH_3\cdot \longrightarrow C_2H_6 \qquad (\beta\beta)$$

Using the hypothesis of a quasi-steady state for the atoms and the free radicals, and using the long-chain approximation, this mechanism leads to the following rate law :

$$r^{o}_{\mu H} = k_3 \, (k_1/k_{\beta\beta})^{1/2} \, (\mu H)^{3/2}$$

where (μ H) is the concentration of neopentane.

In Figure 6, the values of log $\left[r^{o}_{\mu H}\right]$, obtained by the extrapolation of the curves at no extent of reaction, have been plotted as functions of log $\left[(\mu H)\right]$. The initial order, n^{o}, increases with the neopentane concentration and reaches 3/2 at a partial pressure of about 100 Torr. According to other studies (26), this phenomenon can be interpreted by adding the following steps to the previous mechanism :

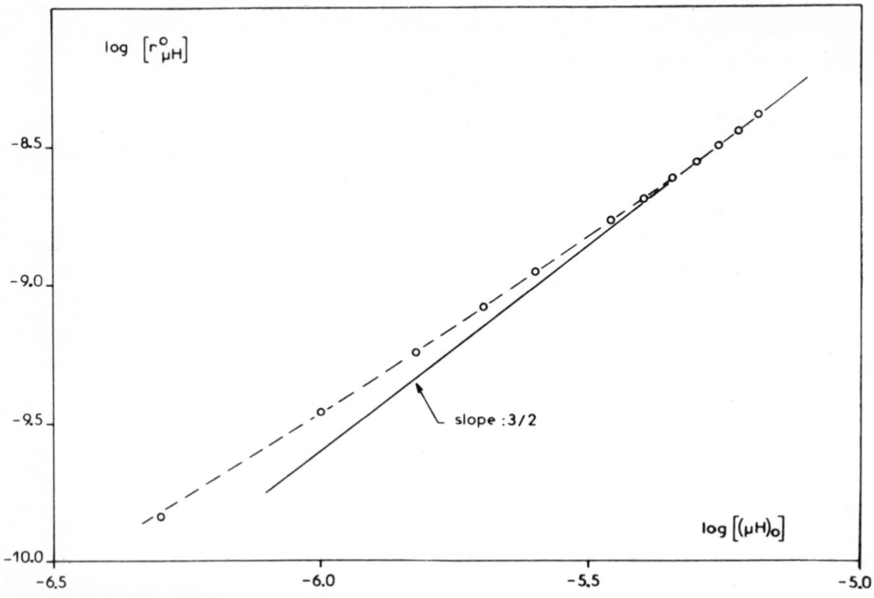

Figure 6. Order of the reaction at 794°K and X = 0

$$CH_3\cdot + CH_3\cdot \rightleftharpoons C_2H_6^* \qquad (\vec{t}) \text{ and } (\overleftarrow{-t})$$

$$C_2H_6^* + \text{neo-}C_5H_{12} \longrightarrow C_2H_6 + \text{neo-}C_5H_{12} \qquad (d)$$

$$C_2H_6^* + N_2 \longrightarrow C_2H_6 + N_2 \qquad (d')$$

where $C_2H_6^*$ is a molecule of ethane vibrationaly excited and which can be deactivated by collisions with molecules of neopentane or nitrogen. This interpretation leads to the following expression of $k_{\beta\beta}$:

$$k_{\beta\beta} = \varphi^{-1} \cdot k_t$$

where $\varphi = 1 + \frac{C_1}{(\mu H)+C_2(N_2)}$ and $C_1 = k_{-t}/k_d$, $C_2 = k_{d'}/k_d$

C_2 is characteristic of the relative deactivating efficiency of nitrogen and neopentane. It has a very low value (26) : $5x10^{-2}$ at 500° C. C_1 is characteristic of the ability of $C_2H_6^*$ to give CH_3. in the presence of the surrounding deactivating molecules. In our experimental conditions its value (26) was found to be about 10^{-6} mole/cm^3.

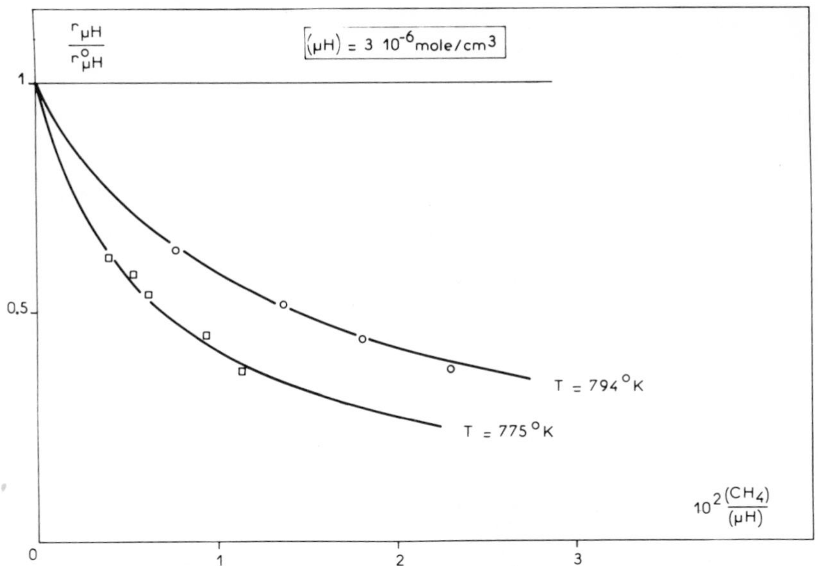

Figure 7. Self-inhibited pyrolysis of neopentane.
—— theoretical model.

The total activation energy E^o derived from the experimental value of $k^o = k_3 k_1^{1/2}/k_t^{1/2}$ was found to be about 50 kcal/mole.

Self Inhibition of Neopentane Pyrolysis

Isobutene and methane are the only main products of the neopentane pyrolysis. Methane has no kinetic effect ; therefore the produced isobutene might be responsible for the strong self inhibition exhibited on the curves of Figures 5 and 7 when neopentane is cracked without additives at low extent of reaction. According to a theoretical mechanism "μH, YH" proposed by NICLAUSE et alii (27), the pyrolysis of neopentane in the presence of isobutene can be represented by the following reaction steps, valid at least when the extent of reaction is zero :

Initiation :

$$\text{neo-}C_5H_{12} \longrightarrow (CH_3)_3C\cdot + CH_3\cdot \quad (1)$$

$$(CH_3)_3C\cdot \longrightarrow i\text{-}C_4H_8 + H\cdot \quad (x)$$

$$H\cdot + \text{neo } C_5H_{12} \longrightarrow H_2 + \text{neo } C_5H_{11}\cdot \quad (y)$$

$$H\cdot + i\text{-}C_4H_8 \longrightarrow H_2 + i\text{-}C_4H_7\cdot \quad (y')$$

Propagation (I) :

$$\text{neo-}C_5H_{11}\cdot \rightarrow \text{i-}C_4H_8 + CH_3\cdot \qquad (2)$$

$$CH_3\cdot + \text{neo-}C_5H_{12} \rightarrow CH_4 + \text{neo-}C_5H_{11}\cdot \qquad (3)$$

Propagation (II) :

$$\text{neo-}C_5H_{11}\cdot \rightarrow \text{i-}C_4H_8 + CH_3\cdot \qquad (2)$$

$$CH_3\cdot + \text{i-}C_4H_8 \rightarrow CH_4 + \text{i-}C_4H_7\cdot \qquad (4\,\beta)$$

$$\text{i-}C_4H_7\cdot + \text{neo-}C_5H_{12} \rightleftharpoons \text{i-}C_4H_8 + \text{neo }C_5H_{11}\cdot \qquad (\overrightarrow{5}) \text{ and } (\overleftarrow{4}\,\mu)$$

Termination:

$$CH_3\cdot + CH_3\cdot \rightarrow C_2H_6 \qquad (\beta\beta)$$

$$CH_3\cdot + \text{i-}C_4H_7\cdot \rightarrow \text{methyl-2 butene 1} \qquad (\beta y)$$

$$\text{i-}C_4H_7\cdot + \text{i-}C_4H_7\cdot \rightarrow \text{2, 5-dimethyl 1, 5-hexadiene} \qquad (yy)$$

The calculation of the rate of formation of methane is made possible by using the quasi-steady state approximation for the atoms and free radicals and using the following approximations :

$$\frac{(k_{\beta\beta})^{1/2}}{k_{yy}} << 2\,\frac{(k_1\,k_{\beta\beta})^{1/2}}{(\mu H)^{1/2}} < k_3 << \frac{k_2}{\mu H}$$

These approximations show that the isobutenyl radical

$$\left[\begin{array}{ccc} CH_2 \text{ ======= } & \overset{\displaystyle C}{|} & \text{ ======= } CH_2 \\ & CH_3 & \end{array}\right]\cdot$$

is stabilized by resonance and is much less reactive than the methyl radical when reacting with neopentane.

Thus, as the elementary step (5) is negligible compared to step (4 β) we can say :

$$r^o_{\mu H,\,YH} = r^o_{\mu H}\,\frac{1 + a_2\,(YH)/(\mu H)}{1 + b_2\,\varphi^{1/2}\,(YH)/(\mu H)^{1/2}}$$

with $a_2 = k_{4\beta}/k_3$, $b_2 = k_{4\beta}/2\,(k_1\,k_t)^{1/2}$

and where (YH) is the concentration of isobutene.

The kinetic law, $r^{o}_{\mu H, YH}$, accounts, at least qualitatively, for the self-inhibited feature of the reaction (Figure 7). But the experimental results do not agree with the law quantitatively, however they can be interpretated, using the same formulation, by the equation :

$$r_{\mu H} = r^{o}_{\mu H} \frac{1 + a'(CH_4)/(\mu H)}{1 + b'(CH_4)/(\mu H)}$$

which is compatible with $r^{o}_{\mu H}$ when $(CH_4) = 0$. The expression $r^{o}_{\mu H}$ can be linearized :

$$\frac{r^{o}_{\mu H}}{r^{o}_{\mu H} - r_{\mu H}} = \frac{1}{1 - a'/b'} + \frac{1}{b' - a'} \frac{(\mu H)}{(CH_4)}$$

and its agreement with the experimental results is clearly demonstrated in Figure 8.

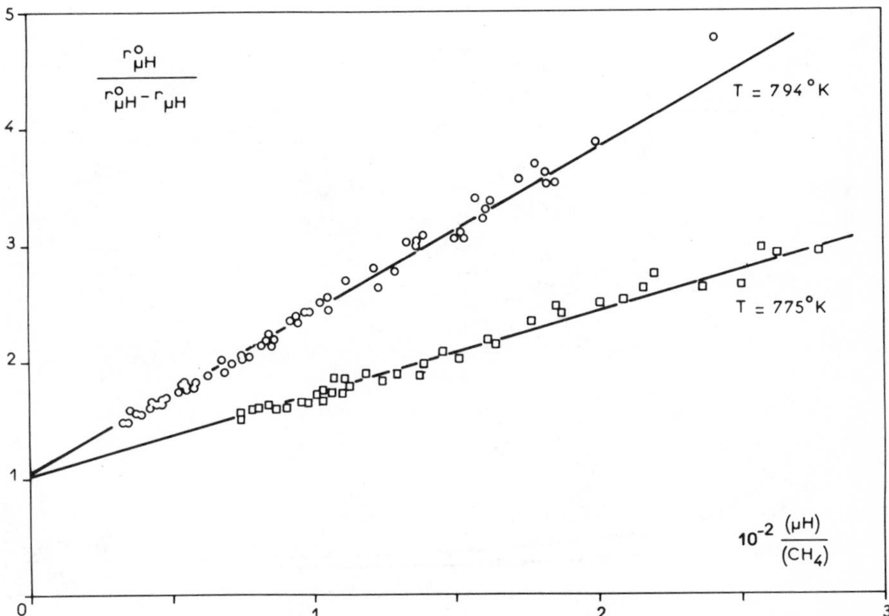

Figure 8. Self-inhibited pyrolysis of neopentane: concentration of neopentane between 9×10^{-6} and 9×10^{-8} mol/cm³. ——theoretical model.

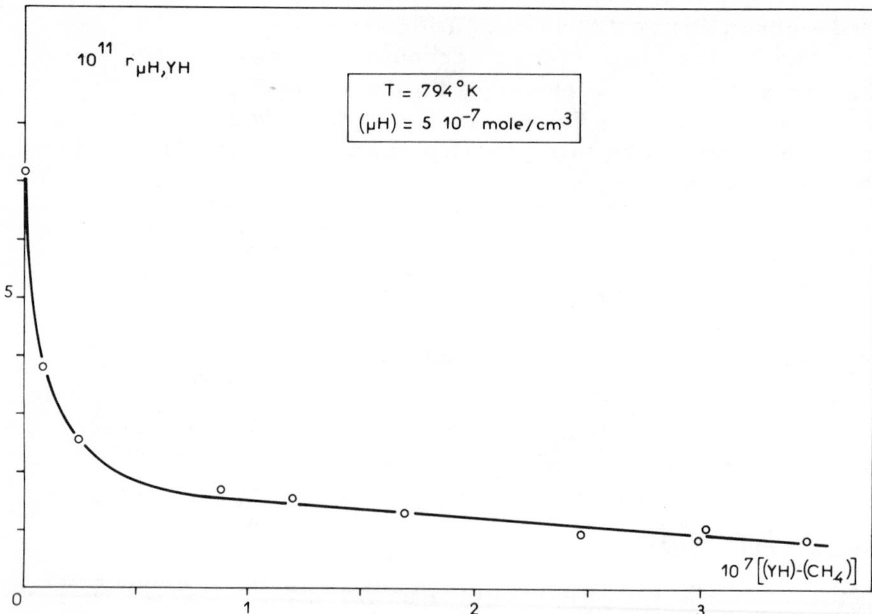

Figure 9. Pyrolysis of neopentane inhibited by added isobutene at $\tau = 60$ sec. —— theoretical model.

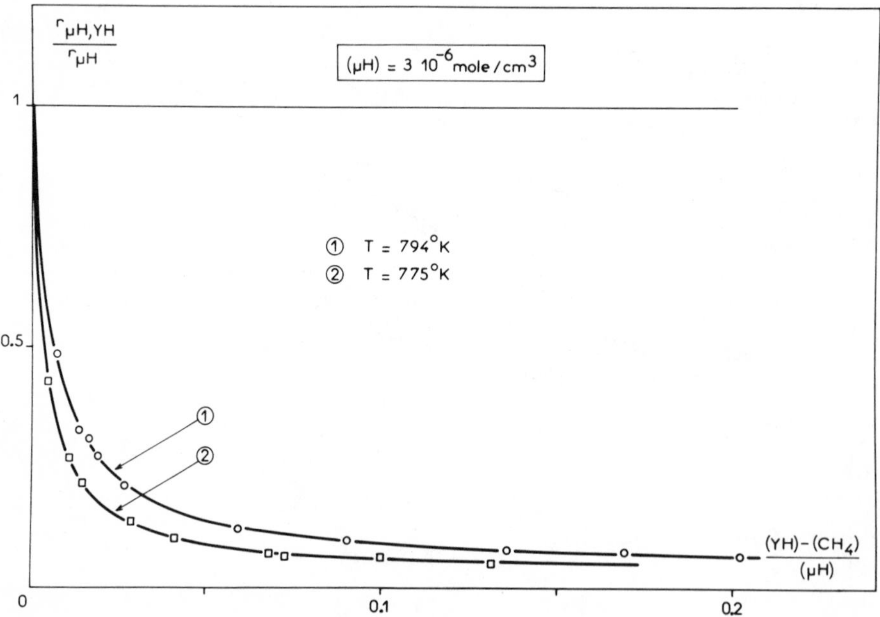

Figure 10. Pyrolysis of neopentane inhibited by added isobutene. ——theoretical model.

At the following operating conditions :

$$502°\,C \leqslant \text{Temperature} \leqslant 521°\,C,$$
$$9\times10^{-8} \leqslant \text{neopentane concentration} \leqslant 9\times10^{-6}\ \text{mole/cm}^3,$$
$$25 \leqslant \text{mean residence time} \leqslant 160\ s,$$
$$\text{extent of reaction} \leqslant 5\ \%,$$

the traces on Figure 8 give a graphical determination of the two parameters a' and b' at two different temperatures, and an estimation of the activation energy of b' : $E_{b'} \simeq -42$ kcal/mole.

Pyrolysis of Neopentane with Added Isobutene

The rate of methane production, $r_{\mu H,\,YH}$, strongly decreases when isobutene is added to the ingoing gaseous stream, the other operating conditions (Figures 9 and 10) remaining unchanged. The kinetic law, $r_{\mu H,\,YH}$, which accounts for the experimental results should be compatible with :

- the expression for $r_{\mu H}$ in the pyrolysis of pure neopentane, when $(YH)_o = (YH) - (CH_4) = 0$ in the ingoing stream
- the expression for $r^o_{\mu H,\,YH}$ in the pyrolysis of neopentane with isobutene when $(CH_4) = 0$

These two conditions are incorporated in the equation :

$$r_{\mu H,\,YH} = r^o_{\mu H}\ \frac{1 + a_2\,\dfrac{(YH) - (CH_4)}{(\mu H)}}{1 + b_2\varphi^{1/2}\,\dfrac{(YH) - (CH_4)}{(\mu H)^{1/2}}} \cdot \frac{1 + a'\,\dfrac{(CH_4)}{(\mu H)}}{1 + b'\,\dfrac{(CH_4)}{(\mu H)}}$$

which can be linearized when (μH) remains constant :

$$\frac{r_{\mu H}}{r_{\mu H} - r_{\mu H,\,YH}} = \frac{1}{1 - a_2/b_2\,\varphi^{1/2}\,(\mu H)^{1/2}} + \frac{1}{b_2\,\varphi^{1/2}\,(\mu H)^{1/2} - a_2} \cdot \frac{(\mu H)}{(\mu H) - (CH_4)}$$

In Figure 11 the variation of $r_{\mu H}/(r_{\mu H} - r_{\mu H,\,YH})$ is plotted as a function of $(\mu H)/[(\mu H) - (CH_4)]$ for a neopentane concentration of 3×10^{-6} mole/cm^3. It is apparent that there is good agreement with the previous relationship and the experimental points at the two operating temperatures. The same graphs, for different neopentane concentrations, at a given temperature, are also

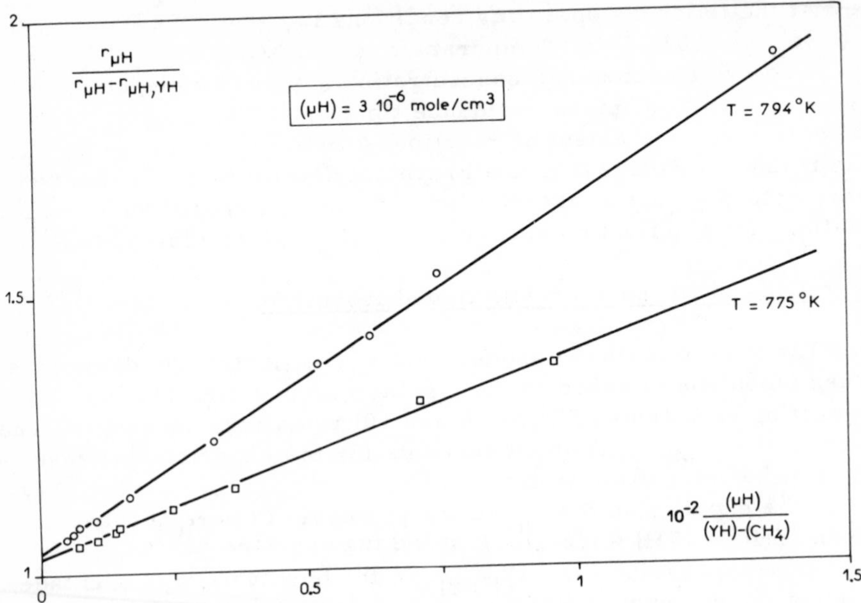

Figure 11. Pyrolysis of neopentane inhibited by added isobutene: effect of the temperature. —— theoretical model.

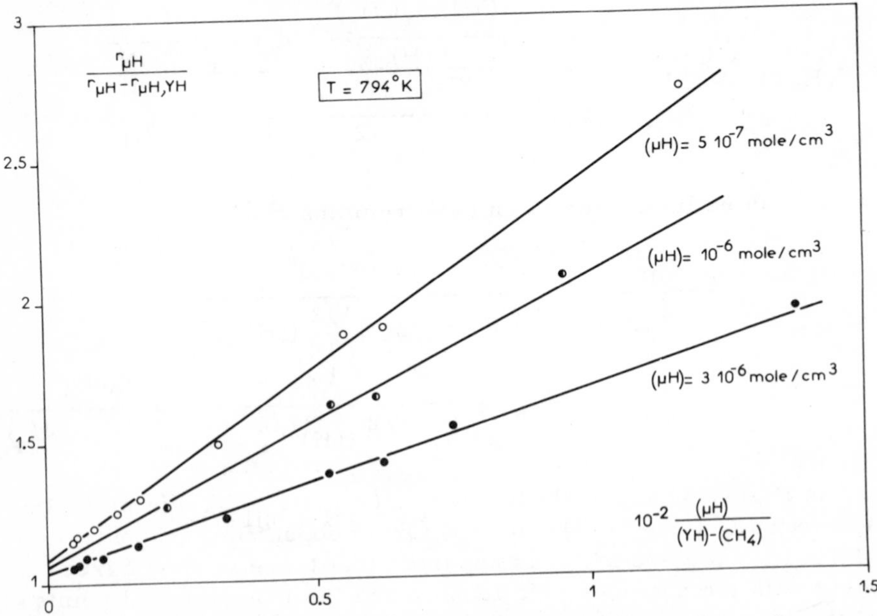

Figure 12. Pyrolysis of neopentane inhibited by added isobutene: effect of the concentration (μH). —— theoretical model.

shown. At a given neopentane concentration, the points are aligned but the slope and the initial ordinate are functions of μH concentrations and vary according to the previous relationship (Figure 12). The kinetic parameters, a_2 and b_2, are obtained at two temperatures from the diagrams. The activation energy of b_2 is about - 36.5 kcal/mole.

Effects of Oxygen and Wall Coating

Neither the addition of small amounts of oxygen, ε_o represents the ratio $(O_2)/(N_2)$ actually achieved in the ingoing gaseous stream, nor the PbO coating of the walls (24) affect the stoichiometric equation of the pyrolysis of neopentane (Figure 13). However oxygen does accelerate the reaction in an uncoated vessel (Figure 14, Curves 1 and 2). This effect is decreased by PbO coating (Figure 14, Curves 2 and 3). Figure 15 shows the variation of $r_{\mu H, YH, O_2}/r_{\mu H, YH}$ as a function of the total concentration of isobutene. This curve presents some rather interesting features: as (YH) increases, the ratio $r_{\mu H, YH, O_2}/r_{\mu H, YH}$ first decreases, goes through a minimum and then increases. This shows that oxygen has a complex action on the rate of the pyrolysis of neopentane in the presence or in the absence of isobutene. Indeed

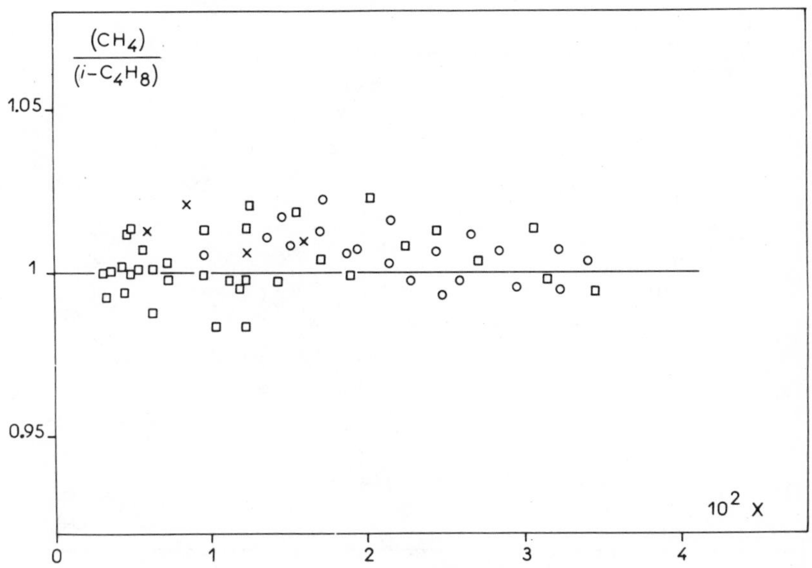

Figure 13. Pyrolysis of neopentane with small amounts of oxygen $(O_2)_o \simeq 2, 3 \times 10^{-8}$ mol/cm³. ○ T = 794°K, uncoated walls; □ T = 777°K, uncoated walls; × T = 775°K, PbO coated walls.

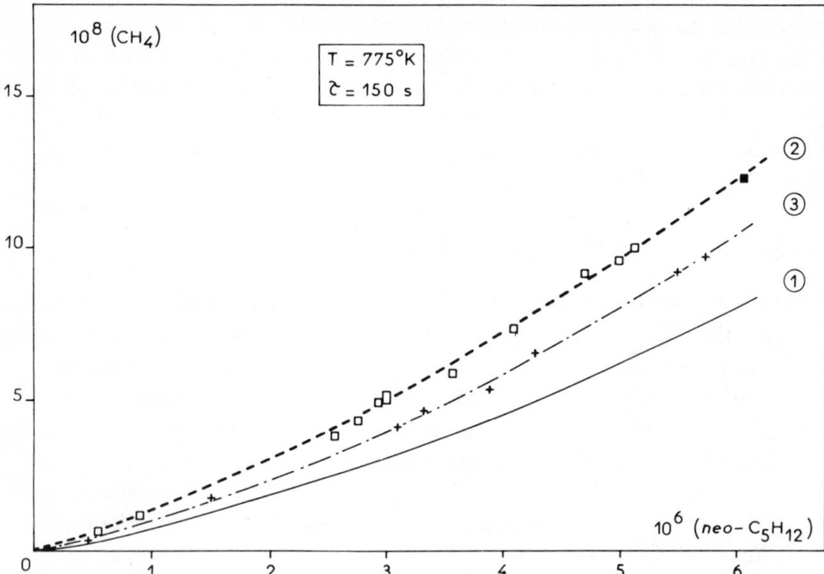

Figure 14. Pyrolysis of neopentane with small amounts of oxygen. ① *reference with less than 2 vpm* O_2*;* ② $(O_2)_o \simeq 2 \times 10^{-8}$ *mol/cm³; uncoated walls;* ③ $(O_2)_o \simeq 2 \times 10^{-8}$ *mol/cm³, PbO coated walls.*

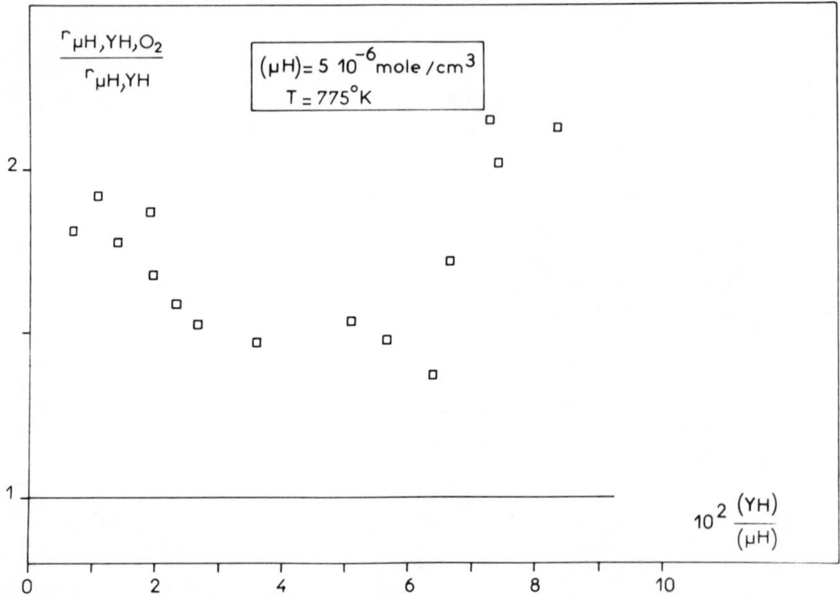

Figure 15. Pyrolysis of neopentane with added isobutene and oxygen. $(O_2)_o \simeq 2 \times 10^{-8}$ *mol/cm³,* $(YH)_o$ *and* τ *variable.*

oxygen acts both as an accelerator and as an inhibitor.

The accelerating effect can be explained by the following :

- new initiation steps :

$$\mu H + O_2 \longrightarrow \mu\cdot + HO_2\cdot$$

$$YH + O_2 \longrightarrow Y. + HO_2.$$

- degenerate branching steps :

$$\mu\cdot + O_2 \rightleftharpoons \mu O_2\cdot$$

$$Y. + O_2 \rightleftharpoons YO_2.$$

$$\mu O_2\cdot + \mu H \longrightarrow \mu O_2H + \mu.$$

$$YO_2. + \mu H \longrightarrow YO_2H + Y.$$

$$\mu O_2H \longrightarrow \mu O. + OH.$$

$$YO_2H \longrightarrow YO. + OH.$$

The inhibiting effect results from :

- the occurence of new homogeneous termination steps with the μO_2. and YO_2. free radicals stabilized by hydrogen bonds.

- the disappearance at the walls of these oxygenated radicals. This explains, at least qualitatively, the experimental findings.

Discussion

The experimental results agree with the kinetic models and the hypothesis of a well micromixed reactor. They are also in good agreement with previous studies conducted in batch reactors (14, 15). Thus one obtains similar values for the main fundamental kinetic parameters, and these are compatible with those already published in the literature (see Table I). The values of the kinetic parameters shown on Table I (this work) were determined by an optimization procedure which indicates the precision of the estimation. However this work shows that the parameters obtained for self-inhibition (pyrolysis of pure neopentane) and inhibition (pyrolysis of a mixture of neopentane and isobutene) at no extent of reaction) are markedly different. We believe that these differences can be explained by the presence, in the reaction mixture, of impurities (essentially ethylenic products). These impurities have two possible sources :

- traces in the neopentane feed (28)
- traces of by-products, from the reaction itself, actually detected by chromatographic analysis. They appear during thermal reaction in addition to the main products (CH_4 and i-C_4H_8). The kinetic law, $r^o_{\mu H,\ YH}$, deduced from the theoretical mechanism is a first approximation for the laws, $r_{\mu H}$ and $r_{\mu H,\ YH}$; it accounts only for the main reaction products CH_4 and i-C_4H_8 but

TABLE - I

KINETIC PARAMETERS			
	This work	Other works	References
n^o	3/2	3/2	(14, 15, 18, 29)
E^o kcal/mole	50	50 51.5 52 55.3	(15) (14, 16) (29) (18)
c_1 Torr	50	50	(26)
c_2	5×10^{-2}		
$E_{(b_2)}$ kcal/mole	- 36.5	- 36 - 33	(15) (30)
$E_{(b')}$ kcal/mole	- 42	- 37.5	(15)
a_2	6 to 8	6 or 7	(15)
a'	3	3	(15)

does not include the kinetic effects of by-products or impurities.

The effects of oxygen are complex. In order to give a quantitative explanation of experimental results, one needs to investigate further :

- the kinetics of the reaction at short mean residence times
- the thermal reaction of isobutene with oxygen
- the reaction at the vessel walls.

Conclusion

The pyrolysis of pure neopentane or neopentane with additives (i-C_4H_8 and O_2) has been studied in a CFSTR . Experiments were interpreted with models based on radical mechanisms and on the use of phenomenological parameters. These models account for the perturbating effects due to the complexity of the reaction medium. Since the reaction rates are of the homographic type, it is particularly easy to estimate the kinetic parameters (by introduction of multiplying factors). Furthermore the different models are compatible because they are based on a mechanistic knowledge of the reaction and they remain valuable in a wide range of operating conditions.

This work shows that a CFSTR can be a convenient means of performing kinetic analysis of a complex gas phase reaction and that it is, in many respects, superior to the batch reactor.

Abstract

The pyrolysis of neopentane, pure, or with added isobutene or oxygen, has been investigated around 500° C in a CFSTR. The main reaction products are methane and isobutene : methane has no kinetic effect upon the rate of neopentane pyrolysis, isobutene an inhibiting one and oxygen an accelerating one (effect of oxygen is reduced when walls are PbO coated). Let us note that small traces of products, either present in the initial reactant or formed in the reaction, might have a strong kinetic influence. We modelled this pyrolysis reactor by a mixed way, using free-radical mechanisms, and phenomenological terms to take perturbing effects into account. Several fundamental kinetic parameters are determined and compared with others already published. This work shows that a CFSTR can be a convenient means of performing kinetic analysis of complex gas phase reactions. It also points out that mixed reactor modelling is suitable to extrapolation.

Acknowledgement

The authors wish to thank Professor M. NICLAUSE for his scientific assistance.

Literature Cited

(1) STEACIE E. W. R., "Atomic and Free Radical Reactions", Reinhold Publ. Corp., New York, (1954).
(2) SEMENOV N. N., "Some Problems of Chemical Kinetics and Reactivity", Pergamon Press, London, (1959).
(3) GORDON A. S., "Review of Kinetics and Mechanism of the Pyrolysis of Hydrocarbons", 5th AGARD Combustion and Propulsion Colloquium, 111, (1963).
(4) NICLAUSE M. and MARTIN R., "Aspects Chimiques et Physicochimiques du Craquage Homogène d'Hydrocarbures Saturés", 36ème Congrès International de Chimie Industrielle, Bruxelles, (1966) and Ind. Chim. Belge, (1967), 2, 621.
(5) PURNELL J.H. and QUINN C. P., "The Pyrolysis of Paraffins", Photochemistry and Reaction Kinetics, ed. by ASHMORE P. G., DAINTON F.S. and SUGDEN T.M., Cambridge University Press, 330, (1967).
(6) BENSON S. W., "The Foundations of Chemical Kinetics", Mc Graw Hill, New York, (1960).
(7) DANCKWERTS P. V., Chem. Engng. Sci., (1958), 8, 93.
(8) ZWIETERING T. N., Chem. Engng. Sci., (1959), 11, 1.
(9) MULCAHY M. F. R. and WILLIAMS D. J., Aust. J. Chem., (1961), 14, 534.
(10) HERNDON W. C., J. of Chem. Ed., (1964), 41, 425.
(11) MATRAS D. and VILLERMAUX J., Chem. Engng. Sci., (1973), 28, 129.
(12) FREY F. E. and HEPP J. E., Ind. Eng. Chem., (1933), 25, 441.
(13) PEARD M. G., STUBBS F. J. and HINSHELWOOD C. N., Proc. Roy. Soc., (1952), A 214, 330.
(14) ENGEL J., Ph D. Thesis, Nancy (1955) ; ENGEL J., COMBE A., LETORT M. and NICLAUSE M., C.R. Acad. Sc. Paris, (1957), 244, 453.
(15) BARONNET F., Ph D. Thesis, Nancy (1970) ; BARONNET F., DZIERZYNSKI M., CÔME G. M., MARTIN R. and NICLAUSE M., Intern. J. Chem. Kinetics, (1971), 3, 197 ; BARONNET F., CÔME G. M. and NICLAUSE M., J. Chim. Phys., (1974), 71, 1214.

(16) ANDERSON K. H. and BENSON S. W., J. Chem. Phys., (1964), 40, 3747.
(17) TSANG W., J. Chem. Phys., (1966), 44, 4283.
(18) HALSTEAD H. P., KONAR R.S., LEATHARD D. A., MARSHALL R.M. and PURNELL J.H., Proc. Roy. Soc., (1969), A 310, 525.
(19) TAYLOR J. E., HUTCHINGS D. A. and FRECH K. J., J. A. C.S., (1969), 91, 2215.
(20) MARQUAIRE P.M., D. E. A., Nancy (1973).
(21) RONDEAU J.A. and ANDRE J.C., Rev. Sci. Instrum., (1975), 46, 692.
(22) CAIRE M.F., D. E. A., Nancy (1972).
(23) KAISER R., "Gas Chromatography", Butterworths, London, (1963).
(24) LEWIS B. and von ELBE G., "Combustion Flames and Explosions of Gases", Acad. Press, New York, (1961) ; SHARMA R.K. and BARDWELL J., Combustion and Flame, (1965), 9, 106.
(25) POWERS D.R. and CORCORAN W.H., Ind. Eng. Chem. Fundam., (1974), 13, 4.
(26) CÔME G.M., BARONNET F., SCACCHI G., MARTIN R. and NICLAUSE M., C. R. Acad. Sc. Paris, (1968), 267 C, 1192 ; HOLE K.J. and MULCAHY M.F.R., J. Phys. Chem., (1969), 73, 177 ; PORTER G. and SMITH J.A., Proc. Roy. Soc., (1961), A 261, 28 ; EUSUF M. and LAIDLER K.J., Trans. Faraday Soc., (1963), 50, 2750.
(27) NICLAUSE M., MARTIN R., BARONNET F. and SCACCHI G., Rev. Inst. Fr. Pétrole, (1966), 21, 1724.
(28) MARQUAIRE P.M., unpublished results.
(29) FRANCK D.J. and SACKETT W.M., Geochim. Cosmochim. Acta, (1969), 33, 811.
(30) RATAJCZAK E. and TROTMAN-DICKENSON A.F., tables of bimolecular gas reactions, N.B.S. (1967).

10

A Kinetic Study on the Formation of Aromatics During Pyrolysis of Petroleum Hydrocarbons

TOMOYA SAKAI and DAISUKE NOHARA

Dept. of Chemical Reaction Engineering, Nagoya City University, 3-1, Tanabedori, Mizuhoku, Nagoya, 467 Japan

TAISEKI KUNUGI

Dept. of Synthetic Chemistry, Faculty of Engineering, University of Tokyo, 7-1, Hongo, Bunkyoku, Tokyo, 113 Japan

For the pyrolysis of paraffinic hydrocarbons at 700 - 800°C, yields of olefins such as ethylene, propylene, butenes, butadiene and cycloolefins increase during the initial stage of the reaction, pass through their maxima, and later decrease; yields of aromatics, hydrogen and methane however increase monotonically throughout the reaction course. Sakai et al. (1) reported previously the result of a kinetic study on thermal reactions of ethylene, propylene, butenes, butadiene and these respective olefins with butadiene at the conditions similar to those of paraffin pyrolysis, directing their attention on the rates of formation of cyclic compounds. Kinetic features of the thermal reactions of these olefins are summarized in Table I combined with the results obtained in later investigations for thermal reactions of cycloolefins (2) and benzene (3).

Thermal reactions of ethylene (4,5) require higher temperatures (750 - 800°C) than the other olefins. Initial reaction products are butadiene, 1-butene, propylene, ethane and acetylene. As the yields of these initial products decrease with increased residence times, cyclic compounds such as cyclopentene, cyclopentadiene, cyclohexene and benzene are produced. In the case of propylene (6,7), the reaction proceeds 2 - 4 times faster than that of ethylene; and ethylene, methane, butadiene, butenes, acetylene, and methylcyclopentene are the main products during the initial step; cyclopentadiene, cyclopentene, benzene, toluene and polycyclic compounds higher than or equal to naphthalene are products of secondary reactions. A remarkable fact for the thermal reaction of propylene is that the yields of five membered ring compounds are larger than those in the case of ethylene.

Different features were observed between the thermal reaction of 1-butene and those of cis- and trans-2-butenes at 640 - 680°C (1). In the former case, the reaction proceeded mainly in three ways; these were pyrolysis to methane and propylene, dehydrogenation to butadiene, and pyrolysis to two moles of ethylene; the ratio of rates for these three reactions are 4 : 3 : 1, respectively. In the latter cases, the main reaction was isomerization between cis- and trans-2-butene, and the selectivities of other reactions than the isomerization were supressed to less than 10%.

Cyclization proceeded in nearly 100% selectivity in the case of thermal reaction of butadiene (1), yielding 4-vinylcyclohexene (VCH) for the first step and ethylene, cyclohexene, cyclohexadiene, and benzene in the secondary steps. Similar highly selective cyclizations were observed for the reactions between butadiene and ethylene, propylene, 1-butene, cis-2-butene, trans-2-butene or isobutylene (1), yielding cyclohexene (HCH), 4-methylcyclohexene (MCH), 4-ethylcyclohexene, cis-4,5-dimethylcyclohexene, trans-4,5-dimethylcyclohexene or 4,4-dimethylcyclohexene, respectively. Based on the above information, it can be said that butadiene plays an important role in the formation of cyclic compounds in pyrolysis conditions.

Next, in order to learn more about the rates of dehydrogenation of cyclohexenes resulting from Diels-Alder reactions between butadiene and olefins, VCH, HCH and MCH were earlier subjected to thermal reactions at 530 - 665°C (2). The main reactions in these cases were reverse Diels-Alder reactions and dehydrogenations. Dehydrogenations which are related to the productions of cyclohexadiene and benzene homologues were 1 : 10 in selectivity as compared to that of the reverse Diels-Alder reaction. An interesting observation related to cyclic compound formation is that, in the case of MCH pyrolysis, cyclohexadiene and cyclopentene are formed at almost the same rates as butadiene and propylene. So that, in this case, about 60% of MCH is employed in the formation of cyclic compounds.

Thermal reaction of benzene (3) proceeded at 800 - 850°C producing biphenyl and hydrogen. Benzene was the most refractory material of the feed stocks employed in these experiments. The addition of 3 - 4 wt% of naphthalene to benzene did not greatly affect the reaction kinetics or product selectivity. However, when ethylbenzene was added in the same small amount, features of the reaction were quite changed. No white crystalline product was obtained, and instead a tarry matter covered the inside of the reactor.

Based on the kinetic data obtained above, a tentative calculation was earlier performed to determine whether or not the rate of cyclic compound formation in actual pyrolysis reactions can be accounted for through the Diels-Alder reactions between butadiene and olefins. The actual rate of cyclic compound formation was much greater than the rate calculated from the concentrations of butadiene and olefins in the actual pyrolysis conditions.

Allyl radicals were tested as materials or substances that cause cyclization with olefins. 1,5-Hexadiene (diallyl) and diallyl oxalate (DAO) were used as the source materials for allyl radicals. These compounds were subjected to the thermal reaction in the presence of ethylene, which resulted in a rapid formation of five-membered cyclic compounds. The role of allyl radical for the formation of aromatics in pyrolysis reaction is however still

Table I. A Summary of Rate Data on Thermal

Reactant	Temperature (°C)	Product primary	Product secondary
C=C	700 - 850	C_4H_6, 1-C_4H_8, C_3H_6, C_2H_2, C_2H_6, H_2	CH_4, H_2, , , 
C=C-C	700 - 850	C_2H_4, C_4H_8, C_4H_6, , C_2H_2, CH_4, H_2	CH_4, H_2, , , , , C_2H_6
C=C-C-C	640 - 680	CH_4, C_3H_6, C_4H_6, C_2H_6, 2-C_4H_8's	—
C-C=C-C (cis-)	640 - 680	trans-C_4H_8, C_4H_6, CH_4, C_3H_6	C_2H_4
C-C=C-C (tr-)	640 - 680	cis-C_4H_8, C_4H_6, CH_4, C_3H_6	C_2H_4
C=C-C=C + C=C-C=C	550 - 750	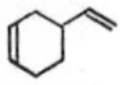	, , , C_2H_4, CH_4, C_3H_6, H_2
C=C-C=C + C=C	510 - 590		—
C=C-C=C + C=C-C	510 - 590		—
C=C-C=C + C=C-C-C	510 - 590	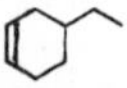	—
C=C-C=C + C-C=C-C (cis-)	510 - 590		—
C=C-C=C + C-C=C-C (tr-)	510 - 590	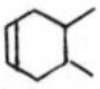	—
C=C-C=C + C=C(C)-C	510 - 590		—
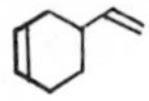	530 - 585	C_4H_6 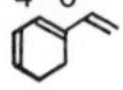	— —
	585 - 665	C_2H_4, C_4H_6 	C_3H_6 
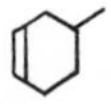	575 - 650	C_3H_6, C_4H_6 CH_4,  C_2H_4,  	— — — —
	700 - 850	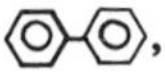, H_2, C_2H_4	Pyrene, CH_4

Reactions of Olefins, Cycloolefins and Benzene

Reaction order	$\log_{10}k$ (ml, mol, sec)				E (kcal/mol)	$\log_{10}A$ (ml, mol, sec)
	500	600	700	800		
1.5	-	-	0.59	1.63	49.6	11.73
1.5	-	-	0.86	2.19	63.2	15.06
1.0	-	-1.88	-0.30	-	61.3	13.47
1.0	-	-0.51	0.95	-	56.5	13.64
1.0	-	-0.50	0.97	-	57.2	13.82
2.0	3.40	4.20	4.84	5.36	24.8	10.41
2.0	2.54	3.47	-	-	28.8	10.68
2.0	2.00	2.96	-	-	29.7	10.40
2.0	0.77	2.48	-	-	52.7	15.67
2.0	0.50	2.27	-	-	54.7	15.97
2.0	2.14	3.14	-	-	30.6	10.80
2.0	1.11	2.59	-	-	45.8	14.06
1.0	-2.09	-0.12	-	-	60.6	15.05
1.0	-3.09	-1.52	-	-	48.4	10.60
1.0	-3.57	-1.44	-	-	65.7	15.01
1.0	-4.02	-2.25	-	-	54.9	11.50
1.0	-3.41	-1.31	-	-	64.9	14.94
1.0	-3.67	-1.69	-	-	61.1	13.61
1.0	-3.94	-1.92	-	-	62.4	13.70
1.0	-3.52	-1.92	-	-	49.55	10.49
1.0(?)	-	-	-2.96	-1.80	55.0	9.40

obscure because the cyclized compounds produced from allyl radical and olefins were exclusively five-membered ring compounds during the initial stage of the reaction (8). However, in combination with the isomerization reactions of substituted cyclopentenes which are expected to proceed easily at pyrolysis conditions, it is likely that allyl radicals are important intermediates causing the aromatization of olefins.

Experimental

Several methods are known for the generation of allyl radical. James and Troughton (9) commented on the photolyses of dicyclopropyl ketone, propylene, 1-butene, and cyclopropane and radiolysis of diallyl ketone. McDowell and Sifniades (10) and James and Kambanis (11) obtained the allyl radical from photolyses of crotonaldehyde and diallyl oxalate at low temperatures, respectively. Al-Sader and Crawford (12) obtained allyl radicals by thermolysis of 3,3'-azo-1-propene. Some of the above materials may however have several disadvantages such as preliminary isomerization of reactants, concomitant generation of other reactive species, complexity of the successive and competing reactions, and difficulty in raw material preparations. At rather elevated temperatures, it is thought that diallyl produces allyl radicals solely during the very initial stage of the pyrolysis. DAO was employed as the source material for allyl radicals for reactions to be investigated at lower temperatures.

The purity of diallyl sample was more than 99.4 wt% by gas chromatographic analysis after distillation with a spinning-band rectification column of 60 stages. The main impurity was 1-hexene. DAO was synthesized and distilled according to a method reported in the literature (13). Polymerization grade ethylene was used without purifications. Impurities in the ethylene sample were methane and ethane in 0.01 and 0.07 mol %, respectively. Nitrogen from a cylinder was deoxygenated by passing it through a reduced copper gauze at 250°C followed by drying in a silica gel column.

A conventional flow-type reaction system was used for reactions at atmospheric pressure. The liquid sample was vaporized at 0°C with the aid of nitrogen and ethylene flows, and the resulting mixtures then entered the reactor. The reactor was an annular quartz cylinder of 200 mm length and 10.6 mm i.d., equipped coaxially with a thermowell of 7.2 mm o.d. The reactor was positioned in an electrically heated brass block of 180 mm length, 18 mm i.d., and 55 mm o.d. The temperature profile of the reactor was measured for each run, and the residence times of the reactants were determined by the method of Hougen and Watson (14). Inlet and outlet gases were analyzed by use of flame-ion-detector gas chromatographs equipped with 50 m capillary columns coated with squalane and di-n-butyl maleate at 50 and 0°C, respectively.

Results and Discussion

In the present investigation, the thermal reactions of diallyl or DAO in excess nitrogen or ethylene were conducted in a flow system. Information concerning the reactions of the allyl radical with olefins were obtained, especially the reactions resulting in the formation of C_5 cyclic compounds. The detailed product distributions are given for varied conversion levels. The primary and secondary products are clearly separated. A reaction scheme is proposed to explain qualitatively the formation of primary and secondary products.

Thermal Reaction of Diallyl. Overall mass balances for the pyrolysis experiments of diallyl in excess nitrogen generally were within 96 - 99 %. It was confirmed that the effect of the quartz surface on the reaction was negligible under the present conditions. The conversion of the reactant has been defined as the ratio of the sum of the peak areas of products to that of all peak areas of the gas chromatogram.

Typical kinetic results are listed in Table II, in which the product distributions are presented as the weight per cent of the individual products in the total products excluding C_{12} and higher compounds.

A kinetic model based on the 3/2-order fits the experimental results. From Arrhenius plots, the overall reaction rate constant was obtained as $k = 10^{13.5} \exp(-44{,}500/RT)$ $ml^{1/2}$ $mol^{-1/2}$ sec^{-1}.

The examination of the product distribution vs. residence time curves at four temperature levels revealed that the same mechanism applied for the reaction within the present experimental conditions, so that, in Figure 1, the product distribution vs. conversion curves were adopted.

From Figure 1, it is clear that the primary products of the thermal reaction of diallyl are ethylene, propylene, 1-butene, butadiene, 1-pentene, cyclopentene, cyclopentadiene, and 1,3,5-hexatriene, and the secondary products are 1,3-cyclohexadiene and benzene. Trace amounts of methane, propane, and 1,4-pentadiene were also found in some experiments. No hydrogen was detected by a nitrogen carrier gas chromatograph with MS 5A column. The formation of C_{12} compounds was noticed at low temperatures. A small amount of liquid product was found in the separator tube after 50 or more experimental runs. The average molecular weight of the liquid product was 428 based on the method of Hill (15).

By extrapolating the curves in Figure 1 to zero conversion, the molar ratios of formation of individual products at the initial stage of the reaction were approximated as listed in Table III, in which the amount of ethylene was taken as unity. From Table III and Figure 1, the radical chain mechanism of the thermal reaction of diallyl is proposed as follows.

Initiation:

1. C=C-C-C-C=C (I) ⟶ 2 C=C-C·

Table II. Typical Experimental Data on Pyrolysis of Diallyl

Temperature, °C	500	540	580	620
Residence time, sec	0.273	0.314	0.386	0.140
Conversion, wt %	0.075	0.437	2.07	6.44
Product distribution[a)], wt %				
Ethylene	9.29	8.16	7.12	8.58
Propylene	30.97	31.12	32.14	32.92
1-Butene	9.94	7.67	8.45	8.24
Butadiene	11.93	12.96	9.27	9.76
1-Pentene	5.51	4.76	3.63	2.80
Cyclopentene	—	3.43	3.25	2.40
Cyclopentadiene	6.73	9.10	8.33	8.02
1,3,5-Hexatriene	13.59	9.85	3.16	0.00
1,3-Cyclohexadiene	6.62	6.62	12.92	17.55
Benzene	0.00	0.42	2.77	4.31
Total[b)]	94.58	94.09	91.04	94.58

a) Formation of C_{12} and higher compounds is excluded.

b) The remainder consists of several minor products such as methane, propane, 1,4-pentadiene, etc.

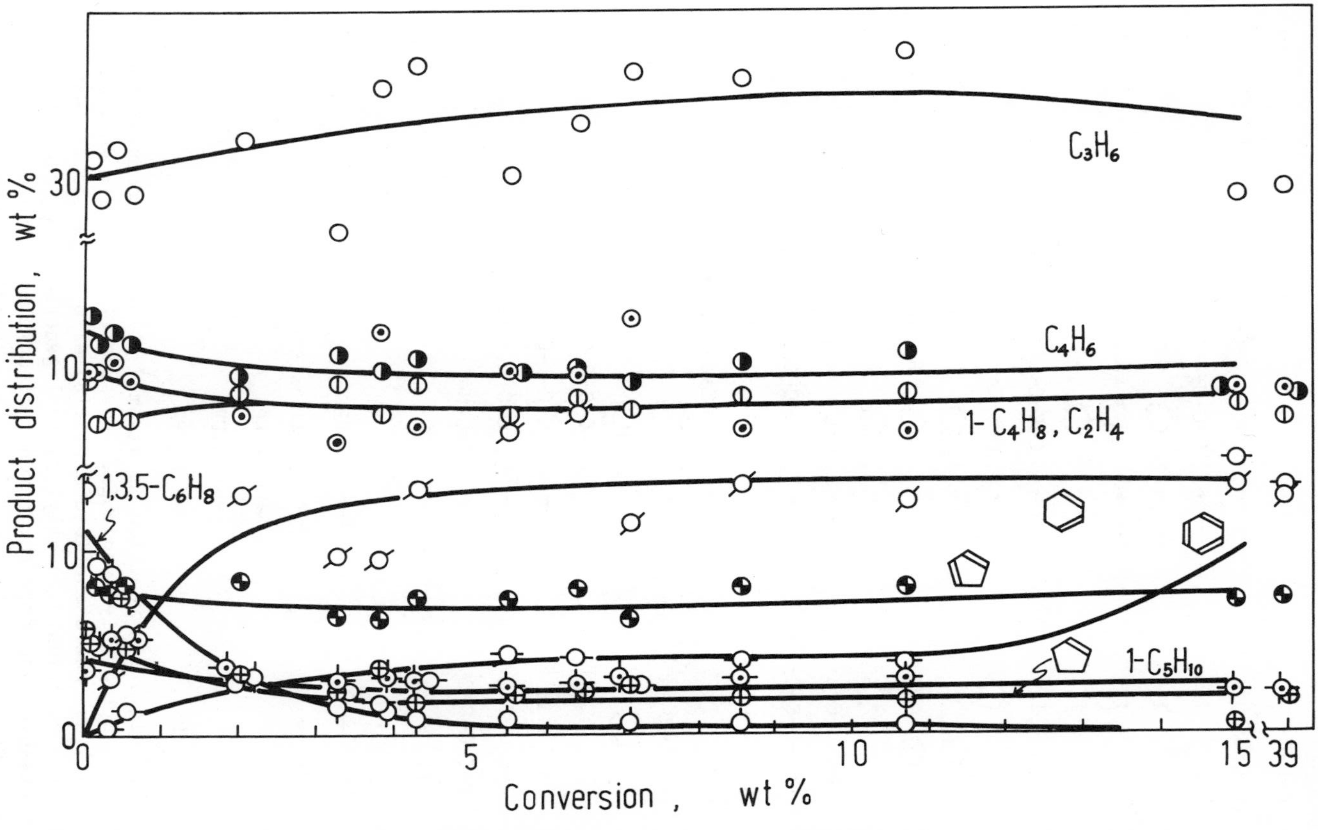

Figure 1. Product distribution vs. conversion for pyrolysis of diallyl

Propagation:

2.	C=C-C· + I	⟶	C=C-C + C=C-Ċ-C-C=C	(II)
3.	C=C-C· + I	⟶	C=C-C-C-Ċ-C-C-C=C	(III)
4.	III	⟶	C=C-Ċ-C-C-C-C-C=C	(IV)
5.	III	⟶	+ C=C-C=C	
6.	III	⟶	+ C=C-C-C·	
7.	III	⟶	+ C=C-C-C	
8.	IV	⟶	C=C-C-C-C· + C=C-C=C	
9.		⟶	C=C-C· + C=C	
10.	+ I	⟶	+ C=C-C-C-Ċ-C	
11.	+ I	⟶	+ C=C-C-C-C-C·	
12.		⟶	+ H·	
13.	C=C-C-C-C·	⟶	C=C-C· + C=C	
14.	C=C-C-C-C· + I	⟶	C=C-C-C-C + II	
15.	C=C-C-C· + I	⟶	C=C-C-C + II	
16.	II	⟶	C=C-C=C-C=C + H·	
17.	C=C-C=C-C=C	⟶		
18.		⟶	+ 2H·	
19.	H· + I	⟶	C=C-C-C-Ċ-C	
20.	H· + I	⟶	C=C-C-C-C-C·	
21.	C=C-C-C-Ċ-C	⟶	C=C-C· + C=C-C	
22.	C=C-C-C-C-C·	⟶	C=C-C-C· + C=C	
23.	II + n I	⟶	P·	

Termination:

24.	H· + R·	⟶	RH
25.	H· + P·	⟶	PH
26.	2R·	⟶	RR
27.	R· + P·	⟶	PR
28.	2P·	⟶	PP

The initiation reaction is assumed to be the decomposition of diallyl into two allyl radicals. Lossing et al. (16), Ruzicka and Bryce (17) and Akers and Throssell (18) also suggested the same initiation reaction. An allyl radical generated by reaction 1 abstracts hydrogen from the parent molecule to produce propylene as in reaction 2. Existence of the 1,5-hexadienyl radical (II) is supported by Ruzicka and Bryce (17) and James and Troughton (9).

Another reaction of allyl radical is an addition to the parent molecule, i.e., reaction 3, to form III, which undergo hydrogen shift to produce IV, i.e., reaction 4.

The appreciable amount of C_5 compounds (1-pentene, cyclopentene, and cyclopentadiene) obtained as primary products leads to the postulation of several reactions of C_9 radicals designated as III or IV. They are reactions 5 - 8. Radicals produced in these reactions undergo either unimolecular scission or bimolecular reaction with I to form 1-pentene, cyclopentadiene, and other primary products as shown in reactions 9 - 15. These schemes are partly in line with the one of Ruzicka and Bryce (17). Disagreements exist in that (a) neither methylcyclopentene nor hydrogen was detected, (b) formation of methane was negligibly small, and (c) all C_5 compounds were obtained as primary products in the present study as compared to findings of Ruzicka and Bryce.

Table III. Molar Ratio of Products at Zero Conversion Relative to Ethylene on Pyrolysis of Diallyl

Ethylene	1
Propylene	2.22
1-Butene	0.46
Butadiene	0.54
1-Pentene	0.20
Cyclopentene	0.22
Cyclopentadiene	0.36
1,3,5-Hexatriene	0.37
1,3-Cyclohexadiene	0
Benzene	0

Among the C_6 compounds produced, only 1,3,5-hexatriene is the primary product. Reaction 16 is proposed for its formation. James and Troughton (9) obtained ethylene and 1,3,5-hexatriene as the primary products in their study on the reaction of diallyl with the ethyl radical at 134 - 175°C. Furthermore, they obtained 1,3-cyclohexadiene as a successive product. Recently Orchard and Thrush (19) reported the thermal isomerization of 1,3,5-hexatriene to 1,3-cyclohexadiene at ca. 400°C and the consecutive formation of benzene at ca. 550°C. In the present work, 1,3-cyclohexadiene (reaction 17) and benzene (reaction 18) were obtained as the secondary products. The hydrogen atom produced in reactions 12, 16 and 18 is considered to react with the parent molecule in two ways as shown in reactions 19 and 20. These radicals decompose to produce the allyl radical and propylene in reaction 21 and the butenyl radical and ethylene in reaction 22.

Reaction 23 is proposed because C_{12} and higher compounds were noticed in the products. P· represents such oligomers. The termination reactions are described by reactions 24 - 28; R· represents all radical species.

The fact that no cyclohexene was detected in the present experiments suggests that the rate of cyclization through Diels-Alder reaction between formed butadiene and olefins is smaller than the rate of cyclization caused by the reaction of allyl radical with olefins at these temperatures.

Thermal Reaction of Diallyl in Excess Ethylene. It was shown in the preceeding section that diallyl decomposed in nitrogen at

Table IV. Typical Experimental Data on Pyrolysis of Diallyl in Excess Ethylene

Temperature, °C	580		620		660		700		660[a)]
Residence time, sec	0.193	0.523	0.352	0.489	0.177	0.377	0.340	0.511	0.266
Diallyl decomposition, %	1.23	13.8	19.7	29.9	40.5	72.1	91.5	100	54.9
Product distribution, mole %									
Propylene	14.5	16.4	14.7	18.3	15.6	17.4	19.0	20.7	15.3
1-Butene + Butadiene	19.0	18.7	15.2	16.8	18.4	21.5	21.1	22.7	17.9
1-Pentene	28.5	28.4	31.3	29.0	30.3	28.5	26.7	23.0	32.8
1,4-Pentadiene	trace	0.8	0.9	0.7	1.5	0.9	1.0	1.2	—
Cyclopentene	38.0	33.9	36.1	32.8	32.4	29.3	26.8	23.6	31.5
Cyclopentadiene	0.0	1.8	1.9	2.4	1.9	2.4	5.5	8.8	2.5

a) Ethylene-nitrogen mixture (37 % ethylene) was used as diluent gas.

500 - 620°C forming allyl radicals as the initial step of the reaction. At the same time, it was reported previously (4) that a thermal reaction of ethylene itself scarcely proceeded below 700°C. Experiments on pyrolysis of diallyl in excess ethylene were conducted at temperatures between 580 and 700°C, and at very low concentrations of diallyl, between 0.2 and 0.7 mol %, so as to cause exclusively the expected reaction of allyl radical with ethylene.

Typical experimental results were listed in Table IV. Overall mass balances of the outlet gas to the inlet gas were within 93 - 98 %. The diallyl decomposition (%) was calculated based on the amount of diallyl in the outlet gas to that in the inlet gas. The product distribution was presented as mole % of each to the total products. Main products were cyclopentene, 1-pentene, 1-butene, butadiene, and propylene. The molar ratio of 1-butene

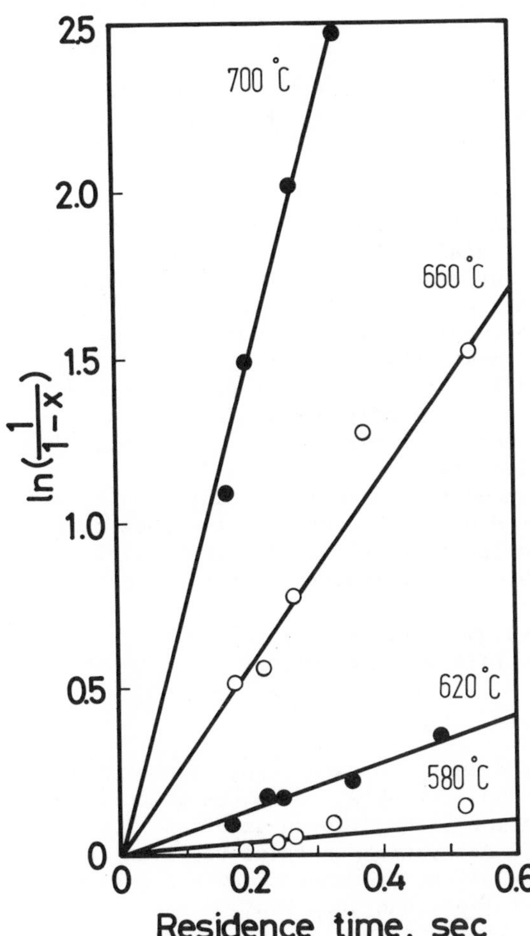

Figure 2. First-order kinetics in decomposition of diallyl in excess ethylene

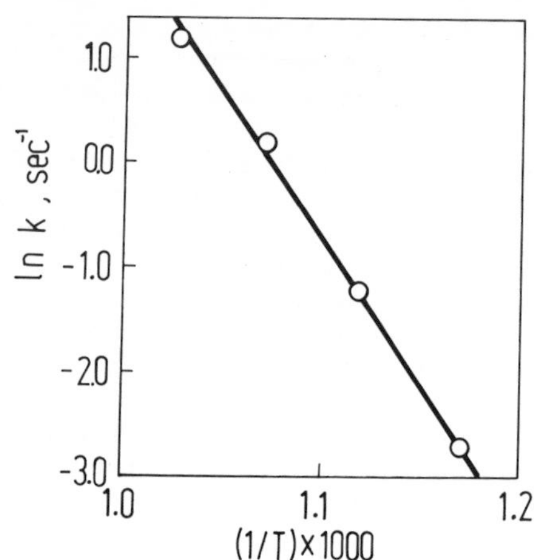

Figure 3. Arrhenius plots for decomposition of diallyl in excess ethylene

to butadiene formation was 0.4 - 0.6. These five components accounted for over 90 % of the total products. Minor products were 1,4-pentadiene and cyclopentadiene. In addition, a trace amount of hydrogen was detected. No methane, ethane, acetylene, propane, allene, cyclopentane or C_8 compounds were detected.

Under the present experimental conditions, first-order reaction kinetics fit the rate of diallyl decomposition, as shown in Figure 2. In the present experiments, the feed ratio of diallyl to ethylene was so small that the reaction of generated allyl radical with diallyl could be neglected. This was confirmed based on the following evidence. First, the reaction order of diallyl decomposition changed from 3/2-order in the case of nitrogen dilution to first-order in the case of ethylene dilution. Second, considerably fewer products were obtained in this case compared with those of nitrogen dilution as shown in Tables II and IV. Therefore, it could be concluded that almost all allyl radicals generated from diallyl reacted with ethylene. At the same time, it could be deduced from the first-order kinetics that allyl radicals were effectively quenched by ethylene under the present reaction conditions.

From Arrhenius plots, shown in Figure 3, the kinetic parameters for the decomposition rate constant of diallyl were obtained as below;

$$k = 10^{12.9} \exp(-55{,}000/RT) , \quad sec^{-1} .$$

The activation energy obtained in the present experiments, i.e., 55.0 kcal/mole, was higher than that obtained in the preceeding section of diallyl decomposition with nitrogen dilution, i.e., 44.5 kcal/mole (500 - 620°C). Activation energies reported in the literature are 44.9 kcal/mole in the presence of excess cyclohexane by Doering and Gilbert (20), and 44.1 kcal/mole at very low partial pressures of diallyl (5×10^{-4} mmHg) in helium at 704-797°C by Homer and Lossing (21). The present activation energy of 55.0 kcal/mole was as large as that reported by Akers and Throssel (18), namely, 56.0 kcal/mole, in their study of diallyl decomposition with the toluene carrier method. The higher activation energy obtained in the present study confirms that ethylene quenches allyl radical as effectively as toluene. However, even the highest activation energy experimentally obtained is considerably lower than the estimated value, 62.2 kcal/mole, reported by Golden et al. (22) in their study on the estimation of the allyl resonance energy.

Examination of the product distribution vs. residence time curves at four temperature levels, i.e., at 580, 620, 660 and 700 °C, revealed that the same mechanism applied in the temperature range investigated. Accordingly, as shown in Figure 4, product distribution versus diallyl decomposition (%) curves were offered for the following discussion.

From Figure 4, it can be said that all the main products which were listed in Table IV are primary products of the reaction. It is ambiguous experimentally whether or not 1,4-pentadiene and cyclopentadiene are primary products. An interesting fact is that the ratio of the amount of cyclopentene to 1-pentene decreases from about 1.3 at the lower diallyl decomposition percentages to

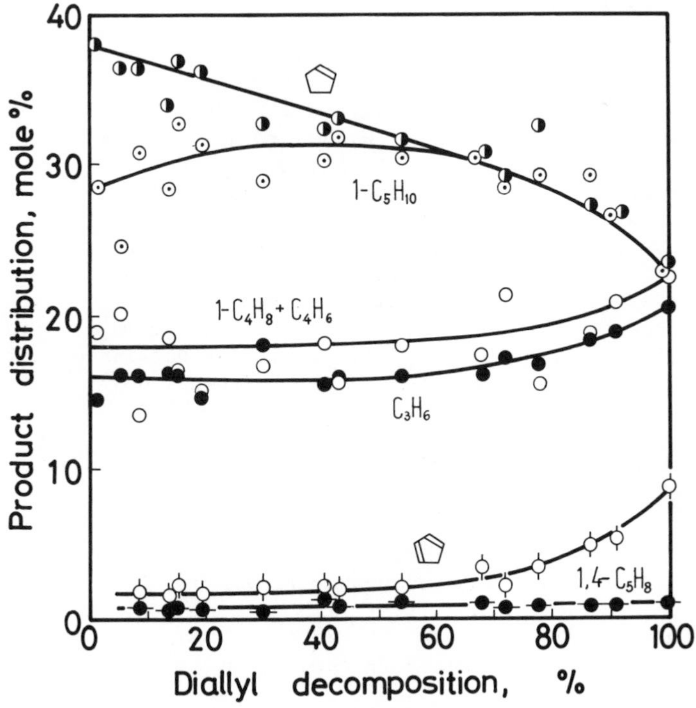

Figure 4. Product distribution vs. decomposition of diallyl in excess ethylene

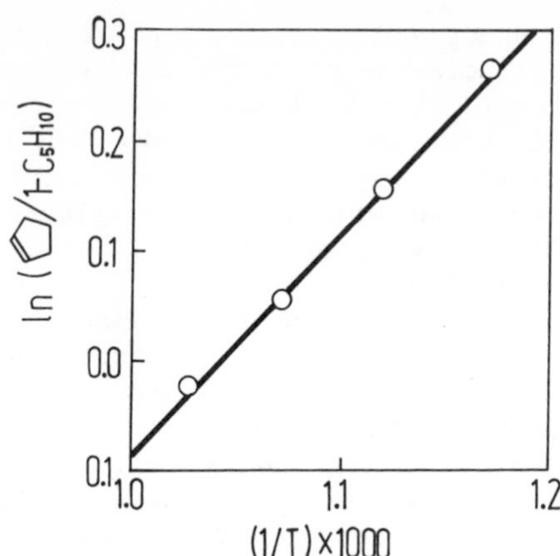

Figure 5. Temperature dependence on the ratio of amount of cyclopentene to 1-pentene

almost unity at higher percentages. A linear relationship occurs between the logarithm of that ratio and the reciprocal temperature as shown in Figure 5. A discussion related to Figure 5 will be given later.

A reaction scheme is proposed below based on product distributions shown in Figure 4 and on energetic as well as kinetic considerations of elementary reactions of the key radicals. The initiation reaction is considered to be a splitting of diallyl to form two allyl radicals. In the present discussion, however, reaction 29 is postulated so that the allyl radicals generated from diallyl are hot radicals. Golden et al. reported that the heat of formation of a stable allyl radical generated from diallyl is 62.2 kcal/mole (22). The activation energy obtained in the present investigation for diallyl decomposition, i.e., 55.0 kcal/mole, may be too small to support the postulate concerning the formation of a hot allyl radical. However, reaction 29 is necessary in order to explain the simultaneous formations of cyclopentene and 1-pentene so as not to conflict with the informations revealed in the literatues (23,24,25,26,27). Details on this point will be described in the course of discussions.

29. C=C-C-C-C=C ⟶ 2 C=C-C·*

A succeeding reaction of hot allyl radical generated by reaction 29 must be either the addition to ethylene, i.e., reaction 30; a hydrogen abstraction from ethylene, i.e., reaction 31; or a quenching to a stable allyl radical, i.e., reaction 32.

30. C=C-C·* + C=C ⟶ C=C-C-C-C·*

31. C=C-C·* + C=C ⟶ C=C-C + C=C·

32. C=C-C·* + M ⟶ C=C-C· + M

4-Pentenyl radicals produced by reaction 30 are also hot radicals. The correctness of this statement will be comprehended from the consideration of the heat content of an allyl radical plus ethylene, that of a stable 4-pentenyl radical, and the activation energy of reaction between allyl radical and ethylene, as will be discussed in detail in the next section. The reverse reaction -30 possibly should be considered as a means of obtaining a deactivated allyl radical.

-30. C=C-C-C-C·* ⟶ C=C-C· + C=C

Next, the cyclization of a hot 4-pentenyl radical, i.e., reaction 33 was postulated.

33. C=C-C-C-C·* ⟶ ⬠·

No direct formation of cyclopentyl radical from an allyl radical and ethylene was considered in this case. It is known that Diels-Alder type addition of an allyl radical to ethylene, i.e., C=C-C· + C=C ⟶ ⬠·, is almost impossible due to the extremely high electronic energy level of the reaction intermediate. In reaction 33, 4-pentenyl radical likely would not be a stable radical, but a hot one, for the following reason. Watkins and Olsen (27) suggested in their study on the photolysis of azo-n-propane in the presence of acetylene that a cyclopentyl radical was formed only from the excited 4-pentenyl radical produced by irradiation, but 1-pentene and 1,4-pentadiene may be formed from the stable 4-pentenyl radical. In this connection, the works of Gordon (23) and Palmer and Lossing (24) concerning a cyclopentyl radical, and that of Shibatani and Kinoshita (25) for 4-pentenyl radical were supposed to have dealt with stable radicals. The former two papers reported that no linear C_5 compounds were formed in their studies on reactions of cyclopentyl radical generated by cyclopentane photolysis at 162 - 445°C (Gordon), and by the thermal decomposition of cyclopentyl methyl nitrite at 300 - 800°C (Palmer and Lossing). On the other hand, the latter paper reported that no cyclized C_5 compound was formed during the pyrolysis of 1-pentene at 480 - 750 °C. In a word, it is impossible to produce any linear (cyclized) C_5 radicals from the stable cyclopentyl (4-pentenyl) radical. That is, only a hot 4-pentenyl radical formed by reaction 30 can form stable 4-pentenyl and cyclopentyl radicals at the same time. The extent of excitation of a 4-pentenyl radical will be discussed in the next section.

Walsh (26) calculated the activation energies of the following reactions, C=C-C-C-C· ⇄ ⬠·, to be smaller than 14 kcal/mole for the forward and to be smaller than 32.5 kcal/mole for the reverse; these values are reported on the estimation of stable radical energies. However, the enthalpy change calculated tentatively from Figure 5 in the present experiment is only 4 kcal/mole instead of 18.5 kcal/mole reported by Walsh. This inconsistency may be caused by the hot 4-pentenyl radical in the present study. Formations of 1-pentene and cyclopentene as main primary products of the reaction can duly be explained by postulation of a hot 4-pentenyl radical as shown in reactions 33, 34, and 35.

34. C=C-C-C-C·* + C=C ⟶ C=C-C-C-C + C=C·

35. (cyclopentyl radical) ⟶ (cyclopentene) + H·

The next discussion concerns the formation of propylene, 1-butene and butadiene which are the other main primary products of the reaction. The hot allyl (or 4-pentenyl) radical generated in reaction 29 (or 30) may abstract hydrogen from ethylene to form propylene (or 1-pentene) and a vinyl radical. From the highly endothermic nature of the vinyl radical formation, the postulation of a hot radical is again reasonable in this step. A similar reaction of a cyclopentyl radical with ethylene, i.e., (cyclopentyl radical) + C=C ⟶ (cyclopentane) + C=C·, was excluded from the whole scheme by the following reasons. Firstly, no cyclopentane was detected experimentally. Secondly, the cyclopentyl radical may have released its energy in the course of cyclization reaction, reaction 33. A hot 4-pentenyl radical itself may partly be quenched by reaction 36.

36. C=C-C-C-C·* + M ⟶ C=C-C-C-C· + M

Vinyl radical generated in reaction 32 undergoes reaction 37, which is considered to be a path for formation of C_4 products. Another, presumably a main, reaction course for C_4 product formations is the addition of ethylene to a hot 4-pentenyl radical, i.e., reaction 38, accompanied with hydrogen shifts (or isomerizations) as exemplified in reactions 39 - 42, and with β-scission in reactions 43 - 46. It is ambiguous whether reactions 45 and 46 should be replaced by reaction 47 because no ethane or propane was detected in the experiments.

37. C=C· + C=C ⟶ C=C-C-C·

38. C=C-C-C-C·* + C=C ⟶ C=C-C-C-C-C-C·

39. C=C-C-C-C-C-C· ⟶ C=C-C-C-C-Ċ-C

40. C=C-C-C-C-Ċ-C ⟶ C=C-C-C-Ċ-C-C

41. C=C-C-C-Ċ-C-C ⟶ C=C-C-Ċ-C-C-C

42. C=C-C-Ċ-C-C-C ⟶ C=C-Ċ-C-C-C-C

43. C=C-C-C-C-Ċ-C ⟶ C=C-C-C· + C=C-C

44. C=C-C-C-Ċ-C-C ⟶ C=C-C· + C=C-C-C

45. C=C-C-Ċ-C-C-C ⟶ C=C-C-C=C + C-C·

46. C=C-Ċ-C-C-C-C ⟶ C=C-C=C + C-C-C·

47. C=C-C-C· ⟶ C=C-C=C + H·

Finally, discussions will be made on the termination reactions. Generally, it is very difficult to follow the scheme of termination reactions in relation to the products obtained in these radical chain reactions. The kinetic results however suggest a chain length of almost unity for the present reaction. The termination reactions are merely speculative based on the main or minor products. They are arranged below as reactions 48 - 56 in the speculative order of effectiveness.

48. 2 C=C-C· ⟶ C=C-C-C-C=C

49. C=C-C· + [cyclopentyl·] ⟶ C=C-C + [cyclopentene]

50. C=C-C· + C=C-C-C· ⟶ C=C-C + C=C-C=C

51. C=C-C-C· + [cyclopentyl·] ⟶ C=C-C-C + [cyclopentene]

52. C=C-C· + C=C-C-C-C· ⟶ C=C-C + C=C-C-C=C

53. C=C-C-C-C· + [cyclopentyl·] ⟶ C=C-C-C-C + [cyclopentene]

54. 2 C=C-C-C· ⟶ C=C-C-C + C=C-C=C

55. C=C-C-C· + C=C-C-C-C· ⟶ C=C-C=C + C=C-C-C-C

56. C=C-C-C· + C=C-C-C-C· ⟶ C=C-C-C + C=C-C-C=C

Although some ambiguity exists in the designation of cyclopentadiene as one of the primary products of the reaction, the formation may be considered partly due to reactions 57 and 58 and partly to reaction 59. The former reactions give relatively small amounts of cyclopentadiene as a primary product. The latter gives it as a secondary one.

57. [cyclopentyl·] ⟶ [cyclopentenyl·] + H_2

58. [cyclopentenyl·] ⟶ [cyclopentadiene] + H·

59. [cyclopentene] + R· ⟶ [cyclopentenyl·] + RH

R· represents any radical present. Stein and Rabinovitch (28) recently reported the existence of reaction 57. Trace amounts of hydrogen present even during the initial stage of the reaction in the experiment confirms the existence of reaction 57. Formations of cyclopentadiene and hydrogen from cyclopentene are the well known dehydrogenation which was reported by Rice and Murphy (29), Vanas and Walters (30), and Mackay and March (31). Increase of cyclopentadiene with diallyl decomposition % in Figure 4 indicates that reaction 59 proceeds as one of the secondary reactions in the whole scheme of the present reaction.

Thermal Reaction of DAO in Excess Ethylene. To get better knowledge about the formation of cyclopentene and 1-pentene, DAO was employed as another source material for allyl radicals. In the case of the pyrolysis of DAO in excess ethylene, the reaction temperature was considerably lowered, and a large amount of diallyl was produced, accompanied with cyclopentene, 1-pentene, propylene, 1-butene, butadiene and cyclopentadiene. All of these except diallyl were the same main products as obtained in the pyrolysis of diallyl in excess ethylene. In other words, the same mechanism regulate both the reactions of DAO and those of diallyl in the presence of ethylene.

The kinetics of formation of diallyl from allyl radicals dur-

Table V. Typical Experimental Data on Pyrolysis of DAO in Excess Ethylene

Temperature, °C	450		470		490	
DAO concentration, mole %	2.84	2.71	2.78	2.83	2.71	2.84
Residence time, sec	2.89	4.65	2.75	3.47	2.60	3.43
DAO conversion, %	2.24	3.44	3.84	4.61	8.82	10.12
Product distribution, moles of products per 100 moles of DAO converted						
Propylene	33.9	34.0	30.8	30.9	23.2	21.2
1-Butene	11.0	13.1	12.4	12.2	10.0	8.8
Butadiene	1.3	1.6	2.0	1.9	2.1	2.1
1-Pentene	11.2	10.8	13.0	13.2	12.3	11.5
Cyclopentene	21.6	21.2	22.3	22.4	20.0	18.6
Cyclopentadiene	6.0	5.5	5.0	4.8	3.2	3.0
Diallyl	27.7	30.1	32.9	34.6	41.9	38.9

ing the pyrolysis of DAO has been extensively studied by Golden et al. (22). With the help of their kinetic data, it is now possible to estimate the concentration of allyl radical in the reaction mixture. Hence it may be possible to get a more profound knowledge about the reactions of allyl radicals with ethylene to produce cyclopentene and 1-pentene.

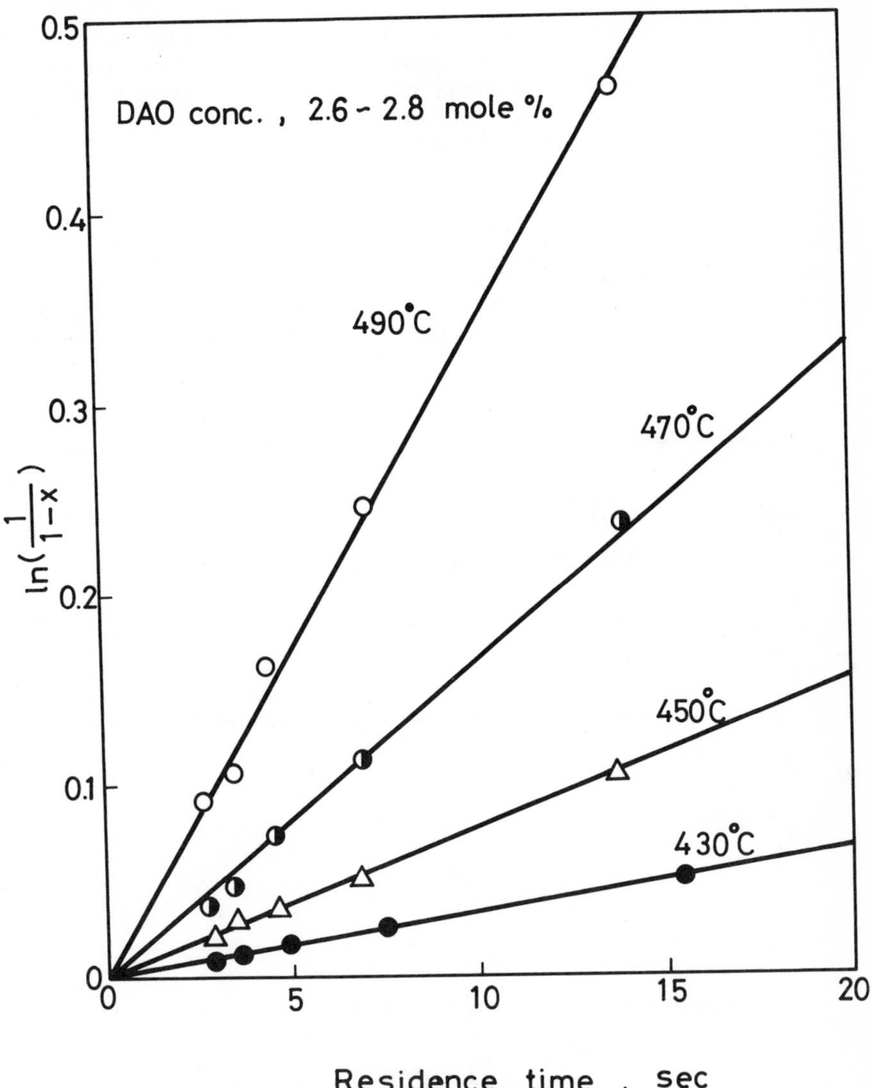

Figure 6. First-order kinetics in decomposition of DAO in excess ethylene

Typical experimental results are listed in Table V. Conversion of DAO, x, was defined based on the stoichiometry of the pyrolysis of DAO to allyl radicals and CO_2 as $x = (1/2)$ (moles of CO_2 formed/moles of DAO fed). Product distribution was calculated on the basis of 100 moles of DAO converted. Fewer products were obtained than those in the reaction of diallyl in excess ethylene. Lower temperatures favored the formation of cyclopentene and 1-butene as compared to those of 1-pentene and butadiene, respectively.

First-order reaction kinetics fit satisfactorily the rate of DAO decomposition as shown in Figure 6. This fact was confirmed also by considering a wide range of DAO concentrations in ethylene at 450°C, as illustrated in Figure 7. From Arrhenius plots, the rate constant of DAO decomposition was obtained as below,

$$k = 10^{10.9} \exp(-43{,}200/RT), \quad sec^{-1}.$$

The correctness of first-order reaction kinetics is in accordance with the fact that the product distribution did not change with the severity of the reaction, as illustrated in Figure 8 for experiments made at 450°C. Consequently, it is possible to obtain a linear relationship between the concentration of products and the residence time of the reaction; the slopes of the straight

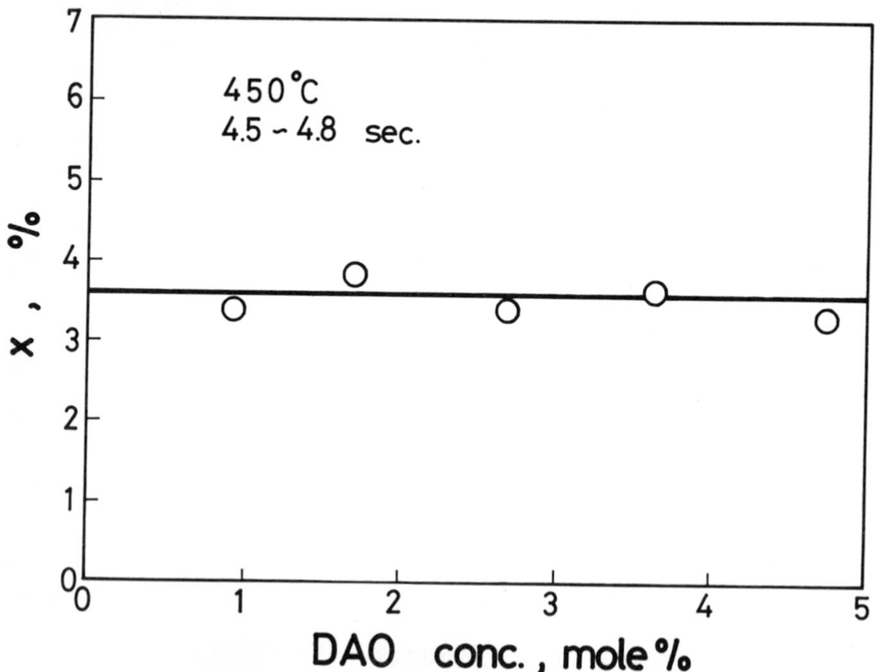

Figure 7. Conversion vs. concentration for decomposition of DAO in excess ethylene

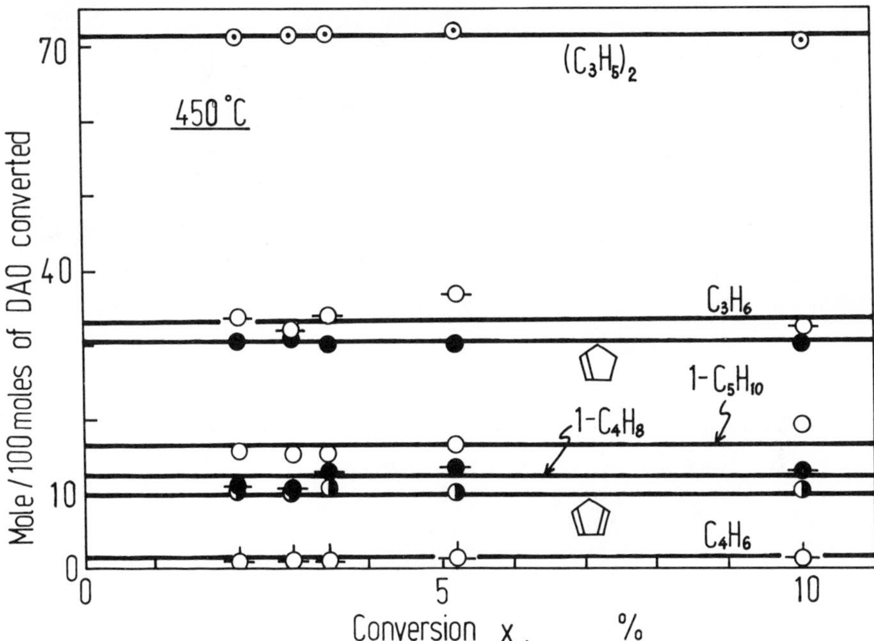

Figure 8. Product distribution vs. decomposition of DAO in excess ethylene

lines obtained in such relationship correspond to the rates of formation of the respective products.

The concentrations of allyl radical in the present experiments were estimated by means of the rate equation below. A second-order rate constant k_0 for recombination of allyl radical to diallyl was approximated to be 5.0×10^9 l/mol·sec by extrapolation from the value reported by Golden et al. (22). The rate of diallyl formation was obtained from the linear relationship between the concentration of diallyl and the residence time as mentioned above. This rate is substituted in the following equation in order to calculate the allyl radical concentration.

$$d[\text{diallyl}]/dt = k_0 [\text{allyl}\cdot]^2$$

The concentration of ethylene and the calculated concentration of allyl radical were employed in the reaction mechanism postulated in the preceeding section to estimate the overall rate constants of the reactions of allyl radical with ethylene to produce both cyclopentene and 1-pentene. Experimental rates of formations of cyclopentene and 1-pentene were of course also used in these calculations.

Experimental evidence indicates a distinct difference between the rate of formation of cyclopentene and that of 1-pentene. Over-

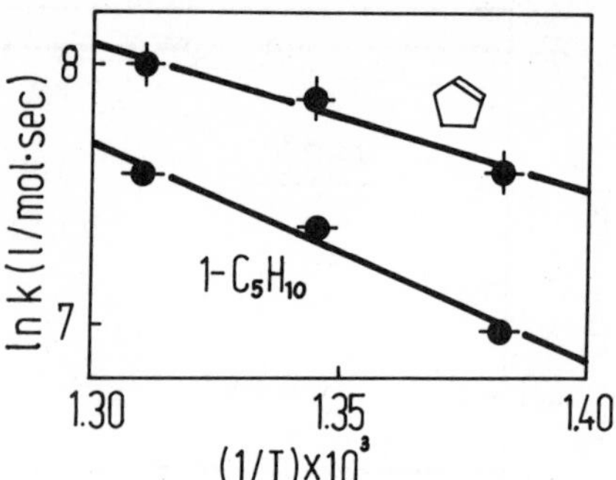

Figure 9. Arrhenius plots for the overall reaction of allyl radical with ethylene to cyclopentene and 1-pentene formation

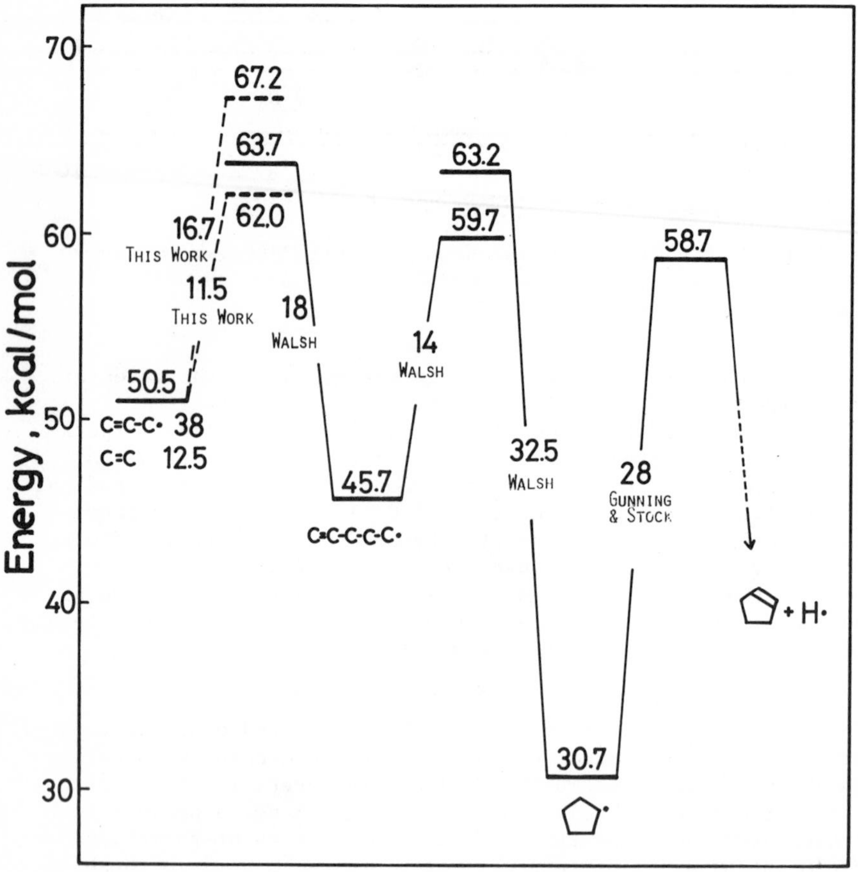

Figure 10. Energy diagram related to reaction of allyl radical with ethylene

$$\text{C=C-C}\cdot{*} + \text{C=C} \underset{-30}{\overset{30}{\rightleftharpoons}} \text{C=C-C-C-C}\cdot{*} \begin{cases} \xrightarrow{33} \text{cyclopentyl}\cdot \xrightarrow{35} \text{cyclopentene} + \text{H}\cdot \\ \xrightarrow[+\text{ C=C}]{34} \text{C=C-C-C-C} + \text{C=C}\cdot \end{cases}$$

all reaction rate constants for the formations of cyclopentene and 1-pentene should be different in the same manner. Moreover, the temperature dependencies of these two overall rate constants are illustrated in Figure 9; the overall activation energy for the formation of cyclopentene and that of 1-pentene were found to be 11.5 and 16.7 kcal/mole, respectively. The difference of these two activation energies, i.e., 5 kcal/mole, corresponds to 4 kcal/mole obtained in Figure 5 for the reaction of diallyl in excess ethylene. Referring to the reaction mechanism postulated above, it can be said that the rate-determining step of the overall reaction is not the addition reaction of allyl radicals to ethylene.

Above all, the important fact is that activation energies of 11.5 and 16.7 kcal/mole were needed for reaction step 33 or 35 and that of 34, respectively. From the heat content of stable allyl radical (38 kcal/mole) and ethylene (12.5 kcal/mole), the energy content of the reaction intermediate which is converted to cyclopentene and 1-pentene is not smaller than 62 - 67 kcal/mole. This high energy content of the intermediate strongly supports the prediction of a hot 4-pentenyl radical in the present study. Step 33 proceeds readily in the case of hot 4-pentenyl radical as stated in the preceeding section. The same situation exists for step 34 because the heat content of vinyl radical is as high as 67 - 69 kcal/mole (32). Thus, the existence of the hot 4-pentenyl radical was reconfirmed during the reaction of an allyl radical with ethylene.

Figure 10 illustrates the heat contents of key radicals such as reported in the literature (23,26,28,33). Activation energies obtained in the present investigation for reactions starting from allyl radical plus ethylene to cyclopentene and 1-pentene formation fit the diagram consistently. This figure strongly supports the conclusions that it is impossible to produce any linear (cyclized) C_5 radicals from stable cyclopentyl (4-pentenyl) radicals, and that, in the case of the reaction of an allyl radical with ethylene, it is possible to produce both cyclized and linear C_5 compounds at the same time.

Abstract

A summary of rate data is given for the systematic study of the formation of cyclic compounds during thermal reactions of olefins or of olefins with butadiene. As a next step in order to investigate cyclization at pyrolysis conditions, the reactions of allyl radicals with olefins were studied kinetically. 1,5-Hexadiene (diallyl) and diallyl oxalate (DAO) were employed as source materials for allyl radical generation. Thermal reactions of pure diallyl and of diallyl or DAO in the presence of excess ethylene were conducted at 430 - 700°C to obtain cyclopentene and other olefins. The rate of cyclopentene formation was relatively large, but C_6 cyclic compounds were not found in appreciable amounts. The reaction mechanism is discussed kinetically for the formation of cyclopentene from an allyl radical and ethylene.

Literature Cited

(1) Sakai, T., Soma, K., Sasaki, Y., Tominaga, H., Kunugi, T., "Advances in Chemistry Series, No.97, Refining Petroleum for Chemicals", p.68, American Chemical Society, Washington, D.C., 1970.

(2) Sakai, T., Nakatani, T., Takahashi, N., Kunugi, T., Ind. Eng. Chem., Fundam., (1972) 11, 529.

(3) Sakai, T., Wada, S., Kunugi, T., Ind. Eng. Chem., Process Design & Develop., (1971) 10, 305.

(4) Kunugi, T., Sakai, T., Soma, K., Sasaki, Y., Ind. Eng. Chem., Fundam., (1969) 8, 374.

(5) Kunugi, T., Sakai, T., Soma, K., Sasaki, Y., Kogyo Kagaku Zasshi, (1968) 71, 689.

(6) Kunugi, T., Sakai, T., Soma, K., Sasaki, Y., Ind. Eng. Chem., Fundam., (1970) 9, 314.

(7) Kunugi, T., Soma, K., Sakai, T., Ind. Eng. Chem., Fundam., (1970) 9, 319.

(8) Nohara, D., Sakai, T., Ind. Eng. Chem., Prod. Res. Develop., (1973) 12, 322.

(9) James, D.G.L., Troughton, G.E., Trans. Faraday Soc., (1966) 62, 145.

(10) McDowell, C.A., Sifniades, S., J. Am. Chem. Soc., (1962) 84, 4606.

(11) James, D.G.L., Kambanis, S.M., Trans. Faraday Soc., (1969) 65, 1350.

(12) Al-Sader, B.H., Crawford, R.J., Can. J. Chem., (1970) 48, 2745.

(13) Vinokurov, D.M., Zabedenii, M.B., Izv. Vysshikh Uchebn. Zavedenii, Khim. i Khim. Tekhnol., (1963) 6, 83.

(14) Hougen, D.A., Watson, K.M., "Chemical Process Principles", p.884, J. Wiley, New York, 1943.

(15) Hill, A.V., Proc. Roy. Soc., Ser. A, (1930) 1279.

(16) Lossing, F.P., Ingold, K.N., Henderson, I.H.S., J. Chem. Phys., (1954) 22, 621.

(17) Ruzicka, D.J., Bryce, W.A., Can. J. Chem., (1960) 38, 827.

(18) Akers, R.J., Throssell, J.J., Trans. Faraday Soc., (1967) 63, 124.

(19) Orchard, S.W., Thrush, B.A., J. Chem. Soc., Chem. Commun., (1973) (1) 14.

(20) Doering, W. von E., Gilbert, J.C., Tetrahedron, (1966) 22, Suppl. 7, 397, footnote 36.

(21) Homer, J.B., Lossing, F.P., Can. J. Chem., (1966) 44, 2211.
(22) Golden, D.M., Gac, N.A., Benson, S.W., J. Am. Chem. Soc., (1969) 91, 2136.
(23) Gordon, A.S., Can. J. Chem., (1965) 43, 570.
(24) Palmer, T.F., Lossing F.P., Can. J. Chem., (1965) 43, 565.
(25) Shibatani, H., Kinoshita, H., Nippon Kagaku Kaishi, (1973) 336.
(26) Walsh, R., Int. J. Chem. Kinetics, (1970) 2, 71.
(27) Watkins, K.W., Olsen, D.K., J. Phys. Chem., (1972) 76, 1089.
(28) Stein, S.E., Rabinovitch, B.S., J. Phys. Chem., (1975) 79, 191.
(29) Rice, F.O., Murphy, M.T., J. Am. Chem. Soc., (1942) 64, 896.
(30) Vanas, D.W., Walters, W.D., J. Am. Chem. Soc., (1948) 70, 4035.
(31) Mackay, G.I., March, R.E., Can. J. Chem., (1970) 48, 913.
(32) Benson, S.W., "Thermochemical Kinetics", pp.214-215, J. Wiley, New York, 1968.
(33) Gunning, H.E., Stock, R.L., Can. J. Chem., (1964) 42, 357.

11

Kinetics of Product Formation in the H_2S-Promoted Pyrolysis of 2-Methyl-2-pentene

DAVID A. HUTCHINGS, KENNETH J. FRECH, and FREDERIC H. HOPPSTOCK

The Goodyear Tire & Rubber Co., Research Division, Akron, Ohio 44316

The discovery, by Ziegler, et. al., (1, 2) of stereospecific polymerization, transformed isoprene into a monomer of great commercial importance. The continuing increase in world demand for synthetic polyisoprene has catalyzed the search for less costly isoprene processes.

Presently, isoprene is being produced via the dehydrogenation of amylenes,(3,4,5) the cracking of propylene dimer, (6,7,8) and by the isolation from refinery C_5 streams.(9,10,11). Isoprene production from isobutylene and formaldehyde appears promising, (12,13,14) with two plants operating successfully in the Soviet Union.

The purpose of this work is to define the mechanistic aspects of propylene dimer cracking in the presence of hydrogen sulfide.

Experimental

All pyrolysis reactions were carried out using the flow reactor described in Figure 1. The reactor was constructed from stainless steel and was adapted to inlet and exit tubing with commercially available stainless steel compression fittings. Liquid hydrocarbon feed was introduced using glass syringes fitted with teflon plungers. The syringes were activated using a commercially available gear pump. When steam was used as a diluent, its rate of introduction was controlled by introducing water in a manner similar to that used for the hydrocarbon. When nitrogen was used as a diluent, the input rate was measured using calibrated rotameters.

The reactor was heated using staged resistance heaters which received their power via proportional band controllers activated by thermocouple sensors

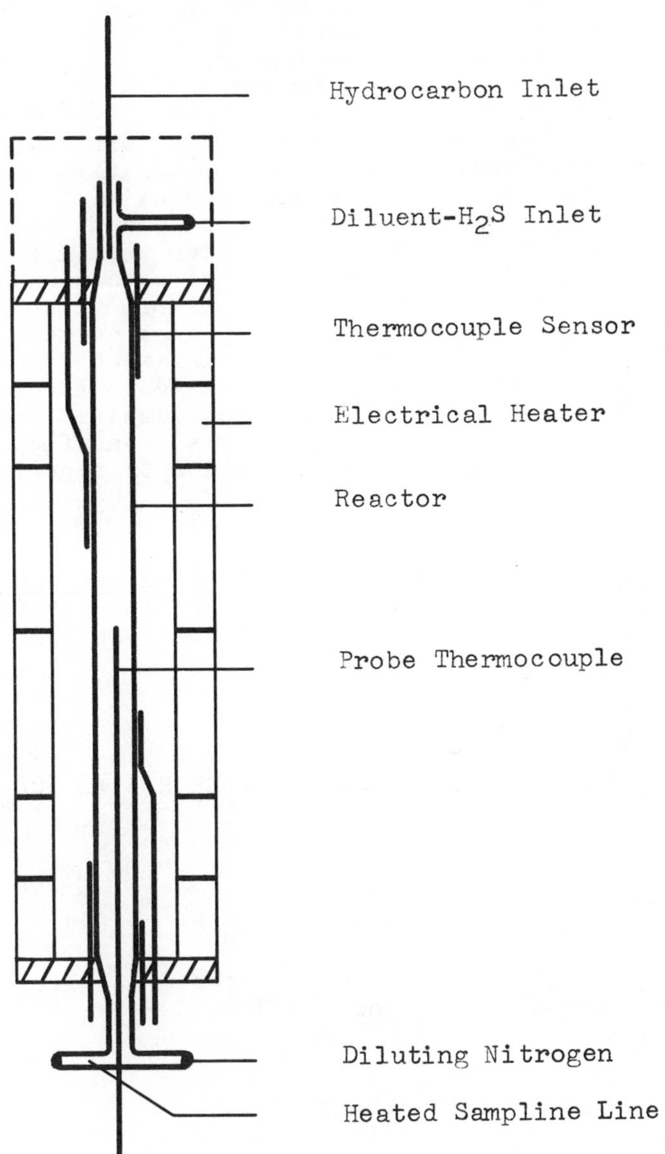

Figure 1. Reactor

mounted on the surface of the tubular reactor. In order to attain an isothermal profile in the reactor, a thermocouple probe technique was used in which the probe was moved along the central longitudinal axis of the reactor. By reading the temperature at fixed positions inside the reactor, a temperature profile along the length of the reactor was obtained. By correlating internal temperature readings with temperature controller settings, an isothermal profile at any desired temperature could be attained within the reactor. Using this technique, a temperature difference along the reactor length of no greater than ± 2°C could be realized with minimal end effects.

During the course of a reaction, products from the reactor passed into the sample loop of a heated gas sampling valve. From the valve, samples were introduced directly into an analytical gas chromatograph.

In this paper, the variables of temperature, pressure (through the use of diluents), and flow rate have been used to gather the kinetic data reported.

Results and Discussion

Thermal Decomposition of 2-Methyl-2-Pentene(2MP2). The following mechanism can be written for the formation of isoprene from the pyrolysis of 2-methyl-2-pentene.

Initiation:

$$CH_3\overset{\overset{CH_3}{|}}{C}{=}CHCH_2CH_3 \xrightarrow{k_1} CH_3\overset{\overset{CH_3}{|}}{C}{=}CHCH_2\cdot + \cdot CH_3$$

Propagation:

$$\cdot CH_3 + CH_3\overset{\overset{CH_3}{|}}{C}{=}CHCH_2CH_3 \xrightarrow{k_p} R\cdot + CH_4$$

$$R\cdot = CH_3\overset{\overset{CH_3}{|}}{C}{=}CH\dot{C}HCH_3 \text{ or } \cdot CH_2\overset{\overset{CH_3}{|}}{C}{=}CHCH_2CH_3$$

$$R\cdot(\alpha) = CH_3\overset{\overset{CH_3}{|}}{C}{=}CH\dot{C}HCH_3$$

$$R\cdot(1-\alpha) = \cdot CH_2\overset{\overset{CH_3}{|}}{C}{=}CHCH_2CH_3$$

Decomposition:

$$\cdot CH_2\overset{\overset{CH_3}{|}}{C}{=}CHCH_2CH_3 \xrightarrow{k_d} CH_2{=}\overset{\overset{CH_3}{|}}{C}{-}CH{=}CH_2 + \cdot CH_3$$

Termination:

$$\cdot CH_3 + R\cdot \xrightarrow{k_t} \text{Termination products.}$$

Based on the preceding mechanism, the following rate law may be derived:

Assumption No. 1

Rate of Initiation = Rate of Termination

$$k_I\left[2MP2\right] = k_T\left[\cdot CH_3\right]\left[R\cdot\right]$$

$$(1)\quad \left[R\cdot\right] = \frac{k_I\left[2MP2\right]}{k_T\left[\cdot CH_3\right]}$$

Assumption No. 2

$$\frac{2\left[\cdot CH_3\right]}{2t} = 0$$

$$\frac{2\left[CH_3\cdot\right]}{2t} = k_d(1-\alpha)\left[R\cdot\right] - k_p\left[\cdot CH_3\right]\left[2MP2\right]$$

$$(2)\quad (1-\alpha)k_d\left[R\cdot\right] = k_p\left[\cdot CH_3\right]\left[2MP2\right]$$

Substituting Eq. 1 in 2:

$$\left[\cdot CH_3\right] = (1-\alpha)^{1/2}\left[\frac{k_d k_I}{k_p k_T}\right]^{1/2}$$

Since the rate of 2MP2 disappearance = $k_p\left[\cdot CH_3\right]\left[2MP2\right]$, the rate law may be defined as follows:

$$\text{Rate of 2MP2 pyrolysis} = (1-\alpha)^{1/2}\left[\frac{k_p k_d k_I}{k_T}\right]^{1/2}\left[2MP2\right]$$

Using thermochemical kinetic techniques (15) described by Benson, the following activation energies have been obtained:

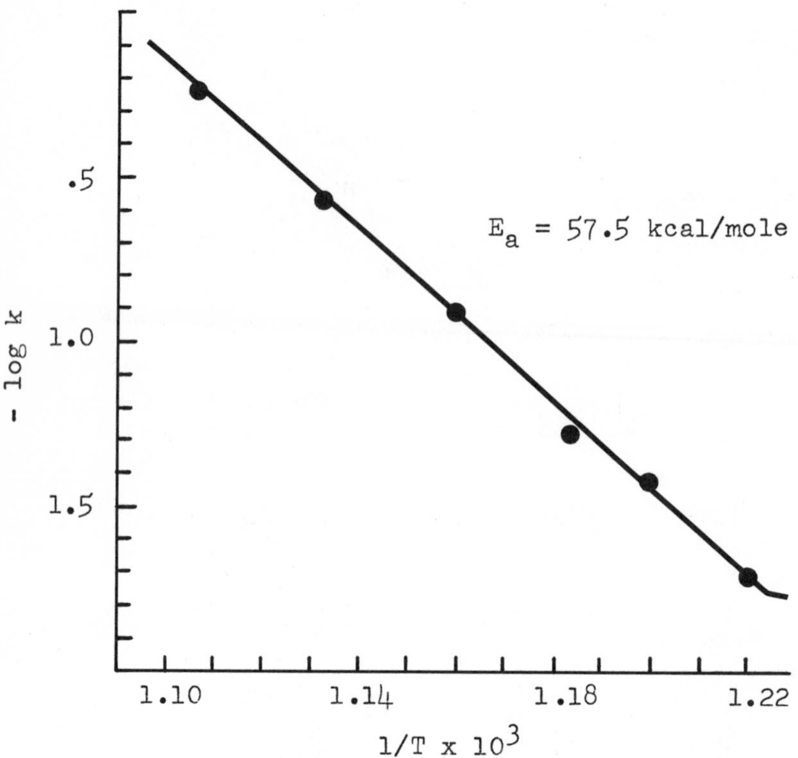

Figure 2. Arrhenius plot of thermal 2MP2 cracking rate data

$$E_d = 38.60$$

$$E_I = 71.17$$

$$E_p = 7.0$$

$$E_T = 0$$

From the rate law:

$$E_{Total} = \frac{E_p + E_d + E_I - E_T}{2} = 58.4 \text{ kcal/mole}$$

An Arrhenius plot of thermal cracking rate data is presented in Figure 2. It will be observed that the activation energy of 57.5 kcal/mole agrees well with the theoretical value.

The following rate has been obtained for the thermal initiation process:

$$k_I = 10^{14.39}\ 10^{71.17/4.6T}$$ (based on thermochemical kinetic techniques)

By comparing rates from Figure 2 with those obtained from the above expression, a kinetic chain length of >600 is determined for the thermal reaction.

<u>H_2S-Promoted Rate Law Derivation.</u> The following mechanism can be written for the formation of isoprene from 2-methyl-2-pentene in the presence of H_2S as a promoter:

<u>The Mechanism of H_2S Promoted 2MP2 Decomposition</u>

Initiation:

$$CH_3C(CH_3)=CHCH_2CH_3 \xrightarrow{k_I} CH_3C(CH_3)=CHCH_2\cdot + \cdot CH_3$$

Radical Transfer:

$$\cdot CH_3 + H_2S \xrightarrow{k_{Trans}} CH_4 + \cdot SH$$

Propagation:

$$HS\cdot + CH_3C(CH_3)=CHCH_2CH_3 \xrightarrow{k_p} H_2S + R\cdot$$

$$R\cdot = CH_3C(CH_3)=CH\dot{C}HCH_3 \text{ or } \cdot CH_2C(CH_3)=CHCH_2CH_3$$

$$R\cdot(\alpha) = CH_3\overset{\overset{\displaystyle CH_3}{|}}{C}{=}CH\dot{C}HCH_3$$

$$R\cdot(1-\alpha) = \cdot CH_2\overset{\overset{\displaystyle CH_3}{|}}{C}{=}CHCH_2CH_3$$

Decomposition:

$$\cdot CH_2\overset{\overset{\displaystyle CH_3}{|}}{C}{=}CHCH_2CH_3 \xrightarrow{k_d} CH_2{=}\overset{\overset{\displaystyle CH_3}{|}}{C}{-}CH{=}CH_2 + \cdot CH_3$$

Termination:

$$\cdot SH + R\cdot \xrightarrow{k_T} \text{Termination products.}$$

Based on the preceding mechanism, the following rate law may be derived:

Assumption No. 1

Rate of Initiation = Rate of Termination

$$(1) \quad k_I\,[2MP2] = k_T\,[\cdot SH][R\cdot]$$

Assumption No. 2

Rate of $\cdot CH_2\overset{\overset{\displaystyle CH_3}{|}}{C}{=}CHCH_2CH_3$ formation =

Rate of $\cdot CH_2\overset{\overset{\displaystyle CH_3}{|}}{C}{=}CHCH_2CH_3$ decomposition.

$$(2) \quad (1-\alpha)k_p\,[HS\cdot][2MP2] = k_d[R\cdot]\,(1-\alpha)$$

Substituting Eq. 1 in 2:

$$[\cdot SH] = \left[\frac{k_d k_I}{k_p k_T}\right]^{1/2}$$

Since the rate of 2MP2 disappearance = $k_p(1-\alpha)\,[HS\cdot][2MP2]$, the rate law may be defined as follows:

Rate of 2MP2 pyrolysis in the presence of H_2S =

$$(1-\alpha)\left[\frac{k_p k_d k_I}{k_T}\right]^{1/2}[2MP2].$$

Activation energies for the above k's have been calculated:

$E_d = 38.6$

$E_I = 71.17$

$E_P = 5.0$

$E_T = 0$

From the rate law:

$$E_{Total} = \frac{E_p + E_d + E_I - E_T}{2} = 57.4 \text{ kcal/mole}$$

An Arrhenius plot of H_2S promoted cracking rate data is presented in Figure 3. The observed activation

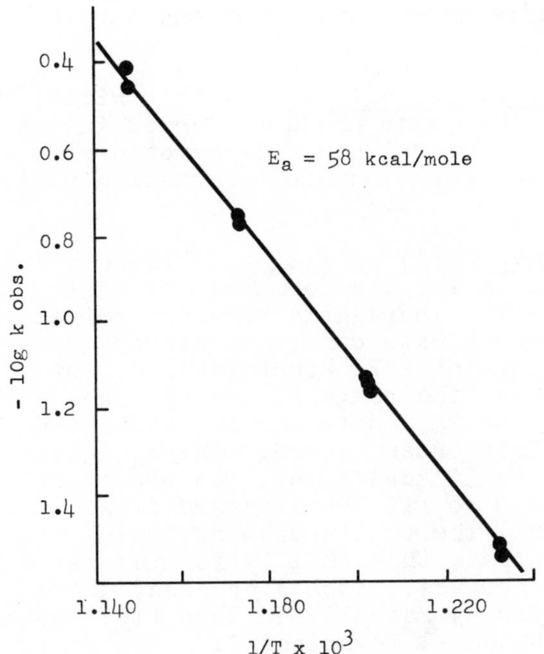

Figure 3. Arrhenius plot of H_2S-promoted 2MP2 cracking rate data

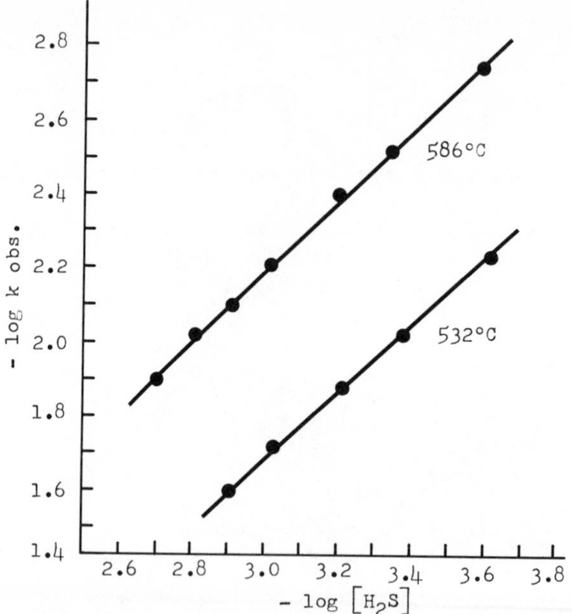

Figure 4. H_2S order data for formation of 2MP1 from 2MP2

energy of 58.0 kcal/mole agrees well with the theoretical value.

By-Product Formation Mechanisms. Table I gives weight percentages of the major products formed in the pyrolysis of 2MP2 under thermal and H_2S-promoted conditions. The mechanisms for by-product formation will now be discussed.

2-Methyl-1-Pentene (2MP1) Formation. Figure 4 represents H_2S order data for the formation of 2MP1 at low conversion levels. These data were determined by calculating first order rate constants for the formation of 2MP1 at a constant 2MP2 concentration, but varying H_2S levels. Over the range of temperatures studied, the slope of the 2MP1 data was one indicating its formation to be first order in H_2S. The E_a calculated from these data is 30 kcal/mole. The order of 2MP1 formation was found to fit the standard first order expression. Thus, the combination of these experimental facts indicate that 2MP1 is forming via a first-order molecular process. Such a process, using HBr, has been postulated by Maecall and Ross (16) for the isomerization of butene-1 to butene-2:

$$CH_3(H)CHC(H)=CH_2 \longrightarrow [CH_3(H)CH-C(H)CH_2 \cdots Br-H] \longrightarrow CH_3(H)C=C(H)CH_3 + H-Br$$

TABLE I

Product Weight Percentages of Total Product for H_2S Promoted and Thermal 2MP2 Cracking

Products	H_2S Promoted	Thermal
	Wt Percent of Total Products	
CH_4	13.0	11.2
C_2	2.55	6.36
C_3	.24	.87
Isobutylene	2.38	5.70
2-Methyl-1-Butene	.41	.596
2-Methyl-2-Butene	3.34	9.30
Isoprene	43.7	45.5
4-Methyl-2-Pentene + piperylenes	13.1	4.40
2-Methyl-1-Pentene	5.01	1.64
Methyl Pentadienes	11.6	8.00
Total Conversion	17.0	14.4

4-Methyl-2-Pentene (4MP2) Formation. A radical chain process may be written for olefin double bond isomerization.

The Mechanism of H_2S-Promoted Isomerization of 2MP2 to 4MP2

Initiation:

$$CH_3\overset{CH_3}{\overset{|}{C}}=CHCH_2CH_3 \xrightarrow{k_I} CH_3\overset{CH_3}{\overset{|}{C}}=CHCH_2\cdot + \cdot CH_3$$

Radical Transfer:

$$\cdot CH_3 + H_2S \xrightarrow{k_{Trans}} CH_4 + HS\cdot$$

Propagation #1:

$$HS\cdot + CH_3\overset{CH_3}{\overset{|}{C}}=CHCH_2CH_3 \xrightarrow{k_{p1}} CH_3\overset{CH_3}{\overset{|}{C}}=CH\dot{C}H-CH_3 + H_2S$$

Propagation #2:

$$CH_3\overset{CH_3}{\overset{|}{\underset{\cdot}{C}}}-CH-CH-CH_3 + H_2S \xrightarrow{k_{p2}} HS\cdot + CH_3\overset{CH_3}{\overset{|}{C}}-CH=CHCH_3$$

Termination:

$$HS\cdot + R\cdot \xrightarrow{k_T} \text{Termination products.}$$

$$R\cdot = (CH_3)_2C{=}CH\dot{C}HCH_3 + \cdot CH_2(CH_3)C{=}CHCH_2CH_3$$

$$R\cdot(\alpha) = CH_3\overset{CH_3}{\overset{|}{C}}{=}CH{-}\dot{C}HCH_3 \equiv CH_3{-}\overset{CH_3}{\overset{|}{C}}{-}CH{-}CHCH_3$$

$$R\cdot(1-\alpha) = \cdot CH_2\overset{CH_3}{\overset{|}{C}}{=}CHCH_2CH_3$$

<u>H_2S-Promoted Rate Law Derivation.</u>

Assumption No. 1

Rate of Initiation = Rate of Termination

$$(1) \quad k_I[2MP2] = k_T[HS\cdot][R\cdot]$$

Assumption No. 2

Rate of $\alpha R\cdot$ formation = Rate of $\alpha R\cdot$ disappearance.

$$(2) \quad k_{p_1}[\cdot SH][2MP2] = k_{p_2}[\alpha][R\cdot][H_2S]$$

Substituting Eq. 1 in 2:

$$[HS\cdot] = \left[\frac{\alpha k_I k_{p_2}}{k_T k_{p_1}}\right]^{1/2} [H_2S]^{1/2}$$

Since the rate of 2MP2 conversion to $\alpha[R\cdot] =$

$$k_{p_1}[\cdot SH][2MP2]$$

The rate law may be defined as follows:

Rate of 2MP2 conversion to 4MP2 =

$$\alpha^{1/2}\left[\frac{k_i k_{p_1} k_{p_2}}{k_T}\right]^{1/2}[H_2S]^{1/2}[2MP2]$$

From thermal kinetic calculations:

$$E_I = 71.17 \text{ kcal/mole}$$

$$E_{p1} = 5 \text{ kcal/mole}$$

$$E_{p2} = 13 \text{ kcal/mole}$$

Then, E_a obs should = 44.5 kcal/mole

Conversion of 2MP2 to 4MP2 data were obtained at several temperatures and H_2S levels (10^{-4} to 10^{-8} molar). Under all conditions, 4MP2 formation data were found to fit a first-order rate equation. Therefore, for any given concentration of 2MP2, the H_2S concentrations were varied from, and the rate of 4MP2 formation determined from single point conversion data. Preliminary data at various H_2S concentrations from such a study are presented in Figure 5. Three points of interest may be noted. First, the log (H_2S) versus log (k) plots show a definite curvature at lower temperatures. At lower H_2S concentrations, the reaction order for H_2S is definitely one half, while at higher H_2S levels the slope approaches one for the lower temperature. The overall trend for the higher H_2S

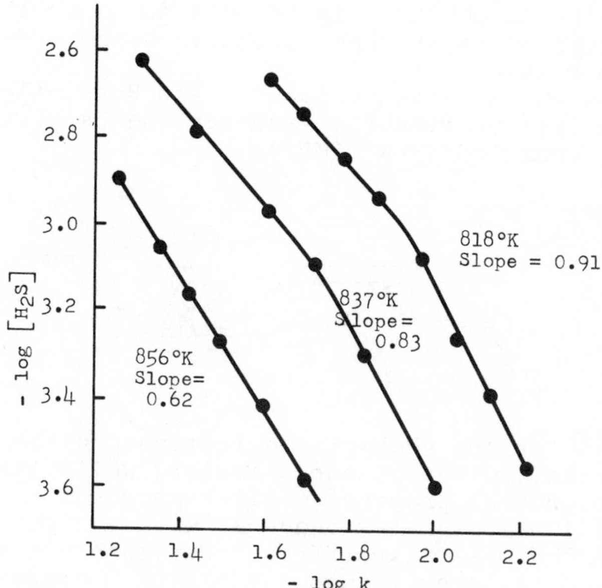

Figure 5. H_2S-catalyzed isomerization of 2MP2 → 4MP2 (order plots)

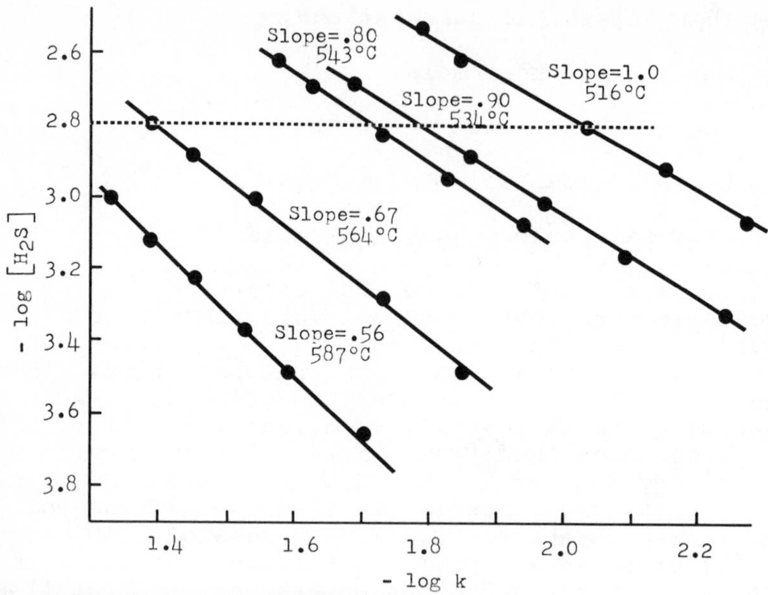

Figure 6. H_2S-catalyzed isomerization of 2MP2 → 4MP2 (order plots)

concentration data is from an order of 1.0 to 0.5 as the reaction temperature is increased. A duplicate set of experiments, Figure 6, confirms the order trend with temperature.

Based on the work of Maccoll and Ross, (16) the following type of reaction sequence can be postulated for 4MP2 formation from 2MP2:

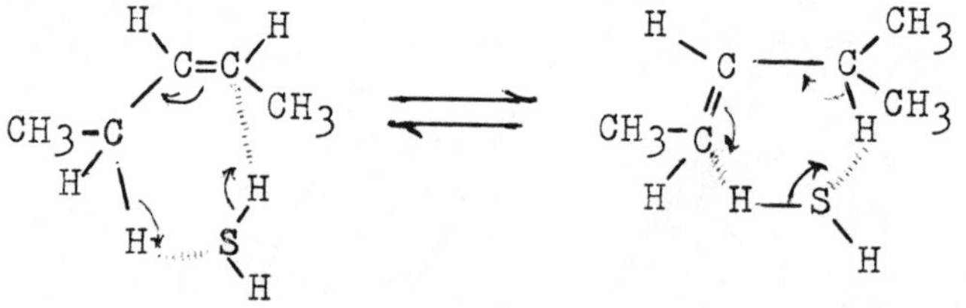

If we consider the above isomerization process to have a low activation energy and A factor, while the radical chain process proposed earlier proceeds via a higher activation energy process, with a higher A factor, then the molecular (first order) process should be observed at lower temperatures with a gradual transition being observed to the radical chain process as the temperature is increased, providing the rates for these processes are comparable. This is the trend which is observed in the data presented in Figure 5. In addition, at lower H_2S levels, the radical chain process would be favored over the molecular process (half order dependency of H_2S for the radical process as opposed to first order for the molecular process). The trend of order from one-half to first-order shown in Figure 5 supports this hypothesis.

The Arrhenius plot presented in Figure 7 was derived from a series of rates (see dotted line in Figure 6) for 2MP2 $\longrightarrow$ 4MP2 isomerization at different temperatures, but at a fixed H_2S level. It will be noted that the E_a calculated from the Arrhenius plot is 36.8 kcal/mole, this being a value between the theoretical E_a's for the molecular and radical chain isomerization processes.

Methyl Pentadienes Formation. Rate data for methyl pentadienes formation in the presence of H_2S are presented in Figure 8. At both temperatures, methyl pentadienes formation is found to be $\sim$ 0.3 order in H_2S. The observed E_a for the formation of methyl pentadienes based on these data is 26.5 kcal/mole. When one considers the possible free radical homogeneous mechanism for methyl pentadienes formation, the following mechanism may be proposed:

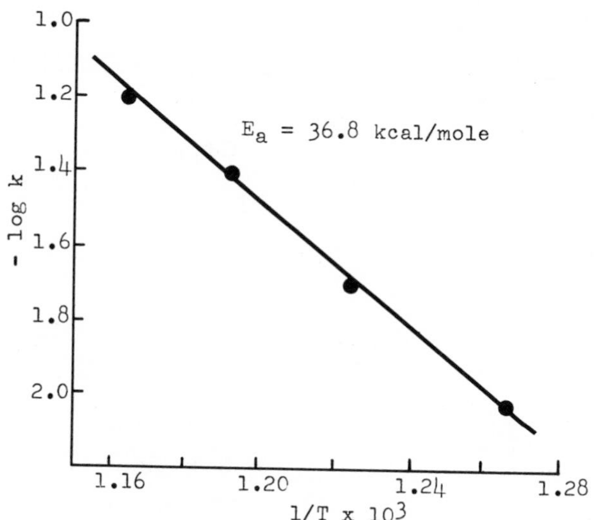

Figure 7. Arrhenius plot of H_2S-catalyzed 2MP2 isomerization to 4MP2 data

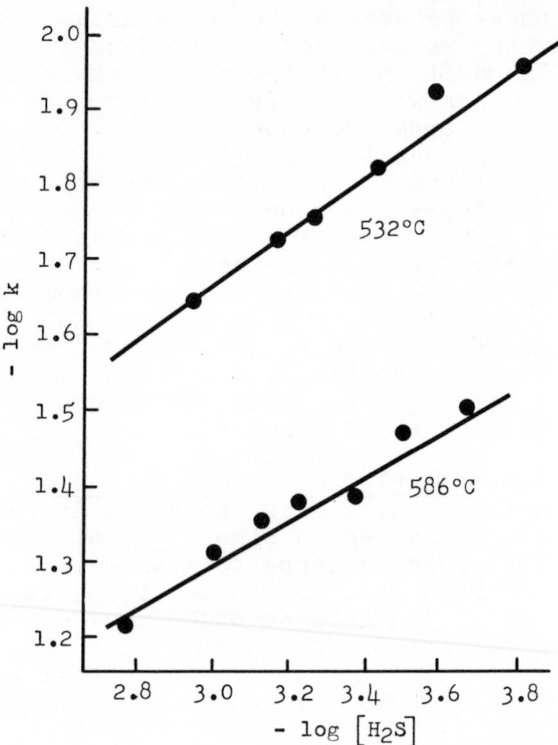

Figure 8. Methyl pentadiene rate data (order in H_2S)

Methyl Pentadienes Formation in the Presence of H_2S

Initiation:

$$\underset{\text{(2MP2)}}{CH_3\overset{\overset{CH_3}{|}}{C}{=}CHCH_2CH_3} \xrightarrow{k_I} CH_3\overset{\overset{CH_3}{|}}{C}{=}CHCH_2\cdot + \cdot CH_3$$

Transfer:

$$H_2S + \cdot CH_3 \xrightarrow{k_{Trans}} HS\cdot + CH_4$$

Propagation:

$$HS\cdot + CH_3\overset{\overset{CH_3}{|}}{C}{=}CHCH_2CH_3 \underset{}{\overset{k_p}{\rightleftharpoons}} H_2S + CH_3\overset{\overset{CH_3}{|}}{C}{=}CH\dot{C}HCH_3$$

Decomposition:

$$CH_3\overset{\overset{CH_3}{|}}{C}{=}CH\dot{C}HCH_3 \xrightarrow{k_d} CH_2{=}\overset{\overset{CH_3}{|}}{C}{-}CH{=}CHCH_3 + H\cdot$$

Termination:

$$\cdot SH + R\cdot \xrightarrow{k_T} \text{Termination products.}$$

$$k_I[2MP2] = k_T[\cdot SH][R\cdot]$$

$$k_p[HS\cdot][2MP2] = k_d[CH_3\overset{CH_3}{\overset{|}{C}}=CH\dot{C}HCH_3] + k_{p\cdot}[H_2S][CH_3\overset{CH_3}{\overset{|}{C}}=CH\dot{C}HCH_3]$$

$$[CH_3\overset{CH_3}{\overset{|}{C}}=CH\dot{C}HCH_3] = \frac{k_p[HS\cdot][2MP2]}{k_p[H_2S] + k_d}$$

Assuming $R\cdot \approx CH_3\overset{CH_3}{\overset{|}{C}}=CH\dot{C}HCH_3$,

$$[HS\cdot] = \frac{k_I^{1/2}}{k_T^{1/2}k_p^{1/2}}\left[k_p[H_2S] + k_d\right]^{1/2}$$

$$R\cdot = \left[\frac{k_Ik_p}{k_T(k_p[H_2S] + k_d)}\right]^{1/2}[2MP2]$$

Rate of Methylpentadiene Formation = $k_d(R\cdot)$

$$\text{Rate} = k_d\left[\frac{k_Ik_p}{k_T(k_p[H_2S] + k_d)}\right]^{1/2}[2MP2]$$

Where $H_2S = 0$

$$\text{Rate} = \left[\frac{k_dk_Ik_p}{k_T}\right]^{1/2}[2MP2]$$

Based on the above mechanism, the reaction leading to methyl pentadienes formation should be first order in 2MP2, have a high activation energy, i.e., $\sim$ 60 kcal/mole, and show a negative 1/2 order dependency on H_2S. Our observed data indicate that a portion of the methyl pentadienes forming reaction may be heterogeneous in origin. If one considers the following low-energy process,

$$CH_3\overset{CH_3}{\overset{|}{C}}{=}CH\dot{C}HCH_3 + H_2S \xrightarrow{k_s} H_2 + \text{(methyl pentadiene)} + \cdot SH$$

the rate equation becomes:

Rate of methyl pentadiene formation =

$k_s [R\cdot][H_2S]$

If $k_p[H_2S] > k_d$,

$$\text{Rate} = k_s \left[\frac{k_I k_p}{k_T k_p}\right]^{1/2} [H_2S]^{1/2} [\text{(2MP2)}]$$

$E_a = \sim 30$ kcal/mole

The Role of H_2S in Modifying the 2MP2 Cracking Product Distributions. Product weight percentages for H_2S-promoted and thermal 2MP2 cracking runs at comparable conversion levels are presented in Table I. It will be observed that for the H_2S-promoted runs, the products derived from H_2S interactions with 2MP2 (2MP1, methyl pentadienes and 4MP2) are major products. In the thermal case, products derived from radical addition reactions predominate. These reactions are as follows:

Methyl Butene Formation:

$$\cdot CH_3 + CH_3\overset{CH_3}{\overset{|}{C}}{=}CHCH_2CH_3 \longrightarrow CH_3\overset{CH_3}{\overset{|}{\dot{C}}}-\underset{CH_3}{\underset{|}{C}}HCH_2CH_3$$

$$CH_3\overset{CH_3}{\overset{|}{\dot{C}}}-\underset{CH_3}{\underset{|}{C}}HCH_2CH_3 \longrightarrow CH_3\overset{CH_3}{\overset{|}{C}}{=}CHCH_3 + CH_3CH_2\cdot$$

Isobutylene Formation:

$$H\cdot + CH_3\overset{CH_3}{\overset{|}{C}}{=}CH\dot{C}H_2CH_3 \longrightarrow CH_3\overset{CH_3}{\overset{|}{\dot{C}}}-CH_2CH_2CH_3$$

$$CH_3\overset{\overset{\textstyle CH}{|}}{\underset{\bullet}{C}}\text{-}CH_2CH_2CH_3 \longrightarrow CH_3\overset{\overset{\textstyle CH_3}{|}}{C}{=}CH_2 + CH_3CH_2\cdot$$

The methyl radicals and hydrogen atoms present in the thermal system can also interact with product isoprene to give rise to additional products. Thus, the role of H_2S appears to be that of a hydrogen and methyl radical scavenger. The resulting •SH radicals, while they can propagate the reaction, cannot participate in the aforementioned radical addition reactions which can lead to unwanted by-products.

Sequential Products. The products whose formation are enhanced by H_2S (2MP1, 4MP2, methyl pentadienes) can react further to produce isoprene at varying efficiencies. These materials, however, also give other by-products:

Compound	Main Reaction
2MP1	isobutylene + ethane or ethylene
4MP2	piperylene + cyclopentadiene
Methyl pentadienes	methyl cyclopentadiene

Conclusions

When 2-methyl-2-pentene is pyrolyzed in the presence of H_2S, the mechanism of isoprene formation is not changed. Hydrogen sulfide serves to modify the by-product mechanism pathways. The major beneficial role of H_2S is as a hydrogen atom scavenger, which decreases the undesirable side reactions resulting from hydrogen atom addition. Aside from these benefits, H_2S works to the detriment of the isoprene formation in several reactions by promoting a number of homogeneous double bond isomerization reactions involving 2-methyl-2-pentene producing olefins which are poorer isoprene precursors.

Acknowledgements

The authors wish to thank Dr. S. W. Benson and Dr. J. E. Taylor for their helpful discussions on the material presented in this paper.

Literature Cited

1. Ziegler, K., Holzcamp, E., Breil, H., and Martin, H., Angew. Chem., 67, 541 (1955); Ziegler, K., Belgium Patent 533,362 (November 16, 1954).
2. Goodrich-Gulf Chemical Co., Belgium Patent 543,292, December 2, 1955.
3. Dow Chemical, British Patent 746,611, March 14, 1956.
4. Pitzer, E. W., U.S. Patent 2,866,790, December 30, 1958.
5. Pitzer, E. W., U.S. Patent 2,866,791, December 30, 1958.
6. Goodyear Tire & Rubber Co., British Patent, July 13, 1960.
7. Goodyear Tire & Rubber Co., British Patent 832,475, August 13, 1960.
8. Goodyear Tire & Rubber Co., British Patent 868,566, May 17, 1961.
9. Kelley, R. T. et al. U.S. Patent 3,012,947, December 12, 1961.
10. Hachmuth, K. H., U.S. Patent 3,038,016, June 5, 1962.
11. King, R. W. et al, U.S. Patent 3,301,915, January 31, 1967.
12. Hellin, M. et al, British Patent 884,804, December 20, 1961.
13. Mitsutani, A., U.S. Patent 3,284,533, November 8, 1966.
14. Mitsutani, A., et al, U.S. Patent 3,284,534, November 8, 1966.
15. Benson, S. W., Thermochemical Kinetics, John Wiley & Sons, New York, 1968.
16. Maecoll, A. and Ross, R. A., J. Am. Chem. Soc., 87, 4997 (1965).

12

Factors Affecting Methyl Pentene Pyrolysis

K. J. FRECH, F. H. HOPPSTOCK, and D. A. HUTCHINGS

The Goodyear Tire & Rubber Co., Research Division, Akron, Ohio 44316

The pyrolysis of propylene dimers to isoprene was investigated as part of the commercialization of the Goodyear Isoprene Process (1,2,3). A significant amount of experimentation has been done to characterize some of the many reaction variables that affect the formation of isoprene during pyrolysis of various methyl pentenes.

The purpose of this paper is to describe, briefly, several interrelated variables that affect the course of methyl pentene pyrolysis. Initially, mechanistic aspects are described for the decomposition of methyl substituted pentenes to isoprene. Next, the high temperature degradation of isoprene and related by-product dienes will be discussed. Finally, the heterogeneous effects associated with reactor metallurgy are detailed.

Experimental

All of the pyrolysis reactions described herein were performed in continuous flow, isothermal reactor systems. One such laboratory reactor system is schematically depicted in Figure 1. Usually, these systems incorporated separate preheat zones, for hydrocarbon and diluent. Catalysts or promoters used were added along with the diluent stream. The heated portions of the preheat and reactor systems were constructed from 316 stainless steel in most cases.

The methyl pentenes were introduced into the apparatus by means of an infusion pump or through calibrated rotameters. Diluents were introduced in the same manner as the liquid hydrocarbons described above. Rotameter systems were employed for the introduction of gases.

In most cases, electrical resistance heaters were used to supply heat to the reactors and preheaters. The

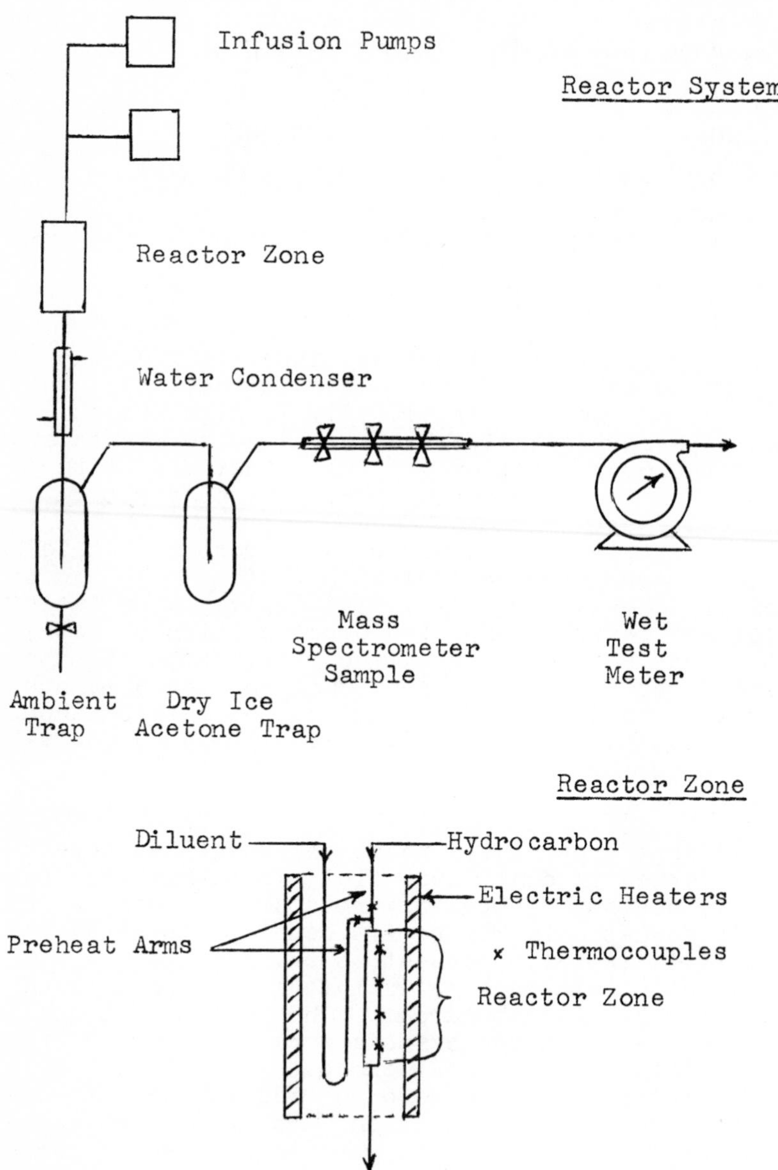

Figure 1. Reactor schematic

heating elements were positioned to obtain optimum isothermal temperature profiles. Temperature was monitored by means of dual-tipped, sheathed and grounded chromel-alumel resistance thermocouples. Temperatures were measured within the gas stream and also externally along the reactor skin. Proportional band temperature controllers were employed to maintain preset conditions.

Sampling techniques included the use of an on-line gas sampling valve, coupled with an analytical gas chromatograph. This procedure was used if the diluent and/or promoter did not interfere with the flame ionization detector system.

An alternative method was the collection of aliquot portions of the total reaction stream using a trap system. For this type of analysis, a water condenser at the exit of the reactor acted as a quench zone. This technique was very effective. The bulk of the reactor effluent stream was collected using an ambient temperature trap. The remainder of the sample was removed via a dry ice-acetone trap or in a mass spectrometer collection tube for later analysis of light gases by low resolution mass spectroscopy. The total volume of these light gases was recorded by passage through a wet test meter. The aqueous and organic portions of the ambient trapped liquids were separated by decantation, the organic portions being combined with the dry ice-acetone trapped material for analysis by gas chromatography.

Factors Affecting Methyl Pentene Pyrolysis

Kinetic studies have shown that methyl pentene pyrolysis occurs via a free radical process having chain lengths in the range of 10^3 (4). As a consequence, product distribution depends on the types of propagating radicals in the system as well as the parent methyl pentene structure.

Product distributions obtained from a number of methyl pentenes are presented in Table I. It will be observed that yields of isoprene and other products vary widely. These results indicate that the following primary factors dictate the yield of isoprene that will be obtained:

1. The ability to form allylic radicals capable of demethylating to form isoprene.
2. The nature of radical addition products to the parent olefin.
3. The steady-state concentration of hydrogen atoms in the system.

TABLE I
Product Distribution-Mole Percent

Component	3MP2	2MP2	2MP1	4MP2
CH_4	23.0	18.3	8.7	4.5
C_2	1.9	4.6	13.7	1.5
C_3	--	.8	2.6	1.4
isobutylene	.3	2.8	10.4	3.5
butadiene	--	.8	.3	--
isoprene	16.7	12.6	7.5	8.5
2-methyl-1 & 2-butene	3.0	3.0	1.9	.8
3-methyl-2-pentene	50.0	--	--	--
2-methyl-2-pentene	--	50.0	.9	5.4
2-methyl-1-pentene	--	--	50.0	1.7
4-methyl-2-pentene	--	--	--	50.0
methyl pentadiene	1.5	2.0	.5	5.2
others	2.2	3.5	2.0	6.3
piperylene	--	--	--	11.0

Conditions Employed:
Temperature: 675°C
Pressure: atmospheric
Residence time: 0.2 seconds
Diluent to hydrocarbon ratio: 6/1

The major initiation reactions for the four methyl pentenes studies are listed below:

I. 3-methyl-2-pentene (cis and trans 3MP2)

$$CH_3\text{-}CH{=}\underset{}{\overset{CH_3}{\overset{|}{C}}}\text{-}CH_2\text{-}CH_3 \xrightarrow{k_i} CH_3\text{-}CH{=}\overset{CH_3}{\overset{|}{C}}\text{-}\underset{\bullet}{C}H_2 + \bullet CH_3$$

II. 2-methyl-2-pentene (2MP2)

$$CH_3\text{-}\overset{CH_3}{\overset{|}{C}}{=}CH\text{-}CH_2\text{-}CH_3 \xrightarrow{k_i} CH_3\text{-}\overset{CH_3}{\overset{|}{C}}{=}CH\text{-}\dot{C}H_2 + \bullet CH_3$$

III. 2-methyl-1-pentene (2MP1)

$$CH_2{=}\overset{CH_3}{\overset{|}{C}}\text{-}CH_2\text{-}CH_2\text{-}CH_3 \xrightarrow{k_i} CH_2{=}\overset{CH_3}{\overset{|}{C}}\text{-}\underset{\bullet}{C}H_2 + \bullet CH_2\text{-}CH_3$$

IV. 4-methyl-2-pentene (cis and trans 4MP2)

$$CH_3\text{-}CH{=}CH\text{-}\overset{CH_3}{\overset{|}{C}H}\text{-}CH_3 \xrightarrow{k_i} CH_3\text{-}CH{=}CH\text{-}\dot{C}H\text{-}CH_3 + \bullet CH_3$$

For all the above examples, initiation involves the homolytic cleavage of the allylic carbon-carbon bond in the molecule. This reaction has an activation energy of approximately 70 kcal/mole and an A factor in the neighborhood of $10^{-14.5} \pm .5$.

In all of the systems described above, the major propagation step involves methyl or ethyl attack on allylic hydrogen positions. For example:

$$\cdot CH_3 + CH_3\text{-}CH{=}\overset{\overset{\textstyle CH_3}{|}}{C}\text{-}CH_2\text{-}CH_3 \xrightarrow{k_p} R\cdot + CH_4$$

The more important allylic radicals for each of the systems investigated are listed below:

I. 3MP2

$$\cdot R_{I_1} = \cdot CH_2\text{-}CH{=}\overset{\overset{\textstyle CH_3}{|}}{C}\text{-}CH_2\text{-}CH_3$$

$$\cdot R_{I_{11}} = CH_3\text{-}CH{=}\overset{\overset{\textstyle CH_3}{|}}{C}\text{-}\dot{C}H\text{-}CH_3$$

$$\cdot R_{I_{111}} = CH_3\text{-}CH{=}\overset{\overset{\textstyle \cdot CH_2}{|}}{C}\text{-}CH_2\text{-}CH_3$$

II. 2MP2

$$\cdot R_{II_1} = \cdot CH_2\text{-}\overset{\overset{\textstyle CH_3}{|}}{C}{=}CH\text{-}CH_2\text{-}CH_3$$

$$\cdot R_{II_{11}} = CH_3\text{-}\overset{\overset{\textstyle CH_3}{|}}{C}{=}CH\text{-}\dot{C}H\text{-}CH_3$$

III. 2MP1

$$\cdot R_{III_1} = CH_2{=}\overset{\overset{\textstyle CH_3}{|}}{C}\text{-}\underset{\cdot}{C}H\text{-}CH_2\text{-}CH_3$$

$$\cdot R_{III_2} = CH_2{=}\overset{\overset{\textstyle \cdot CH_2}{|}}{C}\text{-}CH_2\text{-}CH_2\text{-}CH_3$$

IV. 4MP2

$$\cdot R_{IV_1} = \cdot CH_2\text{-}CH{=}CH\text{-}\overset{\overset{\textstyle CH_3}{|}}{C}H\text{-}CH_3$$

$$\cdot R_{IV_{11}} = CH_3\text{-}CH{=}CH\text{-}\underset{\cdot}{\overset{\overset{\textstyle CH_3}{|}}{C}}\text{-}CH_3$$

For the methyl pentenyl radical systems, the most favorable path to diolefin formation involves allylic carbon-carbon bond scission:

$$\cdot CH_2\text{-}\overset{\overset{\textstyle R}{|}}{C}{=}\overset{\overset{\textstyle R}{|}}{C}\text{-}CH_2\text{-}CH_3 \longrightarrow CH_2{=}\overset{\overset{\textstyle R}{|}}{C}\text{-}\overset{\overset{\textstyle R}{|}}{C}{=}CH_2 + \cdot CH_3$$

$R = CH_3$ or H

Some of the radicals formed in the propagation sequence are incapable of allylic carbon-carbon bond cleavage to yield diolefinic product. For these systems, decomposition is via the much slower carbon-hydrogen scission or disproportionation reactions.

The radical abstraction pathway is depicted as follows:

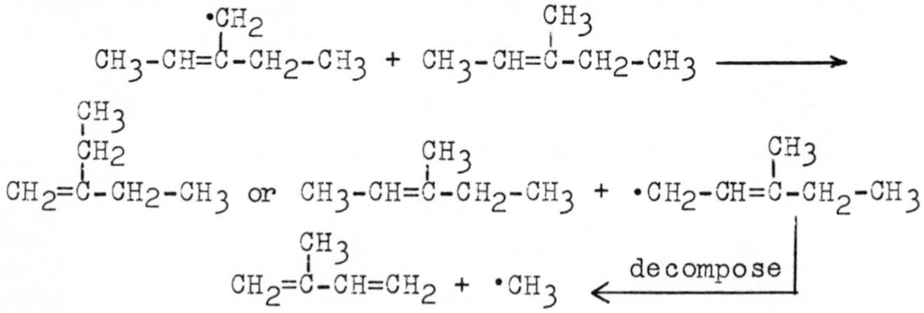

$$CH_3\text{-}CH{=}\underset{}{\overset{\cdot CH_2}{\overset{|}{C}}}\text{-}CH_2\text{-}CH_3 + CH_3\text{-}CH{=}\overset{CH_3}{\overset{|}{C}}\text{-}CH_2\text{-}CH_3 \longrightarrow$$

$$CH_2{=}\overset{\overset{CH_3}{|}}{\overset{CH_2}{\overset{|}{C}}}\text{-}CH_2\text{-}CH_3 \text{ or } CH_3\text{-}CH{=}\overset{CH_3}{\overset{|}{C}}\text{-}CH_2\text{-}CH_3 + \cdot CH_2\text{-}CH{=}\overset{CH_3}{\overset{|}{C}}\text{-}CH_2\text{-}CH_3$$

$$CH_2{=}\overset{CH_3}{\overset{|}{C}}\text{-}CH{=}CH_2 + \cdot CH_3 \xleftarrow{\text{decompose}}$$

The disproportionation sequence is as follows:

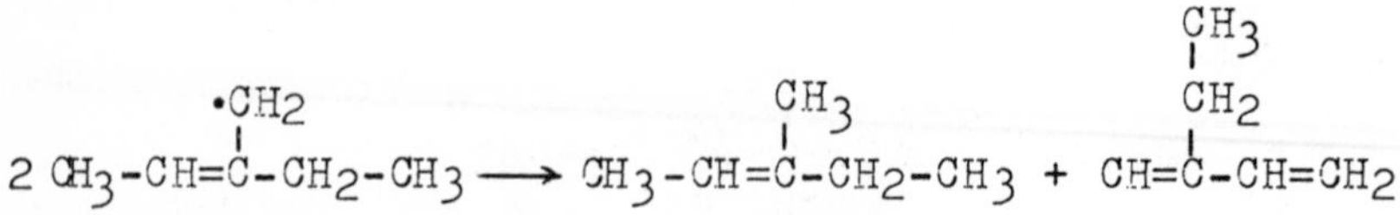

$$2\,CH_3\text{-}CH{=}\overset{\cdot CH_2}{\overset{|}{C}}\text{-}CH_2\text{-}CH_3 \longrightarrow CH_3\text{-}CH{=}\overset{CH_3}{\overset{|}{C}}\text{-}CH_2\text{-}CH_3 + CH{=}\overset{\overset{CH_3}{|}}{\overset{CH_2}{\overset{|}{C}}}\text{-}CH{=}CH_2$$

Reactions of this type are the main pathway for double bond isomerization.

Initial inspection of the allylic radicals formed from 3-methyl-2-pentene, 2-methyl-2-pentene and 2-methyl-1-pentene indicate that these systems are capable of decomposition to isoprene while cis and trans 4-methyl-2-pentene are not. With the latter olefins, cis and trans-piperylene are the major products. The isoprene yield differences between cis and trans 3-methyl-2-pentene, 2-methyl-2-pentene, and 2-methyl-1-pentene can be explained on the basis of radical addition products.

In the case of cis and trans 3-methyl-2-pentene, hydrogen addition leading to the tertiary 3-methyl pentenyl radical is favored:

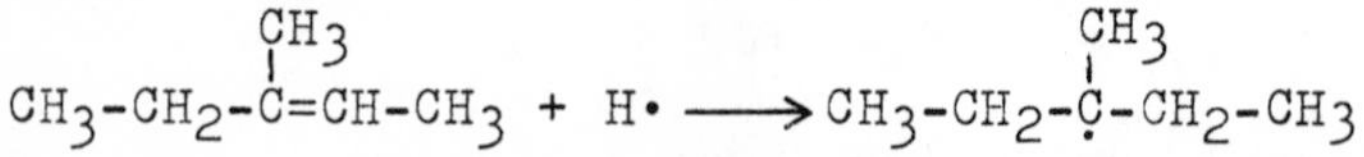

$$CH_3\text{-}CH_2\text{-}\overset{CH_3}{\overset{|}{C}}{=}CH\text{-}CH_3 + H\cdot \longrightarrow CH_3\text{-}CH_2\text{-}\overset{CH_3}{\overset{|}{\dot{C}}}\text{-}CH_2\text{-}CH_3$$

This radical decomposes to give 2-methyl-1-butene plus a methyl radical. The resulting methyl radical can then enter into the normal chain propagation sequence, or add to the parent molecule. As in the case of hydrogen atom addition, methyl radical addition gives predominantly tertiary radical adducts, 2,3-dimethyl-

3-pentyl radicals. Again, only methyl radicals can be formed in this decomposition sequence.

In the case of 2-methyl-2-pentene, hydrogen addition again gives predominantly the tertiary methyl pentenyl radical.

$$CH_3\text{-}\overset{\displaystyle CH_3}{\overset{|}{C}}\text{=}CH\text{-}CH_2\text{-}CH_3 + H\cdot \longrightarrow CH_3\text{-}\overset{\displaystyle CH_3}{\overset{|}{\dot{C}}}\text{-}CH_2\text{-}CH_2\text{-}CH_3$$

Unlike the 3-methyl-3-pentenyl system, the 2-methyl-2-pentenyl radical gives rise to an ethyl rather than a methyl radical.

$$CH_3\text{-}\overset{\displaystyle CH_3}{\overset{|}{\dot{C}}}\text{-}CH_2\text{-}CH_2\text{-}CH_3 \longrightarrow CH_3\text{-}\overset{\displaystyle CH_3}{\overset{|}{C}}\text{=}CH_2 + \cdot CH_2\text{-}CH_3$$

The ethyl radical has two principal options in this system. It can abstract a hydrogen atom, or decompose to ethylene plus a hydrogen atom. Thus, ethyl radicals are a prime source of hydrogen atoms which can participate in a long chain decomposition process in the 2-methyl-2-pentene system:

$$\cdot CH_2CH_3 \longrightarrow CH_2\text{=}CH_2 + H\cdot$$

$$H\cdot + CH_3\text{-}\overset{\displaystyle CH_3}{\overset{|}{C}}\text{=}CH\text{-}CH_2\text{-}CH_3 \longrightarrow CH_3\text{-}\overset{\displaystyle CH_3}{\overset{|}{\dot{C}}}\text{-}CH_2\text{-}CH_2\text{-}CH_3$$

$$CH_3\text{-}\overset{\displaystyle CH_3}{\overset{|}{\dot{C}}}\text{-}CH_2\text{-}CH_2\text{-}CH_3 \longrightarrow CH_3\text{-}\overset{\displaystyle CH_3}{\overset{|}{C}}\text{=}CH\text{-}CH_3 + \cdot CH_2\text{-}CH_3$$

Methyl radical addition to 2-methyl-2-pentene leads to a tertiary radical system which is capable of decomposing to give additional ethyl radicals which also contribute to undesirable decomposition reactions:

$$CH_3\text{-}\overset{\displaystyle CH_3}{\overset{|}{C}}\text{=}CH\text{-}CH_2\text{-}CH_3 + \cdot CH_3 \longrightarrow CH_3\text{-}\overset{\displaystyle H_3C}{\overset{|}{\dot{C}}}\text{—}\overset{\displaystyle CH_3}{\overset{|}{C}H}\text{-}CH_2\text{-}CH_3$$

$$CH_3\text{-}\overset{\displaystyle H_3C}{\overset{|}{\dot{C}}}\text{-}\overset{\displaystyle CH_3}{\overset{|}{C}H}\text{-}CH_2\text{-}CH_3 \longrightarrow CH_3\text{-}\overset{\displaystyle CH_3}{\overset{|}{C}}\text{=}CH\text{-}CH_3 + \cdot CH_2\text{-}CH_3$$

Product distribution data indicate that the major difference between cis and trans 3-methyl-2-pentene and 2-methyl-2-pentene lies in the fact that 2-methyl-2-pentene can more readily generate ethyl radicals and thus produce a higher hydrogen atom flux.

Inspection of 2-methyl-1-pentene pyrolysis products indicates that fragmentations resulting from hydrogen atom addition are increased over those observed for the 2-methyl-2-pentene system. These results reflect the difference in ease of hydrogen addition to a primary versus a secondary olefinic bond.

For 2-methyl-1-pentene:

$$H\cdot + CH_2{=}\overset{CH_3}{\overset{|}{C}}{-}CH_2{-}CH_2{-}CH_3 \longrightarrow CH_3{-}\overset{CH_3}{\overset{|}{\dot{C}}}{-}CH_2{-}CH_2{-}CH_3$$

For 2-methyl-2-pentene:

$$H\cdot + CH_3{-}\overset{CH_3}{\overset{|}{C}}{=}CH{-}CH_2{-}CH_3 \longrightarrow CH_3{-}\overset{CH_3}{\overset{|}{\dot{C}}}{-}CH_2{-}CH_2{-}CH_3$$

The more rapid rate of hydrogen addition to the 2-methyl-1-pentene system, coupled with the slower rate of methyl radical hydrogen abstraction leading to the 2-methyl-1-pentenyl-3 radical,

$$CH_2{=}\overset{CH_3}{\overset{|}{C}}{-}CH_2{-}CH_2{-}CH_3 + \cdot CH_3 \longrightarrow CH_2{=}\overset{CH_3}{\overset{|}{C}}{-}\dot{C}H{-}CH_2{-}CH_3 + CH_4$$

decreases the isoprene yield from the decomposition pathway promoted with a hydrogen atom.

A similar argument can be put forward for isoprene yield effects arising from methyl radical addition to 2-methyl-1-pentene.

Pyrolysis of cis and trans-4-methyl-2-pentene does not yield isoprene. The 4-methyl-2-pentenyl radical decomposes primarily to cis- and trans-1,3-pentadiene:

$$\cdot CH_2{-}CH{=}CH{-}\overset{CH_3}{\overset{|}{C}}H{-}CH_3 \longrightarrow CH_2{=}CH{-}CH{=}CH{-}CH_3 + \cdot CH_3$$

Cis- and trans-4-methyl-2-pentene can contribute to the production of isoprene by first undergoing a high temperature isomerization to the appropriate olefin isomers.

Degradation of Isoprene

Since isoprene was the principal desired product from methyl pentene pyrolysis it was essential to determine its stability under reaction conditions. Isoprene degradation proceeds via second-order rate processes. Isoprene dimer and related compounds formed during the pyrolysis are dependent on the isoprene partial pressure in the reaction system.

$$\text{Rate Isoprene Degradation} = k\left[CH_2{=}\overset{CH_3}{\overset{|}{C}}{-}CH{=}CH_2\right]^2$$

The second order integrated equation is:

$$\frac{1}{a-x} - \frac{1}{a} = kt$$

a = initial isoprene concentration
a-x = concentration of isoprene at time t
t = isoprene residence time.

The rate constants calculated using the integrated second-order rate equation over a wide concentration range are presented in Table II. This agreement can be taken as a positive test for second-order dependency. Figure 2 is an Arrhenius plot derived from second-order rate constants calculated on isoprene conversion data obtained over a wide concentration range.

The activation energy and A factor derived from **Figure 2 indicate:** $k = 10^9 \cdot e^{-31/\theta}$. These values are somewhat higher than those reported in the literature for the dimerization of butadiene (5):

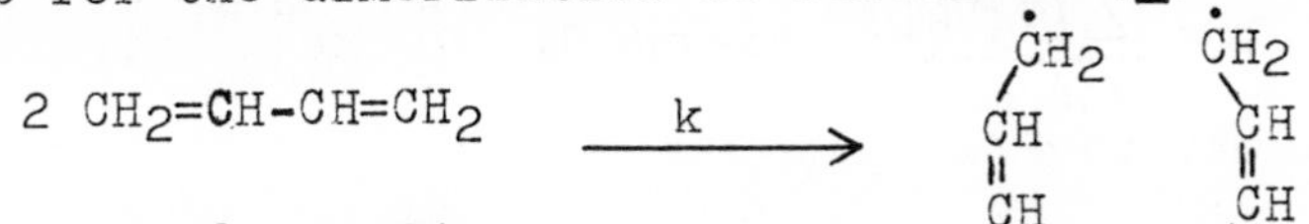

$$2\ CH_2{=}CH{-}CH{=}CH_2 \xrightarrow{k} \dot{C}H_2{-}CH{=}CH{-}CH_2{-}CH_2{-}CH{=}CH{-}\dot{C}H_2$$

$$k = 10^8 \cdot e^{-26/\theta}$$

We consider the isoprene dimer intermediate to be similar, i.e.:

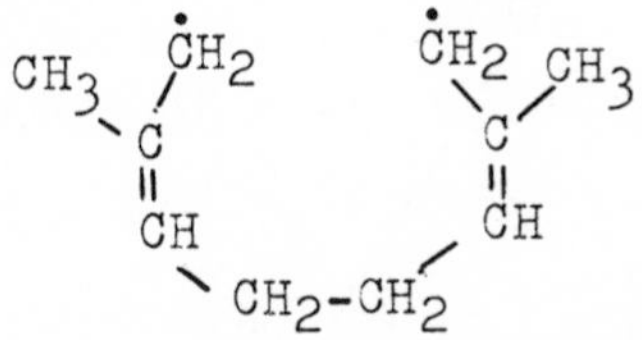

$$\dot{C}H_2{-}C(CH_3){=}CH{-}CH_2{-}CH_2{-}CH{=}C(CH_3){-}\dot{C}H_2$$

The introduction of H_2S to the system supports this hypothesis. Increased methyl butene levels are observed with a reduction in the amount of high molecular weight products (Table III). These data are interpreted in the following manner:

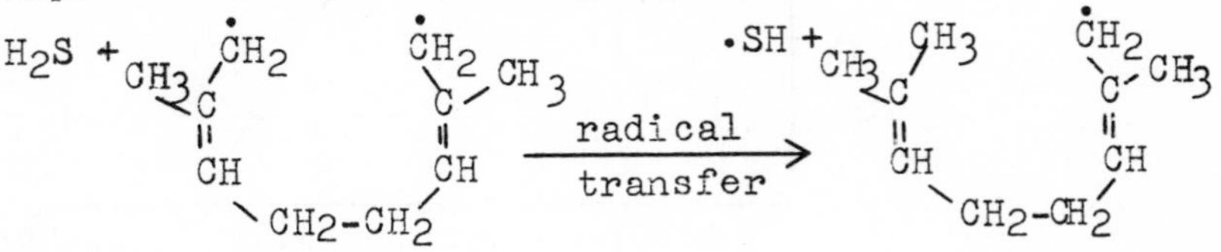

$$H_2S + \dot{C}H_2{-}C(CH_3){=}CH{-}CH_2{-}CH_2{-}CH{=}C(CH_3){-}\dot{C}H_2 \xrightarrow[\text{transfer}]{\text{radical}} \cdot SH + CH_3{-}C(CH_3){=}CH{-}CH_2{-}CH_2{-}CH{=}C(CH_3){-}\dot{C}H_2$$

TABLE II. Isoprene Degradation-Order Determination

Nitrogen/Isoprene Mole Ratio	10:1	9:1	5:1	3:1
1-x	0.909	0.89	0.824	0.724
$[C_5H_8]_0$	12.4×10^{-4}	13.6×10^{-4}	22.7×10^{-4}	$34. \times 10^{-4}$
$[C_5H_8]_1$	11.3 "	12.1 "	18.7 "	26.3 "
$\frac{\Delta[C_5H_8]}{\Delta t = 150°C}$	1.1 "	1.5 "	4.0 "	7.7 "
$k = \frac{1}{C_1} - \frac{1}{C_0}$	82.2	96.5	103.0	97.1

k average = 95

(plot of log $\frac{\Delta[C_5H_8]}{\Delta t \text{ sec}}$ versus log $[C_5H_8]_0$ gives slope equal to 2).

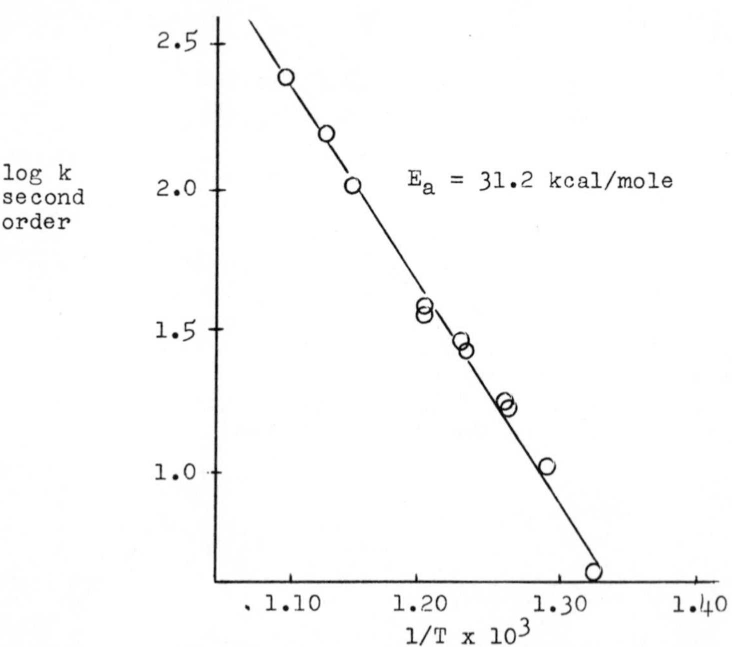

Figure 2. Arrhenius plot of isoprene second-order degradation rate constants

$$CH_3\text{-}C(CH_3){=}CH\text{-}CH_2\text{-}CH_2\text{-}CH{=}C(CH_3)\text{-}\dot{C}H_2 \xrightarrow{\text{decompose}} (CH_3)_2C{=}CH\text{-}\dot{C}H_2 + CH_2{=}C(CH_3)\text{-}CH{=}CH_2$$

$$(CH_3)_2C{=}CH\text{-}\dot{C}H_2 + H_2S \xrightarrow[\text{transfer}]{\text{radical}} (CH_3)_2C{=}CH\text{-}CH_3 + \cdot SH$$

More high molecular weight material is observed than would be predicted from isoprene degradation alone when 2-methyl-2-pentene is pyrolyzed.

Degradation studies with the other diolefins (methyl pentadienes and piperylenes) produced during methyl pentene pyrolysis show only very small quantities of high molecular weight materials under similar test conditions (Figure 3). At temperatures and partial pressures giving 5% heavy material from isoprene, the piperylenes and methyl pentadienes form just 0.5% higher molecular weight material.

TABLE III

Isoprene Degradation Product Variation as a Function of H_2S Concentration

Percent H_2S	0	13.3
Methane	6.8	9.5
C_2	15.7	15.7
C_3	11.8	8.9
Isobutylene	3.5	5.3
Butadiene	3.9	2.9
2-methyl-1-butene	4.0	12.6
2-methyl-2-butene	10.9	32.2
2-methyl pentadiene	22.9	2.7
methyl cyclopentadiene	10.8	3.7
Efficiency to lights	7.2	17.5
Efficiency to heavies	92.8	82.5
% Conversion	19.8	19.4

Conditions Employed:
Temperature: 650°C
Pressure: Atmospheric
Residence Time: 0.2 seconds

An explanation for this behavior may relate to the location of the methyl group on the allylic radical. The methyl group stabilizes the di-radical

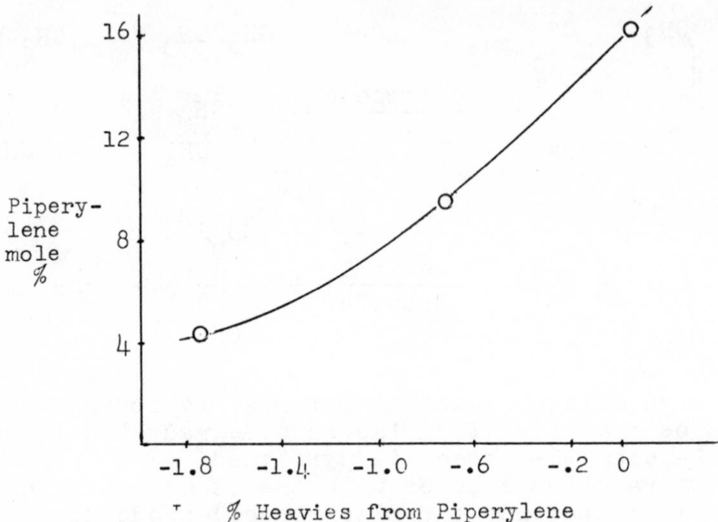

Figure 3. Variation of heavy material with varying piperylene concentration. Conditions: 600°C reactor temperature, atm press., 1.0 sec residence time.

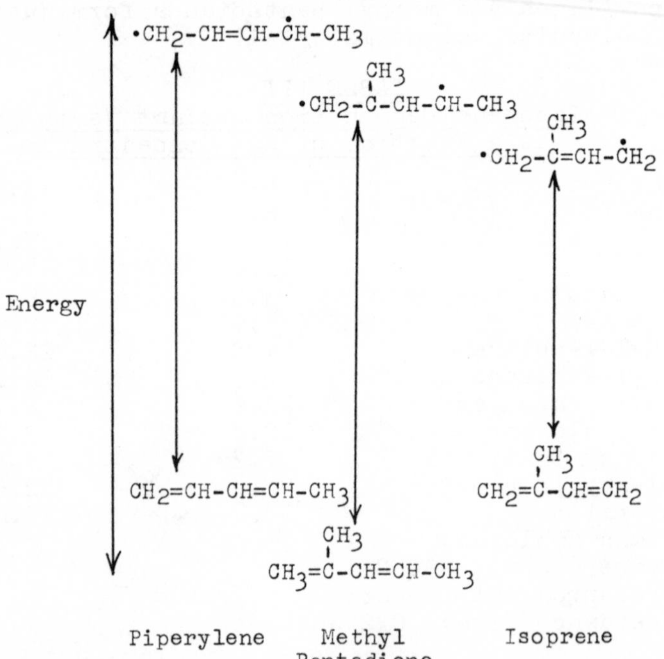

Figure 4. Relative energies: diolefin ground states. Excited diradical states.

structure for isoprene, while in the case of piperylenes and methyl pentadienes the methyl group stabilizes the ground state (Figure 4).

Piperylene forms 0.5% heavy material when reacted alone at 600°C with nitrogen as a diluent. Methyl pentadiene forms 0.7% heavy material and isoprene forms 4.8% heavy material under these conditions. When piperylene or methyl pentadiene is mixed with isoprene in varied ratios, under similar conditions, the amount of heavy material formed is greater than would be expected (Figure 5).

Isoprene appears to have the ability to form heavies by attack on itself and other dienes. These differences can be attributed to steric hindrance by the methyl group. The piperylenes and methyl pentadienes have an alternate reaction pathway--that of cyclization to cyclopentene and 1,3-cyclopentadiene and the methyl cyclopentene and methyl cyclopentadienes, respectively.

In the presence of H_2S, the product distribution changes rather dramatically (Figure 6). Table IV shows the product breakdown of piperylene degradation thermally and in the presence of H_2S. For the unpromoted case, the two major reactions are the cyclization to cyclopentadiene and the decomposition to butadiene (Figure 7).

TABLE IV

Piperylene Reaction Products

(Wt % Efficiencies)

Temperature, °C	650	650
% H_2S based on piperylene	0	100
% Conversion	21.8	15.2
CH_4	6.2	2.0
C_2	4.6	1.0
C_3	2.0	1.4
C_4	2.6	2.8
butadiene	28.8	4.3
pentane	.5	1.2
pentene-1	.6	4.0
pentene-2	2.0	10.2
cyclopentene	6.2	49.8
cyclopentadiene	26.7	20.4
benzene	3.4	1.3
toluene	2.0	.8

Operating Conditions:

Temperature: 650°C

Pressure: Atmospheric

Residence Time: 0.2 seconds

Diluent/hydrocarbon mole ratio: 3/1

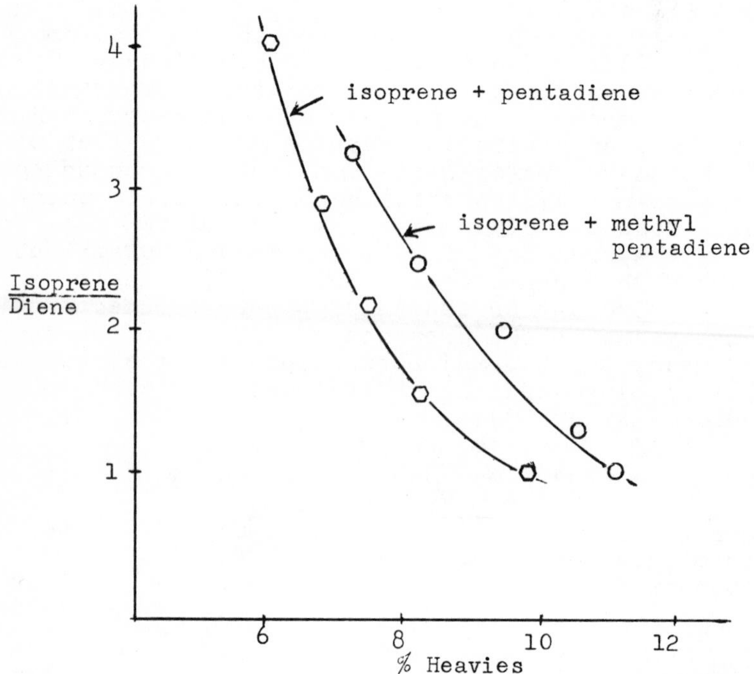

Pentadiene alone - 0.5% heavies
Methyl Pentadiene alone - 0.7% heavies
Isoprene alone - 4.8% heavies

Figure 5. Variation in heavies as a function of diene concentration in a system with a fixed isoprene level. Conditions: N_2/isoprene mole ratio 16.6, residence time 1.0 sec, 600°C reaction temperature.

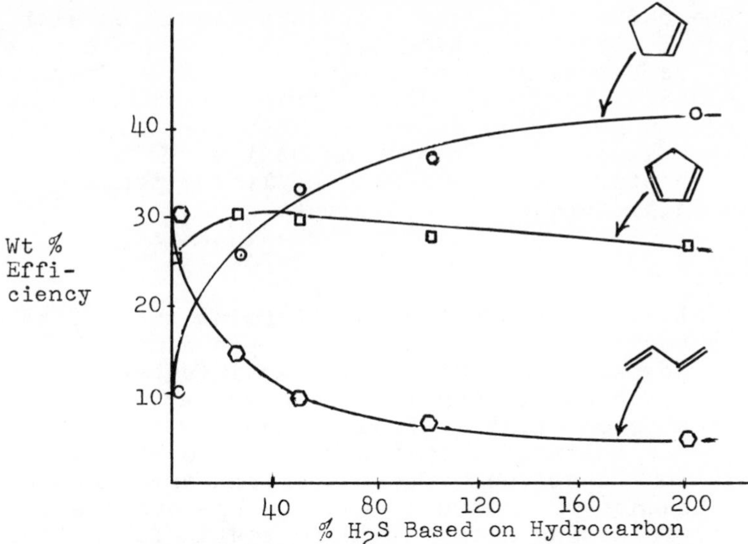

Figure 6. Product wt % efficiencies as a function of H_2S level: piperylene degradation

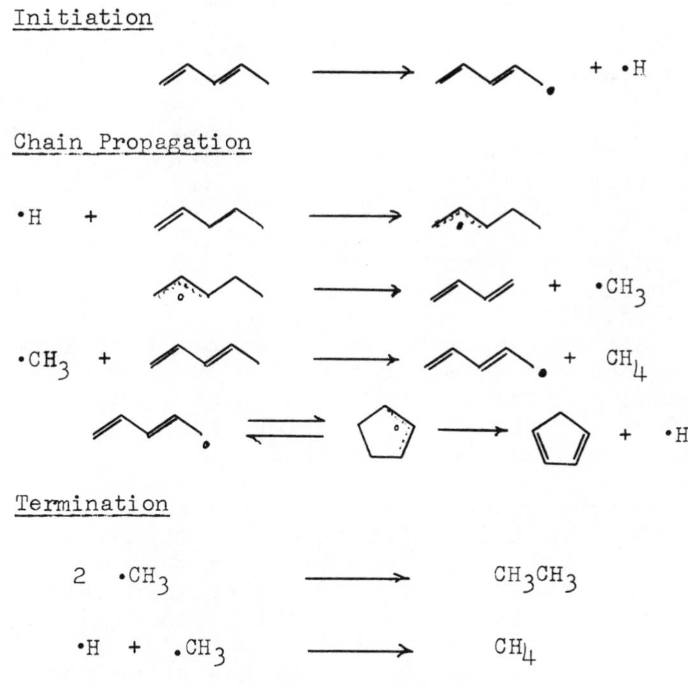

Figure 7. Unpromoted decomposition of piperylene

The reduction in the butadiene formation with increasing H_2S concentration reflects the change in the radical decomposition reaction (Figure 8). Hydrogen atom adducts abstract hydrogen from the H_2S or other hydrocarbon molecules at rates which are competitive with, or faster than, those for radical decomposition (6). Therefore, products such as linear pentenes are formed in preference to butadiene.

Surface Studies

A study was conducted to determine the effect of various reactor surface materials upon 2-methyl-2-pentene degradation. Of particular interest was the material of construction for the hydrocarbon preheater. This area is of prime interest because at lower temperatures we observed isomerization and other olefin degradation reactions. These problems are compounded by the absence of a diluent and the H_2S promoter.

The following materials were tested for heterogeneous degradation activity on 2-methyl-2-pentene (Table V).

TABLE V

Heterogeneous Activity

Test Sample*	Conversion Level	
Temperature, °C	550	620
Residence Time, seconds	2.0	0.5
1. Aluminum	2.2	--
2. Quartz	3.1	--
3. Electrode Carbon	3.1	--
4. 316 Stainless Steel	2.0	6.4
5. 440 Stainless Steel	2.0	6.9
6. Mild Steel	10.0	34.0
7. 5% Cr/.5% Mo Carbon Steel	10.0	33.0
8. 5% Cr/.5% Mo Carbon Steel "Alonized"**	2.5	9.4

* All materials--pretreated by exposure to 2-methyl-2-pentene for 30 minutes at 700°C.

** Alonizing Process--diffusion of aluminum into steel to form an iron-aluminum alloy, Alon Processing, Inc., Tarentum, Pennsylvania.

It may be noted from these data that a small amount of thermal degradation of 2-methyl-2-pentene occurred with all materials to the extent of 2 to 3% conversion. However, the samples of mild steel and carbon steel containing 5% chromium and 0.5% molybdenum yielded conversion levels in excess of 10%.

The two "activated" samples described above were visually inspected and were found to be decidedly

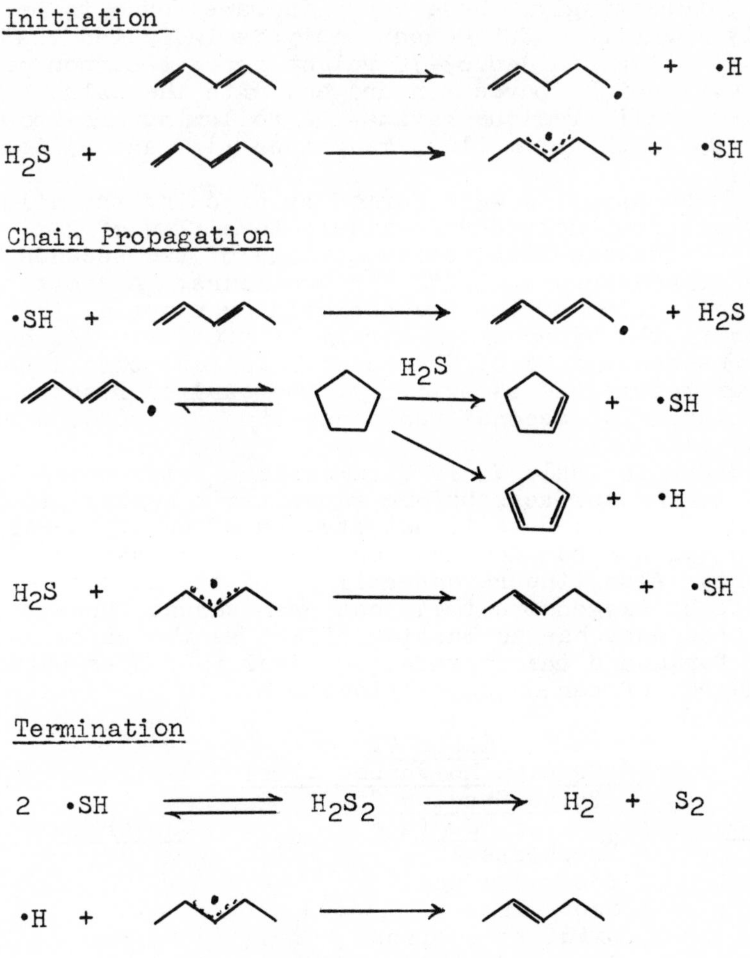

Initiation

→ + •H

H_2S + → + •SH

Chain Propagation

•SH + → + H_2S

⇌ $\xrightarrow{H_2S}$ + •SH

→ + •H

H_2S + → + •SH

Termination

2 •SH ⇌ H_2S_2 → H_2 + S_2

•H + →

Figure 8. H_2S-promoted decomposition of piperylene

different from all other materials in that they were covered with patches of a soot-like material (Figure 9). Interestingly, these deposits were found to be highly magnetic. Subsequent analysis indicated that the deposit contained 63-70 weight percent carbon and 0.05-0.22 weight percent hydrogen, with the balance being metallic residue having the following breakdown: iron (95%), chromium (1%), manganese (1%) and silica (1.2%).

These deposits were formed by exposing the mild and the chrome-moly carbon steels to a flow of 2-methyl-2-pentene at a residence time of two seconds and a temperature of 550°C for two hours. A shortened activation period was achieved by exposing the steels to the hydrocarbon stream at 700°C for 1/2 hour at a residence time of 2 seconds. To test this deposit for heterogeneous activity, the earlier reaction conditions (2.0 seconds residence time at 550°C) were used.

Shown in Table VI is a comparison between the fresh tubing surface, before and after a series of treatments. The data illustrate the effect of accelerated hydrocarbon degradation when the carbon is present. Also, the regenerability of the parent metal surface by oxygen treatment has been shown. However, this procedure has no lasting effect as the carbon again forms and the degradation level increases with the degree of carbon deposition.

TABLE VI

5% Cr/0.5% Mo-Carbon Steel

Pulse Reactor Study

Number	Surface	Conversion
1	Fresh steel	32
2	Sooty surface	50
3	Soot alone	59
4	Oxidized surface	26
5	Oxidized, then resooted	46

Test Sample: 5% Cr/0.5% Mo-Carbon Steel

Conditions: 625°C Temperature
0.5 seconds residence time
2-methyl-2-pentene hydrocarbon

In an effort to solve the problem of carbon formation without resorting to the use of high cost stainless steels, a sample of chrome-moly steel was subjected to the alonizing process (aluminum impregnation). Table VII demonstrates the successful pacification of the surface.

Figure 9. "Activated" turnings after experiment

TABLE VII

Test Sample

"Alonized" 5% Cr/0.5% Mo-Carbon Steel

Number	Elapsed Time	Conversion
1	0-5	25.0
2	5-10	3.4
3	10-15	2.3
4	15-20	2.4
5	20-25	2.4
Control (Non-alonized)	20-25	10.0

Test Sample: Alonized 5% Cr/0.5% Mo-Carbon Steel
Aging Effect for 2-Methyl-2-Pentene Degradation
Conditions: 550°C Temperature
2.0 seconds residence time

Heterogeneous surface effects on the pyrolysis of 2-methyl-2-pentene have been defined for a number of systems. The origin of these effects appears to be related to the ability of the reactor surface to catalyze olefin (or diolefin) decomposition to form a carbon-iron matrix capable of catalyzing further hydrocarbon decomposition.

An interesting sidelight of this work is the use of the "magnetized" carbon described herein as an olefin isomerization catalyst. In the presence of this material we have been able to form thermodynamically less stable methyl pentene isomers at elevated temperatures.

Preliminary work has shown that 4-methyl-1-pentene can be formed from 2-methyl-2-pentene (Table VIII). Additional work needs to be done to define this system more completely.

TABLE VIII

Temp, °C	Residence Time seconds	Mole % Conversion	% Selectivity to 4MP1
400	500	11.0	78.0
450	50	22.5	67.0

Flow reactor packed with "magnetized" carbon.

Summary

Mechanistic aspects have been described for the decomposition of selected methyl-substituted pentenes to isoprene. The observed order of isoprene efficiencies for these systems is: 3MP2 > 2MP2 > 2MP1 > 4MP2.

The ability of isoprene to form heavies by attack on itself and other dienes has been demonstrated. This observation then allows for reactor design considerations that will keep the isoprene partial pressure relatively low.

The nature of the reactor surface has been shown to be related to the catalysis of the olefin (or diolefin) decomposition reaction to form a carbon-iron matrix capable of catalyzing further hydrocarbon decomposition. The use of 316 or 440 stainless steel or Alonized carbon steel will avoid the formation of a carbon-iron matrix.

Literature Cited

1. The Goodyear Tire & Rubber Co., British Patent, 832,475, April 13, 1960.
2. The Goodyear Tire & Rubber Co., British Patent, 841,351, July 13, 1960.
3. The Goodyear Tire & Rubber Co., British Patent, 868,566, May 17, 1961.
4. Hutchings, David A., Frech, Kenneth J., Hoppstock, Frederic H., "The Kinetics of Product Formation in the H_2S-Promoted Pyrolysis of 2-Methyl-2-Pentene", presented April 6, 1975, Philadelphia Meeting, ACS, Division of Petroleum Chemistry; also Chapter 11 of this book.
5. Benson, Sidney W., "Thermochemical Kinetics", pp 94-95, John Wiley & Sons, Inc., New York, 1968.
6. Hutchings, David A., Frech, Kenneth J., Hoppstock, Frederic H., "Gas-Phase Cyclization of Piperylene to Cyclopentene and Cyclopentadiene Using H_2S as a Catalyst", presented April 9, 1972, Boston Meeting ACS, Division of Petroleum Chemistry.

13

Reactor Surface Effects During Propylene Pyrolysis

M. A. GHALY and B. L. CRYNES

School of Chemical Engineering, Oklahoma State University, Stillwater, Okla. 74074

The thermal decomposition of propylene involves a series of primary and secondary reactions leading to a complex mixture of products. Studies showed that the distribution of pyrolysis products varies considerably with the pyrolysis conditions and the type of reactor used. There is agreement among the studies on propylene pyrolysis that the three major products of pyrolysis are methane, ethylene, and hydrogen. However, there is disagreement on the types and amounts of minor or secondary product species. Ethane, butenes, acetylene, methylacetylene, allene, and heavier aromatic components are reported in different studies, Laidler and Wojciechowski (1960), Kallend, et al. (1967), Amano and Uchiyama (1963), Sakakibara (1964), Sims, et al. (1971), Kunugi, et al. (1970), Melloutte, et al. (1969), conducted at different conversion and temperature levels. Carbon was also reported as a product in the early work of Hurd and Eilers (1943) and in the more recent work of Sims, et al. (1971).

The overall rate of propylene pyrolysis has been reported to be first order in some studies and as 3/2 order in others. Most studies conducted at low temperatures (up to 650C) tend to favor 3/2 order, while studies conducted at higher temperatures indicated mostly a first order rate. Clearly, the overall order of reaction is at best only a pseudo or an apparent order representing a combination of many elementary steps.

One feature of the pyrolysis reaction which is not fully understood is the role of reactor surface. Although the thermal decomposition of propylene has been at least implicitly assumed in some cases to be a completely homogeneous gas-phase reaction, the influence of reactor surfaces has been demonstrated frequently. Factors such as the material of construction, surface-volume ratio of the reactor, and chemical treatment of reactor surface often affect the rate of reaction and/or the product distribution. Many previous investigations are of limited value in determining the role of surface in the pyrolysis reaction. The literature on surface effects contains predominantly qualitative information and often contradictions and anomalies. Early studies that reported heterogeneous surface effects were completed several years ago before reliable analytical methods were available thus restricting the accuracy of product analysis and detection of minor products. Most recent studies have been made using batch reactors or tubular glass or quartz reactors usually at conditions different from those of commercial interest. Recent studies of the influence of reactor walls on pyrolysis reactions have been conducted by Crynes (1969) and Herriott (1971). Their studies unequivocally demonstrated the influence of reactor walls on propane pyrolysis in reactors constructed of various metals and with certain gas treatments. The material of construction, reactor history, and the type of wall treatment were shown to be important variables. Another recent study by Taylor, et al. (1972) on propylene pyrolysis in a "wall-less" reactor, i.e., a "completely" homogeneous system, showed that the product species, product distribution, and reaction rate in the absence of a surface were significantly different from those obtained after a stainless steel surface has been introduced into the reaction mixture.

With respect to theories of wall heterogeneous effects in hydrocarbon pyrolysis reactions, the literature is almost void. Rice and Herzfeld (1951) have presented some theoretical arguments but with some severely simplifying assumptions. Polotrak, et al. (1959) proposed mechanisms involving both chain initiation and termination as heterogeneous processes. More elaborate theoretical work on the interaction between hydrocarbons (paraffins and olefins) and metal oxide surfaces was done by Semenov (1958) and Kasansky and Pariisky (1965) in which the authors tried to explain the heterogeneous effects (activity) of the surfaces in terms of electronic conductivity.

A recent study by Tsai and Albright (1975) clearly indicated some of the important surface reactions occuring in reactors constructed of different metals during the pyrolysis of light paraffins. These reactions include formation of carbon, removal of carbon, oxidation of metal surfaces, reduction of surface oxides, formation of metal sulfides, and destruction of metal sulfides. The above authors stated the important need for information on the interaction of ethylene, propylene, and other olefins with the surface oxides in order to clarify the level of surface oxides present on the reactor walls when various hydrocarbon feedstocks are used.

There is a need to determine numerous factors about the mechanisms of hydrocarbon pyrolysis, including the role of reactor surfaces. The present investigation was made in tubular flow reactors constructed of several metals, and data have been obtained on the influence of reactor surfaces on the product distributions, feed conversions and carbon formation while pyrolyzing propylene.

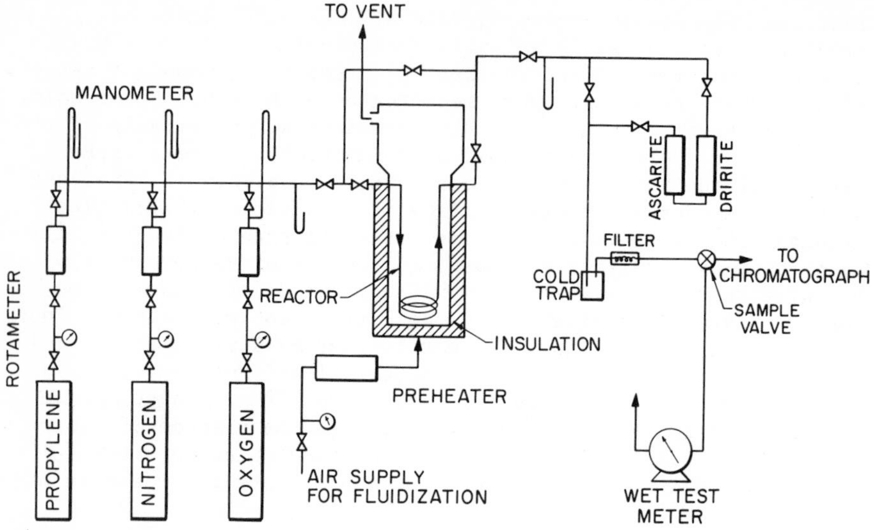

Figure 1. Experimental flow system

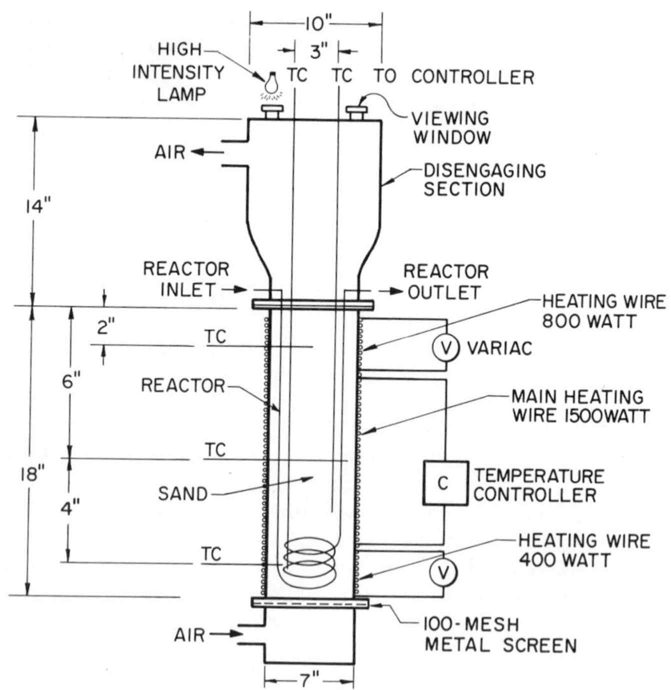

Figure 2. Reactor details

Experimental Details

The flow system used in this investigation is shown in Figure 1. The reactors were constructed from 0.635 cm O.D. (1/4-inch O.D., 20-gauge) tubing. The tubing was coiled to a diameter of 12.7 cm, and about 4.52 to 4.62 m of tube were immersed in a fluidized sand bath to obtain the desired temperatures. A total of seven reactors were used in this study. Reactors were constructed of 304 stainless steel, low-carbon steel, nickel, inconel, and incoloy, all commercially available.

The thermocouples immersed in the sand bath (Figure 2) were accurate to within 1C; and at steady-state, the temperatures inside the fluidized sand bath at various radial and axial positions did not differ by more than 3C in all runs. Thermocouples were also placed in the reactor inlet and exit to measure gas heating and cooling profiles, but these thermocouples were removed during actual runs. The reacting gas was essentially isothermal since less than 8.9 cm were required for heating and cooling, respectively, but the gas temperature was always less than the bath temperature by about 4 to 8C. Sand was fluidized by air that was preheated to about 300C before entering the reactor system. Details of the reactor used in this study are shown in Figure 2.

Product samples were analyzed using a Model 1200 Aerograph gas chromatograph equipped with flame ionization detector. The chromatograph columns were a series arrangement of 1.83cm, 0.32cm O.D. tubing packed with silica gel followed by 1.83cm, 0.32cm O.D. tubing packed with activated alumina. These columns provided for resolution of all hydrocarbons through C_4's. Hydrogen could not be determined by the chromatograph and was calculated from a hydrogen balance.

Reaction Products and Kinetics, 700-850C

A series of runs was made in a reactor constructed from 304 stainless steel at temperatures ranging from 700 to 850C and for space times ranging from 0.1 to 3 seconds. Space time was obtained by dividing the reaction zone volume by the feed volumetric rate measured at reaction temperature and pressure. Hydrogen, methane, ethylene, and carbon were the major products formed at these temperatures, while butenes, 1-3 butadiene, and sometimes ethane were the minor products. Carbon yield was determined from a carbon balance. However, carbon was determined later quantitatively in an identical reactor by burning the reactor walls with oxygen and adsorbing the resulting CO_2 on ascarite. Results showed a good agreement between the amount of carbon determined quantitatively and the value determined from an independent carbon balance.

Propylene conversions ranged from 4 to 48% at all temperatures and space times. At low conversion, ethylene, methane, hydrogen, butenes, and butadiene formed in the

TABLE I

KINETIC PARAMETERS OF PROPYLENE PYROLYSIS IN UNTREATED 304 STAINLESS STEEL REACTOR

Temperature (C)	Reaction Order (n)	Rate Constant $[\frac{(\frac{gmol}{cc})^{1-n}}{sec}]$
700	1.4 ± 0.03	17.60
750	1.2 ± 0.04	2.49
800	1.08 ± 0.03	1.31
850	1.0 ± 0.05	1.80

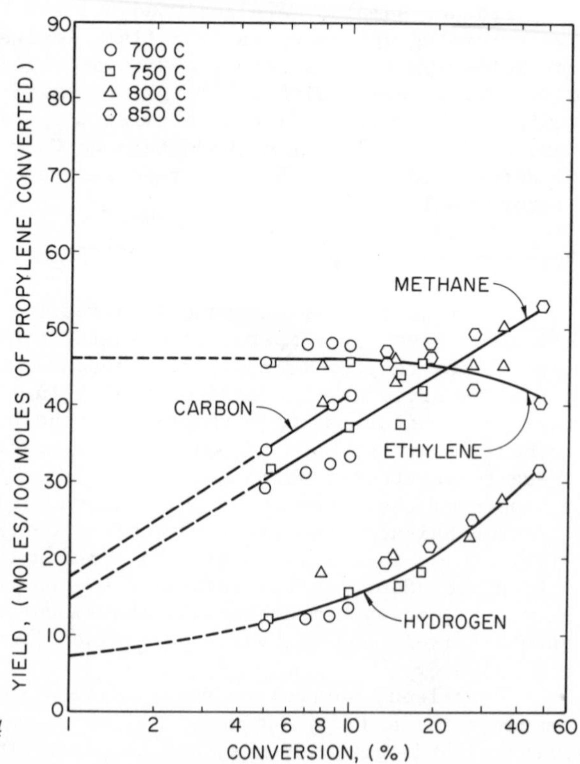

Figure 3. Product yields, 304 stainless steel reactor

approximate ratio of 5:5:2:1:1. As propylene conversion increased, methane and hydrogen yields increased rapidly while ethylene, butenes, and butadiene yields decreased steadily. At these high conversions, the products ratio became approximately 10:13:8:1:1. Carbon yield markedly increased from 34% at low conversion to 95% at high conversion. Total molar yield of products per mole of propylene reacted seemed to increase with conversion and/or temperature levels ranging from 1.15 (700C, 4% conversion) to 1.41 (850C, 48% conversion). Figure 3 shows the effect of conversion on major product distribution. Note that the abscissa of Figure 3 is logarithmic to emphasize lower conversion data.

Hydrogen, ethylene, and methane appear to be primary products, and carbon is possibly a primary product, although not conclusively shown here. Carbon data are somewhat limited relative to that of other species. Few researchers have observed carbon as a primary reaction product. The hydrogen and methane yields are more sensitive to temperature than was ethylene. Both are probably produced by primary as well as secondary reactions (decomposition of butenes and butadiene and also ethylene, the latter at high temperature and conversion).

The conversion-space time data were used to calculate rate constants and overall reaction orders for propylene decomposition. Reaction order was obtained by plotting the logarithm of rate versus the logarithm of propylene concentration according to the eqqation:

$$-r = k_n C^n$$

Reaction rates were obtained by numerical differentiation of conversion-space time curves. The reaction rate constants were calculated at each temperature by numerically integrating the expression

$$k_n = \frac{1}{t\, C_o^{\,n-1}} \int_0^X \left[\frac{1 + (\varepsilon - 1)X}{(1 - X)} \right]^n dX$$

which takes into account the volume expansion as reaction proceeds. The reaction orders and rate constants obtained at the various temperatures are shown in Table I. In order to compare the results from this work to those given in the literature, first and 3/2 order rate expressions were used to obtain rate constants from the experimental data. A comparison of first and 3/2-order rate constants is given in Table II. The values given in the table for this work were those corresponding to zero conversion, and the literature values are from a variety of conditions including batch, flow, and at both low and high temperatures. The values of the first order rate constant are close to the values

TABLE II

COMPARISON OF REACTION RATE CONSTANTS

	1st-order*					3/2 order**		
Temperature (C)	This Work***	Szwarc (1949)	Amano (1963)	Melloutte (1969)	Taylor[+] (1972)	This Work***	Melloutte (1969)	Kunugi (1970)
700	0.035					12.3	0.10	5.8
750	0.10		0.07	0.40		32.5		35.0
800	0.38	0.03	0.48		0.26	112		125
850	2.0	0.10	2.5	3.3	1.1	602		460

* Units on rate constant are $seconds^{-1}$.

** Units on rate constant are $(g\ mole)^{-0.5}\ (sec)^{-1}$.

*** Rate constants are estimated at zero conversion.

\+ Wall-less reactor.

reported by Amano and Uchiyama (1963), at least over the temperature range of 750-850C. The 3/2-order data are generally in agreement with those presented by Kunugi, et al. (1970), again over the range of 750-850C. Mellouttee, et al. (1969) reported a change of reaction order from 3/2 (600-698C) to first order (700-900C) and suggested a change in decomposition mechanism at 700C as a possible reason. The rate constants obtained by Mellouttee are, however, quite different from those obtained in this work at the same temperatures. The pyrolysis of propylene is obviously a complex reaction system, and any attempt to represent the rate of the reaction by a simple overall nth order expression over anything but a very narrow range of conditions is not justified.

Results in Untreated Reactors

A total of twenty runs were made in 5 reactors constructed of different materials at a temperature of 750C and different space times ranging between 0.5 and 2 seconds. The reactors used in these runs are shown in Table III. Stainless steel Reactor 2 was chosen as a reference reactor, and the results (conversions and product distribution) obtained in this reactor were the basis by which the comparative behavior of other reactors were judged. Results obtained under the same experimental conditions but in different reactors were sometimes significantly different as will be shown below. The influence of wall material was judged by the change in propylene conversion (reaction rate) and/or a shift in product distribution. When comparing the different wall materials behavior to that of the reference wall material, i.e., 304 stainless steel, care was taken to distinguish between the transient and steady yields and conversions; the latter was usually obtained in reactors with carbon-conditioned walls.

Stainless Steel (304) Reactor (Reference Reactor). The conversions and product yields obtained at the different space times and at 750C were almost identical to those obtained previously in 304 stainless steel Reactor 1 (used mainly to generate kinetic data), thus indicating the high reproducibility of results. No initial transient behavior of the reactor was noticed since the products concentrations and propylene conversion did not change for samples drawn over a period of 30 minutes in the first run (750C, 1.5 seconds) made in this reactor. No unsteady state behavior was observed in subsequent runs as well. Typical results are shown in Table IV. Propylene conversion ranged from 5 to 18%. Hydrogen yield increased steadily from 12 to 20%, methane yield increased from 31 to 43% and ethylene yield remained practically constant at 45% as conversion increased with space time. Carbon yield increased substantially from 38 to 60% with increase in conversion, while butenes yield decreased from 15 to 11% and butadiene yield decreased from 11 to 8% with increase in conversion. The results suggest that secondary decomposition of products

TABLE III

REACTOR DIMENSIONS AND PROPERTIES

Reactor	Material	Submerged Length, cm	Submerged Volume, cm^3	S/V cm^{-1}	Fe	Composition, % Cr	Ni	Mn	C
1	Stainless Steel (AISI 304)	457	75.0	8.7	7	19	9	2.0 max.	0.08 max.
2	Stainless Steel (AISI 304)	457	75.0	8.7	70	19	9	2.0 max.	0.08 max.
3	Low-Carbon Steel (AISI 1015)	445	73.0	8.7	Bal.	--	--	0.3-0.6	0.13-0.18
4	Nickel	462	75.8	8.7	0.15	--	99.5	0.25	0.06
5*	Inconel 600	452	53.0	10.4	7.2	15.8	76.4	0.20	0.04
6*	Incoloy 801	455	53.3	10.4	45.5	20.5	32.5	0.80	0.04
7*	Incoloy 801	460	54.0	10.4	45.5	20.5	32.5	0.80	0.04

*0.635 cm O.D. (1/4 inch O.D., 18 gage) tubing

TABLE IV

RESULTS FROM 304 STAINLESS STEEL REACTOR
AT 750C, 1 ATM

Space Time (min.)	0.5	1.0	1.5	2.0
Mole Percent				
H_2	0.66	1.52	2.52	3.25
CH_4	1.61	3.71	5.21	7.10
C_2H_4	2.48	4.35	6.24	7.84
C_2H_4	0.05	0.04	0.61	0.95
C_3H_6	93.9	87.5	82.4	77.4
C_4H_8	0.77	1.35	1.71	1.93
C_4H_6	0.57	1.03	1.31	1.56
Conversion (%)	5.3	10.1	14.0	17.8
Total Yield*	1.16	1.25	1.30	1.33
H_2 Yield (%)	12.5	15.3	18.8	19.4
CH_4 Yield (%)	30.6	37.4	39.0	42.4
C_2H_4 Yield (%)	47.1	44.1	46.7	46.8
C Yield (%)	38.0	50.0	55.1	60.0

*Total yield of gaseous products

butene and butadiene takes place as conversion increases, and methane, hydrogen, and probably carbon are products of such decomposition. High carbon yield at conversions as low as 5% suggests strongly that carbon may be formed directly by the interaction of propylene molecule with the metal surface rather than by a secondary coking of some higher molecular weight products. Direct decomposition of paraffins and olefins on metal surfaces to produce carbon was suggested early by Thomas, et al. (1939) and was later confirmed by Tesner (1959) and Tamai, et al. (1968). The absence of a noticeable initial surface activity period does not conclusively prove that stainless steel surface has no heterogeneous effect on propylene decomposition. The formation of carbon in much higher quantities (at low conversion) than those reported in reactors made of glass or quartz under similar conditions strongly indicates a surface activity associated with the metal reactor. Carbon formed during pyrolysis of several hydrocarbons has been shown to possess some catalytic activity as shown by Tesner (1959) and Holbrook, et al. (1968), and it is possible that the carbon formed by interaction of propylene with stainless steel suface induces the same activity towards propylene decomposition as that of the parent surface; hence, no change in surface activity would be noticed with time as more carbon is formed.

Low Carbon Steel Reactor. The low carbon steel (LCS) reactor exhibited a peculiar behavior in the early period of the first run (750C, 1.5 seconds). An increase in activity occurred during the 20 minute period, marked by increased hydrogen and carbon formation, and then the activity dropped steadily until a steady-state was reached after 60 minutes. No initial transient activity period was observed in subsequent runs carried out in the carbon-conditioned reactor. At the point of highest activity, propylene conversion was a high 27%, while hydrogen, methane, ethylene, and carbon yields were 90, 56, 35, and 146% respectively. Using hydrogen yield as a measure of activity, Figure 4 presents the activity profile for the LCS reactor.

The results also indicated that even the steady-state carbon-conditioned LCS reactor was more active than the 304 reference reactor. Figure 5 shows the higher hydrogen and methane concentrations obtained in the LCS reactor as compared to the reference reactor. Carbon yield was much higher in low carbon steel than in the reference reactor, while ethylene yield was slightly lower than that in the reference reactor. Conversions (reaction rates) in low carbon steel were always 25-35% higher than those obtained in the reference reactor.

The same shaped initial activity profile was observed by Crynes (1968) in a LCS reactor during the pyrolysis of propane rather than propylene the higher initial activity of low carbon steel as compared to stainless steel could be attributed to the much higher iron content of the former. Thomas, et al. (1939) study indicated iron to be a strong catalyst in promoting carbon and

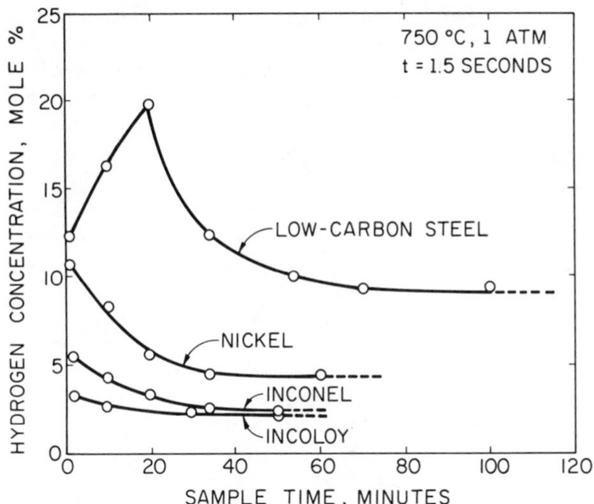

Figure 4. Transient behavior of metal reactors

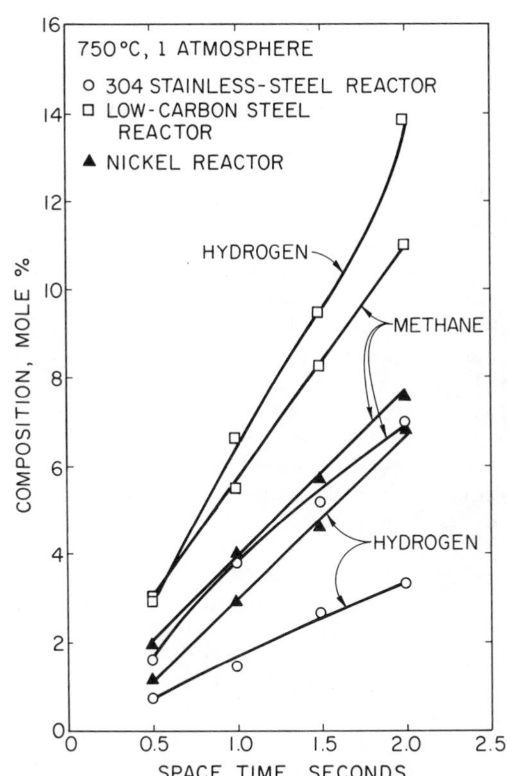

Figure 5. Effect of reactor material on product distribution

hydrogen formation during hydrocarbon pyrolysis, while alloys of iron containing a large amount of chromium were much less active due to the Cr ability to decrease or destroy the activity of iron. The activity of carbon-conditioned low carbon steel was higher than that of stainless steel perhaps because of a more catalytically active carbon layer. Studies by Thomas, et al. (1939) and Tamai, et al. (1968) on carbon formation during the pyrolysis of olefins and paraffins on different metals indicate the possibility of carbon showing a catalytic activity due to carrying iron atoms in the carbon layer perhaps in the form of minute particles.

Nickel Reactor. An initial unsteady state period marked by high activity lasted for 50 minutes in a reactor tube constructed of nickel. The highest activity was obtained at the beginning of the run (750C, 1.5 seconds) and decreased with time as more carbon was deposited on the reactor walls. The transient behavior of the nickel reactor is shown in Figure 4 along with those of other reactors. Highest wall activity corresponded to a hydrogen yield of 74%, methane yield of 47%, ethylene yield of 41%, and carbon yield of 122%; conversion was slightly higher than that obtained in the reference reactor.

At steady-state, conversions were identical to those obtained in the 304 reactor, but products distribution was different indicating a reactor wall more active (selective) than stainless steel but less active than low carbon steel as shown in Figure 5 by the levels of hydrogen and methane obtained in the carbon-conditioning nickel reactor.

The initial wall activity diminished as less active carbon builds upon the walls decreasing available active sites. The initial activity of nickel was clearly lower than that of low carbon steel, but higher than that of stainless steel. The results are in general agreement with the conclusions of Tamai, et al. (1968) and Buell and Weber (1950). The former indicated that nickel had a lower "affinity" to olefins than iron, while the latter concluded that the nickel content in austenitic steel alloys is primarily responsible for their activity (carbon formation) when compared to the less active chrome steel alloys. The carbon-conditioned nickel walls were less active than those of low carbon steel reactor probably because the catalytic activity of the base metal did not penetrate through the carbon layer as effectively as it did with low carbon steel.

Inconel and Incoloy Reactors. Inconel showed an initial activity period that lasted for 30 mintues. During this period, propylene conversion remained constant, but the product distribution changed markedly. The highest wall activity was obtained in the first two minutes of the first run (750C, 1.5 seconds) and then decreased steadily with time. Highest activity corresponded to a hydrogen yield of 43%, methane yield of 51%, and carbon yield of 95%. Steady-state conversions and product distribution in the inconel reactor were similar to those of the reference reactor. Figure 4 shows Inconel transient behavior as measured by hydrogen yield.

The lower initial activity of Inconel as compared to that of low carbon steel or nickel clearly indicates that alloys containing iron and nickel are in general less effective than the pure metals in interacting with propylene.

Incoloy reactor showed an initial activity profile similar to that of Inconel reactor. Steady-state activity was reached after 40 minutes, and the results were similar to those obtained in the reference reactor. The initial activity profile of Incology is shown in Figure 4.

Results in Surface-Treated Reactors

Reactor surfaces were treated with oxygen and hydrogen sulfide since these gases are either used or have been suggested in industrial pyrolysis applications. Oxygen treatments consisted of passing 550-600 cc/min of oxygen through the reactor for times up to 60 minutes with the reactor bath at temperatures ranging from 600 to 800C. In hydrogen sulfide treatment, 250 cc/min of the gas was passed through the reactor for 20-60 minutes, while the reactor bath was at 750C. The reactors were thoroughly purged with helium after each treatment. All propylene runs following a surface treatment were carried out at 750C and 1.5 seconds space time at essentially atmospheric pressure.

Stainless Steel (304) Reactor

a. Oxygen Treatment: The course of the reaction was found to be rather different after the stainless steel reactor (Reactor 2) was treated with oxygen. Figure 6 indicates the results of a run made at 750C after the reactor walls had been treated for 30 minutes with oxygen at a temperature of 600C. An initial activity period was observed during which propylene conversion and product composition changed steadily with time until a steady-state was reached after 80 minutes. The change in wall activity was more drastic in the first 15 minutes of the run. Hydrogen yield decreased from 40 to 29%, methane and ethylene yields remained unchanged, butenes and butadienes yields increased moderately, and carbon yield decreased from 73 to 63%. The results suggest that the oxygen-treated surface directly catalyzed the secondary decomposition of butenes and butadienes. After 80 minutes, values of conversion and product composition approximated those obtained in the untreated reference reactor. The initial reactor activity was found to change with oxygen treatment temperature rather than treatment time. The highest surface activity was obtained for a treatment temperature of 600C while the lowest surface activity was obtained for a treatment temperature of 800C. This change in initial activity seems to indicate a change in the structure of a

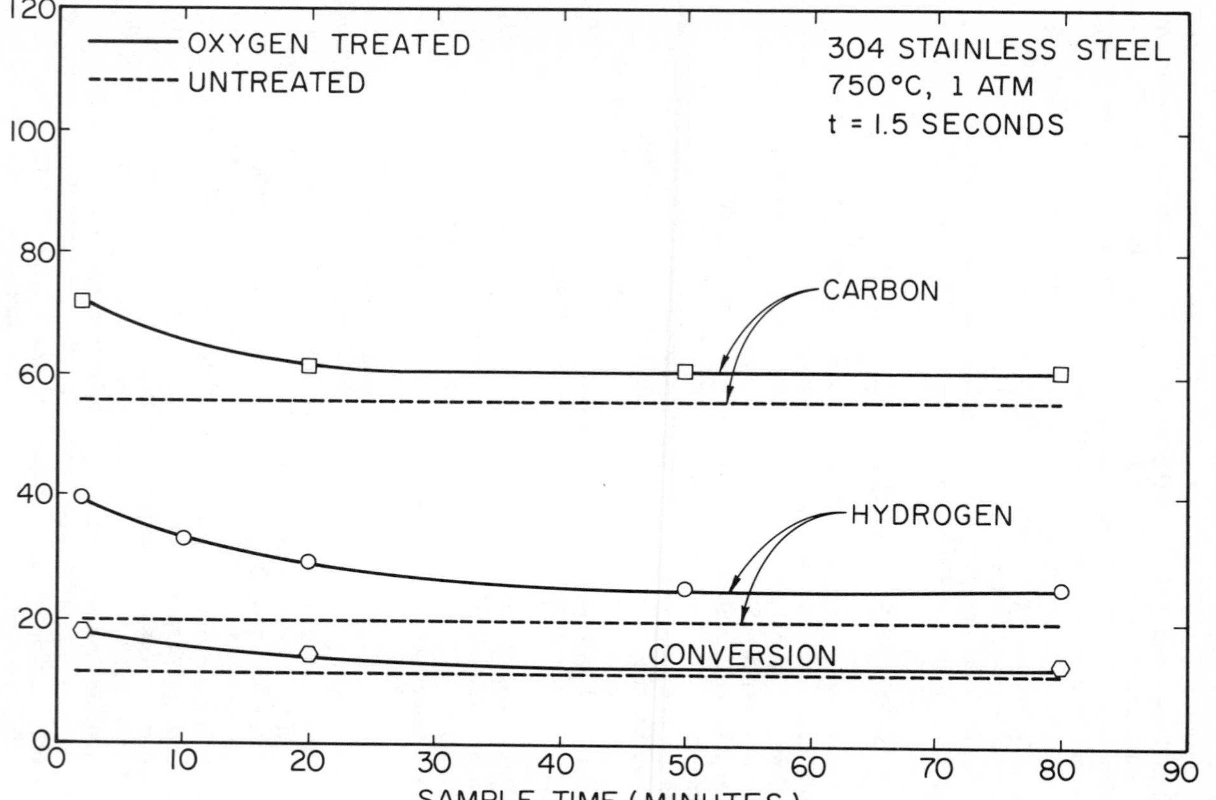

Figure 6. Transit behavior of oxygen-treated stainless steel reactor

catalytic surface species (probably an oxide) when the surface is contacted with oxygen at different temperatures.

b. Hydrogen Sulfide Treatment: Treating the reactor with H_2S for 20 minutes eliminated the surface activity effects of previous oxygen treatments. In the early period of one run (2-20 minutes) the surface was actually less active than that of the untreated reference reactor as indicated by slightly lower hydrogen, methane, and carbon yields. However, as the reaction proceeded and carbon was built up, the activity increased until a steady-state conversion and product composition similar to that in the reference reactor were obtained. Subsequent treatments with oxygen for as long as 60 minutes could not restore wall activity. This behavior was similar to that indicated by Crynes and Albright (1969) during propane pyrolysis in surface-treated reactors.

During sulfiding with H_2S, the surface oxide layer is probably converted to sulfide either entirely or at least to a sufficient depth (Farber and Ehrenberg, 1952), and a durable protective sulfide layer is formed. This sulfide layer is more passive than either the untreated 304 stainless steel surface or the carbon layer formed during pyrolysis. Thus there is an observed initial increase in activity with carbon build-up as the surface type and its activity stabilize.

Low Carbon Steel Reactor.

a. Oxygen Treatment: After treating the LCS reactor with oxygen, an active reactor surface is produced, and a transient activity profile, similar to that obtained with 304 stainless steel, is noticed. A comparison of the transient activity obtained in both the untreated and oxygen-treated LCS reactor is shown in Table V. The table shows that both the activity profile and activity level produced by the two surfaces were considerably different. The rise-fall activity profile was absent in the oxygen-treated reactor, and the activity produced by the oxygen-treated surface was generally lower than that of the untreated surface. Values of initial conversions and product yields shown for the oxygen-treated LCS reactor were obtained after the reactor was treated with oxygen for 20 mintues at 750C. A steady-state activity was eventually reached for the oxygen-treated surface after periods of 50 to 70 minutes. This carbon-conditioned reactor activity was approximately similar to that of the carbon-conditioned untreated LCS surface.

The results suggest that the decrease in initial surface activity of low carbon steel when treated with oxygen is probably due to the formation of a protective oxide film which catalytically is less active than the base metal; i.e., iron. Many literature sources indeed indicate the formation of an oxide layer consisting mainly of Fe_3O_4 under conditions similar to those used in this study. This black iron oxide usually forms a coherent layer which is not easily reduced.

TABLE V

COMPARISON OF INITIAL SURFACE ACTIVITY IN UNTREATED AND TREATED LOW CARBON STEEL REACTOR

	Untreated			Oxygen-Treated*			H_2S-Treated**		
Sample Time (Min)	2	35	70	2	40	60	2	20	60
Temperature (C)	750	750	750	750	750	750	750	750	750
Space Time (Sec)	1.5	1.5	1.5	1.5	1.5	1.5	1.5	1.5	1.5
Conversion (%)	22.7	24.6	20.0	17.2	14.4	14.0	12.5	13.0	14.0
H_2 Yield (%)	62	78	52	27	20	19	15	16	18
CH_4 Yield (%)	43	50	48	39	37	37	33	34	37
C_2H_4 Yield (%)	44	38	41	47	46	46	46	46	45
C Yield (%)	103	128	102	62	54	54	43	46	52

* Oxygen at 750C for 20 minutes

** Hydrogen sulfide at 750C for 20 minutes

b. Hydrogen Sulfide Treatment: Treating the LCS reactor surface with H_2S eliminated the surface activity effects of previous oxygen treatments and produced a more passive surface which exhibited essentially the same conversion and product distribution obtained in the 304 reference reactor. Results obtained in the LCS reactor after treating the walls with H_2S are shown in Table V. Subsequent treatment with oxygen for a rather long period did not restore the original surface activity.

Nickel Reactor. The nickel reactor exhibited a peculiar initial activity profile after the walls had been treated with oxygen. The same peculiar profile was always obtained regardless of the oxygen treatment temperature or duration. Figure 7 shows that the activity of walls dropped steadily in the first 15 minutes of the run, then suddenly the wall activity increased sharply over the next 25 minutes as indicated by the sharp increases in hydrogen and methane yields. Propylene conversion also increased by 30% during that 25 minute period. A sharp decrease in activity then followed and a steady state activity level was finally reached after 90 minutes of run.

The initial activity of the reactor within the first 10 minutes or so was always less than the comparable activity of the untreated nickel surface. However, the product compositions obtained during the later period of decreased activity were approximately similar to those obtained during transient activity behavior of untreated nickel surface. One probable explanation of the above peculiar activity behavior is that the carbon formed on the reactor surface, together with hydrogen, in the early experimental period may have caused, at least, a partial reduction of the nickel oxide surface layer to form the more catalytically active nickel, and as more reduction took place more nickel was formed leading to higher activity. A point must then be reached when no more nickel formation takes place, and the activity eventually declines in a manner similar to that obtained in an untreated nickel reactor. The reduction of nickel oxide to nickel under the influence of hydrogen and/or carbon at high temperatures is sometimes referred to as nickel wildness.

Since nickel is severely attacked by H_2S at high temperatures at even short periods of time, no attempt was made to condition the nickel reactor with H_2S.

Inconel and Incoloy Reactors. The activity of inconel and incoloy reactors increased significantly after the surfaces had been treated with oxygen. Results in these two reactors were practically similar to those obtained in oxygen-treated 304 stainless steel. Again because of the high nickel content of these reactors, no H_2S treatment was attempted.

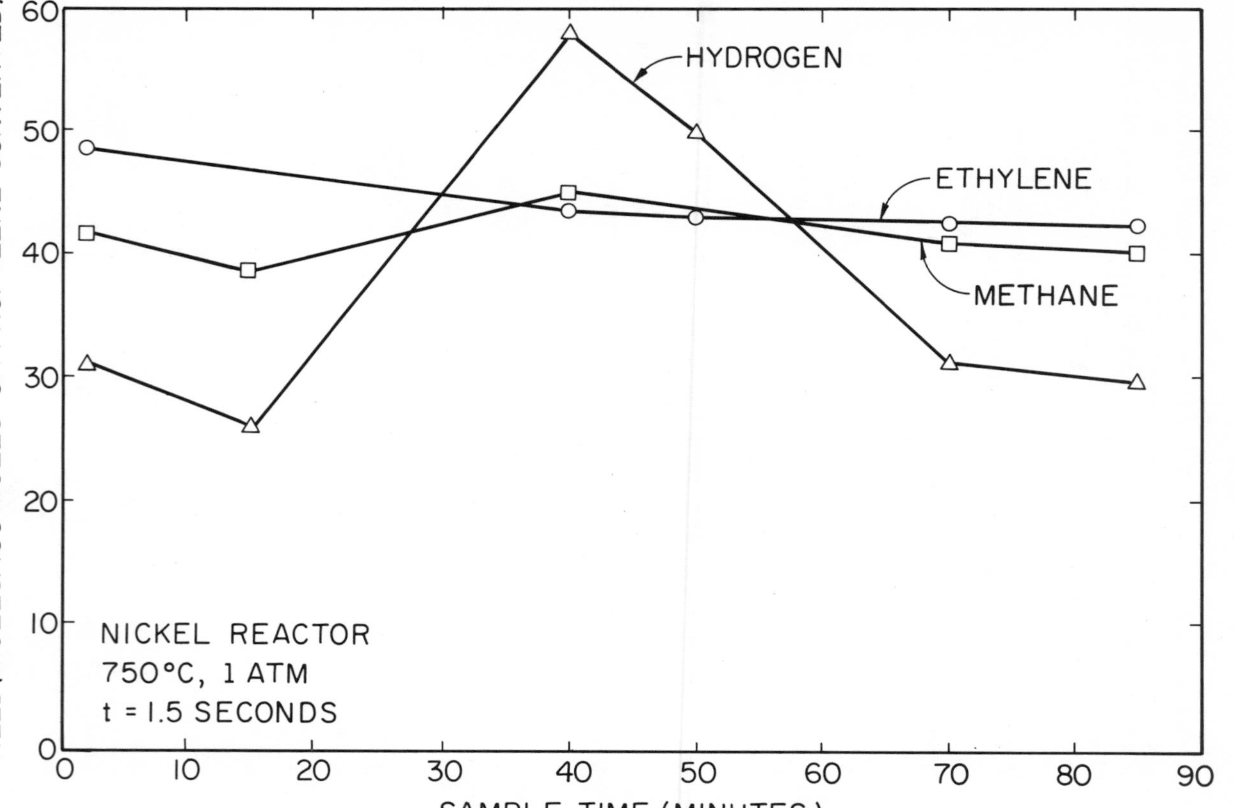

Figure 7. Transit behavior of oxygen-treated nickel reactor

Summary

Simple nth order kinetics are not adequate for representing propylene pyrolysis over any but a relatively short temperature and/or conversion range. A transition does occur from an overall propylene pyrolysis order of 1.5 to 1.0 as the temperature of reaction increases from 700 to 850C.

Significant wall activity was observed in reactors constructed of low carbon steel, nickel, and other alloys. Such activity may result in excessive carbon and hydrogen formation as well as increased conversion. Low carbon steel is an especially active reactor material with a rather unusual and long transient profile.

Carbon appears to be an initial reaction product along with hydrogen, methane, and ethylene.

Not all carbon-conditioned reactors produced identical results. The carbon layer seems to have a catalytic activity whose level depended somewhat on the type of surface present underneath. Both the type of carbon formed and migration and mixing of metal forms within the carbon probably contribute to the increased wall activity.

Reactor surfaces which are treated with oxygen produced strong surface effects. However, the level of activity of the oxidized surface as compared to that of the untreated surface depends largely on the type of metal or alloy. The initial activity of 304 stainless steel, incoloy, and inconel increased after oxygen-treatment of the surface.

The nickel reactor produced a rather unusual activity profile after it was treated with oxygen. Reduction of nickel oxide to nickel followed by carbon build-up is a probable explanation of the unusual activity profile shape.

Hydrogen sulfide treatment of reactor surfaces produced a relatively durable, passive, metal sulfide layer. The oxide layer is probably converted to sulfides either entirely or at least to a sufficient depth that very little catalytic oxides are exposed.

Certain reactor walls (treated and untreated) are rather effective in promoting the formation of carbon and hydrogen and to a lesser extent methane, probably through both primary and secondary reactions. Transient behavior is always associated with active walls and the transient activity profile is dependent on the type of wall. Complex chemical and physical changes are undoubtedly occurring on the reaction walls resulting in such transient behavior. All of the active wall species cannot now be characterized, but general reactions such as reduction and carburization from reactant and products could conceivably cause an initial buildup of catalytically active wall species. Activity passes through a maximum and then diminishes as carbon or carbonaceous products build upon the walls and/or additional reaction products decrease the available active species.

The strong surface effects observed after oxygen treatment of the reactors are probably caused by oxygen adsorption and/or reaction of the oxygen with the reactor walls to form complex metal oxides. Transformations in the structure of these oxides can occur due to change in temperature, exposure to a reducing atmosphere, or attack by another chemical such as hydrogen sulfide.

All of this is, of course, speculative at present and will remain so until more details become available about the wall active species, the elementary reaction steps, and the physical transformations occurring within a pyrolysis reactor. However, precise control of reactor treatment times and conditions still provides excellent conditions for investigating the effect of such treatments during pyrolysis reactions. As research workers continue to identify the effects and reactions of both the homogeneous and heterogeneous pyrolysis of hydrocarbons, and as these reactions mechanisms become quantitative then such efforts will lead to better control of conversion, product yields and in carbon laydown in commercial tubes.

Abstract

Propylene pyrolysis was studied in tubular flow reactors of various materials of construction. The influence of the reactor wall material on both conversion and product yield distribution was assessed over a temperature range of 700 to 850C and for conversions up to 48%. Neither a first order nor three-halves order overall reaction model adequately represent the data over this temperature range. Low carbon steel, nickel, and other alloy reactor materials exhibited a catalytic wall effect as evidenced by product yield changes and/or increased conversions. In a 304 stainless steel reference reactor, carbon appears to be a primary or initial reaction product. Reactor surfaces treated with oxygen showed strong initial surface activity that diminished as carbon was built on reactor walls. Hydrogen sulfide "passivated" the reactor surfaces by forming a protective metal sulfide film which prevented excessive carbon and hydrogen formation by secondary reactions.

Acknowledgement

A portion of this work was supported by the Petroleum Research Fund of the American Chemical Society, and part was supported by institutional research funds of the Oklahoma State University.

Nomenclature

C reactant concentration, compatible units

C_o initial concentration of propylene in feed, compatible units

k specific rate constant, compatible units, depending on order, $\frac{(\text{gmole/cc})^{1-n}}{\text{sec}}$

n reaction order

R gas constant, compatible units

r rate of formation, gmoles/cc sec

T temperature, K

t space time, seconds

X conversion of propylene, moles reacted per mole of feed

ϵ expansion factor, moles of products (excluding carbon) per mole of propylene reacted.

Literature Cited

1. Amano, A., and Uchiyama, M., J. Phy. Chem., (1963), 67, 1241.
2. Buell, C., and Weber, L., Petroleum Processing, (April, 1950), 387.
3. Crynes, B. L., Ph.D. Thesis, "Surface Effects During Propane Pyrolysis in Tubular Flow Reactors," Purdue University (1968).
4. Crynes, B. L., and Albright, L. F., Ind. Eng. Chem. Proc. Des. and Dev., (1969), 8, 25.
5. Faber, M., and Ehrenburg, D., J. Electrochem. Soc., (1952), 99, 427.
6. Herriott, G., Eckert, R., and Albright, L., "Kinetics of Propane Pyrolysis," 68th National Meeting AIChE, Houston, Texas (February, 1971).
7. Holbrook, K., Walker, R., and Watson, W., J. Chem. Soc. (B) Phy. Org., (1969), 1089.
8. Hurd, C. D., and Eilers, K., Ind. Eng. Chem., (1943), 26, 776.
9. Kallend, A., Prunell, J., and Shurlock, B., Proc. Roy. Soc. London, (1967), 300A, 120.

10. Kasansky, V., and Pariisky, G., "Proceedings Third International Congress on Catalysis," Vol. 1, 368, Amsterdam (1965).
11. Kunugi, T., et al., Ind. Eng. Chem. Fundam., (1970), 9, 314.
12. Laidler, K., and Wojciechowski, Proc. Roy. Soc. London, (1960), 259A, 257.
13. Melloutte, H., Maleissye, J., and Delbourgo, R., Bull. Soc. Chemi. Fr., (1969), 8, 2652.
14. Polotrak, V., Leitis, L., and Voevodskii, V., Russ. J. Phy. Chem., (1959), 33, 379.
15. Rice, F., and Herzfeld, K., J. Phy. Coll. Chem., (1951), 55, 975.
16. Sakakibara, Y., Bull. Chem. Soc. Japan, (1964), 37, 1262.
17. Semenov, N., "Some Problems in Chemical Kinetics and Reactivity," Vol. 1, Princeton University Press, Princeton (1958).
18. Sims, J., Kershenbaum, L., and Shroff, J., Ind. Eng. Chem. Proc. Des. Dev., (1971), 10, 265.
19. Szwarc, M., J. Chem. Phys., (1949), 17, 284.
20. Tamai, Y., et al., Carbon, (1968), 6, 593.
21. Taylor, J., et al., Proc. Am. Chem. Soc., Div. Pet. Chem., B47, Boston (1972).
22. Tesner, P., "Seventh International Symposium on Combustion," Butterworth, London (1959).
23. Thomas, C., Egloff, G., and Morrell, J., Ind. Eng. Chem., (1939), 31, 1090.
24. Tsai, C., and Albright L., "Surface Reactions Occurring During Pyrolysis of Light Paraffins," ACS Meeting, Philadelphia (April, 1975).

14

Surface Effects During Pyrolysis of Ethane in Tubular Flow Reactors

JOHN J. DUNKLEMAN* and LYLE F. ALBRIGHT

School of Chemical Engineering, Purdue University, West Lafayette, Ind. 47907

Surface reactions producing coke, carbon oxides, and hydrogen occur during the pyrolysis of ethane, propane, and other hydrocarbons (1,2,3). Such surface reactions can be expected to be of relatively greater importance when smaller diameter (and hence higher surface-to-volume (S/V) ratio) reactors are used. The relative importance of surface reactions also increase as the operating pressures for pyrolysis decrease; lower pressures decrease the ratio of surface to mass of hydrocarbons in the reactor. As compared to commercial units, laboratory units use much smaller diameter reactors, and frequently they operate at lower pressures. Hence surface reactions are relatively more important in laboratory units. This factor is an important, if not the major, reason why there are often significant differences in the products obtained for ethane pyrolysis in the laboratory as compared to commercial units.

Coke formation in a pyrolysis unit is highly undesirable for several reasons: decreased yields of desired products, increased resistance of heat transfer to reactants, and possibly even eventual plugging of the reactor. Production of carbon oxides also reflects decreased yields of desired products. Oxidation of the inner surface of a 304 stainless steel (SS) reactor was found to result in increased coking and other surface reactions (1,4). Treatment of the surface with hydrogen sulfide, however, decreases the importance of surface reactions. The relative importance of nickel, iron, chromium, or various alloys of the three in the metal surface has been investigated to only a limited extent; nickel is generally considered to be a good catalyst for surface reactions including those producing coke. Olefins and acetylene are known to be precursors for coke(5,6). The pyrolytic coke formed is often impregnated with metal atoms (such as iron or nickel) that are catalytic in nature (6,7).

The material of construction and the past history of the surface may have a significant effect on these surface reactions. Crynes and Albright (1) indicated that different materials have some effect on the yields of desired products, but to date there

* Current address: Exxon Research and Engineering Co., Baytown, Texas.

has been no systematic investigation of different materials of construction. Pyrolyses experiments in quartz (or Vycor glass) and metal reactors hence may result in significantly different results; the rates and probable mechanism of coke and of carbon oxide formation probably are quite different in such reactors.

In the present investigation, extensive data were obtained for the pyrolysis of ethane in several reactors. Major differences in ethylene and coke yields were noted depending on the reactor used and its past history.

Experimental Details

The pyrolysis reaction system used was essentially the same as the one described earlier (1,4). Tubular reactors were bent with several loops and positioned inside a fluidized-sand bath that could be controlled to within $\pm$ 2.5°C in the range from 700° to 900°C. Sample ports were provided on the metal reactors, and product samples were taken at axial positions approximately 30, 60, and 100% of the total length. Each of the four reactors used had an internal diameter of approximately 0.46 cm and about 107 cm was positioned inside the sand bath. Two reactors were constructed from 304 stainless steel, one from Incoloy 800, and one from Vycor glass. Typical analyses of these metal alloys before and after pyrolysis experiments are reported by Tsai and Albright (2).

The gas chromatographic unit employed was the same one used earlier (1,4), but the method of calculating the composition of the gaseous product was modified somewhat (8). Hydrogen, carbon monoxide, carbon dioxide, and hydrocarbons up to n-butane were all carefully measured. The analytical results are considered to be more accurate especially at lower conversions of the feed paraffin than those reported earlier (1,4).

Experimental Results

The results of 48 runs using ethane as the feedstock clearly indicate that the material of construction of the reaction and the past history (and pretreatment) of the reactor have a significate effect on the composition and yields of the product. Bath temperatures for these runs were varied from 750° to 900°C; all runs were at atmospheric pressure except for two runs at 1.25 atmospheres; and 50% steam (by volume) was used in the feedstream for some runs. Ethane conversions investigated varied from about 20 to 70%.

Hydrogen and ethylene were generally the major products. Methane was always formed in significant amounts, up to perhaps 7% at higher conversions. Propylene, butane, and propane were minor products. Coke was always a fairly significant product, and in a few cases with highly oxidized metal reactors was a major product. Carbon monoxide and carbon dioxide were always produced in metal reactors when steam was used as a diluent with the feed or when the inner walls of the metal reactor were in an oxidized state. Although only preliminary results were obtained, yields of acetylene and heavier (C_4 and higher) hydrocarbons were less in the present investigation than those normally found in commercial tubular reactors (having much lower S/V ratios).

Figure 1 shows the differences in results for one set of comparative runs made in the Vycor, 304 stainless steel, and Incoloy reactors. All runs were made with a bath temperature of 800°C and with dry ethane feedstock at atmospheric pressure. Major differences occurred in these runs in the amounts of ethylene and hydrogen produced; little or no differences were noted, however, for methane, unreacted ethane, or minor components. Material balance calculations indicate that coke formation varied significantly as will be discussed in more detail later in this paper. More ethylene, less hydrogen (see Figure 1), and less coke were produced in the Vycor reactor as compared to the metal reactors. Experimental runs at 750°, 815°, 850°, and 900°C both with and without steam, although comparatively few in number confirmed the general trends found at 800°C.

The pretreatment of the metal reactors effected the product composition. For example, as shown in Figure 1, one of the runs in the Incoloy reactor having a reduced inner surface had at a given ethane conversion a product composition similar to that for the Vycor reactor. Yet a second run in which the Incoloy reactor had a more oxidized surface resulted in a product containing less ethylene and more hydrogen; the product for the second run was relatively similar to that for the stainless steel reactor.

For ethane pyrolysis in metal reactors, the composition of the product gases often changed with time, especially at the start of a run. Such changes were, however, minimal at most, when the Vycor reactor was used. Figure 2 shows results obtained in both the Vycor and the Incoloy reactors at 800°C. A higher fraction of ethylene was obtained in the run with dry ethane feed in the Vycor reactor than in any of the three runs in the Incoloy reactor. The highest yields of ethylene were obtained in the Incoloy reactor when the ethane feed was dry; in this run the reactor was new and untreated (and hence in a relatively reduced state). Adding steam to the ethane feed, decreased the yield of ethylene in the product and increased the yields of hydrogen and coke produced. Carbon oxides and particularly carbon monoxide were produced in significant amounts whenever steam was used. The composition of the product gas changed appreciably during especially the first half hour of the steam run employing a fairly reduced (and clean) metal reactor; the steam was obviously oxidizing the reactor surface during this initial phase of the run. At the end of this run, significant amounts of coke were present in the reactor.

In the case of steam runs in the H_2S-treated Incoloy reactor, considerably fewer surface reactions occurred than in the untreated (but reduced) reactor as shown in Figure 2. This conclusion is

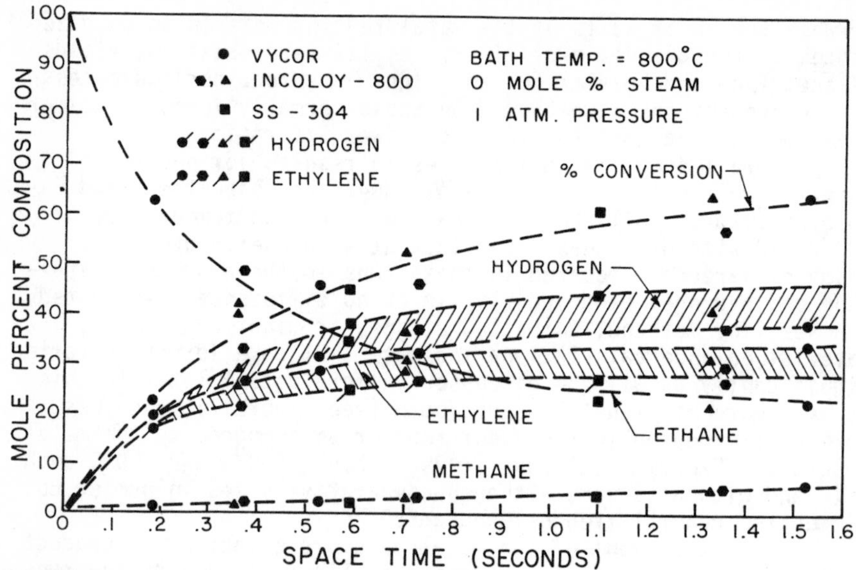

Figure 1. Product composition as a function of space time for runs in different reactors at 800°C and using dry ethane feed

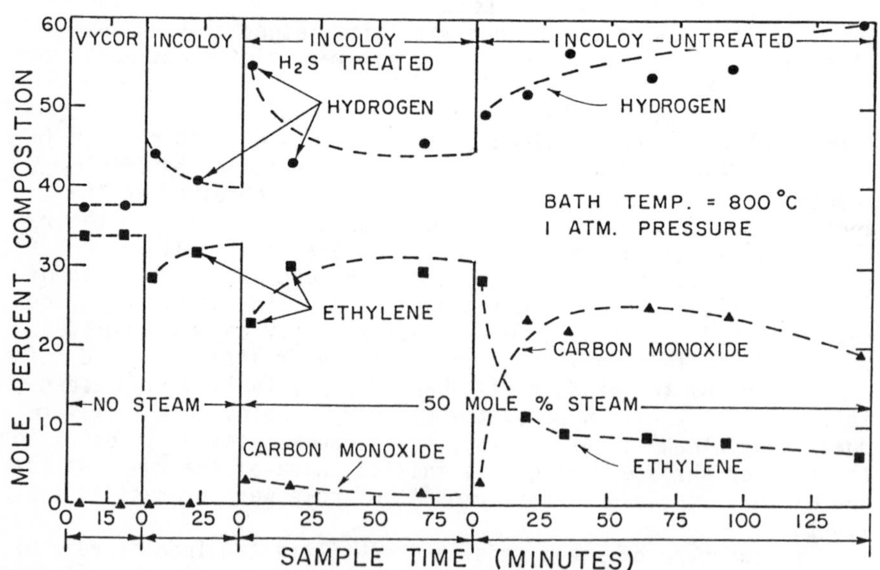

Figure 2. Effect of reactor surface and steam in pyrolysis of ethane at 800°C

based on higher ethylene yields and lower yields of hydrogen, carbon oxides, and coke for the H_2S-treated reactor. Clearly, H_2S-treatment has a highly beneficial effect relative to improved ethylene yields; pure H_2S is, however, corrosive and if used at high temperatures, above 800°C, can seriously corrode or even plug a reactor as was confirmed in this investigation. Unless specified otherwise, H_2S-treatments were at 750° to 800°C for 30-60 minutes. In all cases, such treatments minimized surface reactions in several subsequent runs.

All results obtained in this investigation for comparable runs indicated that whenever the ethylene yields were lower, the combined yields of coke and carbon oxides were higher as were also the yields of hydrogen. Carbon oxides were always formed in significant amounts when steam was added to the ethane feedstream for experiments in the two metal reactors tested but not for those in the Vycor reactor. These findings suggest that part of the ethylene initially decomposes on the surfaces to form carbon (or coke) and hydrogen. Part of the coke deposited on the metal surfaces then reacts with steam (when it is present) to form carbon oxides and more hydrogen. To test this postulate, the experimental data for each run were corrected based on the assumption that all coke, CO, and CO_2 formed were produced from ethylene as indicated by the following three reactions:

$$C_2H_4 \rightarrow 2C \text{ (coke)} + 2H_2 \qquad (1)$$

$$C + H_2O \rightarrow CO + H_2 \qquad (2)$$

$$C + 2H_2O \rightarrow CO_2 + 2H_2 \qquad (3)$$

Using these equations, the amount of ethylene was increased and that of hydrogen was reduced. This correction technique resulted in essentially identical products for a given temperature and partial pressure of ethane for each of the following:

(a) All comparable runs in metal and Vycor reactors. The corrected data for the runs shown in Figure 1 are reported in Figure 3. Relatively small differences occurred between the corrected results (Figure 3) and the experimental results for the Vycor reactor (see Figure 1). Larger differences occurred in the case of the metal reactors.

(b) At all time periods for runs in metal reactors.

Figure 3 indicates only a relatively small scatter of data, as compared to Figure 1; in these two graphs, ethane conversion and the product compositions for hydrogen, ethylene, ethane, and methane were plotted versus space time (defined as the volume of the reactor divided by the volumetric flowrate of the feedstream at reaction conditions). The solid curves shown in Figure 3 are predictions based on the mechanistic (or kinetic) model to be discussed later in this paper. When, however, the "corrected"

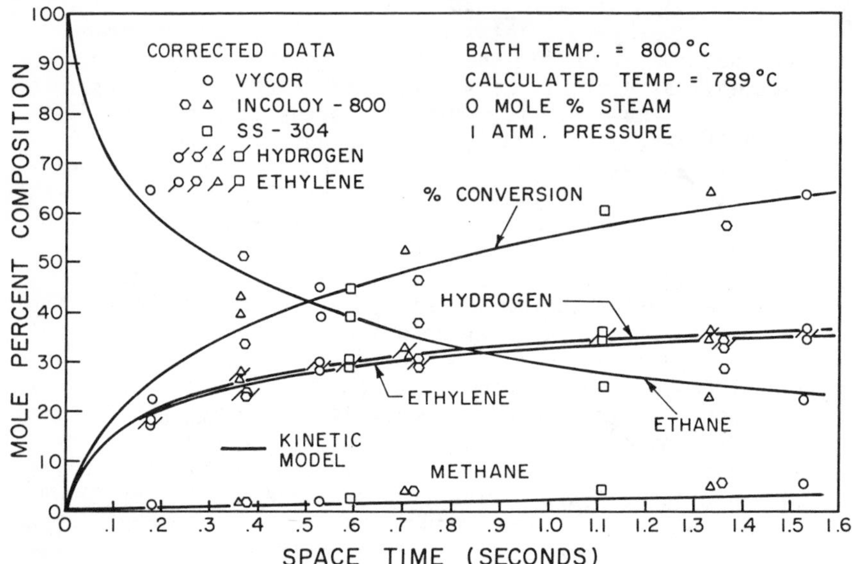

Figure 3. Corrected product composition as a function of space time at 800°C and using dry ethane feed

product compositions were plotted versus ethane conversions, considerably less scatter of data were noted. Figure 4 and 5 indicate examples for uncorrected and corrected results for runs at 800°C (bath temperature) and using a 50:50 mixture of ethane and steam. The 304 stainless steel reactor had in one run an active surface (i.e. a relatively oxidized surface) and had in the other runs a fairly non-active surface (i.e. a relatively reduced surface). Active surfaces resulted in considerably more surface reactions.

Figure 6 and 7 indicate typical ethylene yields as a function of ethane conversion for runs at 800°C with and without steam respectively. These yields (based on the moles of ethane that reacted) decreased with increased ethane conversion and in the following order: Vycor, Incoloy, and 304 stainless steel. Higher yields were also noted in the metal reactors with dry ethane as compared to wet ethane runs; the opposite was the case however, for the runs in the Vycor reactor. Corrected results indicate, however, higher ethylene yields with wet ethane feeds regardless of the reactor used. The corrected results agree well with the mechanistic model also shown in Figures 3,6, and 7; this model will be discussed later in the paper.

Surface reactions including carbon (or coke) deposition (on the reactor surface) varied significantly in the reactors investigated. Table I and Figure 8 show carbon results for several runs in the 304 stainless steel, Incoloy, and Vycor reactors.

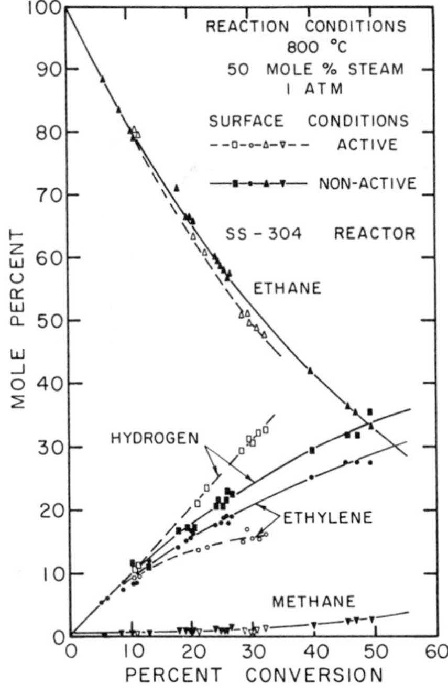

Figure 4. Product composition in both oxidized (active) and nonoxidized (nonactive) 304 stainless steel reactors at 800°C

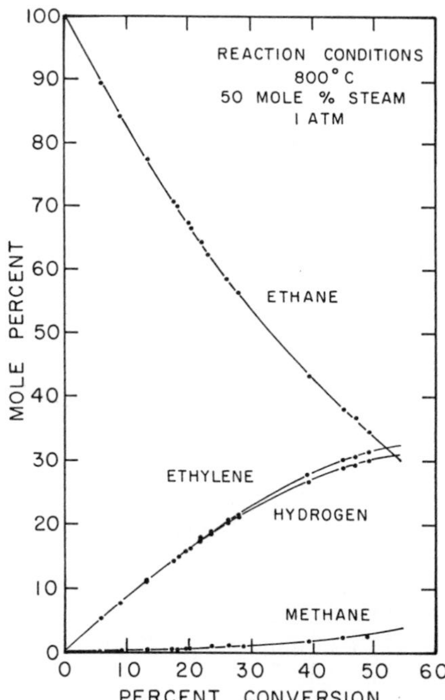

Figure 5. Corrected product composition as a function of ethane conversion for wet ethane feed at 800°C

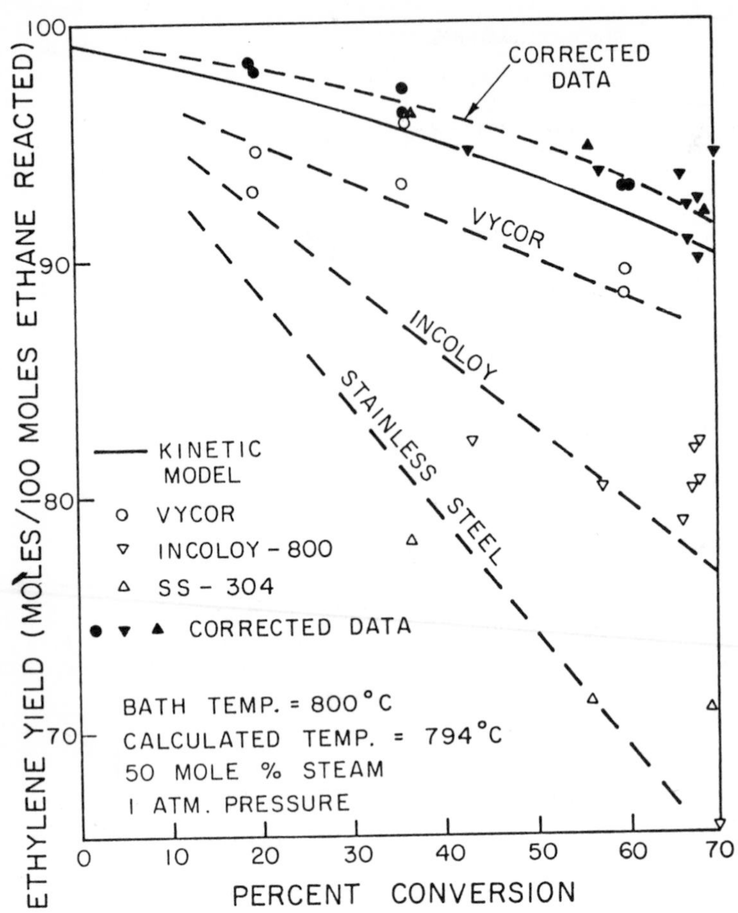

Figure 6. Ethylene yields for runs at 800°C using wet ethane feed in different tubular reactors

Significant findings are outlined below:

(a) Changes in the rate of carbon deposition were always noted at the beginnings of runs in the two metal reactors and perhaps even for runs in the Vycor reactor. In general, the initial rates were higher than those obtained after a short period of operation; apparently the rates then approached more or less steady-state values.

(b) For both metal reactors when they were relatively new, yields of carbon increased very substantially when ethane-steam mixtures were used as compared to dry ethane feed (see Runs 26 vs 27 and 44 vs 45-1). Furthermore, carbon yields (defined here as weight of car-

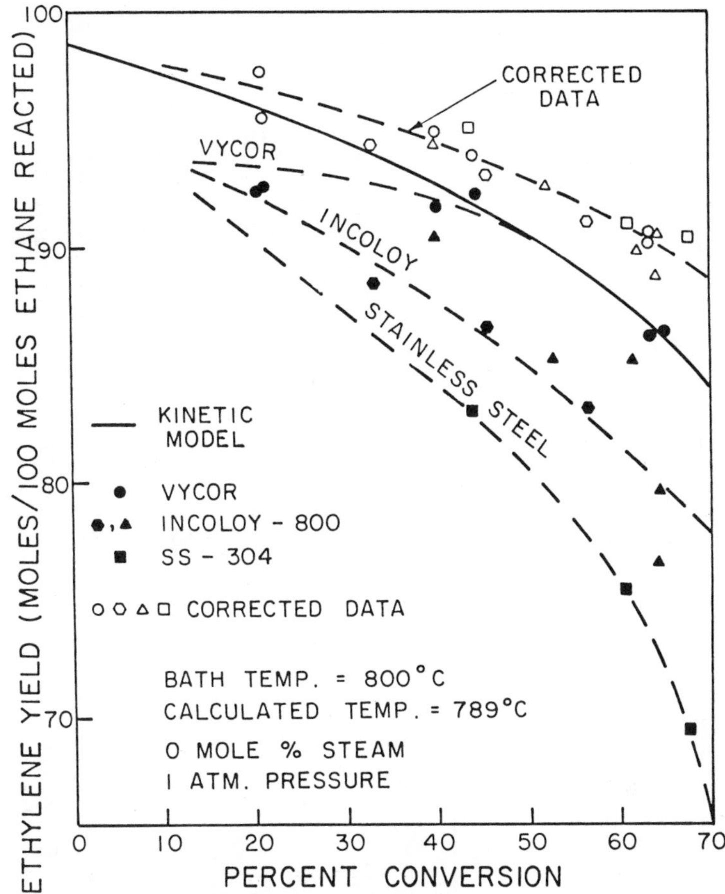

Figure 7. Ethylene yields for runs at 800°C using dry ethane feed in different tubular reactors

bon produced per weight of ethane that reacted) increased substantially during at least the first two hours of the runs (Runs 27 and 45-1) using steam as a diluent of the ethane. Carbon deposition in Run 28, which was at the same operating conditions as Run 27, also had high levels of coke formation. Steam had little if any effect, however, on carbon yields for runs in the Vycor reactor reactor (see Runs 40-1 and 40-2).

(c) Yields of carbon increased steadily with increased ethane conversions (see Figure 8).

(d) For runs both with and without steam, H_2S-treated metal reactors showed a large decrease in the rate of carbon

Table I Carbon Deposition During Ethane Pyrolysis:
Sand-Bath Temperature, 800 $^{\circ}$C; Ethane Conversion, ~ 65%; Atmospheric Pressure

Run No.	Reactor	Mole % Steam	Rate of Carbon Deposition, g/min. X 10^3 (Average at 1.5 sec. space time)	Carbon Yield*	Comments on Reactor
40-1	Vycor	0	1.9	2.8	—
40-2	Vycor	50	1.9	2.9	—
26	Incoloy	0	12	5	New, untreated reactor
27	Incoloy	50	27**	50**	Run after 26
29	Incoloy	50	12	9	Reactor treated before run with H_2S
30	Incoloy	50	15	13	Run after 29
44	SS-304	50	73**	71**	New, untreated reactor
45-1	SS-304	0	16	10	New, untreated reactor
45-2	SS-304	0	8.5	4	Reactor treated before run with H_2S
46	SS-304	50	17	16	Run after 45-2

* For metal reactors, first data point of run was not included since it was high and atypical
** Average value during run; rate of carbon deposition increased as run progressed

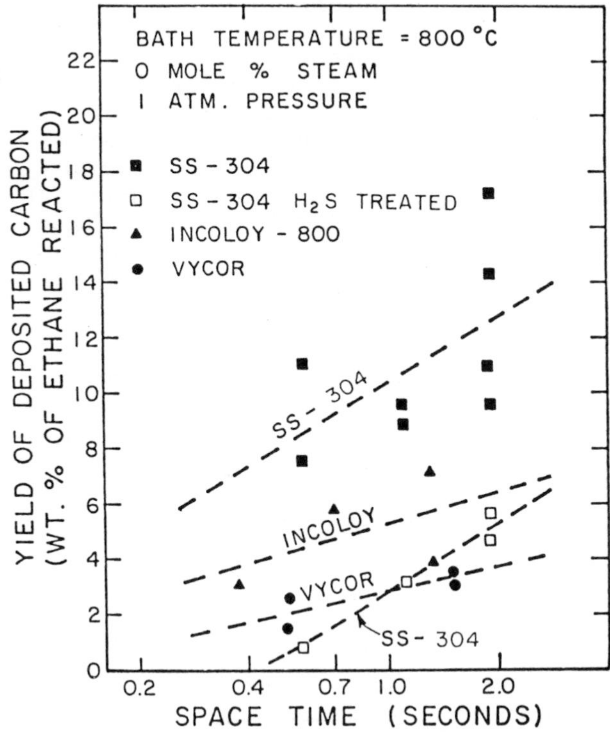

Figure 8. Carbon yields for runs at 800°C using dry ethane feed in different tubular reactors

formation; the results of Runs 27 and 29 and of 45-1 and 45-2 indicate typical comparisons.

(e) When metal reactors that had been treated with H_2S were used for pyrolysis of ethane-steam mixtures, the rates of carbon deposition increased slowly with time. An increase of about 25% occurred in the case of the Incoloy reactor after several hours of operation (see results of Runs 29 and 30). Crynes and Albright (1) also noted a slow rate of increase of surface activity in 304 stainless steel reactors when the reactors were contacted with steam.

(f) As a general rule, the rate of carbon deposition (and other surface reactions) for both dry and wet runs was in the following order for the three reactors investigated:

304 stainless steel > Incoloy > Vycor

In the small diameter reactors (that have a high S/V ratio) of this investigation, carbon yields were much larger as a rule than yields in commercial reactors.

Figure 9 was constructed based on the major gas-phase reaction step; namely $C_2H_6 \rightarrow C_2H_4 + H_2$, and on the two major surface reactions, namely Reactions 1 and 2 reported earlier. Experimental data points for runs in both metal reactors are also plotted on the graph. It is obvious that:

(a) Large percentages of the ethylene often decomposed to carbon and hydrogen in the small laboratory metal reactors when steam was present. Even in metal reactors that had been treated with H_2S, up to 20 to 40% of the ethylene reacted in some cases (see Figure 9). 3 to 4% ethylene reacted in the Vycor reactor.

(b) A higher fraction of the carbon deposited on the surface reacted to produce carbon monoxide in the Incoloy reactor as compared to the 304 stainless steel one.

Corrected pyrolysis results indicated in all cases that ethylene yields increase whenever the reaction temperature increased, the partial pressure of ethane in the feedstream was lowered, or the ethane conversions were less. Lower partial pressures of ethane were realized by use of steam as a diluent and for runs at one atmosphere pressure as compared to runs at 1.25 atmospheres. In one run, about 10% hydrogen was added as a diluent to a wet ethane feed. Similar results were obtained with and without hydrogen except a small decrease of the ratio of carbon dioxide to carbon monoxide occurred in the hydrogen run.

It should be emphasized that pyrolysis runs made in small diameter reactors such as those used here accentuate the relative importance of surface reactions as compared to those made in commercial reactors that employ much larger diameter reactors and also higher operating pressures. Extrapolation of the experimental results of this investigation to commercial units must then

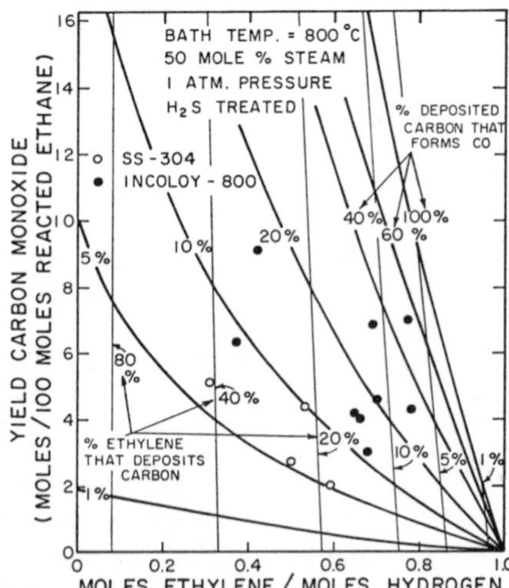

Figure 9. Approximate levels of two major surface reactions during ethane pyrolysis: comparison of experimental results in metal reactors

be done with caution:

(a) The absolute differences in yields of ethylene between various reactors such as shown in Figures 6 and 7 are certainly much greater than those to be found in conventional commercial units. The relative order of yields for the different materials of construction should, however, be identical in both cases.

(b) The relative effect of steam on ethylene yields such as noted in the metal reactors of this investigation is opposite to that found in commercial units. Steam acts to lower the partial pressure of the hydrocarbons; lower partial pressures tend to promote higher ethylene yields as confirmed by correcting the experimental results to a "surfaceless" basis. Steam also promotes surface reactions in metal reactors, as shown in Figure 2; such reactions lower ethylene yields. The partial pressure effect for steam is dominant in most commercial reactors whereas the surface reactions effect was dominant in the small diameter reactors used in this investigation.

Correlation of Ethane Pyrolysis Data

If a mechanistic model is to be developed to represent the ethane pyrolysis data, then all important reaction steps should be included. Clearly surface reactions often fall within this latter category. Since insufficient data have as yet been obtained relative to rate constants and to the effect of surface reactions, it is currently impossible to develop such a model. If the experimental data can be corrected to eliminate the effect of surface reactions, then models using only gas-phase reaction steps could be tested. All corrected data at 750° to 900°C including that shown in Figures 3,5,6, and 7 were tested in this way.

Somewhat simpler and somewhat modified models as compared to the one shown in Table II were initially tested. The results of these initial tests were promising, and by a trial-and-error procedure it was found that the model shown in Table II agreed closely with the corrected data as shown in Figures 3,5,6, and 7: this agreement was within experimental accuracy over the entire range of conditions (both temperature and pressure) investigated. Several features of this model are as follows:

(a) It is basically an 'a priori' model since the parameters, energies of activation and frequency terms, for all reactions of importance during ethane pyrolysis were found in the literature (10-16). For several reactions, ranges of values for the parameters are reported. It was hence necessary in such cases to choose average values or to evaluate critically in order to determine the best values. Kondratiev's compilation (10) with critical remarks was most valuable.

Table II. Rate Constants for Reactions in Mechanistic Model.

No.	Reaction	Forward Reaction A	Forward Reaction E	Reverse Reaction A	Reverse Reaction E
		Rate Constants: A = Frequency Factor, sec^{-1} or cc-g $mol^{-1}sec^{-1}$	E = Activation Energy, kcal-g mol^{-1}		
Ethane Reactions					
1.	$C_2H_6 \rightleftarrows 2\ CH_3\cdot$	1.0×10^{16}	86.0	3.8×10^{13}	0.
2.	$CH_3\cdot + C_2H_6 \rightleftarrows CH_4 + C_2H_5\cdot$	1.0×10^{12}	10.6	1.0×10^{11}	11.0
3.	$CH_3\cdot + H_2 \rightleftarrows CH_4 + H\cdot$	3.2×10^{12}	10.2	1.3×10^{14}	12.6
4.	$C_2H_5\cdot \rightleftarrows C_2H_4 + H\cdot$	3.8×10^{13}	38.0	4.0×10^{13}	1.6
5.	$H\cdot + C_2H_6 \rightleftarrows H_2 + C_2H_5\cdot$	1.3×10^{14}	9.7	3.0×10^{11}	10.8
Propane Reactions					
6.	$C_3H_8 \rightleftarrows C_2H_5\cdot + CH_3\cdot$	1.2×10^{17}	85.0	4.2×10^{13}	0.
7.	$C_2H_5\cdot + C_3H_8 \rightarrow C_2H_6 + C_3H_7\cdot$	7.5×10^{10}	10.0		
8.	$CH_3\cdot + C_3H_8 \rightleftarrows CH_4 + C_3H_7\cdot$	8.1×10^{11}	10.3	2.5×10^{11}	19.5
9.	$H\cdot + C_3H_8 \rightleftarrows H_2 + C_3H_7\cdot$	3.2×10^{13}	7.8	6.3×10^{12}	14.6
10.	$C_3H_7\cdot \rightleftarrows C_2H_4 + CH_3\cdot$	2.0×10^{12}	34.5	3.3×10^{11}	7.8
11.	$C_3H_7\cdot \rightleftarrows C_3H_6 + H\cdot$	1.0×10^{14}	37.0	2.5×10^{14}	3.8
Termination Reactions					
12.	$H\cdot + C_2H_5\cdot \rightarrow C_2H_6$	4.5×10^{13}	0.		
13.	$CH_3\cdot + C_2H_5\cdot \rightarrow C_2H_4 + CH_4$	1.5×10^{12}	0.		
14.	$C_2H_5\cdot + C_2H_5\cdot \rightarrow C_4H_{10}$	1.7×10^{13}	0.		
15.	$C_2H_5\cdot + C_2H_5\cdot \rightarrow C_2H_6 + C_2H_4$	2.6×10^{12}	0.		
16.	$C_3H_7\cdot + CH_3\cdot \rightarrow C_4H_{10}$	2.0×10^{13}	0.		
17.	$C_3H_7\cdot + C_3H_7\cdot \rightarrow C_3H_6 + C_3H_8$	6.3×10^{13}	0.		
Secondary Reactions					
18.	$C_4H_{10} \rightarrow C_2H_5\cdot + C_2H_5\cdot$	7.9×10^{17}	86.0		
19.	$H\cdot + C_2H_4 \rightleftarrows H_2 + C_2H_3\cdot$	8.5×10^{12}	7.2	7.9×10^{12}	7.4
20.*	$CH_3\cdot + C_2H_4 \rightarrow CH_4 + C_2H_3\cdot$	1.5×10^{12}	10.0		
21.	$C_2H_5\cdot + C_2H_4 \rightarrow C_2H_6 + C_2H_3\cdot$	3.2×10^{11}	13.0		
22.	$C_2H_3\cdot + C_2H_4 \rightleftarrows C_4H_6 + H\cdot$	5.0×10^{11}	7.4	6.3×10^{12}	11.0
23.	$C_2H_5\cdot + C_2H_4 \rightarrow C_4H_9\cdot$	1.0×10^{10}	5.5		
24.	$CH_3\cdot + C_3H_6 \rightleftarrows C_4H_9\cdot$	3.2×10^{11}	7.4	4.0×10^{14}	37.0
25.	$CH_3\cdot + C_3H_6 \rightarrow C_2H_4 + C_2H_5\cdot$	3.2×10^{10}	7.5		
26.	$C_4H_9\cdot \rightarrow C_4H_8 + H\cdot$	3.2×10^{15}	46.0		

* "A" value disagrees with literature value by a factor of 10, but the reaction is of minor importance for modeling ethane pyrolysis data.

(b) The model shown in Table II also represents reasonably well, propane pyrolysis data, as will be discussed in the next chapter of this book (9).

(c) This model includes steps for production of most minor products. No provision **is, however**, made in the model for production of acetylene or C_5-C_6 hydrocarbons that are known to be present in small amounts.

In testing the model, two major assumptions were made; these were the plug-flow assumption (the absence of radial concentration gradients) and an isothermal profile throughout the reactor. Calculations by Dunkleman (8) confirmed that these assumptions were closely approached in the reactor.

The model shown in Table II gives rise to rate equations that are a highly non-linear, coupled set of ordinary differential equations with constant coefficients (or rate constants) at a given temperature. Since values of the constants for the various equations often vary by several orders of magnitude, it has been virtually impossible until recently to solve such "stiff" systems of differential equations without the simplifying pseudo steady-state assumption. The integration method of Gear (17), however, has proven most successful for rigorous integration of the equations, and it was used in this investigation.

For each run, an axial temperature profile was first calculated using the best available values for heat transfer coefficients and knowing the heats of reactions. This profile was found to be quite flat, and an average temperature value was then used for solving the equations. The average gas temperatures were calculated to be as follows:

Bath Temp., ^{o}C	Average Gas Temperature, ^{o}C	
	Dry ethane	50:50 ethane and steam
750	747	749
800	789	794
850	822	828
900	843	853

Figures 10 and 11 for runs using dry ethane and an ethane-steam mixture respectively show good agreement between the model and all experimental data obtained in this investigation. Even better agreement is obtained when the temperature values were changed slightly; obviously as shown in Figure 10, the kinetics of pyrolysis are highly sensitive to temperature. There is, of course, some uncertainity in estimating the "best" gas temperature to use.

(a) Heat transfer calculations used to calculate the gas temperatures are subject to some errors and require certain simplifications.

(b) The gas temperature in the tubular reactor varies to at

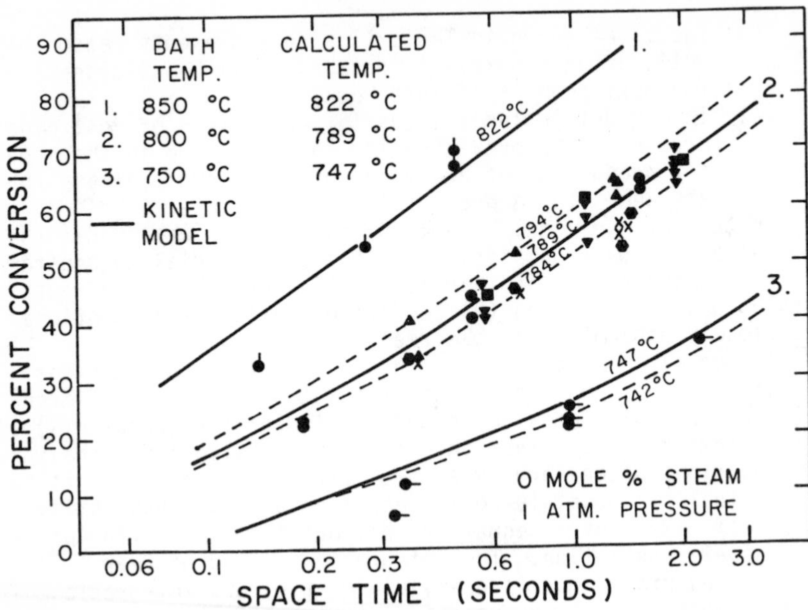

Figure 10. Predicted and experimental ethane conversions with dry ethane feed

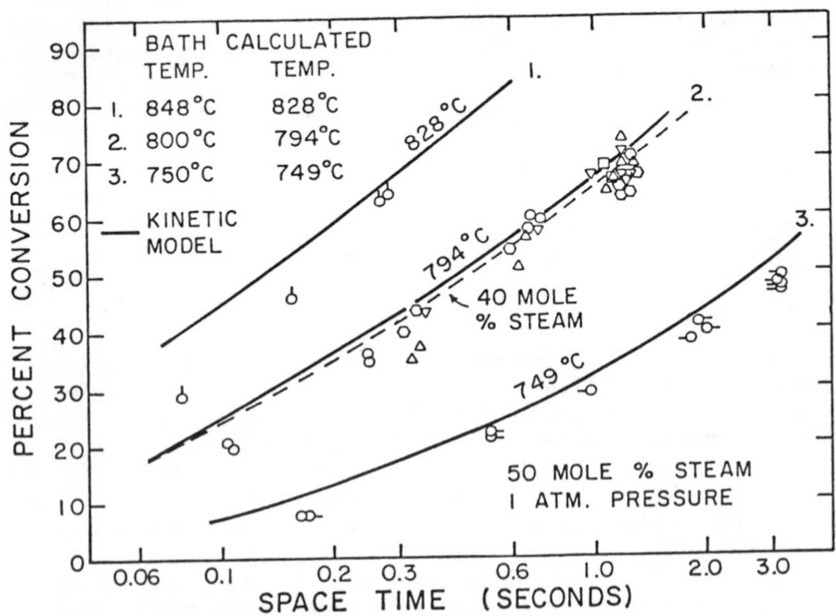

Figure 11. Predicted and experimental ethane conversions with wet ethane feed

least some extent, about $\pm$ 5°C from the average, in the axial direction. The reactor inlet serves as a preheat zone, and there is some uncertainity in calculating the exact residence times. It is not surprising then that the experimental values especially at shorter times are low since some preheating is occurring at such times. Much better agreement in general occurs at higher times.

(c) There is also some uncertainity in measuring and maintaining the desired sand bath temperatures.

Some experimental errors may also occur relative to the exact amount of steam added to the ethane feed stream. Possibly slightly less than the reported amounts of steam may sometimes have been present. Such errors would have small effects on both the calculated ethylene yields (see Figure 11) and on the residence time.

Even better agreement was obtained with the model when the product gas composition was calculated as a function of ethane conversion.

The model reported in Table II is much more reliable than other models reported earlier (16,18,19). These latter models do not fit experimental data well except over limited conditions, primarily at ethane conversions in the 50-70% range (8).

Discussion of Results

The technique developed for correction of ethane pyrolysis data to a "surfaceless" basis is considered to be of major importance. For a hypothetical reactor whose S/V = 0, there would obviously be no surface reactions, and only gas-phase reaction would be occurring. The two factors that strongly support the hypothesis that the corrected results are essentially those to be expected in a surfaceless reactor are as follows:

(a) Excellent fit of predicted results of the 'a priori' model (see Table II) and the corrected results.

(b) Close agreement of corrected results for all comparable runs regardless of the tubular reactors used; the experimental results were often very different.

It is, of course, recognized that both improved techniques for correcting experimental data and improved mechanistic (or kinetic) models can be developed in the future. Considering first the correction technique, other olefins (besides ethylene), diolefins, acetylene, and aromatics which were formed in small amounts (4) all probably contributed to some extent to coke formation in the present investigation. Hence the ethylene yields based on the correction technique used are thought to be too high.

A question that needs to be answered is why were acetylene and heavier hydrocarbons present in only relatively small amounts in the product streams from the laboratory units used in this investigation as compared to streams from commercial units. It is postulated that these compounds were destroyed (forming carbon and hydrogen) in larger fractions because of the relatively high S/V ratio in the laboratory units. When additional quantitative information on the coking abilities of these various hydrocarbons become available, the correction technique can be improved.

The predictions of the present model are thought to be more accurate than the corrected results. The small differences between the predicted and corrected results are probably caused mainly by the small errors in the correction technique used. It is, of course, realized that additional gas-phase reactions are occurring besides those shown in Table II; these latter reactions probably are of relatively minor importance at least for ethane pyrolysis. The model shown in Table II is postulated to be a good representation of all important gas-phase reactions occurring during ethane pyrolysis, and it should be useful in providing a better understanding of these reactions and their relative importance. Because of the mechanistic nature of the model, it probably will predict with good accuracy pyrolysis results up to at least 1000°C and pressures of several atmospheres; such temperatures and pressures likely are the upper limits of commercial interest for ethane pyrolysis.

Small differences in the kinetics of pyrolysis may have occurred in different reactors; see, for example, the results shown in Figures 1 and 3. Such differences could have been caused by one or more factors. One factor that likely is of some importance is small differences in the gas temperature; heat transfer obviously depends to some extent on the materials of construction. Small differences in residence times at reaction conditions could also have occurred in the various reactors that did not have exactly the same internal volumes. In addition, surface reactions may affect the kinetics of the reactions. Some initiation or termination of free radicals may have occurred at the reactor surfaces. Probably at least some hydrogen free radicals were formed during coke formation.

The following procedure is suggested for predicting the product composition that will be obtained in large diameter tubular reactors (such as are used in commercial units):

(a) Experimental data obtained in a small diameter reactor with a large S/V ratio) is first corrected to a surfaceless basis (i.e., a reactor whose S/V = 0).

(b) An interpolation should then be made between the experimental results and the corrected results to the S/V ratio of the commercial reactor. More experimental data are needed to determine the best relationship between the composition of the product and the S/V ratio; until such data are obtained, a straight line relationship can be assumed. If different pressures are used in the laboratory and the commercial units, the ratio of surface to mass of hydrocarbons should be used instead of the S/V ratio.

The results of this investigation and the procedure for extrapolating the results to different size reactors strongly imply that small but nevertheless significant differences in pyrolysis results may occur in commercial units as the material of construction for the tubes is varied. Small increases in ethylene yields and increased time between decoking seems probable if a material of construction that results in less surface reactions as compared to Incoloy 800 is chosen. If the time between decoking could be increased by a factor of two, annual ethylene production for a given pyrolysis unit likely would often increase by 3 to 5%. Furthermore, in such a case, lower operating expenses per given weight of ethylene would likely be realized.

Since different materials of construction were found to result in significantly different yields of ethylene, further investigations are recommended to find materials that will not only have the desired physical and corrosion-resistant properties but will also result in lower levels of surface reactions. Suitable materials that are more inert than Incoloy 800 may be available. More investigations on surface treatments including H_2S are also recommended. Research in this area is now in progress at Purdue University.

Literature Cited

1. Crynes, B. L. and Albright, L. F., Ind. Eng. Chem. Process Design and Develop. 8, 25 (1969).
2. Tsai, C. H. and Albright, L. F., Chapter 16, This Book, 1976.
3. Brown, S. M. and Albright, L. F., Chapter 17, This Book, 1976.
4. Herriott, G. E., Eckert, R. E., and Albright, L. F., AIChE Journal 18, 84 (1972).
5. Bernardo, C. A. and Lobo, L. S., J. Catalysis 37, 267 (1975)
6. Lobo, L. S. and Trimm, D. L., J. Catalysis 29, 15 (1973).
7. Frech, K. J., Hoppstock, F. H., and Hutchings, D. A., Chapter 12, This Book (1976).
8. Dunkleman, J. J., Kinetics and Surface Effects of the Pyrolysis of Ethane and Propane in Vycor, Incoloy, and Stainless steel Tubular Reactors from 750° to 900°C, PhD thesis, Purdue University, 1976.
9. Dunkleman, J. J. and Albright, L. F., Chapter 15, This Book (1976).
10. Kondratiev, V. N., "Rate Constants of Gas Phase Reactions," NSRDS - COM - 10014, U. S. Department of Commerce (1972).
11. Bensen, S. W. and O'Neal, H. E., "Kinetic Data on Gas Phase Unimolecular Reactions," NSRD - NBS 21, U.S. Department of Commerce (1970).
12. Trotman - Dickenson, A. F. and Milne, G. S., "Tables of Bimolecular Gas Reactions, "NSDS 9, U. S. Department of Commerce(1967).
13. Ratajczak, E. and Trotman-Dickenson, A. F., "Supplementary Tables of Biomolecular Gas Reactions," (supplement to "Tables

of Bimolecular Gas Reactions" by A. F. Trotman-Dickenson and G. S. Milne, NSRDS - NBS 9 (1967) University of Wales Institute of Science and Technology, Cardiff, Wales (1970).

14. Benson, S. W. and Haugen, G. R. J. Phy. Chem., 71, 6 (1967).
15. Kunugi, T., Sakai, T., Soma, K., and Y. Sasaki, Ind. Eng. Chem. Fund., 8,3, 379 (1969).
16. Pacey, P. D. and Purnell, J. H. Ind. Eng. Chem. Fund., 11, 2 233 (1972).
17. Gear, C. W., Comm. of ACM 14, 3, 176 (1971).
18. Snow, R. H., Peck, R. E. and von Fredersdorff, C. G. AIChE Journal, 5, 3, 304 (1959).
19. Zdonik, S. B., Green, E. J. and Hallee, L. P. Oil and Gas J. 65, 26, 96 (1967).

15

Pyrolysis of Propane in Tubular Flow Reactors Constructed of Different Materials

JOHN J. DUNKLEMAN* and LYLE F. ALBRIGHT

School of Chemical Engineering, Purdue University, West Lafayette, Ind. 47907

Recent results of Dunkleman and Albright (1) who pyrolyzed ethane have shown that the composition of the product often varies significantly depending on the material of construction of the reactor and on the type of pretreatment of the inner surface of the reactor. Considerably higher yields of ethylene were obtained in a laboratory Vycor reactor as compared to an Incoloy 800 reactor and especially a 304 stainless steel (SS) reactor. Oxidized inner metal surfaces promote the production of coke (or carbon) and carbon oxides, but sulfided surfaces suppress such production.

When olefins are the desired products of propane pyrolysis, ethylene, propylene, methane, and hydrogen are the major products. Brown and Albright (2) found that both ethylene and propylene react to a significant extent on the surface of metal reactors. In general, 304 SS steel reactors result in more olefin reactions and in the formation of more coke and hydrogen than occurs on an Incoloy 800 surface. Ethylene is generally more reactive on the surface than propylene. Metal oxides on the surfaces of these reactors react with hydrocarbons forming carbon oxides, water, and hydrogen; such reactions result in the reduction of the surface. Metal sulfides on the surface, however, minimize surface reactions.

Preliminary investigations (3,4,5) have shown that surface reactions are important during the pyrolysis of propane at temperatures of commercial importance. Until now, however, experimental data were not available to make direct comparisons between reactors constructed of different materials or between propane and ethane relative to surface reactions. Such information has now been obtained in the present investigation.

Experimental Details

The same laboratory apparatus and tubular reactors used for pyrolysis of ethane (1) were employed in this investigation. The gas chromatograph and the method of product analysis used by Herriott (6) were modified, however, in order to obtain a more accurate analysis of the gaseous products up to butane. This was

* Current address: Exxon Research and Engineering Co., Baytown, Texas.

achieved by a combination of techniques including lowering the temperature of the chromatographic column, using Porapak-Q as the column packing, and increasing the sensitivity and chart speed of the recorder. These improvements enabled the peak areas for the individual hydrocarbon products to be more accurately measured. Although helium was retained as the carrier gas, the analysis of hydrogen was improved by use of an atom balance around the flow system (7). Because of these improvements, the earlier results (3,4) were found to be less accurate especially at low propane conversions as will be discussed in more detail later in this paper.

Furthermore, the improved analytical techniques used here enabled the analysis of carbon monoxide and carbon dioxide which were present when steam was added to the propane feed for experiments in the metal reactors; such analyses had not been made earlier. In general, relatively little carbon dioxide was produced during a pyrolysis run.

Material balance calculations based on both the inlet and the exit gaseous streams to the tubular reactor indicated the amount of coke deposited on the reactor wall. On several occasions, an oxygen burn-out technique was used in an attempt to check the amount of carbon formed based on the above calculations. In the burn-out technique, pure oxygen was passed through the hot reactor and the amounts of carbon oxides formed were measured. Less carbon was in each case burned out of the reactor than indicated by the material balance. Some of the carbon was, however, later found to have been blown from the reactor, and hence not burned, as indicated by carbon eventually found in the relatively cool exit line.

Results

Significant differences were noted in the composition of the resulting gaseous product when propane was pyrolyzed in Vycor, 304 SS, and Incoloy 800 reactors. Comparative runs were made both at 750° and 800°C; in some cases, the propane was fed to the reactor dry, and in other cases, it was premixed with steam in a 1:1 mole ratio. Table I shows the results of experimental runs in the three reactors at 800°C and using wet propane as feed.

Yields of particularly ethylene, hydrogen, and coke often varied significantly for comparative runs. In such cases whenever higher ethylene yields occurred, yields of both hydrogen and coke were less. Furthermore, the composition of the product gases frequently changed appreciably during the first few minutes of runs in metal reactors. For the runs in the Incoloy 800 and the 304 SS reactors, shown in Table I, ethylene yields increased substantially during the initial phases of the runs. In general, the ethylene yields varied in the three reactors as follows:

Vycor glass > Incoloy 800 and 304 SS

Table I
Product Composition and Carbon Yield
for Propane Pyrolysis in Different Reactors
(800°C, 50% steam in feed)

Reactor	Time During Run, min.	60% Propane Conversion			85% Propane Conversion		
		Composition of Product Gas, %		Coke Yield, wt.%	Composition of Product Gas, %		Carbon Yield, wt.%
		Ethylene	Hydrogen		Ethylene	Hydrogen	
Model*	--	24.0	12.5	0.0	33.0	11.0	0.0
Vycor	10-60	23.8	14.2	0.2	33.6	10.0	1.4
Incoloy 800**	10	19.9	16.9	0.8	23.9	30.2	8.7
	60	—	—	—	27.4	20.1	4.4
304SS**	10	—	—	—	24.0	27.3	7.0
	40	—	—	—	26.0	19.2	3.1

* The mechanistic model was described earlier (1).

** Both metal reactors were treated with H_2S for 15 minutes and then used for several hours of pyrolysis of wet ethane before being used for propane pyrolysis.

Insufficient data were, however, obtained to compare the yield results in the two metal reactors under essentially identical levels of surface oxidation. If there were any differences in the results for these two reactors, these differences were probably, at most, small.

Figure 1 shows the product composition for a run in a Vycor glass reactor maintained at 800°C and using wet ethane feed; the major products (ethylene, propylene, hydrogen, and methane) and ethane are shown as a function of space time. Propane conversion is also shown. The results of all propane runs including the one used to prepare Figure 1 indicated the following:

(a) The mole fractions of both hydrogen and ethylene in the gaseous product stream passed through maxima at relatively low propane conversions, perhaps about 30%. Obviously, both propylene and hydrogen were reactive, entering into secondary reactions, especially at higher conversions.

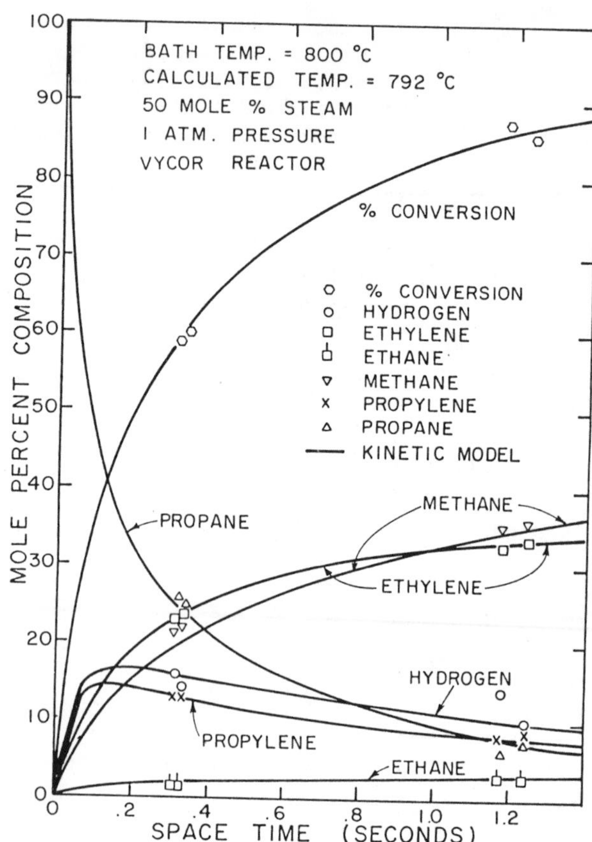

Figure 1. Product composition for propane pyrolyses at 800°C bath temperature and using wet ethane feed in Vycor reactor

(b) At propane conversions up to perhaps 70-80%, the mole fractions of ethylene in the gaseous product streams were greater than those of methane. A crossover occurred, however, at higher conversions; apparently considerable ethylene was destroyed by secondary reactions at higher propane conversions. Methane was however relatively unreactive at the conditions investigated.

For runs at 800°C and at fairly low propane conversions, the fractions of propylene and hydrogen in the gaseous product exceeded those of ethylene and methane. These findings differ from the results reported by other investigators (3,4) who are thought to have used less accurate analytical techniques. Still other investigators (8, 9) report results similar to those found in this investigation. It is not clear though if propylene and hydrogen compositions are greater at lower temperatures, such as 750°C or less. For the results at 800°C, apparently propyl radicals decompose more rapidly into propylene and hydrogen radicals than into ethylene and methyl radicals.

The results of the present investigation for propane pyrolysis and those of the earlier investigation (1) describing ethane pyrolysis were compared at operating conditions of commercial importance. Clearly surface reactions in the case of propane pyrolysis are of lesser importance as compared to those for ethane pyrolysis. Specific comparisons were made using the results of ethane pyrolyses at about 65% ethane conversions and those for propane pyrolyses at about 85-90% propane conversions. The lesser importance of surface reactions for propane pyrolyses is based on lower levels of coke formation and smaller yields of carbon oxides. Table I of the present investigation and Table I of the earlier paper (1) describing ethane pyrolysis results show three specific comparisons of the carbon (or coke) yields at 800°C using wet paraffins as feedstocks. In the Vycor reactor, carbon yields were lower for propane pyrolysis by a factor of about two. In the metal reactors, the factor was even greater, especially when steam was used as a diluent.

Surface reactions are thought to be relatively unimportant when propane is pyrolyzed in a Vycor glass reactor. This conclusion is based on two factors. First, surface reactions in the Vycor reactor were of fairly minor importance for ethane pyrolysis (1) and are considered to be even less significant for propane pyrolysis. Secondly, as was shown in both Table I and Figure 1 and as will be described later in this paper, there was good agreement between the experimental results for the Vycor reactor and the predictions of the mechanistic model described earlier (1).

The pyrolysis results obtained using the Vycor reactor were used to make various comparisons. Increased temperatures and decreased partial pressures of entering propane both result in increased ethylene yields. Such findings are consistent with the general trends reported by many previous investigators for pyrolyses of other hydrocarbons and specifically by those (3,4) who have investigated propane pyrolyses in metal reactors. The pretreatments of metal surfaces also affected product composition. Less surface reactions occur on H_2S-treated metal surface as compared to untreated surfaces. Oxygen-treated surfaces that have metal oxides on the surface tend to promote undesired surface reactions and to produce considerably more carbon oxides.

Modeling of Propane Pyrolysis Data

Several mechanistic models involving numerous free-radical gas-phase reactions were tested and compared to propane pyrolysis results obtained in the Vycor glass reactor. Data obtained in metal reactors were not compared since surface reactions were obviously of much greater importance in these latter reactors. The model developed as a result of this preliminary testing is presented in Table II of the previous chapter of this book (1). Good agreement occurred between the predicted values and all experimental results obtained in the Vycor reactor (at both 750° and 800°C with and without steam as a diluent). Figures 1 and 2 show

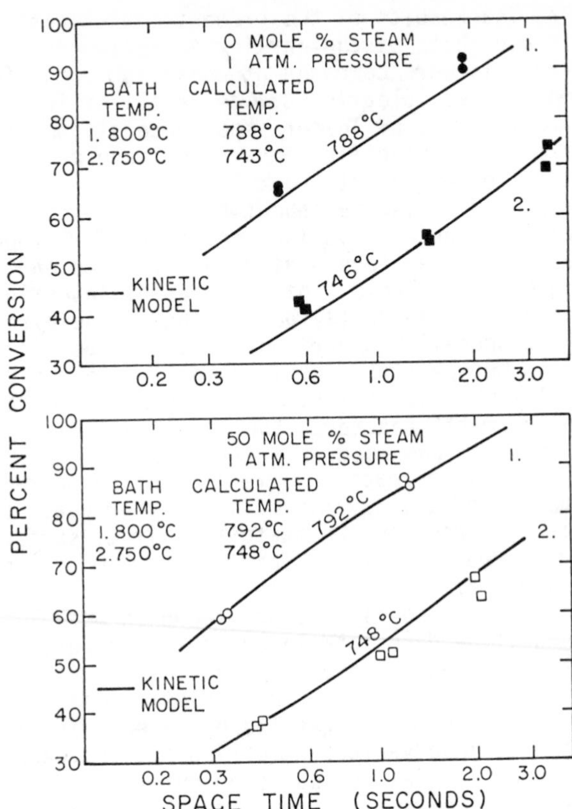

Figure 2. Comparison of kinetic (or mechanistic) model and experimental data for propane pyrolyses in Vycor reactor

comparisons for the product composition and for propane conversions respectively. Good agreement was also noted when the isothermal data of Kinney and **Crowley** (8) for both propane and ethane pyrolyses were compared with the predictions of the model.

The mechanistic model (of Table II of the previous chapter) was developed by successive approximations. First, only those reactions assumed to be most important were used in the model, and later secondary reaction steps were added until good predictions were obtained of the experimental results. In the previous chapter of the book, the parameters for the various reaction steps are reported. The major reactions for pyrolyses of both ethane and propane are grouped in Table II (1). When ethane is pyrolyzed, the propane reactions are of relatively minor importance and eliminating them from the model has little effect on the ethane pyrolysis predictions. When propane is pyrolyzed, however, significantly better predictions result when the ethane reactions are included and the entire set of reactions are employed.

Table II
Corrected Product Composition Results
(800°C, 50:50 Mixture of Ethane and Steam)

Reactor	Incoloy (oxidized)		Incoloy (reduced)		Model
Sample	1		4		-
Conversion, %	83		81		83
Space time, sec.	0.95		0.96		0.98
Gas Composition	Exp. Data	Corrected	Exp. Data	Corrected	
H_2	30.2	15.4	20.1	10.9	11.3
CH_4	25.6	28.6	29.8	31.7	32.2
C_2H_6	2.3	2.6	2.3	2.4	3.0
C_2H_4	23.9	35.9	27.4	34.0	32.6
C_3H_8	8.0	8.9	10.0	10.6	9.6
C_3H_6	7.5	8.4	9.6	10.2	8.7
C_4H_{10}	0.1	0.1	0.0	0.0	1.0
CO	2.1	-	0.6	-	-
CO_2	0.3	-	0.1	-	-
Carbon Yield, %	11.4	-	5.8	-	-

Reactor	304 SS (oxidized)		304 SS (reduced)		Model
Sample	1		3		-
Conversion, %	77		73		74
Space time, sec.	0.68		0.69		0.61
Gas Composition	Exp. Data	Corrected	Exp. Data	Corrected	
H_2	27.3	15.1	19.2	13.0	13.6
CH_4	24.5	26.4	26.2	27.3	26.9
C_2H_6	1.9	2.0	1.9	2.0	2.6
C_2H_4	24.0	33.1	25.9	30.3	29.1
C_3H_8	12.0	12.9	15.2	15.8	15.3
C_3H_6	9.7	10.4	11.2	11.6	10.6
C_4H_{10}	0.0	0.0	0.0	0.0	0.9
CO	0.6	-	0.3	-	-
CO_2	0.1	-	0.0	-	-
Carbon Yield, %	9.2	-	4.3	-	-

The agreement between the values predicted by the model and the experimental data are somewhat poorer, however, for propane pyrolysis as compared to ethane pyrolysis. Such a finding is not surprising since pyrolysis of propane is more complicated, involving more gas phase reactions. More uncertainities also exist in the kinetic parameters given in the literature for some of the propane pyrolysis reactions. It was necessary to make judicious choices for parameters for especially reactions 6,7, and 10 in order to obtain good fits with the experimental data. The values for the parameters used were always within the limits reported in the literature, except for reaction 20. In this latter case, a value that differed by about 10 was finally selected in order to obtain better agreement of the predicted values with the data.

Even though the model contains 39 reaction steps, it is still relatively incomplete for propane pyrolysis. Reaction steps to produce acetylene, diolefins, aromatics, and other heavier (C_4 and higher) hydrocarbons are of at least minor importance. Furthermore, reactions to differentiate between n-propyl and isopropyl radicals likely are needed. Reactions involving allyl radicals also occur to some extent. Although additional reactions have been proposed (7), reliable parameters (activation energies, E, and frequency factors, A) for such equations are not yet available.

Although the model used here for propane pyrolysis can certainly be improved in the future, it is much superior to other models that have been used to predict propane pyrolysis data obtained at conditions approximating those of commercial plants (4, 10).

Correction of Composition Results to Surfaceless Basis

Attempts were made to correct the product composition results obtained for propane pyrolysis in different reactors to a surfaceless basis. Assuming a perfect correction procedure could be developed, the corrected composition of the product stream would be identical for all runs at the same operating conditions regardless of the reactor used or the past history of the reactor. The correction technique used earlier (1) for ethane pyrolysis and which assumes that ethylene is the only hydrocarbon reacting at the surface was found to be less successful in the case of propane pyrolysis. Table II shows several comparisons of experimental data, corrected values, and values predicted using the mechanistic model (1) for runs made at 800°C with wet ethane feed in the two metal reactors. When essentially steady-state operation was obtained after 1-2 hours of operation (at which time the inner reactor surface was in a relatively reduced condition), fairly good agreement, within about 10% on a relative basis, was obtained between the corrected values and those predicted by the mechanistic model. Based on the mechanistic model, the ethylene values were overcorrected but the hydrogen values were not corrected enough. Somewhat poorer agreement resulted between the corrected and predicted values when more oxidized reactors were used; the surface of the reactor was relatively oxidized in some runs during start-up.

Although ethylene was certainly a major precursor for coke and carbon oxides during propane pyrolyses, there were probably also other precursors. Propylene that is formed in substantial amounts during propane pyrolysis certainly reacts to some extent to form coke (2). Acetylenes, diolefins, and aromatics likely are of minor but nevertheless of greater importance as coke precursors for the following two cases: propane pyrolysis as compared to ethane pyrolysis or pyrolysis in small laboratory reactors with high S/V ratios as compared to pyrolysis in commercial units. Before a completely successful correction technique can be developed for propane pyrolysis, quantitative data will be needed concerning the relative reactivities of various carbon precursors.

Since the analytical equipment used was not able to analyze with good accuracy C_4 and heavier hydrocarbons, the yields of carbon reported in this investigation and in the earlier one (1) dealing with ethane pyrolyses are really measures of the carbon that formed both solid coke and also the heavier hydrocarbons. Some hydrogen was of course, present in these heavier hydrocarbons, but such hydrogen was not taken into account in the calculations made to determine elemental hydrogen in the product streams. Failure to include any heavier hydrocarbons in procedures (7) employed to calculate the yields of gaseous components causes yields of hydrogen to be slightly too low and those of ethylene and propylene to be slightly too high.

The rates of propane disappearance were slightly higher in oxidized reactors as compared to reduced reactors (see for example the propane conversion results shown in Table II). A similar finding was reported earlier by Crynes and Albright (3). Such a finding can be explained by increased concentrations of free radicals in the gas phase as a result of increased carbon formation in the oxidized reactors. As hydrocarbons, probably adsorbed, decompose on the surface to form carbon and hydrogen, obviously carbon-hydrogen bonds are broken. Possibly some hydrogen radicals formed escape into the gas phase. The hydrocarbons that decompose may even include propane in the case of the oxidized reactors.

Propane and Ethane Pyrolysis

A key finding of this investigation was that the condition of the inner surface of the reactor changed when the feedstock was switched to propane after ethane or vice versa. This conclusion is based on runs made using a bath temperature of 800°C, with a feed mixture containing 50 mole % steam, and with both 304 SS and Incoloy reactors. The following phenomena occurred in both metal reactors that were initially used for pyrolysis of ethane:

(a) After switching from an ethane to a propane feed, the yields of ethylene increased and the yields of those products (carbon, carbon oxides, and hydrogen) obtained as a result of the surface reactions decreased during the first hour of the propane portion of the run; essentially steady-state product compositions were then obtained.

(b) When the feed was next switched back to ethane, the initial yields of ethylene were significantly higher and those of the products obtained because of surface reactions were lower as compared to yields before the propane portion of the run. Carbon yields during ethane pyrolysis were 13% and 3.8% respectively in the Incoloy reactor before and immediately after the propane phase of the run. In the 304 SS reactor, they were 16% and 7.1% respectively.

Clearly the activity of the surface relative to promotion of surface reactions differed depending on whether ethane or propane was used as a feedstock. This finding is not surprising since Tsai and Albright (5) report that dynamic equilibria occur during pyrolysis relative to the following: oxidation of surface with steam vs reduction of surface with gaseous components; and carbon (or coke) deposition vs removal of coke with steam. The present results are, however, the first to show conclusively that surface activity depends on the feedstock; surface activity must be related in some way to the levels of both surface oxides and surface carbon.

As further comparison of propane and ethane pyrolyses, four different mixtures of propane and ethane were pyrolyzed in the Incoloy reactor employing a 800°C bath temperature and one atmosphere total pressure. These mixtures had propane-to-ethane ratios of 0.04:1, 0.29:1, 1:1, and 2.97:1; the last three mixtures were premixed with steam on a 1:1 basis. In estimating the conversions of both propane and ethane during pyrolysis of the mixture, it was assumed that no propane was formed as a result of the pyrolysis of ethane and that no ethane was formed from pyrolysis of propane; these two assumptions are certainly not quite accurate since small amounts of both ethane and propane were formed during the pyrolysis of the pure paraffins.

Conversions of both ethane and propane at a given residence time varied significantly as a function of the propane-to-ethane ratio in the feedstream. Ethane conversions decreased as the ratio increased (or as the amount of propane in the mixture increased). Propane conversions on the other hand increased slightly as more ethane was added to the mixture. Figure 3 shows results for a space time of 0.9 seconds.

The kinetic model that was successful for prediction of the pyrolysis results for both pure propane and pure ethane was unable to predict results for propane-ethane mixtures. It was, of course, appreciated for these runs in the Incoloy reactor that surface reactions may be quite important, and hence the model would predict more olefins than are present in the exit product. The model, however, predicts that ethane conversions would increase, and not decrease as was found, as the amount of propane

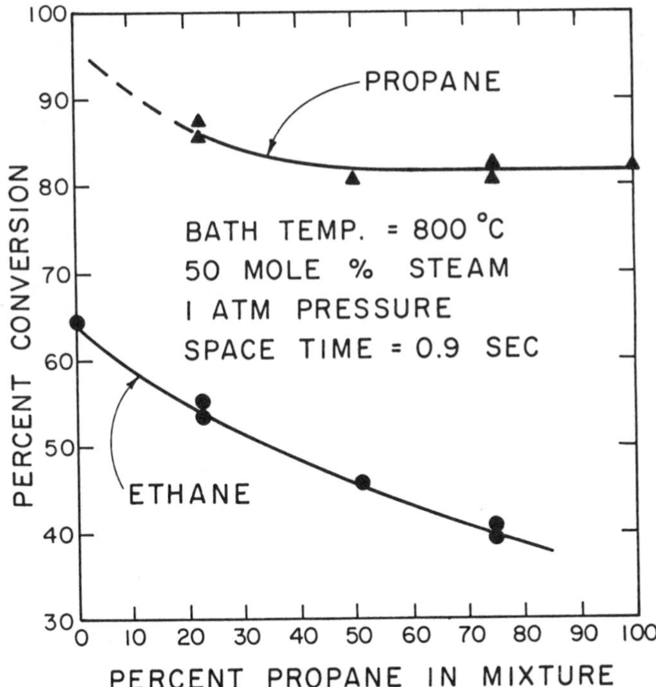

Figure 3. Conversions for propane–ethane mixtures at 800°C bath temperature and using incoloy reactor

in the feedstream increased. The model also predicts propane conversions would be slightly retarded by the presence of large amounts of ethane. The reason for failure of the model to predict pyrolysis data for mixtures is not known. One possibility is that the termination steps may be incomplete; termination steps at the Incoloy surface or gas-phase termination reactions between ethyl and propyl radicals may be important when mixtures are used, but would be of little importance for runs with either pure propane or pure ethane.

Discussion of Results

The present results significantly clarify the role of surface reactions relative to propane pyrolysis. Although the differences in the composition of the product stream obtained in different reactors were not as large as they often were with ethane pyrolyses, nevertheless, the differences were still significant in many cases.

Two factors must be considered in extrapolating the results to commercial reactors. First, as also discussed earlier (1), the small diameter tubular reactors used in this investigation accentuate the relative importance of surface reactions as compared to larger diameter reactors used commercially. A specific technique for extrapolating to larger diameter reactors has also been proposed earlier. Second, as a reactor is used, coke forms on the surface and such coke partially masks the metal so that it has less effect on surface reactions. Brown and Albright (2) have recently reported, however, that coke formation was faster on coke surfaces as compared to Vycor glass surfaces. Apparently the coke acts to some extent to adsorb olefins and other hydrocarbons; these adsorbed hydrocarbons seem to decompose rather rapidly to coke. Furthermore, iron and nickel particles are frequently present in the coke formed (11, 12). In fact, coke formed in this investigation was attracted to a magnet, seemingly indicate the presence of such particles. These particles are in at least some cases both magnetic and catalytic in character. It seems safe to conclude that the coke formed on the surface of commercial pyrolyses tubes only partially masks the inner metal surfaces relative to their effect on surface reactions.

The results obtained in the Vycor reactor are of special interest since surface reactions are of relatively minor importance in this reactor. The results of this reactor then are most useful in clarifying the mechanism of the gas-phase reactions.

The results of the present investigation that indicate that the surface activity of a reactor changes as the feed stream was switched from propane to ethane strongly suggest that changes of surface activity would, in addition, occur if higher molecular weight feeds were used. It seems safe to conclude that the importance of surface reactions varies with the feed used.

The mathematical model used to correlate the propane pyrolysis data obtained in the Vycor reactor can likely be used for at least moderate extrapolations to temperatures and pressures beyond those investigated.

Acknowledgment

Professor Pál Siklós of Technical University, Budapest, Hungary made valuable contributions to the modeling of the pyrolysis data.

Literature Cited

1) Dunkleman, J. J. and Albright, L. F., Chapter 14, This Book (1976).
2) Brown, S. M. and Albright, L. F., Chapter 17, This Book (1976).
3) Crynes, B. L. and Albright, L. F., Ind. and Eng. Chem. Proc. Design Develop., 8, 1, 25 (1969).
4) Herriott, G. E., Eckert, R. E., and Albright, L. F., AIChE J., 18, 1, 84 (1972).
5) Tsai, C. H. and Albright, L. F., Chapter 16, This Book (1976).
6) Herriott, G. E., "Kinetics of the Pyrolysis of Propane at 700 to 850°C," Ph.D. dissertation, Purdue University, 1970.

7) Dunkleman, J. J., "Kinetics and Surface Effects of the Pyrolysis of Ethane and Propane in Vycor, Incoloy, and Stainless Steel Tubular Flow Reactors from 750° to 900°C", Ph.D. dissertation, Purdue University, 1976.
8) Kinney, R. E. and Crowley, D. M., Ind. and Eng. Chem. 46, 1, 258 (1954).
9) Buekens, A. G. and Froment, G. F., Ind. and Eng. Chem. Proc. Design Develop. 7, 3, 435 (1968).
10) Zdonik, S. B., Green, E. J. and Hallee, L. P., Oil and Gas J. 65, 26, 96 (1967).
11) Lobo, L. S. and Trimm, D. L., J. Catalysis 29, 15 (1973).
12) Frech, K. J., Hoppstock. F. H., and Hutchings, D. A., Chapter 12, This Book (1976).

16

Surface Reactions Occurring During Pyrolysis of Light Paraffins

CHUNG-HU TSAI and LYLE F. ALBRIGHT

Purdue University, Lafayette, Ind. 47907

Coke and carbon oxides, both undesirable by-products, are always formed to some extent in commercial pyrolysis units. The carbon oxides are produced when part of the coke reacts with steam that is used as a diluent with the hydrocarbon feedstock. Most, if not all, of these undesired products are formed by surface reactions that reduce the yields of olefins and other desired products. Coke also acts to increase heat transfer resistances through the tube walls, and most pyrolysis units must be periodically shut down for decoking of the tubes. During decoking, pure steam or steam to which a small amount of oxygen (or air) is added is fed to the reactor, and the coke is oxidized to produce carbon oxides.

At pyrolysis conditions, steam and oxygen react with iron, nickel, and chromium to form oxides of these metals (1,2). Crynes and Albright (3), in their 304 stainless steel reactor, found that metal oxides were formed on the inner wall of the reactor used for propane pyrolysis. These metal oxides apparently often promote secondary and undesired reactions that reduced the yield of products. More recently, Dunkleman, Brown, and Albright(4,5) have reported more evidence confirming the undesirability of these metal oxides.

Several gaseous components present during most commercial pyrolysis runs react with or at the surface. For example, hydrogen reduces the surface oxides (6), desulfurizes coke (7), and reacts with the coke itself to produce methane (8). Cleaning coke from the surface may act to promote more coke formation, but reduction of surface oxides presumably often decreases the rate of coke formation. Carbon monoxide also is a reducing agent for metal oxides and is sometimes employed during the manufacture of steel.

Hydrogen sulfide and various sulfur-containing hydrocarbons result in complex surface reactions. Treating the inner surface

*Present address: The Lummus Co., Bloomfield, New Jersey

of a 304 stainless steel tubular reactor with hydrogen sulfide reduced the subsequent formation of coke during propane pyrolysis (4, 9, 10). Hydrogen sulfide often acts to form metal sulfides (11, 12). More information is certainly needed as to the complete role of the hydrogen sulfide that is sometimes added in small amounts when ethane or other light paraffins are used as feed-stocks.

In the present investigation, considerable information has been obtained to clarify the role of the surface reactions that occur during pyrolysis. Reactions investigated include the formation and destruction of metal oxides and metal sulfides. Information has also been obtained relative to coking and decoking.

Experimental Details

Tubular reactors constructed from 304 stainless steel and Incoloy 800* were used to investigate surface reactions that occur during the pyrolysis of light hydrocarbons. Flows to the reactor of various gases including oxygen, hydrogen, carbon monoxide, and hydrogen sulfide were controlled to within about 2% on a relative basis using conventional metering equipment (including pressure regulators, differential manometers, and metering valves). The gas stream to the reactor could be bubbled, if desired, through water maintained at temperatures varying from about $0^{\circ}C$ to $90^{\circ}C$; by this technique, the desired amount of steam could be added to the gas stream entering the reactor. The inlet line to the reactor was heated to prevent condensation of water.

The reactors used were 1.09 cm I.D. and about 80 cm long, and 45.7 cm (equivalent to 42.8 cc) of the reactor was positioned inside a Hoskins electric furnace, type FC-301. The ends of the furnace were insulated with fire brick. Five chromel-alumel thermocouples were attached to the outer surface of the reactor at various positions. Temperature variations along the reactor were in general less than $90^{\circ}C$ when the temperature was controlled in the 700° to $900^{\circ}C$ range. The temperatures reported in the result section are the maximum temperatures for the various runs.

The exit gas from the reactor was cooled, and it was then analyzed in a gas chromatograph employing three columns. Carbon monoxide, carbon dioxide, sulfur dioxide, water, nitrogen, olefins, and methane were determined on a relative basis to within about 2-3%. Analysis of hydrogen was probably accurate on a

*Incoloy 800 obtained from Huntington Alloys is an alloy reported by them to contain 30-35% nickel, 19-23% chromium, 1.5% (max) manganese; 1.0% (max) silicon, and the remainder mainly iron. 304 stainless steel contains 8-12% nickel, 18-20% chromium, and the remainder primarily iron.

relative basis to within about 10% (A thermal conductivity cell and helium were used, and hydrogen sometimes resulted in reverse peaks).

The exit gas stream was cooled to almost $0^{o}C$ using the ice bath, and most of the water vapor was condensed. The remaining gas was metered using a soap-bubble meter.

The general experimental procedure employed was to flow first one gas or mixtures of gases to the reactor at a controlled rate; the reactor temperature was adjusted to a desired level. After a desired period of time, the flow of the first gas was terminated, and a second gas flow was started. Frequently helium was used to flush the first gas from the reactor before the second one was started. Chromatographic analyses were made in most cases at frequent intervals, and the results were used to make material balances of all atoms into and out of the reactor. This general technique indicated which gases resulted in the production or destruction of coke, surface oxides, or surface sulfides.

Experimental Results

Oxidation of the Inner Surface of the Reactor. When oxygen was passed through unoxidized reactors (i.e. reactors with relatively few metal oxides on the inner surface), the following reaction occurred:

$$(\text{metal})_{\text{surface}} + O_2 \rightarrow (\text{metal oxides})_{\text{surface}} \tag{1}$$

When coke (or possibly metal carbides) were present on the surface, the following reaction also occurred:

$$(C)_{\text{surface}} + O_2 \rightarrow 2CO \text{ (or } CO_2) \tag{2}$$

In order to study the surface oxidations, a series of experimental runs were made in which the surface was first oxidized and then reduced. This procedure was repeated numerous times in both reactors investigated.

Starting either with a reduced and clean reactor or with a coke-covered reactor, the rate of oxidation was initially relatively high in all cases, and the rate then decreased becoming essentially zero after all carbon on the surface was converted to carbon oxides and after a surface layer of metal oxides was formed. The following describes the phenomena noted when a new (and non-coked) Incoloy reactor was oxidized at $800^{o}C$. With an inlet flow of oxygen of 30 std. cc/min., the exit flow was initially 19 cc/min, indicating that 11 cc/min were being reacted on the surface. After about 20 minutes, the inlet and exit flows were almost identical indicating that the rate of oxidation was very low. Graphical integration of the graph of the rate of surface oxidation versus time indicated that 33 millimoles O_2/sq.

meter had reacted (see Table I). The surface area was calculated based on the assumption that the inner surface of the reactor was smooth, and it made no allowance for surface roughness or porosity that developed as the reactor was repeatedly oxidized and reduced.

Key findings are summarized in Table I for runs in both the Incoloy 800 and 304 stainless steel reactors. First, the amount of oxygen that reacted with the surface increased as the reactor tube was used, i.e. as oxidation-reduction sequences were repeated. Although all runs were not directly comparable since somewhat different operating conditions were sometimes used, the amounts of oxygen that reacted (calculated as millimoles O_2/sq. meter) increased by a factor of about 12 as a result of the first 30 oxidation-reduction sequences in the stainless steel reactor and by factors of about 4 because of the first eight sequences in the Incoloy reactor. The largest increases in the amount of oxygen reacted occurred in both reactors as a result of the first few oxidation-reduction sequences.

As the reactors were repeatedly oxidized and then reduced, the time required to complete the oxidation of the surface with oxygen increased significantly. For runs in relatively new reactors at 800°C, about one third of an hour resulted in complete oxidation. Longer times, up to several hours, were required for older reactors that had significantly roughened surfaces.

The following first-order kinetic equation correlated reasonably well the oxidation results of this investigation after induction periods of several minutes that were sometimes noted:

$$\frac{d\,(\text{metal oxides})}{dt} = k_{ox}\left[\begin{pmatrix}\text{metal}\\ \text{oxides}\end{pmatrix}_{sat} - \begin{pmatrix}\text{metal}\\ \text{oxides}\end{pmatrix}\right] \qquad (3)$$

where k_{ox} = first-order rate constant

$(\text{metal oxides})_{sat}$ = concentration of metal oxides when surface is completely oxidized (second from last column in Table I)

(metal oxides) = concentration of metal oxides at time t

In general for relatively unused (or new) reactors, a constant value of k_{ox} occurred during the entire run, but for older reactors, two values were as a rule noted (see Table I). Whenever oxygen was used as the oxidant, the k_{ox} value for the initial portion of the run was always larger, and perhaps it represented the oxidation of the inner and more easily oxidized layer of the surface. The k_{ox} value for the second phase of the run probably represented the oxidation of the metal layers somewhat below the

Table I. Oxidation of Inner Surfaces of Reactors

Reactor	Temp., °C	Prior Oxidation Runs	Prior Sulfiding Treatments	Oxidant Used	Time for Complete Oxidation, hours	Total Oxygen Reacted with Surface, Millimoles/ sq. meter	k_{ox}, * min.$^{-1}$
Incoloy 800	800	0	0	oxygen	0.33	33	0.24 (0.24)
	800	5	0	oxygen	0.67	60	0.31 (0.15)
	800	8	0	oxygen	1.0	128	____**
	800	1	0	steam-He	1.0	17	0.06 (0.013)
304 SS	810	3	0	oxygen	0.33	125	____**
	800	20	0	oxygen	0.33	383	0.64 (0.18)
	800	34	1	oxygen	1.0	2030	____**
	800	38	2	oxygen	5.0	4090	0.021 (0.021)
	800	40	2	steam-N_2	24***	8300	0.0008 (0.002)

* Values are reported for initial (and latter) portions of run.

** k_{ox} values could not be calculated since initial surfaces were coke covered.

*** Oxidation with steam was not complete even at end of run.

surface; diffusion of oxygen through pores to these layers was probably occurring. An additional explanation is that iron, chromium, and nickel are transition metals each with more than one oxide. The initial stages of oxidation may have involved the formation of lower oxides. As a reactor was repeatedly oxidized and then reduced (the reduction step will be described later), the values of (metal oxides)$_{sat}$, also expressed as millimoles O_2/sq. meter, also increased often very substantially. Furthermore, the rate of surface oxidation increased, but as a rule, k_{ox} decreased.

Increased temperature was found in runs made in both the Incoloy and 304 stainless steel reactors to result in increased values of both k_{ox} and (metal oxides)$_{sat}$. For runs in the Incoloy reactor at 700°, 800°, and 900^oC, the k_{ox} values for the initial (and latter) stages of the runs were 0.20 (0.09), 0.31 (0.15), and 0.54 (0.54) min^{-1} respectively and the values of (metal oxides)$_{sat}$ were 34, 60, and 91 millimoles oxygen/sq. meter. These runs were made after 3, 5, and 4 oxidation-reduction sequences respectively.

The k_{ox} value for a run in a 304 stainless steel reactor that had been oxidized and reduced 38 times was only 0.021 min^{-1}, but (metal oxides)$_{sat}$ was 4090 millimoles oxygen/sq. meter. (A values much greater than those noted for the Incoloy 800 reactor). This 304 stainless steel reactor, however, had a highly porous and corroded surface as indicated by subsequent inspection of the reactor.

Runs made using steam indicated that metal oxides were produced as follows:

$$(\text{metal})_{surface} + H_2O \rightleftarrows (\text{metal oxides})_{surface} + H_2 \qquad (4)$$

Tests with steam were made using mixtures of essentially 50% steam and the remainder helium or nitrogen. Such mixtures have a steam content similar to steam-hydrocarbon mixtures used as feedstocks in many pyrolysis units. For reactors that had no coke deposits on their surface, the rate of oxidation of the surface can be calculated from the rate of hydrogen formation. When coke was present on the walls of the reactor, the coke was also oxidized by the steam as follows:

$$(C)_{surface} + H_2O \longrightarrow CO \text{ (or } CO_2) + H_2 \qquad (5)$$

The amount of hydrogen produced in general first increased for several minutes at the start of a run, and then decreased as the steam flow continued. At least a partial reason for the increase in the rate of reaction during the first several minutes of the run is because some (or most) of the steam was adsorbed on the reactor walls during this period of time.

The most rapid rates of oxidation of the Incoloy reactor

Table II. Comparison of Hydrogen and Carbon Monoxide for Reduction of Surface Oxides

Reactor	Temp., °C	Prior Oxidation Runs	Prior Sulfiding Treatments	Reducing Gas	Time for Reduction hours	Oxygen Removed, Millimoles/ Sq. Meter Surface	Carbon Deposited, millimoles/ sq.meter	k_{red}, min.$^{-1}$
Incoloy 800	700	4	0	H_2 - He	5	49***	--	0.027(0.006)
	800	3	0	H_2 - He	5	56***	--	0.031(0.006)
	900	5	0	H_2 - He	5	58***	--	0.024(0.005)
	700	7	0	CO	12	150	3300	0.011*
	800	6	0	CO	12	170	3500	0.010*
	900	8	0	CO	12	130	4100	0.003*
304 SS	750	15	0	H_2	12	330	--	0.13(0.021)
	700	27	0	H_2 - He	12	380	--	0.047(0.028)
	900	30	0	H_2 - He	12	735	--	0.045(0.032)
	800	35	1	H_2 - He	11	610	--	0.017(0.008)
	800	39	2	H_2 - He	12	1020	--	0.011(0.011)
	801	7	0	CO	10	760	3800	0.007**
	762	15	0	CO	11	910	8800	0.003**
	699	28	0	CO	12	1570	7500	0.006**
	901	31	0	CO	12	1360	9200	0.009**
	808	34	1	CO	12	2030	23300	--

When two k_{red} values occurred during hydrogen runs, initial (and latter) values are both shown.

* For initial period of run.

** After maximum rate of reduction.

*** Increased amounts of oxygen had been on surface for higher temperature runs in this series.

with steam occurred during the first 20 minutes of the run, and a small but significant rate of reaction persisted over the final 40 minutes of the run. In the case of the old stainless steel reactor (that had been subjected to 40 oxidation-reduction sequences) between 8 and 22% of the inlet steam reacted (by Reaction 4) forming hydrogen and surface oxides during the first 1.5 hours of the run. Conversions then remained fairly constant for an additional 5.5 hours, and they finally decreased from 22% to about 8% in the period of 7 to 24 hours of the run. The amount of surface oxides formed in this run (8300 millimoles oxygen /sq. meter) was higher than in any previous run. The reactor tube broke at the end of this run, as will be described later. Values of k_{ox} in the time period following the maximum rate of oxidation with steam were lower, by a factor of 3-8, as compared to k_{ox} values obtained with oxygen.

Two oxidation runs were made using oxygen in the stainless steel reactor after it was treated with pure hydrogen sulfide at 800°C first for 2 hours after 30 oxidation-reduction sequences and then for 18 hours after 35 sequences. After the first treatment, the amount of oxygen reacted increased by about 20-25%. There was an even higher increase after the second (and longer) treatment as shown in Table I. As will be discussed in more detail later, sulfiding and then oxidizing the surface roughens the inner surface significantly and permits more oxygen reactions on the surface.

Reduction of Surface Oxides. The metal oxides formed on the surface of the metal reactors using either oxygen or steam were reduced with either hydrogen, carbon monoxide, hydrogen sulfide, ethylene, or propylene. A summary of the results with hydrogen and carbon monoxide are shown in Table II.

The results of the reduction runs correlated reasonably well using the following first-order kinetic equation:

$$\frac{-d(1-y)}{dt} = k_{red}\,(1-y) \qquad (6)$$

In this equation, k_{red} is a pseudo reaction rate constant and y is the fraction of reducible oxides that are reduced. Hydrogen could reduce only about 20-30% of the surfaces oxides on the 304 stainless steel surface and 65-90% on Incoloy. Carbon monoxide and hydrogen sulfide, however, could reduce essentially all the surface oxides.

Although more information is needed, hydrogen is probably quite effective for the reduction of oxides of iron and nickel but not of chromium (13). When hydrogen was first contacted with a surface that had been oxidized with oxygen, there was a relatively rapid rate of reduction for one third to one half hour, but then the rate decreased rapidly essentially ceasing after 5 to 12 hours. In general, the time to complete the reduction increased as the reactor aged (because of repeated oxidation and reduction).

Two values of k_{red} sometimes occurred during a hydrogen reduction run. The value during the first portion of a run was larger, and it probably represents the reduction of the more easily reducible metal oxides, i.e. those oxides that are reducible because of the type of metal or their location near the inner surface.

The effect of temperature in the 700° to 900° C range needs more investigation relative to hydrogen reductions. Preliminary results indicate that a more reduced surface was obtained at 700°C as compared to 800°C or especially 900°C. The higher amounts of oxygen removed from the surface at 800°C and 900°C (see Table II) were probably caused by the increased levels of surface oxides present at the start of the reduction steps for these two runs. Of interest, k_{red} values at 800°C were higher than those at 700°C and 900°C. Presumably, hydrogen reduction involves both initial adsorption of hydrogen followed by reactions between the adsorbed hydrogen and the surface oxides. Increased temperatures would presumably decrease hydrogen adsorption, and k_{red} apparently depends to some extent on adsorption.

Although carbon monoxide was effective for reducing surface oxides, relatively long periods of time were required to complete the reduction. Two surface reactions occurred:

$$CO + (\text{metal oxides})_{surface} \longrightarrow (\text{reduced metal})_{surface} + CO_2 \quad (7)$$

Part of the carbon monoxide also disproportionated as follows:

$$2CO \longrightarrow C(\text{coke}) + CO_2 \quad (8)$$

For an oxidized reactor, the exit flow rate was significantly lower than the inlet rate for about 6-12 hours indicating significant disproportionation of carbon monoxide (and formation of coke). Even after 12 hours of operation, a small amount of carbon dioxide was often being formed in many runs.

The following procedure was employed to distinguish between Reactions 7 and 8:

$$\begin{bmatrix}\text{Inlet flow}\\ \text{rate, cc/min}\end{bmatrix} - \begin{bmatrix}\text{Outlet flow}\\ \text{rate, cc/min}\end{bmatrix} = \begin{bmatrix}CO_2 \text{ formed by}\\ \text{Reaction 8, cc/min}\end{bmatrix} \quad (9)$$

$$\begin{bmatrix}\text{Outlet flow}\\ \text{rate, cc/min}\end{bmatrix}\begin{bmatrix}\text{Fraction } CO_2\\ \text{in exit stream}\end{bmatrix} - \begin{bmatrix}CO_2 \text{ formed by}\\ \text{Reaction 8, cc/min}\end{bmatrix} = \begin{bmatrix}CO_2 \text{ formed by}\\ \text{Reaction 7, cc/min}\end{bmatrix} \quad (10)$$

Figure 1 indicates the results of a run (made at 699^{0}C after 28 oxidations of the stainless steel reactor and also described in Table II). Over half of 60 cc/min of feed carbon monoxide reacted initially and approximately 90% of the carbon dioxide produced was by Reaction 8 (disproportionation reaction or Boudouerd reaction). After about one hour, almost all of the reaction was by Reaction 8. Then the moles of carbon monoxide that reacted by Reaction 7 first increased and then decreased as shown in Figure 1. After about four hours, Reaction 7 became and remained the predominant reaction for the remainder of the run that lasted for 12 hours. The areas under the curves of Figure 1 were integrated to make material balances. 1570 millimoles O_2/sq. meter were removed from the surface, and 7500 millimoles carbon/sq. meter were deposited (see Table II).

The major fractions of surface oxides for the stainless steel reactor were reduced in each run during the second phase of the reaction, i.e. in the time period during which the rate passed through the maximum. In the case of the run shown in Figure 1, this time period extended from about 1-12 hours. The k_{red} values of this reactor, as shown in Table II, are for the time period after the rate has passed through the maximum. In the case of the run shown in Figure 1, the time period is from 4 to 12 hours. For the Incoloy reactor, however, most of the surface oxides were reduced during the initial time period of a run, i.e. before the rate passed through a maximum which was relatively small. The k_{red} values for the Incoloy reactor are calculated for the initial time period (during the first two or three hours of the run).

Following the carbon monoxide portion of a run, the reactor surface and the deposited carbon were often oxidized with oxygen. The results for a run made at 753^{0}C in the stainless steel reactor are typical. In this run made in the reactor after 13 oxidation-reduction sequences, the exit flow rate was initially 50% greater than the inlet rate (30 cc/min) because of the formation of carbon monoxide from the deposited carbon. As much as 75% of the exit stream was carbon monoxide in the early stages of oxidation. As the run progressed, the carbon monoxide content decreased and the carbon dioxide content increased to a value of 95%. For almost the first hour, essentially all of the entering oxygen reacted. Then after about one hour, the oxidations (of the coke and of the surface) were completed, and the exit gas stream then became essentially 100% oxygen.

The oxidation and reduction procedures were repeated several times using oxygen and then carbon monoxide respectively. Material balances indicated that the unaccounted carbon during the carbon monoxide treatments was recovered as carbon monoxide and carbon dioxide during the following oxidation step. Also, the unaccounted oxygen of the oxidation step was recovered in the carbon dioxide formed by Reaction 7 of the next carbon monoxide treatment.

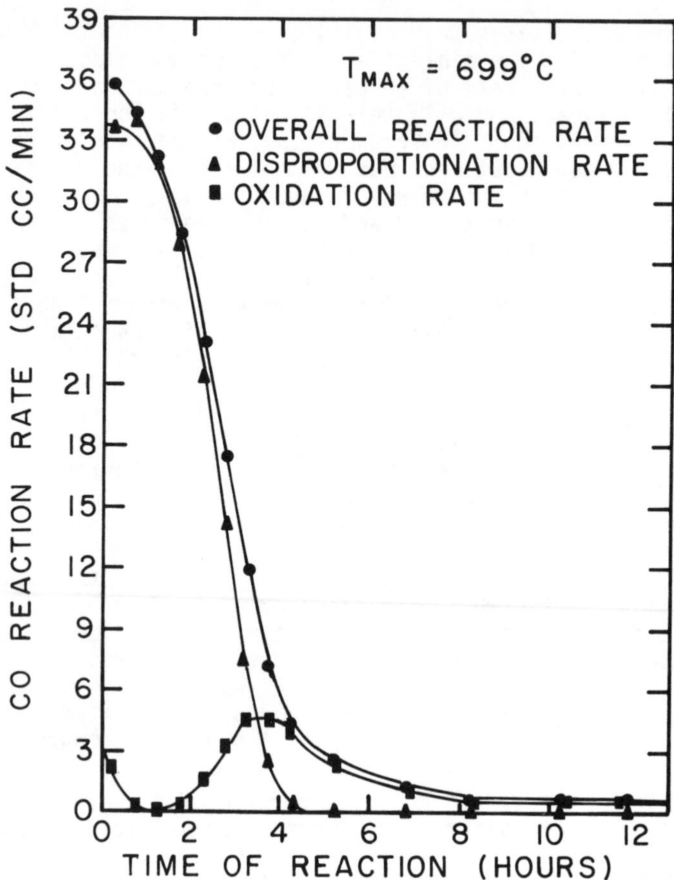

Figure 1. Carbon monoxide reaction in 304 stainless steel reactor at 699°C (flow rate of 60 cc/min)

The amounts of carbon monoxide that reacted during a run by disproportionation and oxidation reactions increased significantly as the reactor was aged (because of oxidation-reduction sequences). Furthermore, the time to complete reductions with carbon monoxide increased. For the stainless steel reactor, approximately 75 to 95% of the carbon monoxide that reacted during the run was by disproportionation. About 94-97%, however, reacted in this manner in the Incoloy reactor.

Figure 2 indicates the effect of changes of the inlet flow rate of carbon monoxide and the effect of reactor age. The total initial rates for both Reactions 7 and 8 were essentially directly proportional to the inlet flow rates. The total time required to complete the carbon monoxide reactions was less in runs with increased inlet flow rates. The total amount of car-

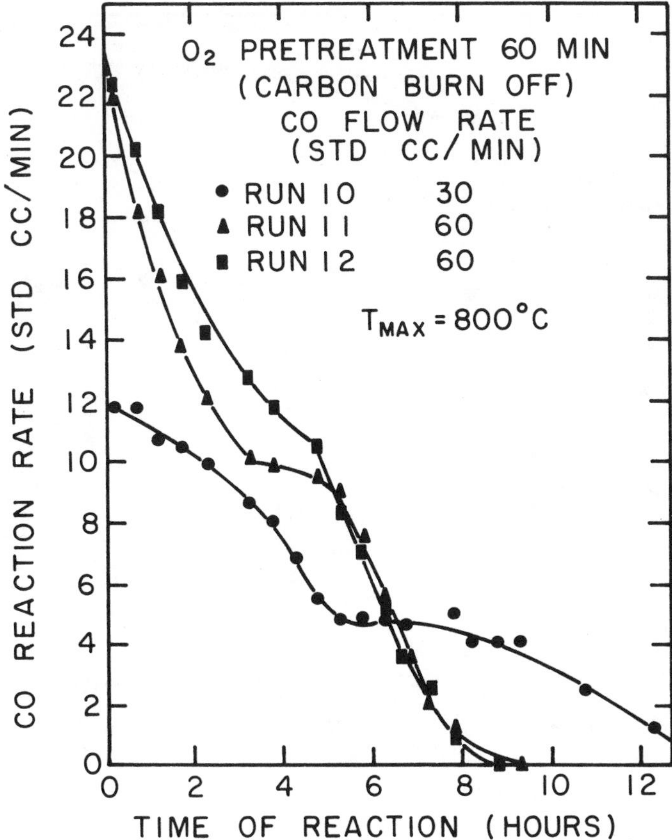

Figure 2. Effect of flow rate and reactor age on carbon monoxide reactions in 304 stainless steel reactor

bon monoxide that reacted during a run was unchanged because of differences in the inlet flow rate.

In the range of 700° to 900°C, increased temperatures resulted in decreased kinetics for both Reactions 7 and 8. Figure 3 shows the results for runs made in the 304 stainless steel reactor. The inner surface of the reactor was cleaned before each run relatively well of all carbon (or coke) deposits by an oxygen pretreatment, a subsequent reduction with hydrogen, and then a 20 minute treatment with oxygen. In the 700°C run, the carbon dioxide content in the exit stream decreased to 1% or less after about 8 hours; in the 800°C run, after 11 hours; and in the 900°C run, the content was still 6.6% after 12 hours. The kinetics of reduction also decreased in the Incoloy reactor with increased temperature as shown in Table II.

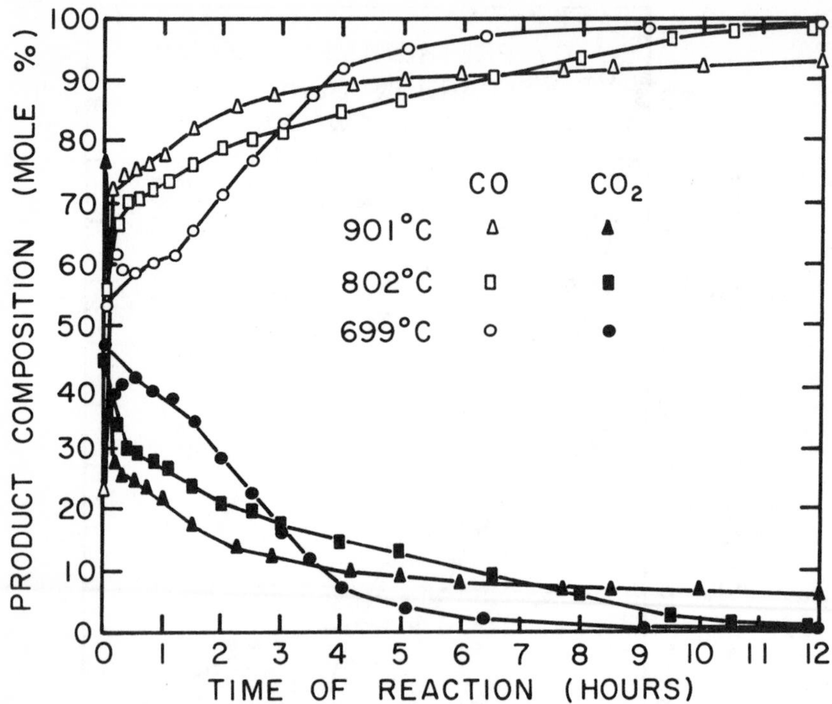

Figure 3. Effect of temperature on product composition of carbon monoxide runs in 304 stainless steel reactor

Both ethylene and propylene were found to reduce effectively the oxidized surface of the Incoloy reactor in the temperature range of 500°C to 800°C. In a run in which the two olefins were fed to the reactor intermittently, carbon oxides, steam, and coke were produced as a result of the surface reactions. Although the results are quite preliminary in nature, the rates of surface reactions increased during the first 5-6 hours which were made at 500°C. A maximum rate then resulted, and after about 16 hours the surface was essentially completely reduced. It had been hoped by feeding the two olefins intermittently to determine the relative reducing abilities of the two. Because of unsteady phenomena noted shortly after switching feeds, a definite conclusion on this matter cannot currently be made; both olefins apparently had at least fairly comparable reducing abilities. It was found, however, that increased temperatures increased the rates of reduction in the temperature range investigated. Furthermore, flushing the reactor with helium while the olefin flows were being switched subsequently resulted in higher rates of olefin reactions; presumably, the helium acted to strip some hydrocarbons from the reactor surface.

Decoking with Hydrogen. Hydrogen was found to be effective, but relatively slow in removing coke formed by carbon monoxide disproportionation. Hydrogen reacted with coke as follows:

$$(C)_{surface} + 2H_2 \longrightarrow CH_4 \qquad (11)$$

The results for a run in the stainless steel reactor that had been oxidized (and then reduced) 15 times demonstrate the nature of the above reaction. In this particular run, the oxidized reactor was treated with carbon monoxide until no further reactions occurred. Calculation indicated that 8800 millimoles of carbon were deposited per sq. meter of surface. When hydrogen was passed at a flow rate of 60 cc/min through the reactor, the exit gas stream was found to contain up to 9.5% methane and the remainder hydrogen; trace amounts of water were noted just after the hydrogen flow was started. Carbon equivalent to 8100 millimoles/sq. meter was removed as methane during 38 hours of hydrogen treatment. In this time period, the methane concentration in the exit stream decreased to less than 0.1%.

Several consecutive runs were made in which the reactor was first contacted with carbon monoxide (until no further reactions occurred) and then with hydrogen (to remove all carbon from the surface). As a result of these treatments, little or no surface oxides were present on the walls of the reactor, and carbon monoxide reacted only by disproportionation. Preliminary evidence indicates that carbon monoxide reactions with the reduced (or at least partially reduced) surfaces were as fast if not more so than with the more oxidized surfaces at 700°, 800°, and probably also 900°C.

Production and Destruction of Metal Sulfides. Hydrogen sulfide was passed on two different occasions through a preoxidized stainless steel reactor (i.e. one that had considerable metal oxides on the surface). On both occasions, both water and sulfur dioxide were produced in significant amounts especially during the initial stages of the run. At least part of the hydrogen sulfide decomposed, perhaps in the gas phase, as follows:

$$H_2S \longrightarrow H_2 + S \qquad (12)$$

Elemental sulfur condensed and solidified in the exit line from the reactor and especially in the cold trap just before the gas chromatographic equipment. Certainly part of the hydrogen so produced was oxidized by metal oxides producing water (reverse step of Reaction 4) and probably some if not most of the sulfur formed was oxidized to SO_2 as follows:

$$S + (\text{metal oxides})_{surface} \longrightarrow SO_2 + (\text{metal sulfides})_{surface} \qquad (13)$$

In addition, the following reaction that may actually be a combination of Reactions 4 (reverse), 12, and 13 was possibly occurring:

$$(\text{Metal oxides})_{\text{surface}} + H_2S \longrightarrow (\text{Metal sulfides})_{\text{surface}} + H_2O + H_2 + SO_2 + S \quad (14)$$

During the first hydrogen sulfide treatment of the reactor at 800°C; the hydrogen sulfide flow rate was 30 cc/min. The exit gas from the reactor contained initially 73% water, 3.7% sulfur dioxide, and the remainder hydrogen sulfide. As the run progressed, the water content decreased to about 15% after one hour and to 2.2% after two hours. The sulfur dioxide content decreased to slightly less than 2% after two hours. As the run progressed, elemental hydrogen was noted in the product stream, being about 3% after almost an hour and 10% after 1.5 - 2 hours; the hydrogen sulfide composition of the product gas stream increased from about 23 to 86% during the run. The k_{red} value for this 2-hour run was 0.028 min^{-1}, and based on the water and sulfur dioxide produced, approximately 1170 millimoles of oxygen were removed per sq. meter of surface. This amount is only about 67% of that removed in comparable runs using carbon monoxide, but it should be emphasized that the hydrogen sulfide flow was terminated while surface reactions were still occurring. Next, a carbon monoxide flow was started, and the rate of carbon monoxide reaction was low; the carbon dioxide concentration in the product gas was less than 1%. Only small amounts of oxidation and disproportionation reactions (Reactions 7 and 8) were noted. Oxygen still on the surface could not be removed by the carbon monoxide treatment in any reasonable length of time. Apparently hydrogen sulfide had "poisoned" the surface for carbon monoxide reactions.

A second hydrogen sulfide run was later made in the stainless steel reactor. First, the reactor was pre-oxidized with oxygen, and the reactor was then partially reduced using a hydrogen-helium mixture; about 450 millimoles of oxygen were removed per sq. meter of surface area. Then hydrogen sulfide was used to further reduce the surface for a total of 18 hours, and an additional 1230 millimoles of oxygen were removed per sq. meter of surface. The total oxygen removed, using both hydrogen and hydrogen sulfide treatments, was essentially the same as removed by comparable carbon monoxide treatments.

After each hydrogen sulfide treatment of the stainless steel reactor, oxygen was passed through the reactor and initially most of the oxygen was adsorbed on the reactor surface. For example, the inlet flow rate of oxygen was adjusted to 13 cc/min. after the second hydrogen sulfide treatment; initially the outlet flow was almost zero for about 50 minutes. Subsequently, the exit flow increased in amount during the next three hours, and the following reaction was important:

$$\text{(metal sulfides)}_{\text{surface}} + O_2 \rightarrow \text{(metal oxides)}_{\text{surface}} + SO_2 \qquad (15)$$

The sulfur dioxide content in the exit stream decreased from about 99% after 50 minutes to 1.5% after about four hours. A total of 750 millimoles of sulfur dioxide were produced per sq. meter of surface area. This value is of the same order of magnitude as the amount of water produced during a hydrogen reduction run of the oxidized reactor. It can be concluded that the concentrations of surface oxides on the oxidized reactor were similar to the concentrations of surface sulfides on the sulfided reactor.

Based on several comparative but nevertheless preliminary runs, the rates of reduction (r) of the surface oxides were generally as follows during the first several minutes of a run:

$$r_{H_2} > r_{H_2S} > r_{CO}$$

After 5-6 hours,

$$r_{H_2S} \text{ and } r_{CO} > r_{H_2} > 0$$

More data are needed relative to the comparison between r_{CO} and r_{H_2S}. The reason why there is a minimum and then a maximum for r_{CO} as a function of time is not yet known.

Analysis and Inspection of the Inner Surfaces. Pyrolysis leads to major changes in the composition and roughness of the inner surface of the metal reactors. This fact was confirmed by inspection of several reactors. The 304 stainless steel reactor used in this investigation and which had been subjected to treatments of oxygen, steam, hydrogen, carbon monoxide, and hydrogen sulfide had broken near its midpoint as it was being positioned in the furnace after the final treatment with steam. The reactor was then cut and examined at several axial positions. This reactor (that had been positioned horizontally in the furnace) was badly corroded on the inner wall, and the internal diameter varied significantly from about 0.71 cm near the midpoint of the reactor to about 1.09 cm at each end. This latter value is the same as the original internal diameter. The outside diameter of the tube had changed only slightly. It was as high as 1.29 cm near the midpoint where the highest temperatures occurred to 1.27 cm. at the end.

The wall thickness of this reactor also varied significantly near the midpoint as a function of the radial position. On the top wall, the thickness was 0.19 cm; at the bottom, it was 0.26 cm. The thickness of the original wall was 0.089 cm, and this

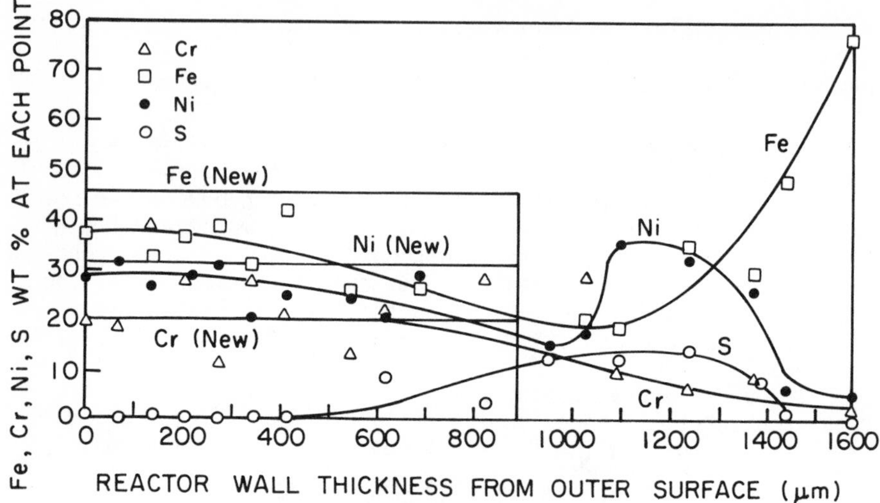

Figure 4. Concentration profile for Incoloy 800 reactor used for pyrolysis of ethane and propane after hydrogen sulfide treatment

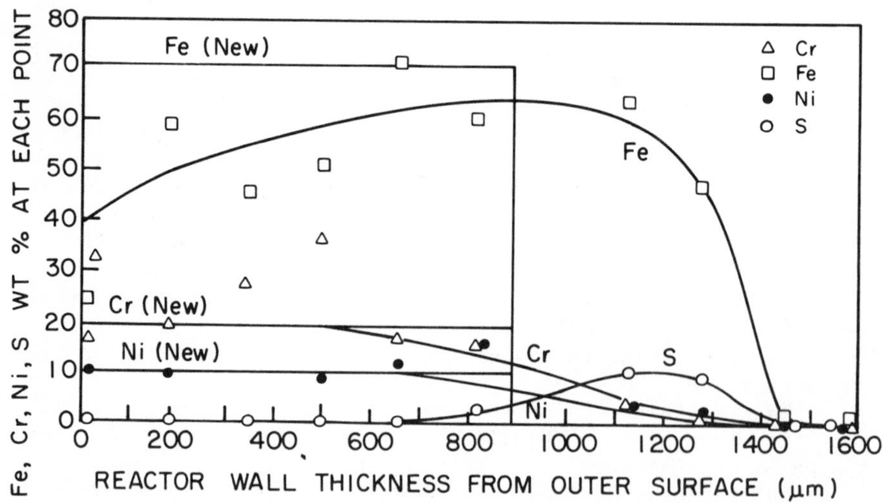

Figure 5. Concentration profile for 304 stain less steel reactor used for pyrolysis of propane

thickness still occurred near the ends of the tubes (which had been heated less). Of interest, variable wall thicknesses have been found for some tube reactors taken from commercial pyrolysis units.

The inner walls of the reactor, especially near the midpoint of the tube were highly porous. Investigation of the cross-sectional areas of the tube walls at various axial positions along the tube using x-ray diffraction analysis indicated as many as four rather distinct layers. The outer layer was essentially the metal alloy of the original tube. The middle two layers contained iron, ferric oxide (Fe_2O_3), nickel, chromium, and chromium oxides (both CrO_2 and Cr_2O_3). The inside layer contained in addition to iron and nickel, iron oxides (Fe_2O_3 and Fe_3O_4) and chromous oxide (CrO).

Three reactors (I.D. = 0.455 cm) used by Dunkleman (4, 10) for pyrolysis of both ethane and propane were cut and sections were examined. These reactors had each been treated two or three times both with oxygen to remove coke and with hydrogen sulfide for a few minutes. His Incoloy reactor had partially plugged as it was treated the last time with hydrogen sulfide. The plugged section was found to consist of metals and metal salts, primarily metal sulfides. Figure 4 indicates the composition as measured by the scanning electron microscope in the plugged areas as a function of the radial position. Figure 5 reports similar results for a 304 stainless steel reactor that had been used for a considerable number of pyrolysis runs. No plugs had developed in this reactor but the internal diameter had decreased significantly especially near bends.

Examination of the inner surface of these reactors clearly indicates that the surfaces were rough and porous. Furthermore, the relative composition of the nickel, iron, and chromium varied significantly in the inner portions of the tubes. Figure 4 indicates high compositions of nickel and especially iron in the material acting to plug the Incoloy tube. Figure 5 shows an inner surface relatively deficient in nickel and chromium. Sulfur had penetrated to considerable depths into the walls of both reactors. The composition of metal and sulfur atoms did not add to 100%, indicating that some materials, including perhaps carbon, were also present.

X-ray diffraction was used in an attempt to clarify some of chemistry that occurs when steam contacts Incoloy 800, 304 stainless steel, and nickel. Small flat-plate samples were analyzed both before and after treatment with 50:50 mixtures of steam and helium at 800°C for 24 hours. Table III clearly indicates significant changes in the composition reported as peak areas; steam treatment decreased the Fe-Cr-Ni peak in both Incoloy and 304 stainless steel, and increased the amounts of several metal oxides (and particularly NiO and Fe_3O_4).

The analyses with the scanning electron microscope and x-ray diffraction must be considered preliminary. It is suggested that

Table III X-ray Diffraction Analysis of Various Materials Before and After Steam Treatment (Values reported are peak areas)

Peak	S-5 (new Incoloy, untreated)	S-6(steam treated Incoloy)	S-7(new nickel, untreated)	S-8(steam treated nickel)	S-9(new 304 SS, untreated)	S-10(steam treated 304 SS)
Fe-Cr-Ni	13.26	4.91	0.00	0.00	12.24	0.00
Fe-Ni	0.00	2.01	0.00	0.00	0.00	0.82
Fe	0.00	0.00	0.00	0.00	0.00	0.00
Cr	0.00	0.00	0.00	0.00	0.00	2.78
Ni	0.00	0.00	40.02	13.06	0.00	0.00
Fe_2O_3	3.11	2.89	2.40	6.24	5.46	1.26
Fe_3O_4	0.00	0.46	0.00	3.60	0.00	5.91
Cr_2O_3	0.00	0.00	0.00	0.00	0.00	0.12
NiO	0.00	4.91	0.00	6.66	0.00	1.95
Si	0.00	0.00	0.00	0.00	0.00	0.00
SiO_2	0.00	0.00	0.00	0.00	0.00	0.00

detailed analyses be made in the future of the surface as a function of the level of oxidation, reduction, and sulfiding of the surface. Such information should clarify how the surface composition varies with past history and how the surface composition affects the surface reactions.

Discussion of Results

Based primarily on the results of this investigation plus other investigations at Purdue University (3,4,5,9,10,14), considerable data have now been obtained to indicate the main surface reactions and their relationship to the gas-phase reactions that occur during pyrolysis. Figure 6 summarizes the key reactions. Although this figure shows the gas-phase reactions for ethane pyrolysis, the types of surface reactions would probably be basically the same regardless of the hydrocarbon feedstock. The relative importance of the surface reactions vary, however, depending on the feedstock used. This point should be obvious since the gas phase provides the reactants for the surface reactions; the composition of the gas phase of course depends on the hydrocarbon feedstock used and on the degree of conversion.

Several dynamic equilibria can be expected to occur once "steady-state" operation of the pyrolysis unit occurs. Both coking and decoking reactions are occurring. In a well operated system, the net formation of coke is low. More or less equilibria can also be expected to occur on the inner surfaces of a reactor tube for both the metal oxides and the metal sulfides that are being simultaneously formed and destroyed. Formation of metal oxides during pyrolysis runs are primarily by means of reactions between the metal surfaces and steam. The metal oxides that are excellent oxidizing agents at the temperatures employed are destroyed (resulting in a reduced surface) by several gaseous materials. In addition to hydrogen, carbon monoxide, hydrogen sulfide, and olefins as demonstrated in this investigation, paraffins and other hydrocarbons are oxidized forming carbon oxides and water. Such a conclusion is based in part on extensive results obtained in related work by Crynes and Albright (3), Herriott, Eckert, and Albright (9) and Mahajan (14). Olefins were also found to react at the surface to produce coke and hydrogen (4,5). Considerable more information is, however, needed on the relative reactivities of various olefins, diolefins, aromatics, etc. at the surface and with various metal oxides.

Although hydrogen sulfide was the only sulfur-containing compound tested in this investigation, presumably mercaptans or organic sulfides would also produce metal sulfides on the surface and sulfur dioxide. The formation and subsequent destruction of metal sulfides is thought to be a major reason for the high porosity of the inner surface and for the thickening of the wall. Nickel in particular is susceptible to sulfur attack and some nickel sulfides are liquids at pyrolysis temperatures.

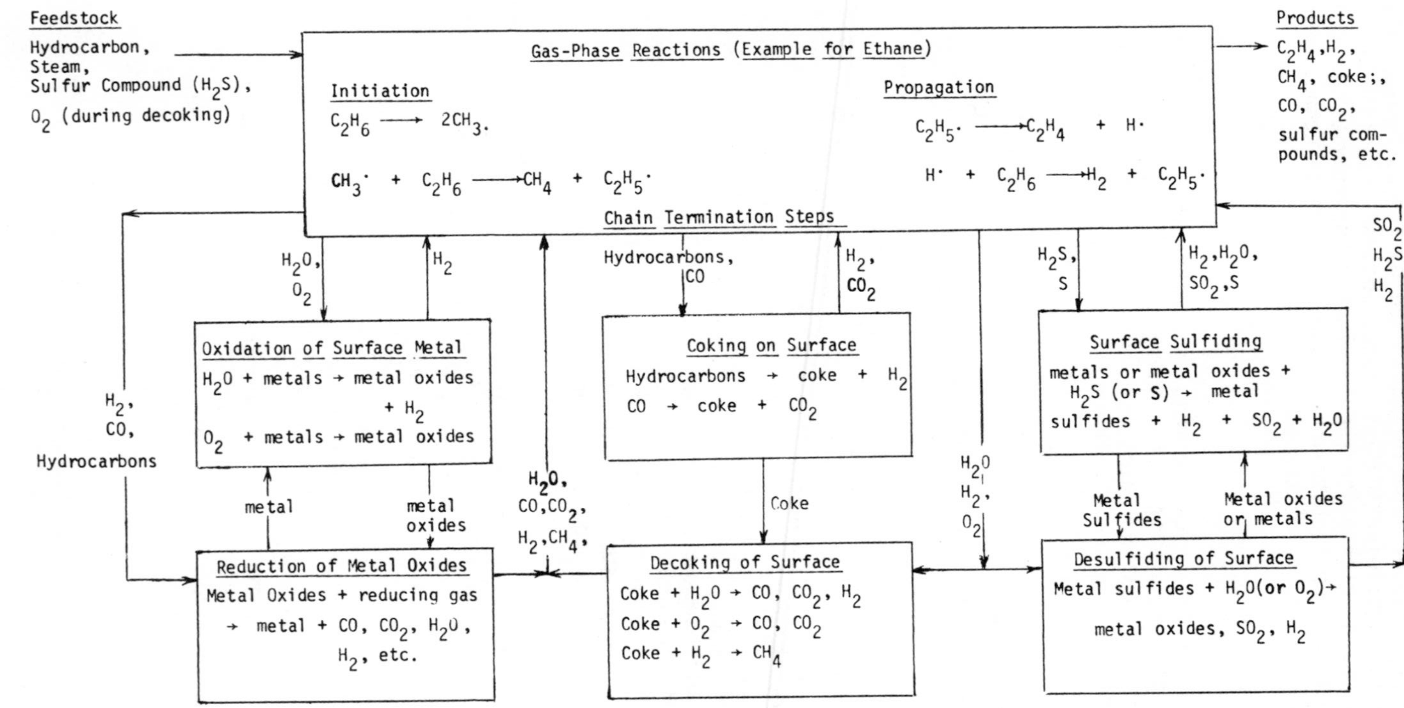

Figure 6. Overall mechanism for pyrolysis including both gas-phase and surface reactions

How to best apply the results of the present investigation to commercial units is not yet known. A small diameter reactor (such as operated in this investigation at atmospheric pressure) has much larger surface-to-volume and surface-to-mass of gas ratios; hence the relative importance of surface reactions is much greater in these reactors than in larger diameter commercial reactors operated at several atmospheres of pressure. Information currently being obtained will hopefully indicate improved techniques needed to predict the importance of surface reactions in commercial pyrolysis units. Presumably, complicated mathematical models that can be solved with the aid of computers can be developed in the near future.

Acknowledgment

Purdue University provided both teaching and research fellowships to the senior author.

Literature Cited

1. Fast, J. D., "Interaction of Metals and Gases", Academic Press, N.Y.C. (1965).
2. Lymar, T., ed., "Metals Handbook", American Society for Metals, Cleveland (1961).
3. Crynes, B. L., and Albright, L. F., Ind. Eng. Chem. Proc. Design and Develop., 8, 1, 25 (1969).
4. Dunkleman, J. J. and Albright, L. F., Chapters 14 and 15, This Book (1976).
5. Brown, S., and Albright, L. F., Chapter 17, This Book (1976).
6. Holm, C. F. Vernon, and Clark, A., J. of Catalysis, 11, 305-316 (1968).
7. Kivlen, J. A., Struth, B. W., and Weiss, C. P., U.S. Patent 3,641,190 (1972).
8. Gilliland, E. R., and Harriott, P., Ind. Eng. Chem., 46, 10, 2193 (1954).
9. Herriott, G. E., Eckert, R. E., and Albright, L. F., AIChE J. 18, 1, 84 (1972).
10. Dunkleman, J. J., PhD thesis, Purdue University (1976).
11. Dravnieks, A., and Samans, C. H., J. Electrochem. Soc., 105, 183 (1958).
12. Sorell, G., and Hoyt, W. B., Twelfth Annual Conference, National Association of Corrosion Engineers, New York, N.Y., March 12-16 (1956).
13. Krupkowski, A., Zasady nowoczesney metalurgii w zarysie, PCWS, Warsaw, p. 98 (1951).
14. Mahajan, S., PhD dissertation, Purdue University (1972).

17

Role of the Reactor Surface in Pyrolysis of Light Paraffins and Olefins

STEVEN M. BROWN and LYLE F. ALBRIGHT

School of Chemical Engineering, Purdue University, West Lafayette, Ind. 47907

Surface reactions occurring during pyrolysis still need considerable clarification. These reactions probably include the initiation and termination of free radicals, coking, and the formation of metal oxides, metal carbides, and metal sulfides. Earlier findings at Purdue University (1-4), including those of Tsai and Albright (5), have indicated the importance of these surface reactions during the pyrolysis of light hydrocarbons. More quantitative information is, however, still needed particularly regarding the surface reactions of light paraffins and olefins as a function of the materials of construction of the reactor.

In order to accentuate the surface reactions, the present investigation was limited to relatively low temperatures (450-750°C) and to long space times (5-30 sec.). For such relatively low temperatures, gas-phase reactions and/or initiation of reactions are slight; surface reactions, however, often occur readily at such conditions. The surface reactions were investigated in four tubular flow reactors: 304 stainless steel, 410 stainless steel, Incoloy 800, and Vycor glass. Four light hydrocarbons were pyrolyzed in this study: ethane, propane, ethylene, and propylene. The results were used to investigate conversions, carbon deposition, and product yields as a function of temperature, of hydrocarbon feed, and of the material of construction, age, and pretreatment of the reactor. Pretreatments were performed with oxygen and hydrogen. Steam was added to the feed in some runs.

The equipment and operating procedure employed were similar to those used by Tsai (5), and they are discussed in more detail by Brown (4).

*Present address: Diamond Shamrock Corp., Cleveland, Ohio.

Experimental Results

Valuable new information have been obtained relative to the surface reactions that are often of importance during the pyrolysis of light hydrocarbons. The material of construction, the history and pretreatment of the reactor surfaces, and the operating conditions employed, all had an appreciable effect on these surface reactions that result in the destruction of olefins and in the production of coke and carbon oxides.

Effect of Temperature. Experiments in which the temperature of the reactor was increased from about 450°C to perhaps 700°C were helpful in clarifying the importance of surface reactions in the metal reactors tested. Figures 1, 2, and 3 indicate that for ethane, ethylene, and propylene, appreciable conversions of the dry feed hydrocarbons begin at 450°-475°C in the metal reactors tested, but do not begin until about 550°-575°C in the Vycor reactor. Propane conversions began at about 450°-475°C in all reactors investigated.

The composition of the product gases for runs made in the range of about 450°-600°C differed significantly, depending on the reactor used. Table I shows the products obtained for runs with dry ethane, propane, ethylene, and propylene; the products are arranged in the approximate order of importance. In the Vycor reactor, the expected gas-phase products are predominant; for ethane pyrolysis, ethylene and hydrogen are expected to be the major gas-phase products, and for propane pyrolysis, ethylene, propylene, methane, and hydrogen are expected. In the metal reactors and especially in the 304 stainless steel reactor, surface reactions resulting in coke are of much greater importance. Coke was sometimes a major product whereas it is not in commercial pyrolysis units that have much lower surface-to-volume ratios in the reactor tubes.

Hydrocarbon conversions can, in general, be represented fairly well by first-order reaction kinetics, and the conversion levels for runs made with a constant hydrocarbon flow rate as a general rule increased significantly as the temperature increased. Figures 1, 2, and 3 show typical results for ethane, propane, ethylene, and propylene. Based on first-order kinetics, the activation energies for ethane, propane, ethylene, and propylene were determined in the various reactors tested. In the Vycor reactor, these activation energies were approximately 51, 57, 56, and 66 k cal/g mole respectively. They were much lower in metal reactors especially after the reactor was oxidized. In a relatively new and unoxidized Incoloy reactor, the activation energies were 15, 47, 27, and 26 k cal/g mole respectively.

Two explanations were considered to account for the significant reactions and for the different types of products obtained at lower temperatures (475° - 600°C) in the metal reactors. The first one considered was that free radicals formed in the gas phase are destroyed to a greater extent at the walls of the Vy-

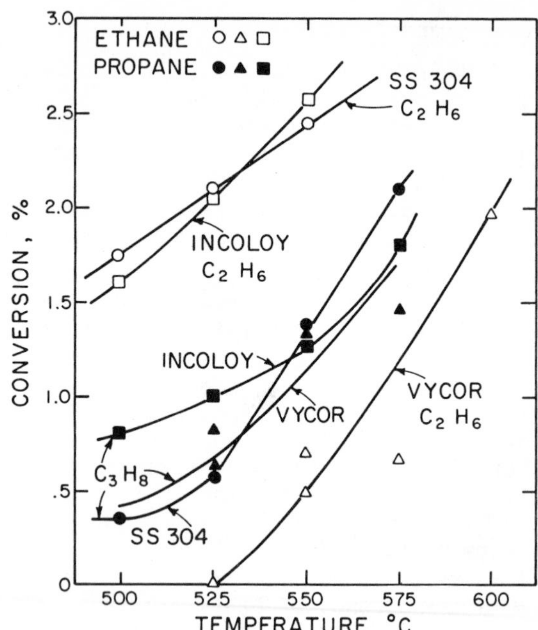

Figure 1. Ethane and propane conversions vs. temperature in several different reactors

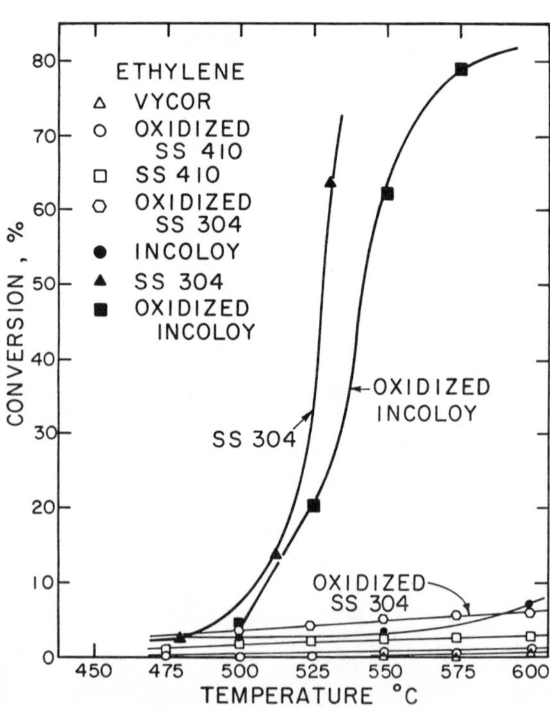

Figure 2. Ethylene conversions vs. temperature in several different reactors

Table I. Major Products for the Pyrolysis of Light Hydrocarbons in Three Reactors.
(450° to 600°C)

Type of Reactor		
Vycor Glass	Incoloy 800	304 Stainless Steel
	Ethane Feed	
Ethylene	Ethylene	Hydrogen
Hydrogen	Hydrogen	Coke
Methane	Methane	Ethylene
Propylene	Propylene	Methane
Coke	Coke	Propylene
	Propane Feed	
Propylene	Hydrogen	Hydrogen
Methane	Propylene	Coke
Ethylene	Ethylene	Propylene
Hydrogen	Methane	Methane
Ethane	Coke	Ethylene
Coke	Ethane	Ethane
	Ethylene Feed	
Ethane	Hydrogen	Hydrogen
Acetylene	Coke	Coke
Hydrogen	Ethane	Ethane
Methane	Methane	Methane
Coke	Acetylene	Acetylene
	Propylene Feed	
Ethylene	Hydrogen	Hydrogen
Methane	Coke	Coke
Hydrogen	Ethylene	Ethylene
Coke	Methane	Methane
Ethane	Ethane	Ethane

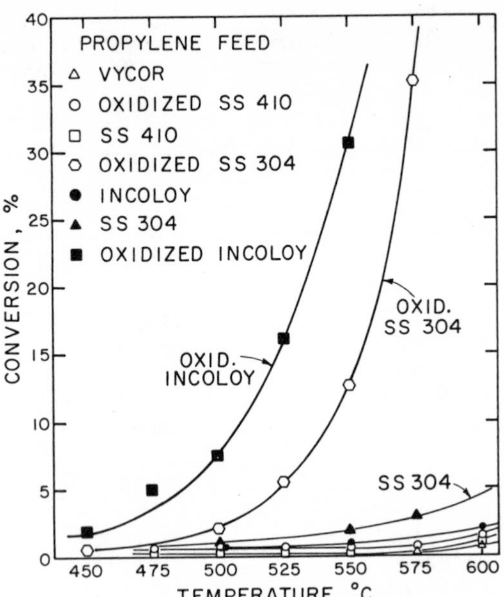

Figure 3. Propylene conversions vs. temperature in several different reactors

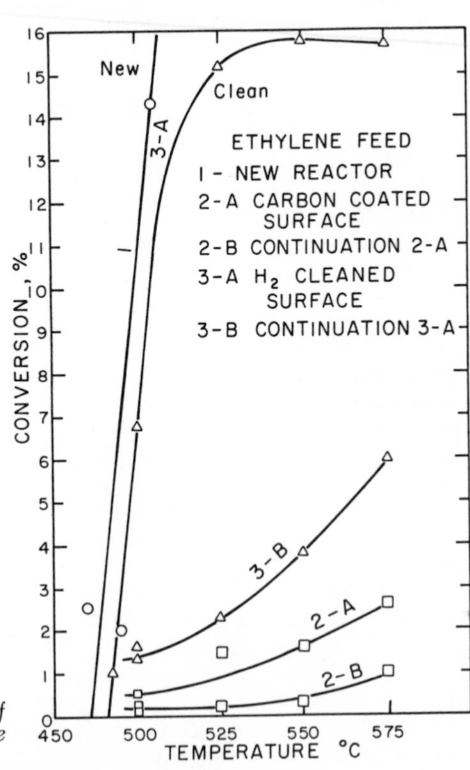

Figure 4. Effects of the past history of stainless steel 304 reactor on ethylene conversion

cor glass reactor as compared to the metal reactors, and hence reduced the reaction rate in the glass reactor. This explanation seems unlikely since metal surfaces are generally considered to be better free-radical terminators than glass surfaces (6). The preferred explanation is that the metal surfaces initiate both decomposition and pyrolysis reactions to a much greater extent than Vycor. The hydrocarbon feed probably reacts breaking carbon-hydrogen bonds and resulting in essentially the following reaction:

$$\text{Hydrocarbon} \longrightarrow \text{Coke} + \text{Hydrogen Radicals } (\text{H}\cdot)$$

To support this explanation, coke was formed (as a major product) on the metal surfaces in the temperature range of 450° to 550°C. It seems quite likely that as the carbon-hydrogen bonds are broken on the reactor surface, hydrogen atoms (or free radicals) form and at least some migrate into the gas phase to initiate gas-phase reactions. As further support of this hypothesis, some products formed in the 450 to 550°C range appear to be products that are formed by gas-phase reactions, namely olefins, hydrogen, and methane.

With propane as the feed hydrocarbon, relatively little difference was noted in the conversions in the different reactors (see Figure 1). Probably gas-phase reactions begin with propane at much lower temperatures than for the other three hydrocarbons; propane is the only hydrocarbon of those tested that has secondary carbon-hydrogen bonds. For propane, the metallic reactors did result in slightly higher coke and hydrogen yields than the Vycor reactor, but the conversions were similar.

As the metal reactors were used in this investigation, they became roughened, and the relative importance of surface to gas-phase reactions increased. This roughening also resulted in higher reaction rates and lower activation energies, hence indicating the increased relative importance of surface reactions.

Deactivation and Activation of Metal Reactors. As coke was formed on a metal reactor, the reactor in general was partly deactivated, but never to the low level of the Vycor reactor; deactivation is defined as lower conversions of the feed hydrocarbon. Figure 4 shows results with ethylene feed in the 304 stainless steel reactor. Conversions were highest in the new reactor (Run 1); a new reactor is defined as one that has been used for reactions less than 1-2 hours. The conversions in both Runs 2-A and 2-B were lower than those of Run 1 because of the coke formed during Run 1: Run 2-B (made as a continuation of Run 2-A) had even lower conversions than Run 2-A. After Run 2-B, hydrogen was passed through the reactor; the hydrogen reacted with the surface carbon forming methane. Run 3-A was made after the rate of reaction between hydrogen and surface coke had decreased to almost zero. Run 3-A indicated an active reactor especially in the early stages of the run (namely up to 525°C). At 550° and 575°C, the reactor was still much more active than in Runs 2-A and 2-B. Apparently, however, the reactor deactivated rapidly during Run 3-A at 575°C, presumably because of the buildup of coke. The reactor was considerably less active in Puns 3-B made after Run 3-A.

Figure 5 gives more details on the deactivation of the 304 stainless steel reactor during Run 3 (using dry ethylene feed) and during Run 4 (using dry propylene feed). The reactor had been cleaned using hydrogen (and hence activated) before the start of both runs. The data points for these two runs are numbered chronologically. Deactivation because of coke formation was significant during both runs as a comparison of data points # 4 and # 6 and of # 3 and # 7 indicate.

Considerable new information has been obtained concerning the role of metal oxides on the surface of metal reactors. The results indicate that as a reactor oxidized with oxygen is used,

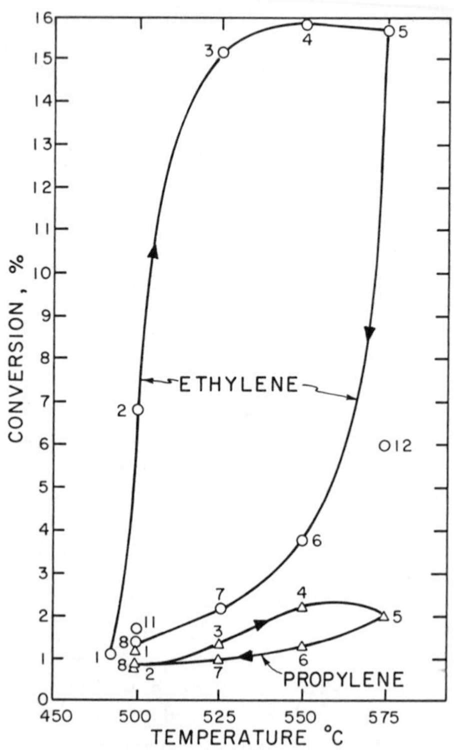

Figure 5. Effects of past use of stainless steel 304 reactor on ethylene and propylene conversions

the activity of the reactor often passes through a maximum. Figure 6 shows how propylene conversions changed in both oxidized 304 stainless steel and oxidized Incoloy reactors. The data points are again numbered chronologically. In both reactors, conversions in the oxidized reactors were initially very low with dry propylene feed at temperatures up to about 525 to 550°C. Then at 550-575°C in the Incoloy reactor and at 575-600°C in the 304 stainless steel reactor, significant reactions were noted. Products were obtained of what was apparently both gas-phase free-radical reactions and surface decomposition reactions; coke and carbon oxides were both formed in significant amounts. The production of carbon oxides indicated that the propylene was reacting with surface oxides on the reactor surface. The level and the types of surface oxides were changing (e.g. Fe_2O_3 was probably converted to Fe_3O_4 or FeO), hence changing the surface activity.

When the reactors were then cooled to 475° to 525°C, significant propylene reactions were still noted. Clearly the reac-

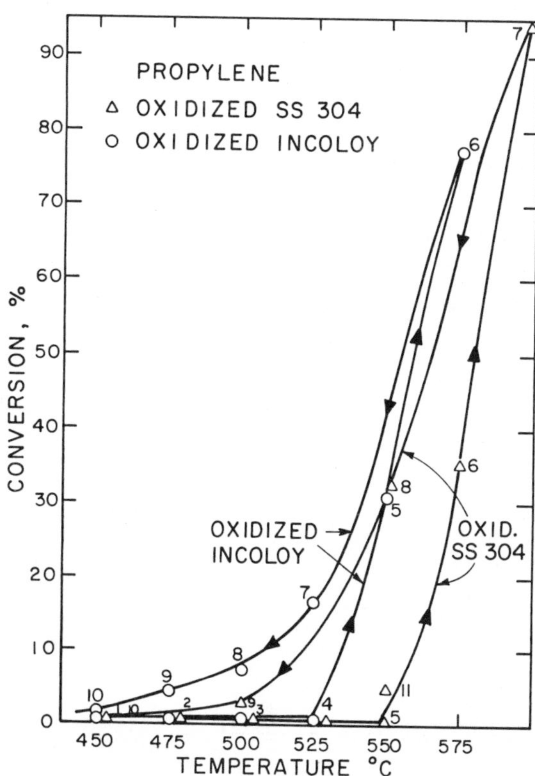

Figure 6. Effects of oxidation of stainless steel 304 and Incoloy reactors on propylene conversions

tors were then in a much more active state than during the initial phases of the runs. With continued use of the reactor and with dry propylene feed, both reactors then rather slowly deactivated. This latter deactivation presumably occurred because of the formation of coke.

Figure 7 shows the results of a propylene run at 500°C in an oxidized Incoloy reactor. In the first 75 minutes using dry propylene, the conversions passed through a maximum. Wet propylene was then used as a feedstock for the next 60 minutes; propylene conversions were slightly higher and they perhaps decreased somewhat during this period. Of considerable interest, the subsequent phase of the run with dry propylene indicated that the reactor had been significantly activated during the wet propylene phase of the run, presumably because of the formation of metal oxides on the surface (5). Continued use of dry propylene resulted in a rather rapid deactivation of the reactor; the reactor

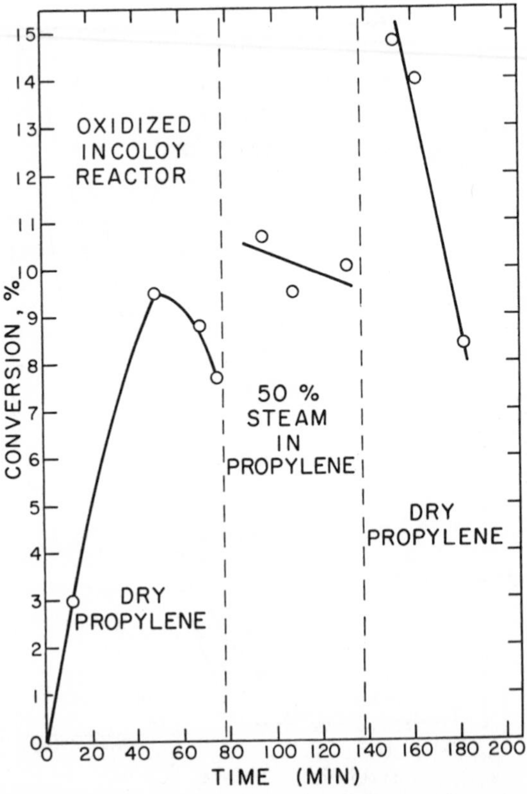

Figure 7. Effects of past history and steam in Incoloy reactor on propylene conversions at 500°C

was still very active after about 40 minutes of dry propylene feed. When, however, steam diluent was added to ethylene feed in the metallic reactors, the activity remained the same or decreased slightly with time. Clearly the effects of a steam are complicated and more study is needed.

Flushing a coke-coated reactor with either helium or nitrogen for about 30 minutes was found on several occasions to activate the reactor in the subsequent run. Certain hydrocarbons were desorbed during the flushing as will be discussed in more detail later.

Materials of Construction. The materials of construction of the reactor clearly affect the level and probably to some extent the type of surface reactions. Surface reactions are in general much less important in Vycor reactors at least in the range of approximately 450° to 550°C (see Figures 1, 2, and 3), as indicated by the lower hydrocarbon conversions in these reactors. The general level of activity of the Incoloy 800 and the 304 stainless steel reactors depends in some complex manner on the level of the surface oxides and apparently on the hydrocarbon feedstock. Similar conversions were obtained with dry ethane and with dry propane in relatively new (and unoxidized) Incoloy and 304 stainless steel reactors (see Figure 1). Yet as shown in Figure 2, ethylene was much more reactive in the unoxidized 304 stainless steel reactor than in the unoxidized Incoloy reactor. With ethylene, oxidized Incoloy gave much higher conversion than the reduced new Incoloy; the reverse was found to be true with oxidized and reduced 304 stainless steels.

The 410 stainless reactor showed much lower activity than either of the other two metal reactors. This reactor had activities only slightly greater than those of Vycor at temperatures up to about 650°C. Yet the Vycor reactor had higher hydrocarbon conversions than the 410 stainless steel reactor at higher temperatures (between 650°C and 750°C). The 410 stainless steel reactor probably terminated more free radical reactions than did the Vycor surface.

Characteristics of Surface Reactions. In general, the reactivities of the feed hydrocarbons tested in reduced metallic reactors were in the following order:

Ethylene > Ethane > Propane > Propylene

Such a conclusion is shown for example by comparing the results of Figures 2, 3, and 5. Only the olefins were tested in oxidized metallic reactors, and both were very reactive. The olefin feeds were apparently more strongly affected by the oxidized metallic surfaces than the paraffins.

The amount of coke formed on the inner surface of a reactor was in general closely related to the surface activity. With the long residence times employed and with the relatively small diameter reactor used, coke yields based on the moles of feed hydrocarbon reacted were much higher than those normally found in commercial reactors. The relative comparisons for different reactors and hydrocarbons are, however, thought to be reliable. Figures 8, 9, and 10 show several comparisons. In all cases, the coking abilities of the hydrocarbons were as follows:

Ethylene > Propylene > Propane > Ethane

These results are shown in Figure 8 and by making a comparison of Figures 9 and 10. Lobo et al (7) also found that olefins produced much more coke than paraffins. The greater coking ability of ethylene as compared to propylene is caused primarily by the higher reactivity of ethylene; somewhat higher yields of coke (based on the amount of olefin that reacts), however, occur with propylene as compared to ethylene.

A key finding of this investigation was that significant amounts of olefins and heavier hydrocarbons were often absorbed on the inner surface of the reactor, probably mainly on the coke deposited there. About 0.2 millimoles of ethylene were, for

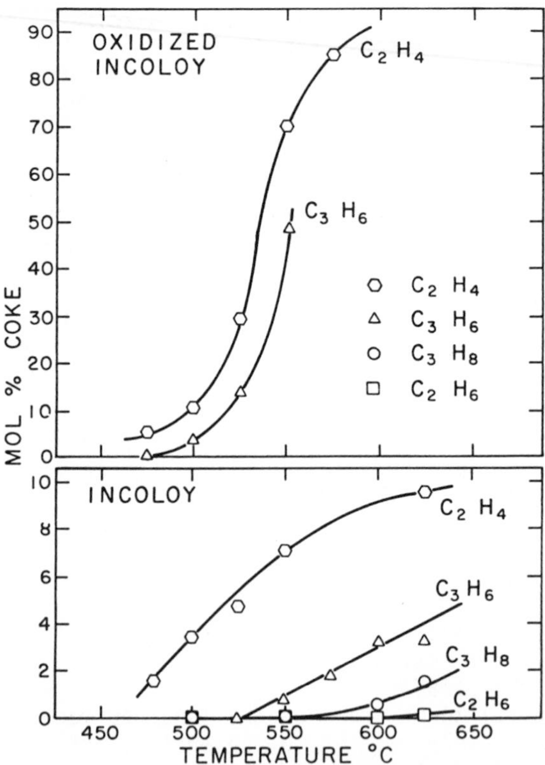

Figure 8. Coke produced from ethane, ethylene, propane, and propylene feeds in reduced-Incoloy and oxidized-Incoloy reactors

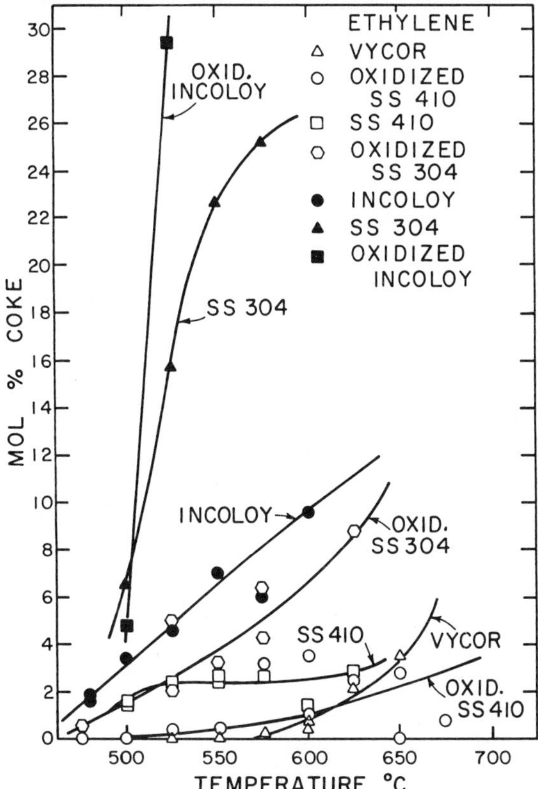

Figure 9. Coke produced from ethylene feed in several different reactors

example, absorbed on the 304 stainless steel reactor that was operated at 530°C. After several runs, helium was used to flush out the reactor; several different hydrocarbons were desorbed and found in the exit helium stream for up to 1.5 hours. These hydrocarbons included olefins and heavier hydrocarbons; these latter hydrocarbons were probably in the C_5 - C_6 range and may have included some aromatics. Adsorbed hydrocarbons presumably are relatively reactive and would likely dehydrogenate fairly rapidly forming coke. Appleby (8) and John (9) have also postulated a similar coking mechanism.

Clearly propylene and ethylene are important during pyrolysis in the production of coke. The high reactivity of ethylene as compared to propylene is thought to be caused primarily by the lower stability of vinyl radicals, as compared to that of allyl radicals. Formation of the C_5 - C_6 intermediates possibly occurs by dimerization and/or trimerization of these radicals.

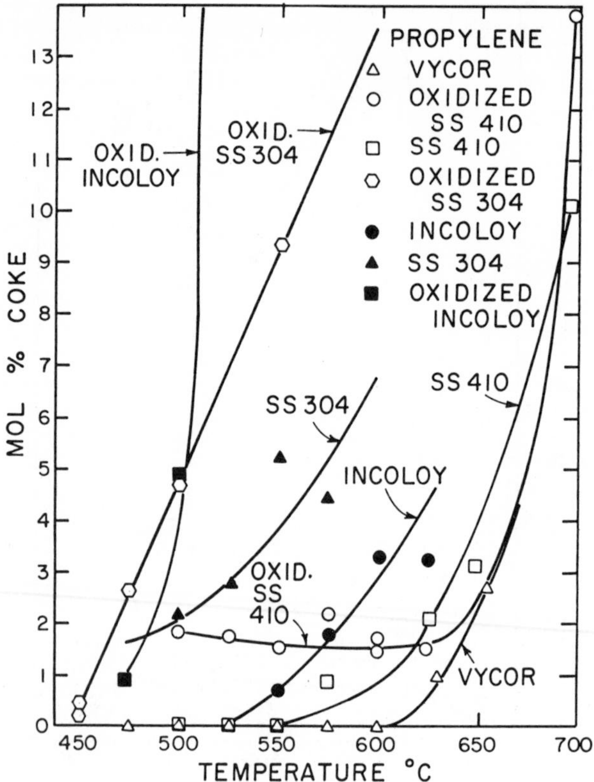

Figure 10. Coke produced from propylene feed in several different reactors

The results of the present investigation are probably applicable to commercial units even though significantly lower temperatures were employed in this investigation. Fewer hydrocarbons would adsorb on the reactor surface at the higher temperatures used commercially, but extrapolation of the low temperature results to higher temperatures seems to indicate that significant adsorption does still occur at temperatures of commercial interest. Presumably, the absorbed hydrocarbons would decompose rapidly into coke and hydrogen at higher temperatures. It is concluded that coke formation is often, if not always, a two-step process involving first adsorption and then surface decomposition (probably dehydrogenation and hydrogenolysis).

A run with dry propylene feed at temperatures between 500°C and 575°C resulted in almost complete plugging of an unoxidized 304 stainless steel reactor after 7 hours of operation. Even though the metal surface was rather completely covered with carbon by the end of the run, propylene conversions were still higher in this reactor than in the Vycor reactor at comparable conditions. Propylene or hydrocarbons formed from propylene may have diffused through the porous carbon layer to the metal surface where they reacted to form more coke. Metal atoms in the coke such as detected by Hoppstock et al (10) and by Lobo et al (7) may also have been a factor.

Miscellaneous Results. After the Vycor reactor was contacted with oxygen for several hours at 800°C, the activity of the reactor was increased slightly. Both ethylene and propylene showed detectable reactions for almost 15 minutes at 500°C which is 50°C lower than for the untreated reactor. Small amounts of carbon oxides were detected during this time indicating that some oxygen had been adsorbed on the inner Vycor wall. After all of the oxygen had desorbed, the activity of the reactor returned to a level similar to that in the untreated Vycor reactor. A steam treatment of the Vycor reactor at high temperatures did not, however, produce any noticeable increase in the reactor activity.

Surface deposits were noted in the Vycor reactor following the experimental runs. These dark brown or black deposits occurred primarily at the exit end of the reactor. These deposits were oily and tarry in nature, and were probably coke and condensed heavy hydrocarbons.

Hydrogen treatment at 800°C of the Vycor reactor, which contained some coke or tarry material, resulted in the production of some methane. This result was surprising since it had been thought that methane formation would occur only in the presence of a metal catalyst such as nickel or iron.

The 410 stainless steel reactor was oxidized upon completion of a run, and a red rust-like powder was formed on the wall. Part of this powder was easily brushed from the reactor.

Discussion of Results

The present results clearly confirm the importance and complexity of surface reactions during pyrolysis reactions. Obviously, the composition of the inner surface of the reactor is of importance relative to the level and types of surface reactions. In addition, valuable new information has been obtained concerning the role of coke in affecting more coke formation. Although the deposition of coke on the walls of a metal reactor decreases the activity of the reactor, it is of interest that the surface activities of coke-covered metal reactors always remained higher than those for the Vycor reactor. Lobo and Trimm (11) have indicated that carbon without contaminants is inactive. Based on this finding, metal contaminants were presumably present in the coke formed. Other investigators (10, 11) have found both nickel and iron contamination of various cokes. Furthermore, coke is sometimes reported to be autocatalytic in nature. The evidence that olefins and other hydrocarbons adsorbed on the surface and presumably on or in the coke would seem to offer at least one explanation for autocatalysis.

Currently the major factors considered in the selection of a tube to be used in a pyrolysis furnace are the physical properties and the expected longevity of the tube. Corrosion and carburization certainly are of importance in this latter respect, but the surface activity should also be considered. Based on the present results, increasing the chromium content and decreasing the nickel content of the metal would reduce coking and other undesired surface reactions. A high chromium stainless steel would likely result in relatively few surface reactions.

Chromizing the inner surface of tube is a possibility that should be considered in attempting to obtain a high-chromium surface. Aluminizing the surface may also produce relatively inert surfaces. The surface chromium and aluminum likely would react with steam or oxygen to form stable oxides that are probably fairly non-reactive relative to undesired surface reactions.

Clearly more investigations are needed to clarify the role of metal oxides on the surface of the reactor. Analysis of the surface for metal oxides and metals would be most helpful in this endeavor. In addition, the roles of the metals (or metal oxides) and of the coke in initiating reactions need clarification. Methods of extrapolating the results of a laboratory unit to commercial units should also be developed.

Acknowledgments

Purdue University and the Procter and Gamble Company provided generous financial support.

Literature Cited

1. Crynes, B. L. and Albright, L. F., Ind. Eng. Chem. Proc. Des. Dev. 8, 1, 25 (1969).
2. Herriott, G. E., Eckert, R. E., and Albright, L. F., AIChE Journal, 18, 1 (1972).
3. Dunkleman, J. J. and Albright, L. F., This Book, Chapters 14 and 15, 1976.
4. Brown, S. M., M. S. Thesis, Purdue University, May, 1976.
5. Tsai, C. H. and Albright, L. F., This Book, Chapter 16 (1976).
6. Nishyama, Y., Bull. Chem. Soc. J., 42, 9, 2494 (1969).
7. Lobo, L. S., Trimm, D. L., and Figuerdo, J. L, Proc. Int. Congr. Catalysis, 5th, 1125 (1973).
8. Appleby, W. G., Gibson, J. W., and Good, G. M. Ind. Eng. Chem. Proc. Des. Dev. 1, 2, 102 (1962).
9. John, T. M. and Wojciechowski, B. M., J. Catalysis, 37, 240 (1975)
10. Hoppstock, F. H., Hutchings, D. A., and Frech, K. J., This Book, Chapter 12, 1976.
11. Lobo, L. S. and Trimm, D. L., J. Catalysis, 29, 15 (1973).

18

Pyrolysis Bench Scale Unit Design and Data Correlation

J. J. LEONARD, J. E. GWYN, and G. R. McCULLOUGH

Shell Development Co., Westhollow Research Center, P.O. Box 1380, Houston, Tex. 77001

Ethylene cracking plants represent large capital investments, and the cost of feedstock to such plants is a major operating expense. Hence, there is considerable justification for evaluating potential feedstocks and for determining optimum operating conditions for each cracker feedstock. One approach to fulfilling such goals is to develop a reactor model that predicts yields for virtually any cracking coil and any operating condition using a bench scale unit to obtain feedstock-dependent parameters required for the model. A further extension is to have the feedstock-dependent parameters as functions of some measurable feedstock properties.

In this paper, the design and operation of a bench scale unit are described. The yield results from the unit are compared to commercial yields with emphasis on the correlation of cracking severity. One approach to reactor modeling is described. Feedstock parameters are obtained from bench unit data and residence time plus profiles of pressure, fluid and metal temperatures, heat flux and methane yields for commercial conditions are calculated using these parameters. This model can then be used in conjunction with a yield selectivity model, not described here, to predict a full spectrum of yields.

Experimental

The bench scale unit was designed to have temperature and pressure profiles and residence times similar to conventional commercial crackers. Thus, comparative data for different feeds can be used directly. However, some sort of empirical approach would be necessary to be able to stipulate what commercial furnace operating conditions the data represent.

A schematic of the bench scale unit is shown in Figure 1. The feed system is fairly conventional. Light feeds are pressured into the unit and liquid feeds are pumped from weighed feed tanks. The radiant coil is divided into four sections of small diameter stainless tubing. The outlet bulk gas temperature of each zone

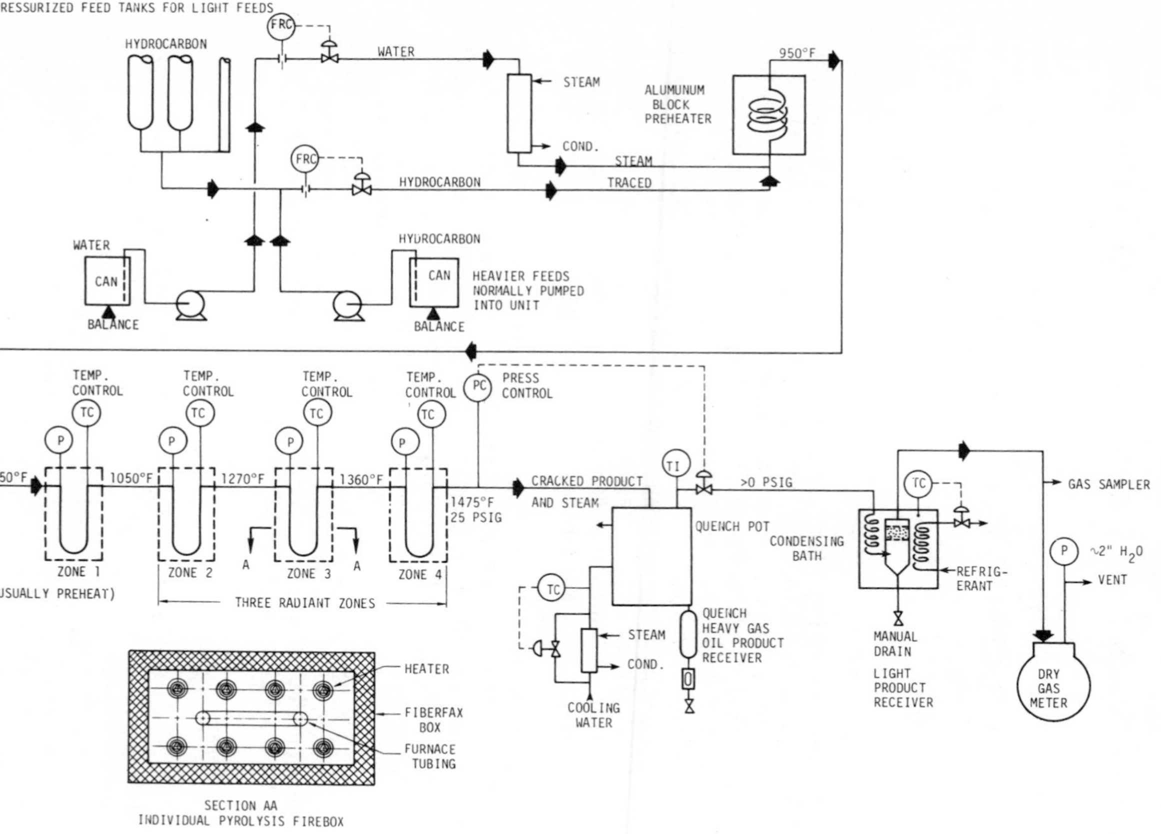

Figure 1. Flow diagram of pyrolysis plant

is measured with thermocouple and the temperature is controlled by four radiant electrical heaters that parallel each straight length of tubing. Pressure is controlled only at the furnace outlet. However, pressure profile across the reactor can be controlled. The hot gases from the radiant zone are passed through a short insulated tube and blown into condensed liquid product for quenching. In this fashion, little difficulty with coking is experienced.

Three product streams are recovered; a heavy oil/water quench liquid, a light oil/water liquid condensed at about 10°C, and a gas stream. Routine product analysis is shown in Table I. The capillary gas/liquid chromotography (GLC) on the product gas is solely for the purpose of obtaining isomer distributions. The capillary GLC on the light oil is primarily for C_5 isomer breakdown and for benzene, toluene, and xylenes (BTX) analysis.

Table I. Product Analysis

Gas
- Duplicate Mass Spectroscopy
- Capillary Gas Chromotography

Light Liquid
- "True Boiling Point" - Liquid Chromotography
- Capillary Liquid Chromotography
- Carbon/Hydrogen/Sulfur

Heavy Liquid
- "True Boiling Point" - Liquid Chromotography
- Carbon/Hydrogen/Sulfur

The TBP-GLC's on the liquids are for calculation of yields of various boiling range liquid products. The C/H/S analyses are used for elemental material balance, calculation of heats of reactions and calculation of hydrogen distribution, i.e., hydrogen content of liquid product. From the material balance and the product analyses, a final yield report is made as shown in Table II. With each yield report, the operating conditions are reported which include tube identification, hydrocarbon and water feed rates and pressure and bulk gas temperature profiles.

Severity of Cracking

There are numerous expressions that are useful for representing severity of cracking.(1) The most commonly used terms in our work, i.e., for liquids cracking, are shown in Table III, and are discussed below.

Methane yield increases monotonically with increased furnace firing, pressure and residence time. It is feedstock dependent, but, because it is a dependent variable, it is useful for comparing yields for similar feeds at equal severities without

Table II. Example Yield Report

%wt		%wt		
0.43	Hydrogen	5.7	C_4 w/o BD	→ C_4 Isomer Breakdown
9.2	Methane	29.6	C_5-450°F	
0.17	Acetylene	10.0	450-615°F	→ C_5 Isomer Breakdown
18.6	Ethylene	3.4	615+	→ C_6 Isomer Breakdown
4.3	Ethane	5.4	Benzene	→ C_7 Isomer Breakdown
0.15	Methyacetylene + Propadiene	3.6	Toluene	
13.6	Propylene	2.7	Xylene	
0.65	Propane	8.5	Hydrogen Content Liquid Product	
4.2	Butadiene	600.	Heat of Reaction, cal/gm	

Table III. Measures of Severity

Methane Yield, Percent Weight

Propylene/Methane Weight Ratio

Hydrogen Content Liquid Product, Percent Weight

$$\int_0^\theta K dt$$

explicit definitions of furnace operating conditions.

Propylene to methane weight ratio ($C_3^=/C_1$) is inversely proportional to methane weight yield over a wide range of severities encompassing normal operating conditions. In Figure 2, data from a commercial unit are shown with bench scale data from two different geometry coils. Within some reasonable span of residence time and pressure, a single line relationship is obtained between $C_3^=/C_1$ ratio and methane yield. Thus, methane yield from a furnace can be determined by measuring the $C_3^=/C_1$ ratio of a furnace effluent if the relationship has been established beforehand.

The hydrogen content of the liquid product is a useful operational control variable. Though subject to interpretation, generally the lower the hydrogen content of the liquid product the more severe the cracking and the closer the operation is to excessive coke formation.

The fundamental expression of severity given in Table III is integrated conversion parameter expressed as $\int_0^\Theta K dt$, where K is the first order reaction rate constant for some specified hydrocarbon and is a function only of temperature, and Θ is time.(2) This is the expression used in our reactor model to relate bench scale units and commercial cracking coils. For the specified hydrocarbon, conversion is obtained from

$$\text{Ln}\ \frac{C_{FD}}{C_{PR}} = \int_0^\Theta K dt = \overline{K\Theta}.$$

For any given feed, a conversion of say 90%, or $\overline{K\Theta}$ equals 2.3, has the same meaning whether the feed was cracked in a bench unit or a commercial unit. Obviously; however, it is impossible to express the conversion of a multicomponent feed as a single value. Thus, the first order rate reaction constant, K or $Ae^{-B/TR}$, is selected as that for a model compound, n-C_{16}.(3) A tacit assumption is made here that the activation energies for the decomposition of the liquid feedstocks vary over a relatively narrow range.(1) As a numerical integration is performed along the length of the tube, the value of residence time is calculated for the real feed and not the model compound. An in-house empirical correlation of hydrocarbon molecular weight as a function of conversion allows such a calculation.

Reactor Calculation Model

The calculation model integrates along the reactor in small increments. It maintains a detailed heat and material balance, pressure, methane yield, conversion, volumetric expansion and other factors necessary to calculate a true severity profile. The calculations involve three levels of interaction. Each increment of reactor must be in heat flux and temperature balance; each furnace zone must satisfy outlet temperature or overall heat flux specifications; and the overall reactor must satisfy

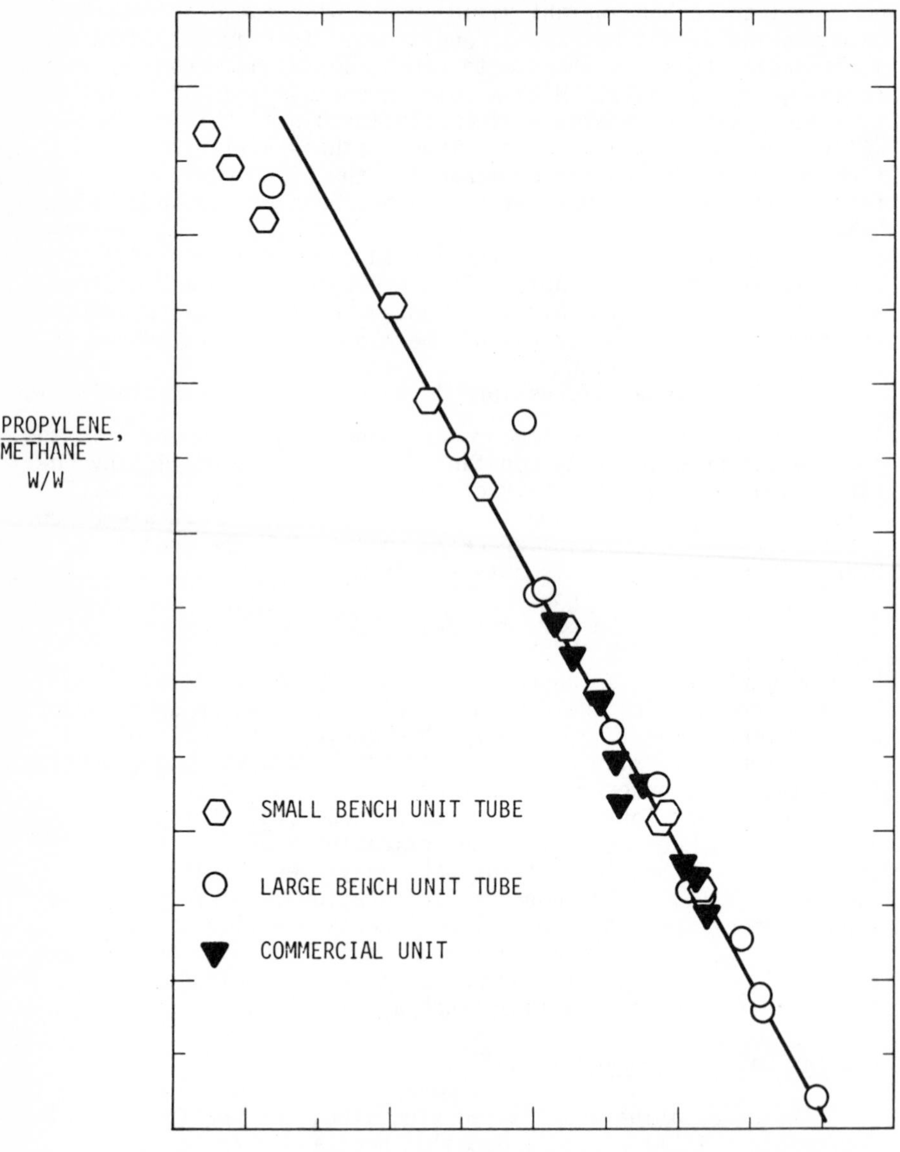

Figure 2. Pyrolysis of a gas oil

specifications of outlet pressure, conversion, etc. The interactions are nested accordingly.

An example calculation procedure where reactor outlet pressure and each zone gas outlet temperature are specified is as follows:

1. Overall Convergence Loop - Inlet pressure is assumed.
2. Zone Convergence Loop - A zone radiation temperature is assumed.
3. Increment Convergence Loop - The average gas bulk temperature of the incremental segment is set equal to inlet gas temperature. An external tube wall temperature is assumed and the heat flux, pressure drop, and bulk gas temperature, T_C, are calculated as discussed in the next section. The wall temperature is adjusted until the calculated bulk gas temperature agrees with the preset value. From the heat flux and reaction heat, (4) the temperature rise along the reactor increment, and hence the outlet temperature of that increment, are calculated. Thus, a better average bulk gas temperature is available and a better estimate of temperature rise can be made by the Runga-Kutta or some similar predictor-corrector technique. The calculation then proceeds to the next segment by this step 3 procedure.
4. If at the end of the zone, the calculated gas temperature does not match the specified temperature, a new radiation temperature is estimated and the calculation is repeated from step 2 until convergence is achieved. The calculation then continues to the next zone.
5. If at the end of the last zone the calculated pressure does not agree with that specified, a new inlet pressure is estimated and the calculation repeated from step 1. On convergence, the profile calculation is complete.

Incremental Yield Calculations

In order to properly calculate the residence time in an increment of reactor, $\Delta\Theta$, it is necessary to know the volumetric flow of the gas. This requires the knowledge of the molecular weight of the gas which is related to the molecular weight of the feed and the methane yield. Thus:

$$MW_{HC} = f(MW_f, C_1)$$

The methane yield, C_1, itself is a function of severity expressed as $\overline{K\Theta}$.

The pressure drop is calculated from the normal friction factor correlation which is a function of Reynold's number. The viscosity value for the Reynold's number is based on the API Technical Data Book.(5) Specific heat for temperature calculations

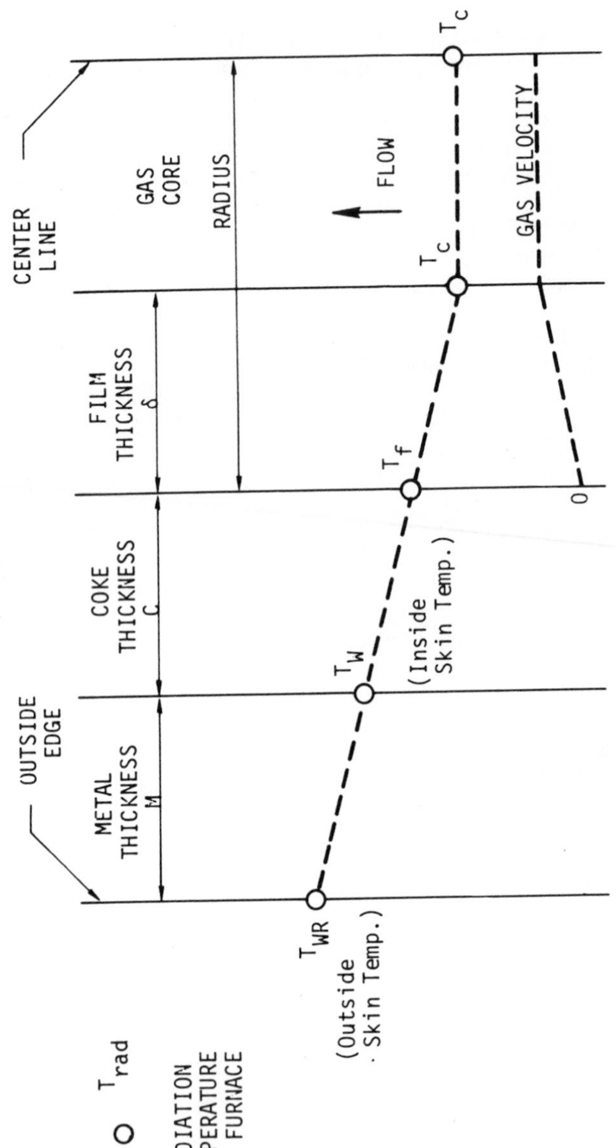

Figure 3. Furnace tube profile

is also based on the API Technical Data Book.

Radiation flux to the furnace coil is given by

$$Q_R = 4.709 \times 10^{-6} \Phi A_0 [T_{rad}^4 - T_{WR}^4]$$

in cal/hr where Φ is the geometry factor between receiver and source. This, in turn, equals the flux across the metal and coke layers, and to the transfer across the turbulent boundary layer given by

$$Q_f = h A_I [T_W - T_C]$$

A cross section of a furnace tube for one small segment of the numerical integration is shown in Figure 3. An assumption is made for this model that the radiant temperature, T_{rad}, has a constant value for each cracking zone. The coke thickness as well as the metal thickness is specified. This model, as presented, does not include any techniques for forecasting coke buildup on the inner tube wall. The fluid film thickness is calculated from an ideal film model where

$$Nu = 0.0243 Re^{0.8} Pr^{0.4}$$

$$Pr = \frac{C_p \mu}{K_C}$$

$Pr \approx 0.9$ for the normal steam/hydrocarbon mixtures in pyrolysis

or $Nu \approx 0.0234 Re^{0.8}$

$$Nu = \frac{hD}{K_C} = \frac{D}{\delta}$$

The temperature drop across the film is assumed linear and the axial temperature across the bulk gas flow is considered uniform. The conversion of feed in the gas film is calculated and added to that of the bulk gas in proportion to the volume of each (see Appendix 1). This simplified model assumed uniform circumferential temperatures.

As mentioned earlier, feedstock-dependent parameters are used in the relationship between conversion and methane yield. Initially, these values are estimated. Calculated values of $\overline{K\Theta}$ using the estimated values are then plotted against measured methane yield and new values for these constants are obtained. This procedure is an off-line iteration, but usually requires only one cycle to converge.

Calculation Features

The output from the computer program is listed in Table IV. The temperature profiles and heat flux profiles are plotted in addition to being tabulated. The presssure, partial pressure, heat flux, and methane yield profiles are tabulated. An example

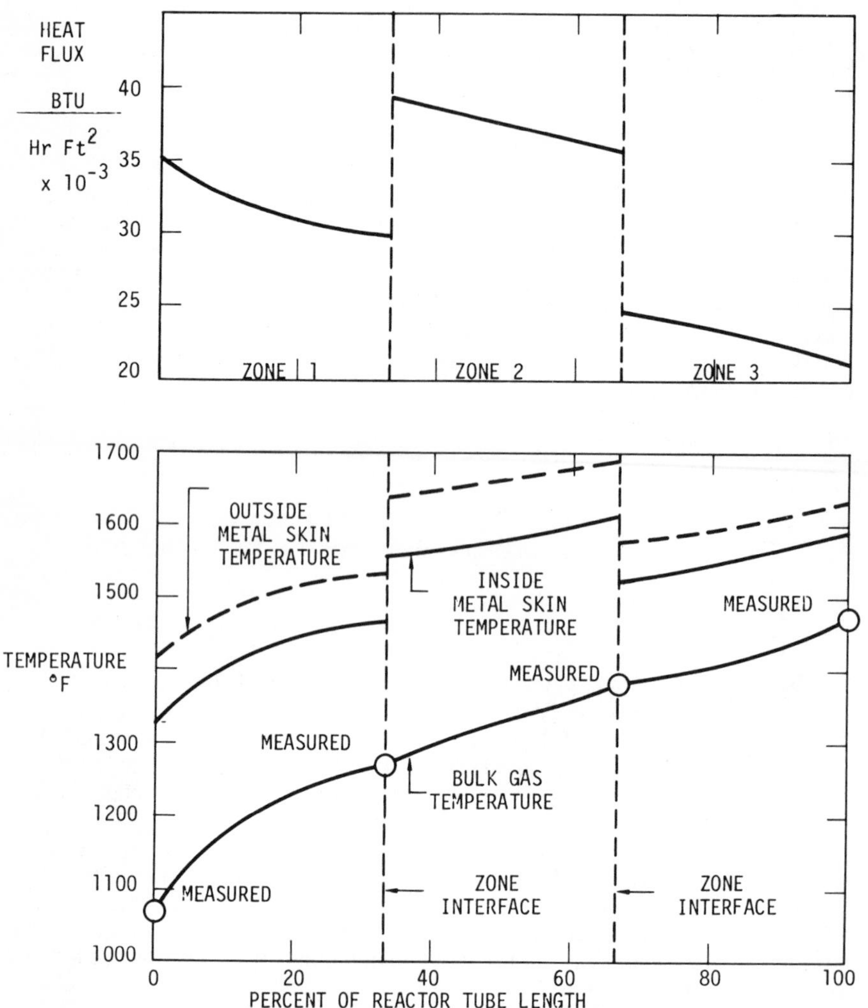

Figure 4. Example of heat flux and temperature profiles calculated by model

of a calculated heat flux and temperature profile for a commercial unit is shown in Figure 4.

Table IV. Computer Output

1. Severity $K\Theta = -\ln(1 - \text{Conv}) = \sum A e^{-B/T} \Delta\Theta$
2. Residence Time, Θ
3. Profiles of:

   Pressure and Partial Pressure
   Gas Temperature
   Metal Temperatures
   Molecular Weight
   Heat Flux
   Methane Yield

4. Heat Duty per Zone
5. Velocities, Reynolds Numbers, etc.

The input required for this model can be grouped into three categories and is shown in Table V. The first grouping describes the properties of radiant coil. The dimensions of the tube must be specified as well as the material of construction. Metal thermal conductivities are catalogued in the program. The location of zone interfaces must be specified as fractional distance along the tube length. The model will consider a one, two or three zone furnace. Unusual tube geometry such as tapered coils can be easily written into the program, but is not now a routine type of calculation.

The only feedstock properties required are the molecular weight, the hydrogen content and the heat of reaction at total conversion (H_R). From this value of H_R, the heat of reaction along the tube length is correlated with conversion. Three feedstock-dependent parameters are used to relate methane yield and conversion. These are obtained from cracking data as discussed later.

The operating conditions that must be specified are few -- the hydrocarbon and steam feed rates, the outlet pressure of the radiant coil and the inlet and outlet temperature of each zone. In the calculation of heat flow from the radiant flame or heating element to the tube wall, a geometry factor must be given. The geometry factor is dependent on the shape and relative orientation of the radiating bodies and implies that each furnace is unique.

Does the estimation of these feedstock-dependent parameters from bench scale data allow a good estimation of commercial furnace performance? We believe so. Test runs were carefully made on a commercial furnace at several severity levels. The same gas oil feed was then cracked in the bench unit under a wide variety of conditions. The feedstock-dependent parameters that relate methane yield and $\overline{K\Theta}$ were then determined by plotting measured methane yield vs. a function of $\overline{K\Theta}$, as shown in Figure V,

Table V. Input

Tube
- Inside Diameter
- Outside Diameter
- Total Length
- Equivalent Length (total resistance to flow)
- Location Zone Interface (fractional distance along tube)
- Metal Conductivity

Feed
- Molecular Weight
- Weight Percent Hydrogen
- Heat of Reaction (heat of formation of products minus feed)

Operating Conditions
- Hydrocarbon Feed Rate
- Water Feed Rate (STM/FD)
- Outlet Pressure
- Outlet Temperature Each Zone
- Geometry Factor

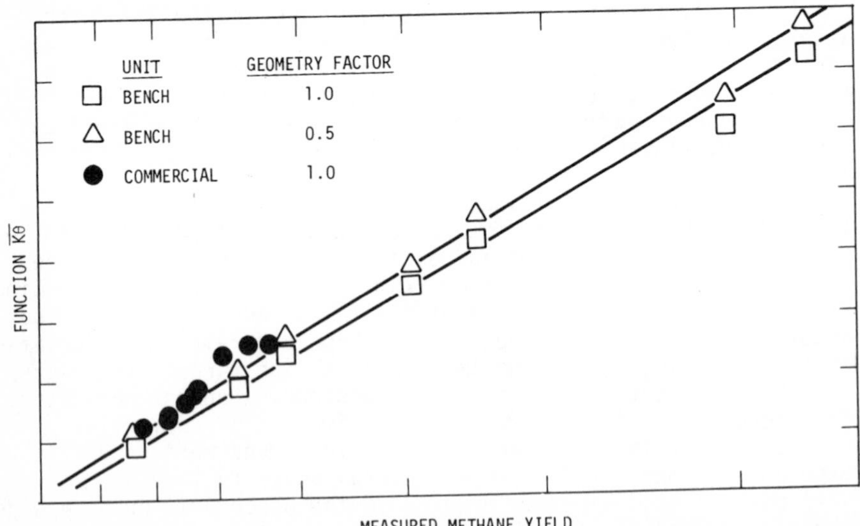

Figure 5. Correlation of pilot unit and commercial furnaces gas oil

for the gas oil data. An adjustment to the geometry factor was made so that the pilot data and commercial data would agree. The geometry factor for the pilot unit was set at 0.5 instead of 1. This alters, slightly, the calculated temperature profile for the pilot unit. Without the correction the mismatch is still only about 4% of the methane yield or less. With the methane-severity correlation determined, methane yields can be calculated for this feed at different operating conditions or different coil designs.

To be sure pilot data could be used to predict commercial performance by the above approach another series of tests was made with a naphtha feedstock. The feedstock-dependent parameters were determined from pilot data, and methane yields were predicted at the commercial test run conditions. Predicted and measured methane yields are shown in Table VI. The agreement appears to be better than one would expect to be able to measure commercial yields.

Table VI. Naphtha Test Runs

Outlet Temp., F	Measured Commercial Methane Yield	Predicted Methane Yields From Bench Data
1480	11.0	11.1
1505	12.5	12.2
1540	13.6	13.6

Sensitivity

The two important features of this approach are the boundary film effect and temperature profile.

The film effect is smaller than has been previously reported. (6) In commercial size tubes, the Nusselt number is large and the film volume becomes insignificantly small. In small bench unit tubes where the film thickness might be significant, the temperature drop across the film is small due to the high transfer surface to volume ratio.

The most important effect is the nonlinear temperature profile. This can vary considerably from frequently assumed linear or smooth profiles.

Conclusions

We have employed some rather simple kinetics, conventional pressure drop and heat transfer formulae, ideal film calculations and some basic empirical correlations to develop a reactor model that quite adequately relates the cracking severity of small bench scale units and commercial furnaces. Thus, evaluation of feedstocks using the bench unit has significantly more meaning. Optimization of operating conditions can be calculated for various cracking coil configurations.

The effective $\Delta\overline{K\Theta}$ for the tube is

$$\Delta\overline{K\Theta} = \frac{q_c \ \Delta\overline{K_c\Theta_c} + q_f \ \Delta\overline{K_f\Theta_f}}{q_t}$$

which upon resolution becomes

$$\Delta\overline{K\Theta} = \Delta\overline{K\Theta_C} \left[\frac{(D-4\delta) + 4\delta \dfrac{(\exp \beta-1)}{\beta}}{D} \right]$$

where $\Delta\overline{K\Theta_C}$ is that calculated from average tube flow and bulk core temperature.

β and δ tend to be counteracting effects. Bench scale units with sufficient δ/D have low temperature gradient effects, β, because of the high surface to volume ratio of the reactor.

Literature Cited

(1) Davis, H.G. and Keister, R.G., "The Advanced Cracking Reactor (ACR): A Process for Cracking Hydrocarbon Liquids at Short Residence Times, High Temperatures, and Low Partial Pressures", Preprints of General Papers, Division of Petroleum Chemistry, Inc., American Chemical Society, (February 1975), Vol. 20, (No. 1), pg. 158; also Chapter 22 of this book.

(2) Zdonik, S.B., Green E.J. and Hallee, L.P., "Ethylene Worldwide-5", The Oil and Gas Journal, (June 26, 1967), pg. 96.

(3) Zdonik, S.B., Green, E.J. and Hallee, L.P., "Ethylene Worldwide-6", The Oil and Gas Journal, (July 10, 1967), pg. 192.

(4) Zdonik, S.B., Green, E.J., and Hallee, L.P., "Ethylene Worldwide-15", The Oil and Gas Journal, (May 27, 1968), pg. 103.

(5) Technical Data Book - Petroleum Refining, American Petroleum Institute, Washington, D.C., (Second Edition), (1970).

(6) de Blieck, J.L. and Goosens, A.G., "Optimize Cracking Coils", Hydrocarbon Processing, (March 1971), pg. 76.

Nomenclature

A	Frequency factor, log hr^{-1}
A_I	Inside tube area, cm^2
A_O	Outside tube area, cm^2
B	Activation energy/1.987, °C
BTX	Benzene, toluene, xylenes
C_1	Methane yield basis feed, %wt.
C_P	Heat capacity, cal/gm/°C
C_{FD}, C_{PR}	Concentration of feed and products respectively
D	Inside tube diameter, cm
h	Coefficient of heat transfer, cal/hr, cm^2,°C
H_R	Heat of reaction, cal/gm
K	Reaction velocity constant, sec^{-1}
K_C	Thermal conductivity, cal/hr, cm^2,°C per cm
$\overline{K\Theta}$	Reaction severity parameter
MW	Molecular weight
Nu	Nusselt number
Pr	Prandtl number
q	Volumetric flow rate, m^3/sec
Q	Heat duty, cal/hr
Re	Reynolds number
t	Time, sec
T_C	Bulk gas temperature, °K
T_R	Temperature, Kelvin
T_{rad}	Radiation temperature, °K
T_W	Inside tube wall temperature, °K
T_{WR}	Outside tube wall temperature, °K
v	Velocity, cm/sec
Z	Length, cm
Θ	Residence time, sec.
Φ	Geometry factor
δ	Film thickness, cm

APPENDIX 1

Estimation of Laminar Sublayer Effect on $\overline{K\Theta}$ Severity Parameter

The quantity of film flow is closely approximated by

$$q_f = \pi \frac{D}{2} \delta v_c$$

with a maximum error for the bench scale unit of about 3 percent of q_f. Total tube flow is

$$q_t = v_c \pi \frac{D}{2} \left(\frac{D}{2} - \delta\right)$$

and core flow is

$$q_c = v_c \pi \frac{D}{2} \left(\frac{D}{2} - 2\delta\right)$$

The film thickness is estimated from the Nusselt number, $Nu = \frac{hD}{K_c}$, (see text) and

$$\delta = \frac{K_c}{h} \quad \text{or} \quad \delta = \frac{D}{Nu}$$

Also, the wall temperature is related to the heat flux, Q_{dZ}, necessary for the heat of reaction and temperature increase in length, dZ (see discussion in text) by

$$Q_{dZ} = h\pi D dZ(T_w - T_c)$$

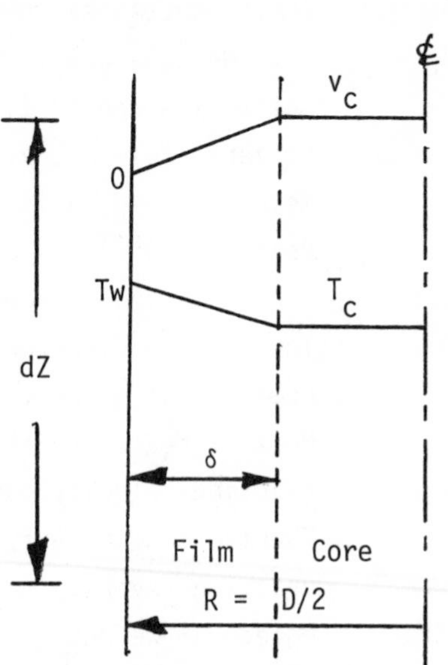

Figure 1-1: Laminar Boundry Layer Concept of Turbulent Flow

To calculate the effective $\overline{K\Theta}$ it is necessary to weigh the local $K\Theta$ by the quantity of flow. For the film

$$\Delta\overline{K_f\Theta_f} = \int_R^{R-\delta} Ktq\,dr \Big/ \int_R^{R-\delta} q\,dr$$

where $q = \pi D v$

$t = dZ/v$

Substituting for v and $K = A \exp\left(\frac{-B}{T}\right)$ gives an effective $\overline{\Delta K_f\Theta_f}$ of

$$\overline{K_f\Theta_f} = 2\Delta\overline{K_c\Theta_c} \left(\frac{\exp \beta - 1}{\beta}\right) \geq 2\Delta\overline{K_c\Theta_c}$$

where

$$\beta = \frac{B(T_w - T_c)}{T_c^2}$$

19

Bench Scale Study on Crude Oil Pyrolysis for Olefin Production by Means of a Fluidized Bed Reactor

TAISEKI KUNUGI, DAIZO KUNII, and HIROO TOMINAGA

Faculty of Engineering, University of Tokyo, Hongo 7-3-1, Bunkyo, Tokyo, Japan

TOMOYA SAKAI

Faculty of Pharmaceutical Science, Nagoya City University, Tanabe-dōri 3-1, Mizuho-ku, Nagoya, Japan

SHUNSUKE MABUCHI and KŌSUKE TAKESHIGE

Tōyō Soda Manufacturing Co., Ltd., Tomita 4560, Shin-nanyo-shi, Yamaguchi, Japan

The need for a diversification in the raw material for olefin production is being recognized in view of the limited supply of naphtha upon which Japanese petrochemical industry has been almost entirely dependent. In this respect, a new cracking technology is expected to be developed which permits processing heavy fractions other than naphtha, including residual and crude oils itself. A fluidized bed reactor where carbon particles play the role of heating media may provide a solution of coking troubles which are inherent in pyrolysis of heavy oils. The coke deposited on carbon particles can be burnt off in an annexed regenerator and carbon particles are recycled at high temperature to provide the heat required for pyrolysis.

In this paper experimental results are presented on crude oil pyrolysis which was carried out by use of a bench scale batch type fluidized bed reactor heated by an electric furnace. Good yields of olefins and aromatics were obtained from several crude oils. Monographs are also presented which provide the means for estimating the products yield based on the characterization factor of the feed stock and the intensity function representing the severity of pyrolysis.

This study laid the groundwork for the development of so-called "K-K process" in a test plant scale of which details were presented in PD 19, 9th World Petroleum Congress. (1)

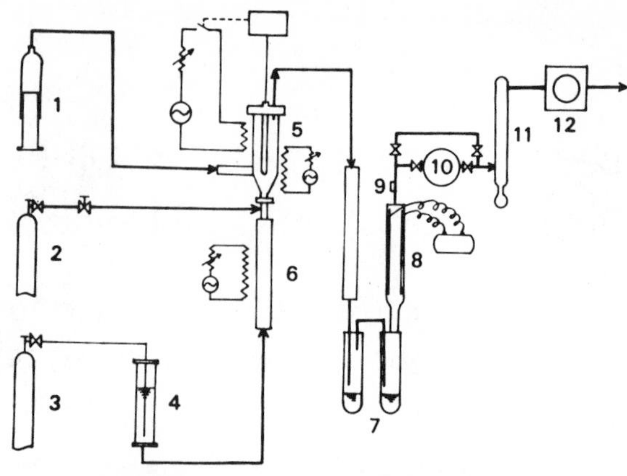

Figure 1. Schematic of experimental apparatus

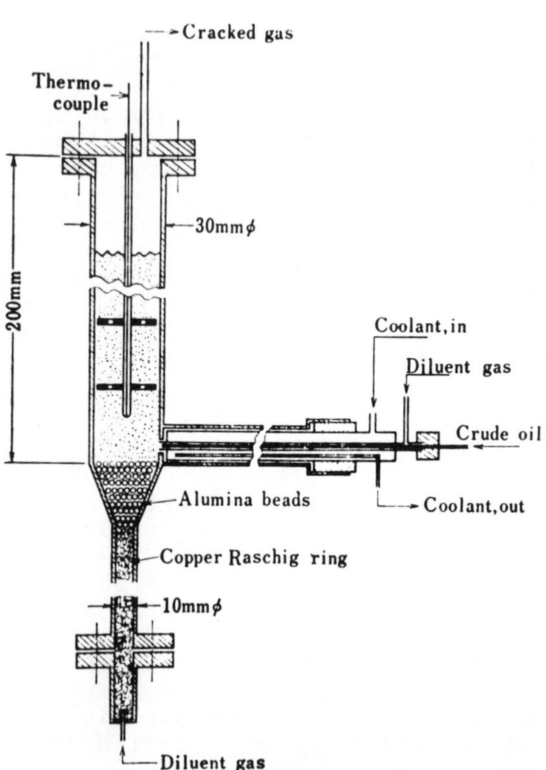

Figure 2. Cross section of reactor

Experimental

Apparatus. The flow scheme of the experimental apparatus is given in Figure 1. Details of the fluidized bed reactor are also shown in Figure 2.

Hydrocarbons were supplied to the bottom of the reactor by use of a syringe, and mixed with a given amount of diluent which was usually steam produced in a steam generator. A stainless steel reactor, 30 mm in inner diameter and 200 mm in length, contained carbon particles of which size distribution ranged from 24 ∿ 32, 42 ∿ 65, or 65 ∿ 100 mesh depending on the space velocity or the residence time of the reactant. An electric furnace was employed to heat the reactor, giving a fairly constant temperature profile (±5°C) throughout the fluidized bed which is 100 mm in height.

Cracked products leaving the top of reactor were cooled with water indirectly to separate gas and liquid, the former was further purified using a Cottrell type demister.

Analysis. Gaseous and liquid products were analyzed respectively by the gas chromatography for their hydrocarbon compositions.

Feed Stocks. Seria, Mukhanovskaya, Minas, Gach Saran, Kuwait, Khafji and Duri crudes were subjected to the pyrolysis. Several distillates and the topped residue from Kuwait crude were also employed to cover the wide variety of hydrocarbon compositions of feed stock with which the pyrolysis results should be closely correlated. General properties of these feed stocks are given in Tables I and II, respectively.

Results and Discussions

Observations on Pyrolysis of Kuwait and Minas Crudes.

Kuwait Crude. Kuwait crude, one of the typical Middle East crude oils, was subjected to the pyrolysis. The experimental conditions covered the ranges of temperature 700 ∿ 900°C, of residence time 0.2 ∿ 1 sec, and of steam dilution ratio 0.5 ∿ 2 by weight to crude oil. These operating variables were examined relative to their effect on the yields of olefins and aromatics.

In this experiment, the best yield of ethylene of 26 wt% based on the initial crude was obtained at 850°C and 0.5 sec, whereas the best one for propylene was 15 wt%, which was attained at a lower temperature, namely 750°C, and 0.6 sec. Total aromatics yield including

Table I. Crude Oil Evaluation Data

	Seria	Mukhanovskaya	Minas	Gach Saran	Kuwait	Khafji	Duri
Sp.gr., 15°/4°C	0.8359	0.8430	0.8462	0.876	0.8681	0.886	0.9017
Conradson C. (wt%)	0.19	2.63	2.76	5.4	5.9	6.75	8.3
Sulfur (wt%)	0.08	1.05	0.10	1.62	2.53	2.93	0.17
Hempel Dist. (°C)							
IBP	29	25	24	23	22	22.5	28
10 (vol%)	110	107	144	116	107	131	132
20 (vol%)	143	152	222	168	167	191	190
30 (vol%)	182	203	276	230	229	256	253
40 (vol%)	226	255		284	278	295	293
50 (vol%)	254	294			294		
60 (vol%)	280						

Table II. Properties of Kuwait Crude Fractions

	Light Naphtha	Heavy Naphtha	Kerosene	Gas Oil	Total Distillate		Topped Crude	
Sp.gr., 15°/4°C	0.6702	0.7483	0.7895	0.8416	0.7596(25°C)		0.956	
ASTM Dist. (°C)						vol%	Dist. (1 mmHg)	°C (corrected to 760 mmHg)
IBP	31	104	148	190	Light naphtha	10.9	IBP	265
10 (vol%)	46	117	176	250	Heavy naphtha	11.0	5 %	—
50 (vol%)	66	134	197	292	Kerosene	12.6	10 %	329
90 (vol%)	98	161	225	342	Gas oil	12.6	20 %	375
EP	125	181	240	375			30 %	417
Rec. (vol%)	97.5	98.0	97.8	98.7	(Topped crude	52.6)	40 %	450
Res. (vol%)	1.0	1.1	2.0	1.2	Total:	99.7	50 %	492
							60 %	537
							70 %	—
Type Analysis							Type analysis	
P (vol%)	96	60	59	83			Saturates (wt%)	34.6
N (vol%)	0	25	23				Aromatics (wt%)	45.8
O (vol%)	0	0	0	0			Resins (wt%)	16.4
A (vol%)	4	15	18	17			Asphaltenes	3.2
Sulfur (wt%)		0.04	0.02				Flash point (°C)	138
Basic N. (ppm)		0.04					Pour point (°C)	+12.5
F-1 (clear)		46.5					Viscosity, at 50°C	222
Smoke Point (mm)			26.5				Sulfur (wt%)	3.68
Aniline Point (°C)		55.8	61.8				Carbon residue (wt%)	8.8
							Ash (wt%)	0.02

benzene and naphthalene ranged between 8 ∿ 20 wt% based on the crude. Typical product yields are summarized in Table III.

Since these data are promising, further study was made to obtain kinetic informations. Olefins yield are most strongly affected by two operating variables, which are pyrolysis temperature and residence time. At a fixed temperature there exists an optimum residence time which gives the best yield of the respective olefin. The optimum residence time depends on the pyrolysis temperature; the higher the temperature, the shorter the optimum residence time. The optimum set of reaction conditions varies with olefins. These results are reasonably interpreted by assuming a consecutive reaction mechanism in which olefins, which are the primary products of pyrolysis, are lost by succeeding reactions such as hydrogenolysis into lower olefins, condensation into aromatics, etc. This is verified by the experimental works on thermal reactions of olefins including those of the present authors. (2,3)

Incidentally, a desired product pattern in a given range would be obtained by a proper choice of conditions of crude oil pyrolysis.

Increasing the steam to crude ratio from 0.5 to 2 results in a relatively small increase in olefin yield. As a result, the space time yield of olefins may be somewhat reduced by ratios of steam to crude greater than 1. The use of hydrogen as the diluent acts to effect increased yields of ethylene, but with the sacrifice of the yields of higher olefins such as propylene, butenes, and butadiene because of hydrogenolysis.

Minas Crude. Minas crude, which is classified as paraffinic, was subjected to pyrolysis to compare the results with those from Kuwait crude which is of mixed base. Some typical data are shown in Table IV.

The experiment indicated that Minas crude was more susceptible to pyrolysis than Kuwait crude, and thus gave $C_1 \sim C_4$ cracked gas yields of 60 ∿ 70 wt% based on the initial crude at 750 ∿ 850°C and 0.2 ∿ 1 sec. At the similar reaction conditions, Kuwait crude gives $C_1 \sim C_4$ cracked gas yields of 50 ∿ 60 wt%. Olefin yields from Minas crude are also higher than those from Kuwait crude. Within the experimental conditions covered by this study, the maximum yield of ethylene from Minas crude is 33 wt% at 850°C and 0.2 sec, and that of propylene is 16 wt% at 750°C and 0.2 sec.

Benzene yields from Minas crude are somewhat higher than those from Kuwait crude, while the yields of toluene are a little lower and those of naphthalene

Table III. Typical Cracking Data (Kuwait Crude)

Experimental No.	64		75		72	
Temperature (°C)	750		800		850	
Residence Time (sec)	0.63		0.51		0.50	
Steam/Oil (wt/wt)	0.96		1.03		1.01	
Product Yield	(wt%)	(vol%)*	(wt%)	(vol%)*	(wt%)	(vol%)*
CH_4	11.0	(27.3)	12.8	(23.3)	15.0	(29.9)
C_2H_2	0.04	(0.1)	0.4	(0.6)	0.6	(0.6)
C_2H_4	19.7	(28.0)	24.7	(31.8)	26.8	(30.3)
C_2H_6	3.3	(4.4)	2.7	(3.3)	2.0	(2.2)
C_3H_6	15.2	(14.4)	11.2	(9.6)	5.8	(4.4)
C_3H_8	0.6	(0.5)	0.4	(0.3)	0.1	(0.1)
C_4H_6	5.8	(4.3)	4.9	(3.3)	3.2	(1.9)
$C_4H_8+C_4H_{10}$	4.8	(3.4)	2.7	(1.8)	1.8	(1.0)
C_6H_6	5.5		5.3		7.4	
$C_6H_5CH_3$	3.3		2.2		3.3	
$C_6H_4(CH_3)_2$	5.1		1.6		4.3	
$C_{10}H_8$	1.3		3.9		6.8	
H_2	0.7	(15.2)	1.0	(22.1)	1.6	(24.9)
CO	0.2	(0.8)	0.4	(1.4)	0.6	(1.5)
CO_2	0.2	(0.6)	0.5	(1.5)	0.7	(1.9)
H_2S	0.8	(1.0)	1.1	(1.0)	1.0	(0.9)
$C_1 \sim C_4$ Hydrocarbons	60.4		59.8		55.3	
Non-hydrocarbon Gases	1.9		3.0		3.9	
B, T, X & N**	15.7		13.0		21.8	
Other Liquid Products	12.5		13.3		11.4	
Coke	7.4		7.7		8.9	
Mass Balance (wt%)	97.9		96.8		101.3	

* Molar composition of gaseous product
** Naphthalene

Table IV. Typical Cracking Data (Minas Crude)

Experimental No.	152		138		155	
Temperature (°C)	750		800		850	
Residence Time (sec)	0.50		0.48		0.47	
Steam/Oil (wt/wt)	1.00		1.08		1.06	
Product Yield	(wt%)	(vol%)*	(wt%)	(vol%)*	(wt%)	(vol%)*
CH_4	10.7	(24.2)	14.0	(28.2)	16.7	(28.3)
C_2H_2	0.3	(0.4)	0.4	(0.4)	1.6	(1.6)
C_2H_4	27.3	(35.6)	31.0	(35.8)	33.3	(32.3)
C_2H_6	2.5	(3.1)	2.5	(2.8)	1.8	(1.6)
C_3H_6	13.8	(12.1)	11.3	(8.7)	5.1	(3.3)
C_3H_8	0.4	(0.4)	0.3	(0.2)	0.1	(0.1)
C_4H_6	7.6	(5.2)	0.6	(3.7)	3.9	(1.9)
$C_4H_8+C_4H_{10}$	5.6	(3.6)	3.4	(1.9)	2.0	(1.0)
C_6H_6	5.2		7.7		9.4	
$C_6H_5CH_3$	2.7		3.5		2.5	
$C_6H_4(CH_3)_2$	1.9		2.2		1.7	
$C_{10}H_8$	1.2		1.7		3.3	
H_2	0.7	(14.3)	0.9	(16.6)	1.6	(26.3)
CO	0.2	(0.6)	0.4	(1.0)	0.9	(1.9)
CO_2	0.1	(0.4)	0.2	(0.6)	0.7	(1.6)
H_2S	0.06	(0.1)	0.06	(0.1)	0.09	(0.1)
$C_1 \sim C_4$ Hydrocarbons	68.2		69.5		64.5	
Non-hydrocarbon Gases	1.1		1.6		3.3	
B, T, X & N	11.0		15.1		16.9	
Other Liquid Products	17.3		12.1		15.6	
Coke	1.9		3.5		5.7	
Mass Balance	99.5		101.8		105.9	

* Molar composition of gaseous product

are particularly lower by a factor of 2 ∿ 3.

At a given set of pyrolysis conditions, the differences between product yields from Kuwait and Minas crudes obviously suggest the dissimilar hydrocarbon compositions of the two crudes. On the other hand, very similar general features of the changes in product yields as a function of pyrolysis conditions have been observed with the both crudes. Consequently, the most important two factors, hydrocarbon composition and pyrolysis condition, were examined in more detail to establish quantitatively their cause-and-effect relationships with product yield. These are described in the next section.

Prediction of Product Yield from Crude Oil Pyrolysis.

Product Yield and Characterization Factor of the Feed Stock. Because the primary products yields of pyrolysis are naturally related with hydrocarbon composition of the feed stock, the former can be estimated based on the latter if any experimental equation is provided. The detailed analysis of crude oil for its hydrocarbon composition is, however, extremely complicated and beyond the common practice. Hence, modified characterization factor defined by the next equation (1) was examined as a possibility to be employed for representing the feed stock property to fulfil our purpose of prediction.

$$K = \sqrt[3]{T_B}/d \qquad (1)$$

where, T_B: 50% point by Hempel or Engler distillation, °R, d: specific gravity, @ 60°F.

Seven crude oils shown in Table I, of which K values range from 11.40 for Duri to 12.28 for Minas, and Kuwait distillates and residue in Table II having K values of 11.64 ∿ 12.65, were pyrolyzed at 800°C, 0.5 sec, and steam to hydrocarbon dilution ratio of 1:1. The yields of coke, liquid product, non-hydrocarbon gases, and C_1 ∿ C_4 hydrocarbons are plotted against K values in Figure 3. Olefin yields are also presented in Figure 4 in the same way.

Quasi-linear relationships are recognized between the product yields and K values of the feed stocks. The same applies to the experimental results of pyrolysis performed at another sets of conditions, namely at 800°C, 0.2 or 1 sec, and steam dilution 1:1.

The relationship that olefin yields increase with the K value is consistent with the general knowledge that paraffinic (low d) hydrocarbons with high boiling points (large T_B) are most susceptible to pyrolysis and thus give high olefin yields. As a result, Figures 3

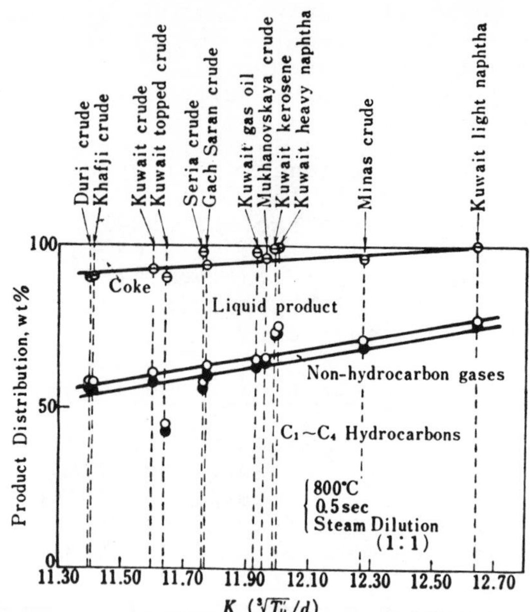

Figure 3. Product distribution from various feeds plotted against their K values (800°C, 0.5 sec, steam dilution 1:1)

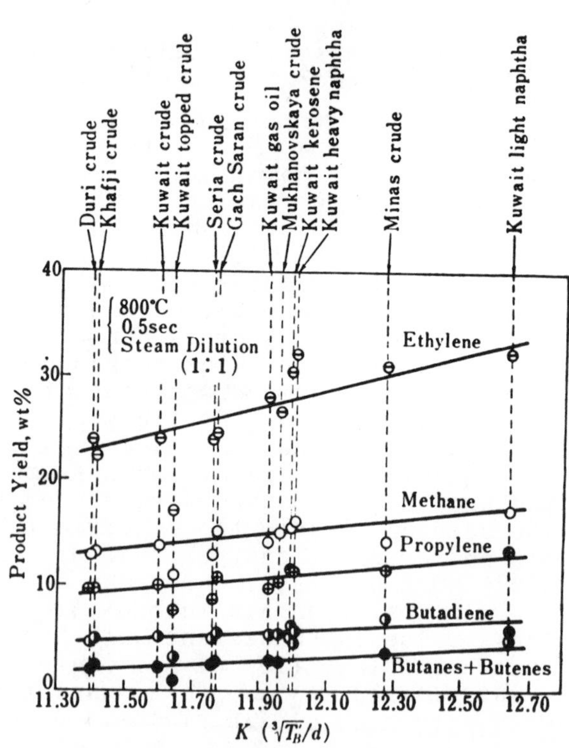

Figure 4. Olefin yield from various feeds plotted against their K values (800°C, 0.5 sec, steam dilution 1:1)

and 4 may provide a useful means for the prediction of the product pattern obtained by crude oil pyrolysis.

Another approach was made for estimating the yield of coke, which is one of the key factors which determine the heat balance of the process. The coke yield is considered to be in close correlation with Conradson residual carbon content of the crude oil. This is experimentally confirmed as is shown in Figure 5, which gives the following equation.

$$Y_C = C_R + 1.5 \qquad (2)$$

where, Y_C: coke yield at 800°C, 0.5 sec, steam dilution 1:1, wt% on crude, C_R: Conradson residual carbon, wt%.

Product Yield and Intensity Function of the Pyrolysis Condition. Kinetic considerations on the Arrhenius expression of the rate constant suggest that, in order to obtain a given conversion of reactant, reaction temperature and residence time are interchangeable to some extent. The intensity function proposed by Linden (3) for pyrolysis of hydrocarbons (4) is, accordingly, reasonable in principle.

$$I_F = T \cdot \Theta^{0.06} \qquad (3)$$

where, I_F: intensity function, T: temperature of pyrolysis, °F, Θ: residence time, sec. This index was adopted for interpretation of our experimental results ranging from 750 to 850°C in temperature and from 0.2 to 1 sec in residence time, where steam dilution ratio was maintained at 1:1.

In Figures 6 and 7, the product yields from Kuwait crude are plotted against I_F, and Figures 8 and 9 show the data with Minas crude.

These figures indicate that the yields of methane and coke, which are stable at the pyrolysis temperature, increase in linear proportion to I_F. While the yield values of olefins, which are the intermediate products of pyrolysis, when plotted against I_F form complicated relationships.

A maximum is clearly shown for each of the following propylene, ethylene, and butadiene. Comparison of Figures 7 and 9 denotes that the maximum yield of olefin is larger with crude oil of the higher K value, and it is attained by pyrolysis at conditions having a lower I_F value.

In summary by use of Figures 3 and 4 the product yield of the pyrolysis of any crude oil at a standard condition can be estimated from the K value of the crude. The change in the product yield as a function of pyrolysis conditions represented by I_F can be predicted by use of one of Figures 6 ∿ 9 or their interpo-

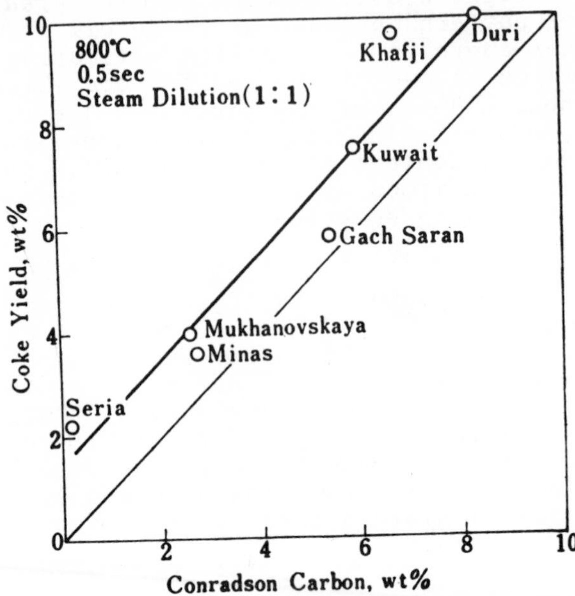

Figure 5. Coke yield plotted against Conradson residual carbon content (800°C, 0.5 sec, steam dilution 1:1)

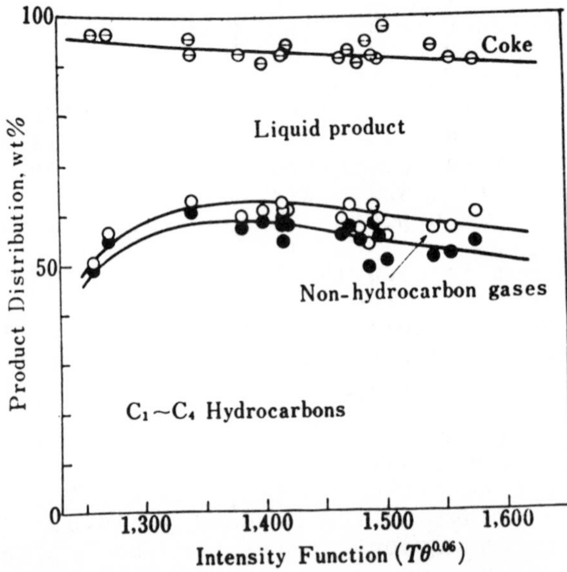

Figure 6. Product distribution from Kuwait crude plotted against intensity function (steam dilution 1:1)

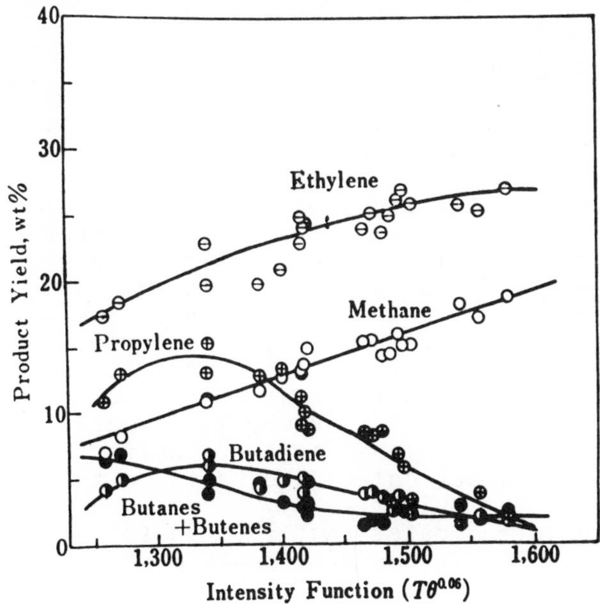

Figure 7. Olefin yield from Kuwait crude plotted against intensity function (steam dilution 1:1)

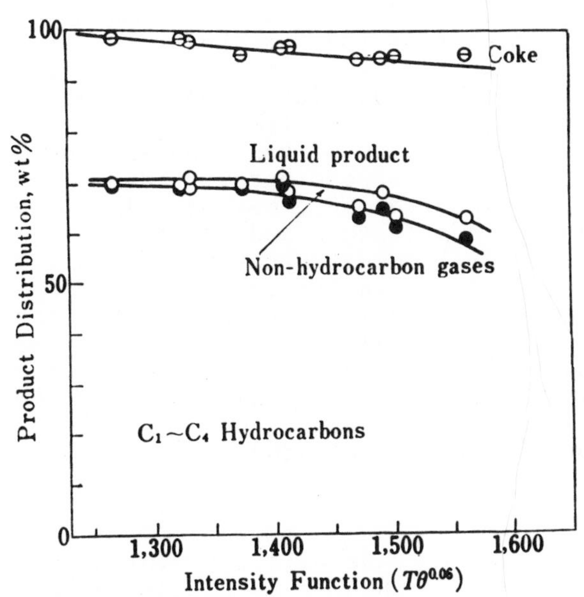

Figure 8. Product distribution from Minas crude plotted against intensity function (steam dilution 1:1)

lation.

Distribution of Sulfur and Heavy Metals into the Pyrolysis Products. Since the crude oils employed in this study contain fairly large amounts of sulfur and heavy metal compounds, the behaviors of them in the course of the crude oil pyrolysis should be examined.

Sulfur. In the experiments of Kuwait crude oil pyrolysis, sulfur distributions into the respective products, gas, liquid, and coke, were determined, and plotted against I_F as is shown in Figure 10.

Hydrogen sulfide in the gaseous products was analyzed by use of a cadmium acetate aq. solution (JIS K 2302-1963). Mercaptans and thiocarbonyl, which may possibly be present in the gaseous products, were not determined. Total sulfur in the liquid products was analyzed by a conventional method. Sulfur in the coke was not analyzed but calculated based on the material balance as to the sulfur.

Figure 10 indicates that from 30 to 45 parts of the sulfur in Kuwait crude is cracked into hydrogen sulfide and appears as such in the gaseous products. The evolution of hydrogen sulfide increases in proportion to I_F. While the sulfur in the liquid products decreases in reverse proportion to I_F, and the sulfur in the coke consequently remains at a fairly constant level.

Table V summarizes the sulfur inspection data for Kuwait, Khafji, and Gach Saran crude oil pyrolysis. The sulfur compounds in Khafji crude are known to be rather refractory and thus remain largely in the coke and liquid products. While the sulfur in Gach Saran crude is more susceptible to the pyrolysis.

Heavy Metals. Distributions of vanadium and nickel in the liquid products and coke were determined by use of an atomic absorption spectroscopic method. In Table VIgiven are the data on Kuwait crude and its topped crude pyrolysis. The table suggests that these heavy metals are mostly concentrated in the coke. Accordingly, in the long run of a commercial cracking plant, the accumulation of these heavy metals in the coke may possibly exert some adverse effects on pyrolysis of hydrocarbons in the reactor and/or combustion of coke in the regenerator, etc. These will remain as open questions until the completion of pilot plant study.

Summary

A study of pyrolysis of crude oils for olefin production was made by use of a bench scale fluidized bed

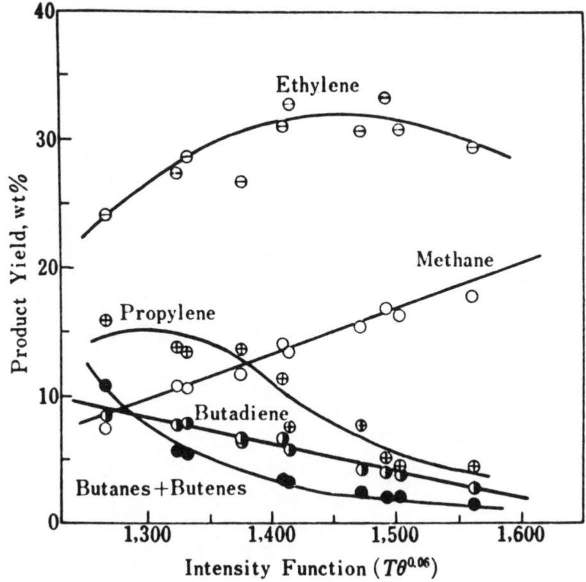

Figure 9. Olefin yield from Minas crude plotted against intensity function (steam dilution 1:1)

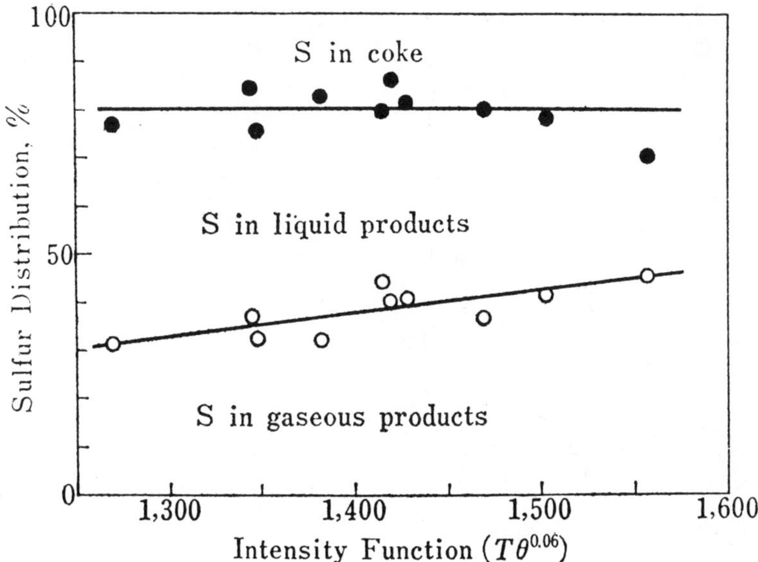

Figure 10. Sulfur distribution into the pyrolysis products (Kuwait crude, steam dilution 1:1)

Table V. Sulfur Distribution into the Pyrolysis Products

Crude Oil	Kuwait				Khafji		Gach Saran	
Pyrolysis Conditions								
Temperature (°C)	750	800	800	850	800	800	800	800
Residence Time (sec)	0.63	0.60	0.99	0.52	0.50	1.05	0.57	1.00
Steam Dilution	0.96	0.88	1.04	0.94	0.95	1.02	0.85	0.93
Sulfur in Crude (wt%)	2.53				2.93		1.62	
Sulfur Concentration in Pyrolysis Products (wt%)								
Gas (H_2S)	0.96	0.97	0.90	0.99	0.78	0.80	0.73	0.68
Liquid	3.13	2.62	2.86	2.44	4.35	4.31	1.95	2.46
Coke	8.2	8.2	5.5	6.1	8.5	7.0	4.0	1.2
Sulfur Distribution (%)								
Gas (H_2S)	32.4	40.7	36.7	41.3	24.0	25.9	42.5	42.5
Liquid	43.6	40.7	43.3	36.8	48.2	47.4	41.4	51.5
Coke	24.0	18.6	20.0	21.9	27.8	26.7	16.1	6.4

Table VI. Heavy Metals in the Pyrolysis Products

<table>
<tr><th>Feed Stock</th><th>Kuwait Crude</th><th colspan="2">Kuwait Topped Crude</th></tr>
<tr><td>Pyrolysis Conditions</td><td></td><td></td><td></td></tr>
<tr><td>Temperature (°C)</td><td>800</td><td>800</td><td>800</td></tr>
<tr><td>Residence Time (sec)</td><td>0.51</td><td>0.20</td><td>0.97</td></tr>
<tr><td>Steam Dilution</td><td>1.01</td><td>1.09</td><td>1.08</td></tr>
<tr><td>Heavy Metals in Feed (ppm)</td><td></td><td></td><td></td></tr>
<tr><td>V</td><td>38</td><td colspan="2">45</td></tr>
<tr><td>Ni</td><td>9</td><td colspan="2">9.5</td></tr>
<tr><td>Heavy Metals in Products (ppm)</td><td></td><td></td><td></td></tr>
<tr><td>V in Liquid</td><td>12</td><td>14</td><td>9</td></tr>
<tr><td>in Coke</td><td>128</td><td>118</td><td>110</td></tr>
<tr><td>Ni in Liquid</td><td>1.4</td><td>2.1</td><td>1.4</td></tr>
<tr><td>in Coke</td><td>47</td><td>40</td><td>37</td></tr>
</table>

reactor containing carbon particles. Seven crude oils were tested. Kuwait and Minas crudes as well as Kuwait distillates and residue were examined in detail. Pyrolysis conditions were 750∿850°C in temperature, 0.2∿1 sec in residence time, 0.5∿2 in steam to hydrocarbon dilution ratio by weight. The product yields were plotted against the modified characterization factor ($K = \sqrt[3]{T_B}/d$) of the respective feed. The changes in the product yields by progress in pyrolysis severity were represented in terms of the intensity function ($I_F = T \cdot \Theta^{0.06}$). These monographs provide the means for estimating the pyrolysis results of any crude oil, with known 50% distillation point (T_B, °R) and specific gravity (d, @ 60°C), at a given reaction temperature (T, °F) and residence time (Θ, sec) with steam dilution.

Literature Cited

(1) Kunii, D., Kunugi, T., Ichihashi, T., Nakamura, R., Tamaoki, H., Takenouchi, Y., Matsuura, T., and Yamanaka, T., 9th World Petroleum Congress, PD 19, No.3 (1950).

(2) Kunugi, T., Tominaga, H., and Abiko, S., 7th World Petroleum Congress, (1967) 5, 239.

(3) Sakai, T., Soma, K., Sasaki, Y., Tominaga, H., and Kunugi, T., Advances in Chemistry Series, (1970) (97) 68.

(4) Linden, H.R., et al., Proc. Amer. Gas Assoc., (1948) 30, 315.

20

An Industrial Application of Pyrolysis Technology: Lummus SRT III Module

J. M. FERNANDEZ-BAUJIN and S. M. SOLOMON
C-E Lummus, Bloomfield, N.J.

The pyrolysis of hydrocarbons for the production of olefins and diolefins has been extensively investigated during the last four decades by industrial and academic institutions. Many investigators are still working on the development of kinetic models which can be used to predict yield patterns as a function of the operating conditions for a given hydrocarbon. Such models have not been completely accurate even for such a simple hydrocarbon as ethane.(1) Of course, the complexity of such a modelling technique is increased several fold when it is applied to heavier hydrocarbons and to a mixture of them, i.e., naphthas and gas oils. As a result, the development of techniques to predict industrial yield patterns generally is based on a combination of theoretical and empirical methods.

PYROLYSIS TECHNOLOGY – BRIEF REVIEW

A hydrocarbon system undergoing pyrolysis is a most complex mixture of molecules and free radicals which react simultaneously with one another in a multitude of ways. Based on established theories supported by experimental data, the production of olefins and diolefins – that is, pyrolysis selectivity towards olefins and diolefins – has been found to be favored by short residence times and low hydrocarbon partial pressures (2, 3, 4). Pyrolysis reactor selectivity has been expressed as a function of the following two parameters:

1. Residence Time, θ
2. Hydrocarbon Partial Pressure, P_{HC}

The specific functions of residence time and partial pressure must describe the history of the feedstocks pyrolyzed and the history of the products produced in passing through the coil.

Bulk residence time, θ_B, is defined by Equation (1)

$$\theta_B = \int_0^{V_T} \frac{dVc}{V} \qquad (1)$$

where V_T = Total volume of the pyrolysis coil,

V_c = Volume of the pyrolysis coil as a function of its length, and

V = Volume of gas per unit time passing through the pyrolysis coil. This volume changes with coil length.

The bulk residence time, θ_B, which is the more commonly used parameter to correlate pyrolysis selectivity, expresses only the length of time that a mass of gas spends in the pyrolysis coil. Bulk residence time is not adequate to represent the combined effects of temperature and chemical reaction which take place in the pyrolysis coil as witness by the failure to correlate the available experimental data as a function of bulk residence time especially when data from pyrolysis coils of different configurations and, therefore, different temperature profiles are considered.[3]

In an effort to correlate the existing experimental data from bench scale, pilot plant and commercial reactors, the definition of average residence time, θ_A, shown in Equation (2) has been developed.

$$\theta_A = \frac{1}{a_o}\int_0^{\theta_B} a \; d\theta_b \qquad (2)$$

where

θ_B = Bulk residence time,

θ_b = Gas residence time along the length of the pyrolysis coil

a = Feedstock conversion along the length of the pyroysis coil,

a_o = Conversion of the feedstock at the pyrolysis coil outlet.

Conversion, a, is empirically defined as a function of the molecular weight of the feedstock and the composition of the cracked gas. The average residence time, expressed as a function of feedstock conversion, is a measure of the average time the reaction products spend in the coil.

As with residence time, the correlating definition of hydrocarbon partial pressure must consider the partial pressure history in the coil. For a given feedstock cracked at a fixed conversion and gas outlet pressure in a pyrolysis coil of a given configuration, the average hydrocarbon partial pressure, P_{HCA}, is determined by:

+ The pressure drop through the pyrolysis coil, and

+ The dilution steam-to-hydrocarbon ratio.

The arithmetic average of inlet and outlet hydrocarbon partial pressure alone does not properly reflect the pressure history of the gas in the pyrolysis coil.
The definition of average hydrocarbon partial pressure, P_{HCA}, which has been developed, is shown in Equation (3).

$$P_{HCA} = \frac{1}{a_o}\int_o^{a_o} P_{HC} \, da \qquad (3)$$

where

P_{HC} = Hydrocarbon partial pressure along the coil length,

a = Feedstock conversion along the length of the pyrolysis coil, and

a_o = Conversion of the feedstock at the pyrolysis coil outlet.

The average hydrocarbon partial pressure, expressed as a function of feedstock conversion, is a measure of the average hydrocarbon partial pressure of the reaction products in the coil.

Pilot plant, prototype and commercial data on pyrolysis selectivity collected by Lummus have been correlated as a function of the average residence time and the average hydrocarbon partial presure. A method of correlation which has proved highly successful is illustrated in

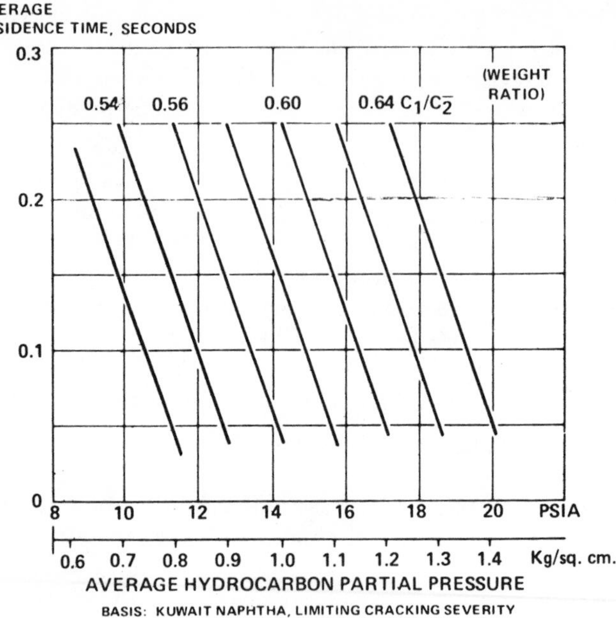

Figure 1. Typical pyrolysis selectivity chart

Figure 1. Each sloped line represents the loci of all possible combinations of average residence times and hydrocarbon partial pressures which are consistent with a fixed pyrolysis yield pattern, i.e., constant pyrolysis selectivity lines. For liquid feedstocks, the methane-to-ethylene ratio found in the pyrolysis reactor effluent has been used as a good overall indicator of pyrolysis reactor selectivity. Low methane-to-ethylene ratios correspond to a high total yield of ethylene, propylene, butadiene and butylenes. Consequently, the yields of methane, ethane, aromatics and fuel oil are reduced. Therefore, each constant pyrolysis selectivity line shown in Figure 1 is identified with a fixed methane-to-ethylene ratio. This specific selectivity chart applies to a Kuwait heavy naphtha which is pyrolyzed to achieve a constant degree of feedstock dehydrogenation, i.e., a constant hydrogen content in the effluent liquid products, which in this case corresponds to the limiting cracking severity.

The results of this correlation work have shown that, while the average residence time and hydrocarbon partial pressure are both key factors in determining pyrolysis selectivity, the average hydrocarbon partial pressure is somewhat more important than previously realized by investigators.

The results of one of the many extensive experimental naphtha pyrolysis compaigns performed with the objective of verifying the exact slope of the constant selectivity lines will be discussed next. A theoretical analysis which supports the selectivity chart concept follows thereafter.

PROFILE CONTROLLED PYROLYSIS UNIT AND FEEDSTOCK PROPERTIES

The experimental work was conducted in the profile controlled pyrolysis pilot reactor, PCP. This unit was developed in 1967 and its performance has closely reproduced the pyrolysis yields obtained in commercial-size coils using naphthas and gas oils as feedstocks. [5, 6]

The unit consists of a small diameter coil which is electrically heated in a multiple zone heater. The hydrocarbon feed and the steam are preheated to the desired inlet temperature of the commercial coil before they are fed to the pyrolysis coil. The pyrolysis coil has been designed to allow the control of the process gas temperature along the length of the pyrolysis coil. Since the coil used in the pilot plant has essentially no pressure drop, a technique has been developed to also control the hydrocarbon partial pressure profile in the coil, and therefore, accurately simulate commercial coil performance. The pyrolysis heater effluent is rapidly quenched and sent to an elaborate sampling system where the different fractions are collected at optimal temperature and pressure. The effluent is analyzed routinely for the presence of 45 chemical compounds and a detailed hydrocarbon breakdown including C_4's, C_5's and pyrolysis gasoline is obtained.

The properties of the naphtha feedstock utilized in the experimental campaign are presented in Figure 2. The source of the feedstock was Kuwait crude. This naphtha feedstock was synthesized by blending 35 volume percent light Kuwait and 65 volume percent heavy Kuwait. This blend does not occur in the virgin crude and as such is somewhat heavier than the true virgin feed. The isoparaffins were about 60% of the total paraffins and the alkylcyclohexanes were about 60% of the naphthenes.

COMPARISON OF PYROLYSIS COILS OF DIFFERENT GEOMETRIES – EXPERIMENTAL RESULTS

Two commercial pyrolysis coils of different geometry were designed to operate on a constant pyrolysis selectivity line. These coils will be referred to in this work as Coil 1 and Coil 2. Pyrolysis Coil 1 consisted of

SOURCE OF NAPHTHA	KUWAIT	
Specific Gravity, 15.5°/15.5°C, 60°/60°F	0.743	
Gravity, °API	59	
Sulfur, wt%	0.044	
Hydrocarbon Analysis, wt%		
Paraffins	62.9	
Naphthenes	23.9	
Aromatics	13.2	
Aromatic Breakdown, wt%		
Benzene	0.76	
Toluene	1.36	
Xylenes & Ethylbenzenes	2.74	
Alkylbenzenes	8.34	
Normal Paraffins/Total Paraffins, Wt. Ratio	0.41	
Alkylcyclopentane/Alkylcyclohexane, Wt. Ratio	0.7	
ASTM Distillation,	°F	°C
IBP, vol%	116	47
10	186	85
30	265	129
50	311	155
70	336	169
90	361	183
EP	372	189
Mean Average Boiling Point	279	137

Figure 2. Feedstock properties

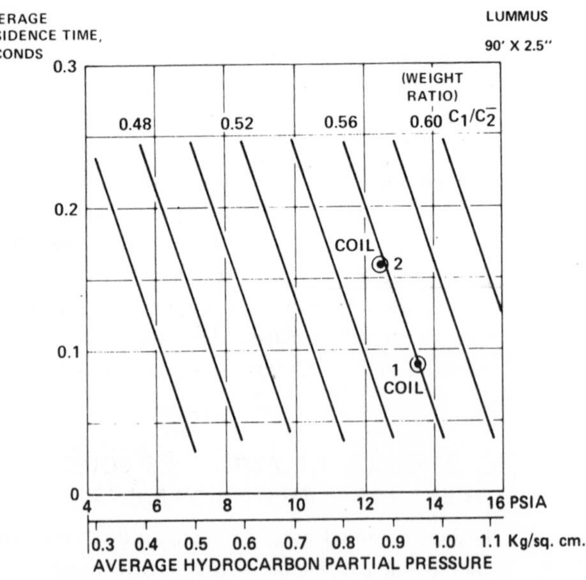

Figure 3. Pyrolysis selectivity chart

a single tube diameter (about 2.25 inches inside diameter), four passes and a single tube per pass. Each tube was about 29 ft. long. The total length of the coil was 116 ft. Pyrolysis Coil No. 2 was similar to the coil used in the commercial SRT® III furnaces for cracking naphthas and gas oils. This coil is referred to in the literature as the "swage" pyrolysis coil. (7)

This coil, among other features, had varying diameters which range up to 5 inches at the outlet. The total length of the pyrolysis coil was 160 ft. which was 38% longer than that of Coil No. 1.

Coil No. 1 was characterized by lower average residence time and higher average hydrocarbon partial pressure while Coil No. 2 was characterized by higher average residence time and lower average hydrocarbon partial pressure. Pyrolysis Coil No. 1 had a constant surface-to-volume ratio along its length while Coil 2 had a variable ratio from inlet to outlet.

Both pyrolysis coils were designed to satisfy the following operating criteria:

1. Identical pyrolysis selectivity toward the production of olefins, i.e., same pyrolysis selectivity line as shown in Figure 3, at the same feestock conversion.

2. Identical cross-over temperature, i.e., gas inlet temperature to the pyrolysis coil.

3. Identical dilution steam-to-oil ratio.

4. Identical maximum tube wall metal temperature at the start of run.

The capacity of Coil 2 is 4.3 times that of Coil 1. The gas temperature profiles of Coils 1 and 2 are shown in Figure 4. While the axial temperature profile in Coil 1 is approximately linear, the temperature profile of the SRT III coil is approximately hyperbolic.

The residence time profile for Coils 1 and 2 are shown in Figure 5. The average residence time of the SRT III coil is longer than that of Coil 1. The SRT III pyrolysis coil (Coil No. 2) had the lower hydrocarbon partial pressure profile as shown in Figure 6.

The typical operating process characteristics of both coils will be discussed later in this paper.

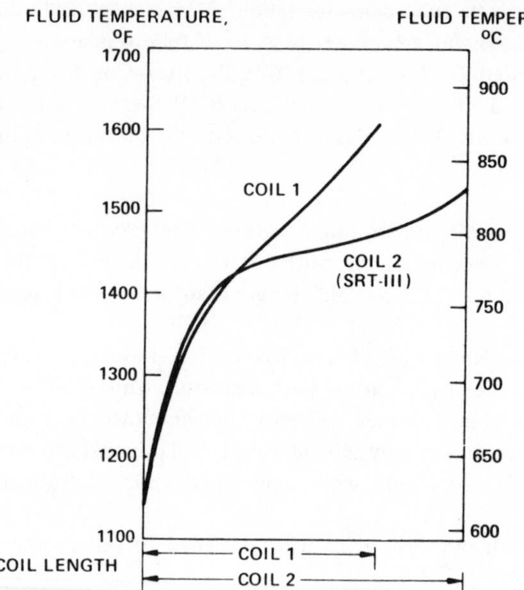

Figure 4. Axial pyrolysis temperature vs. coil length

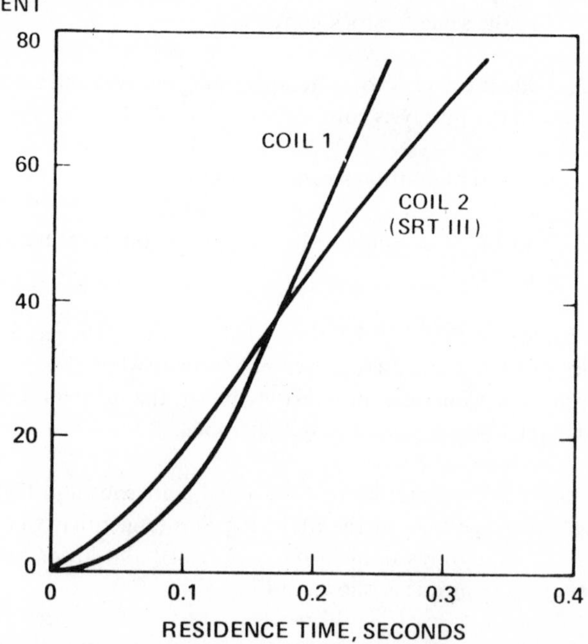

Figure 5. Conversion vs. residence time

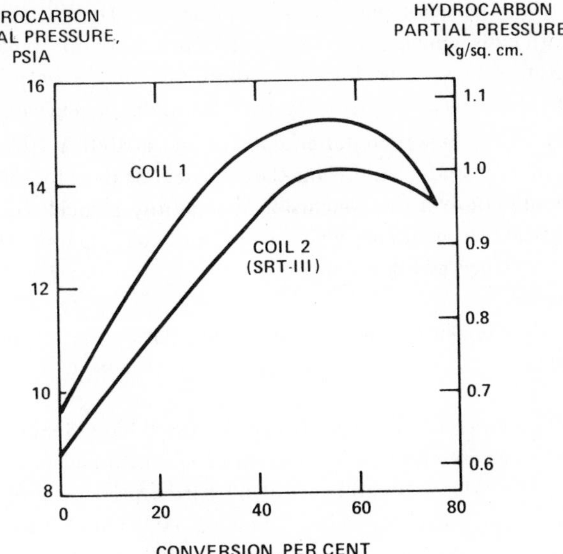

Figure 6. Hydrocarbon partial pressure vs. conversion

To determine the pyrolysis yields obtained from both coil designs, a series of experiments were conducted over a wide range of cracking severities. The cracking severity region beyond the maximum propylene yield was extensively investigated. The experimental results have shown that the pyrolysis yields obtained from Coil 1 were essentially identical to the yields from the SRT III coil. In Table I, the pyrolysis

TABLE I

COMPARISON OF PYROLYSIS YIELDS FROM DIFFERENT COIL CONFIGURATIONS OPERATING ON THE CONSTANT SELECTIVITY LINE

Coil No.	1	2 (SRT III)	%
Hydrogen, wt. %	0.88	0.86	+2.3
Methane	15.90	15.61	+1.9
Ethylene	28.32	27.82	+1.8
Ethane	3.82	3.78	+1.1
Propylene	11.06	10.97	+0.8
Butadiene	4.91	5.29	- 7.2

$$\% = \frac{(\text{Yield Coil 1} - \text{Yield Coil 2})}{\text{Yield Coil 2}} \times 10^2$$

yields from two pilot plant runs are presented. These runs were made with both coils which, as previously described, have different configurations. Both runs were performed at approximately 28 wt. percent ethylene; the differences between the two sets of data being well within the accuracy of the experimental equipment and analytical techniques employed in this study. Adjusting the two sets of data to 28% ethylene yield would confirm the conclusion that pyrolysis yields of both coils are identical. The methane yields from both coils are plotted as a function of ethylene yield in Figure 7.

These results consitute one of many confirmations of the slope of the constant pyrolysis selectivity lines previously discussed.

The two commercial coil designs just discussed have identical pyrolysis selectivity despite their differences in geometrical and process characteristics. The axial gas temperature and partial pressure profiles constitute the major differences. The effect of axial temperature and axial partial pressure profiles on pyrolysis reactor selectivity are taken into account in the definitions of average residence time and hydrocarbon partial pressure. Therefore, when pyrolysis coils of different geometries and thus different temperature and partial pressure profiles are com-

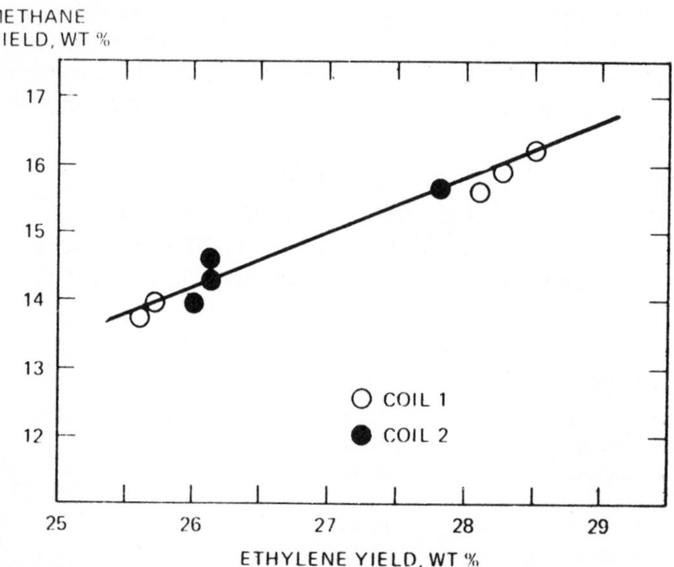

Figure 7. Pyrolysis yields: comparison of pyrolysis coils with different configurations operating on the same selectivity line

pared, the choice of the most selective coil is made by the location of their average residence times and average hydrocarbon partial pressures in the pyrolysis selectivity chart.

KINETIC CONSIDERATIONS SUPPORT EXPERIMENTAL AND COMMERCIAL RESULTS

The pyrolysis selectivity chart shown in Figure 3 and confirmed by the experimental results of this pyrolysis study is also supported by kinetic considerations.

The mass balance for a reactant can be written in a general form applicable to any type of reactor (8) as shown in Equation (4),

$$\begin{pmatrix}\text{Mass of Reactant}\\ \text{fed to the}\\ \text{Volume Element}\end{pmatrix} - \begin{pmatrix}\text{Mass of Reactant}\\ \text{leaving}\\ \text{Volume Element}\end{pmatrix}$$

$$- \begin{pmatrix}\text{Mass of Reactant}\\ \text{Converted in the}\\ \text{Volume Element}\end{pmatrix} = \begin{pmatrix}\text{Accumulation of}\\ \text{Reactant in the}\\ \text{Volume Element}\end{pmatrix} \quad (4)$$

The above equation applies to a time element Δt and a volume ΔV. In the case of the ideal tubular reactor shown in Figure 8, the mass balance equation can be written for a volume element ΔV extending over

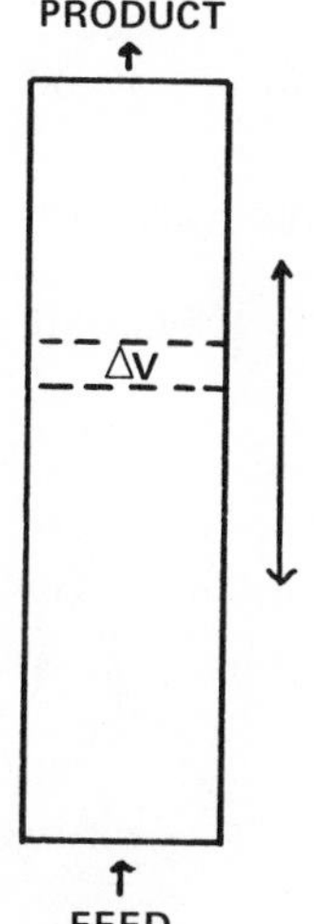

Figure 8. Ideal tubular reactor

the entire cross section of the pyrolysis coil. This assumes no variation in properties or velocity in the radial direction which is consistent with the high Reynolds numbers observed in the commercial pyrolysis coils. In the absence of axial mixing, the reactant can enter the element ΔV only by bulk flow of the stream. Hence,

$$\begin{pmatrix}\text{Mass of Reactant}\\ \text{fed to Volume}\\ \text{Element}\end{pmatrix} = F\,(1\text{-}a)\,\Delta t \qquad (5)$$

where

F = Mass feed rate of reactant to reactor

a = Conversion of this reactant at the entrance to the volume element.

If the conversion leaving the element is $a + \Delta a$, then,

$$\begin{pmatrix}\text{Mass of Reactant}\\ \text{leaving Volume}\\ \text{Element}\end{pmatrix} = F\,(1\text{-}a - \Delta a)\,\Delta t \qquad (6)$$

The fourth term of Equation (4) is zero because the operation is at steady state. The third term is

$$\begin{pmatrix}\text{Mass of Reactant}\\ \text{Converted in the}\\ \text{Volume Element}\end{pmatrix} = \Gamma \Delta V \Delta t \qquad (7)$$

where Γ = Rate of reaction.

Then, the mass balance equation can be rewritten as follows:

$$F\,(1\text{-}a)\,\Delta t - F\,(1\text{-}a - \Delta a)\,\Delta t - \Gamma\Delta V\Delta t = 0$$

or

$$F\Delta a = \Gamma\Delta V$$

Dividing by ΔV and taking the limit as $\Delta V \longrightarrow 0$, the following expression is obtained:

$$\frac{da}{dV} = \frac{\Gamma}{F} \qquad (8)$$

Where Γ varies with longitudinal position (volume) in the reactor.

During the pyrolysis of hydrocarbons for the production of olefins and diolefins, primary and secondary reactions have been postulated. The primary reactions reflect the decomposition of the reactant essentially by free radical mechanisms. The secondary reactions become important when the reactant conversion through primary reactions has reached high levels. The postulated important secondary reactions are hydrogenation and condensation reactions. In the case of liquid feedstocks, the primary reactions can be represented as first order and the secondary reactions as second order.

For first order reactions, Equation (8) can be written as follows:

$$F \frac{da_R}{dV} = \Gamma_R = k_p C_R$$

where

$$C_R = \frac{N_R}{V}$$

$$F \frac{da_R}{dV} = k_p \left(\frac{N_R}{V}\right) \qquad (9)$$

where

a_R = Conversion of reactants. This assumes that the initial conversion of reactant occurs through free radical mechanisms which can be expressed as first order reactions. Primary reactions.

N_R = Moles of reactants.

k_p = Reaction velocity constant of primary reactions.

For the second order reactions, the following equation can be written:

$$F \frac{da_p}{dV} = \Gamma_p = k_s C_p^2$$

$$F \frac{da_p}{dV} = K_s \frac{N_p^2}{V2} \qquad (10)$$

where a_p = Formation of secondary products relative to moles of reactants.

N_p = Moles of primary products.

ks = Reaction velocity constant of secondary reactions.

Pyrolysis selectivity is a function of the relative rates of the primary reactions (olefins forming reactions) to the secondary reactions (olefins consuming reactions).

The relative rates of reactions can be expressed as follows:

$$\frac{F\,\dfrac{da_R}{dV}}{F\,\dfrac{da_p}{dV}} = \frac{K_p}{K_s}\,\frac{N_R}{N_p^2}\,V$$

At the conditions practiced in hydrocarbon pyrolysis,

$$V = \frac{N_{HC}RT}{P_{HC}}$$

where N_{HC} = Total number of moles of hydrocarbons,
Then:

$$\frac{\dfrac{da_R}{dV}}{\dfrac{da_p}{dV}} = \frac{K_p}{K_s}\,\frac{N_R R N_{HC}}{N_p^2}\,\frac{T}{P_{HC}} \tag{11}$$

where T = System temperature

P_{HC} = Hydrocarbon partial pressure

When the entire length of the pyrolysis coil is considered,

$$\int_0^{a_R} da_R = \int_0^{a_p} \left[\frac{K_p\,N_R\,R\,N_{HC}\,T}{K_S\,N_p2\,P_{HC}}\right] da_p$$

The integration of the above expression is quite complex. To facilitate its solution, the average values of the parameters in parathensis are used as shown below:

$$\int_0^{a_R} da_R = \left[\left(\frac{K_P}{K_S}\right)_{AVG} \frac{N_{RA}\ N_{HCA}\ T_A}{N_{PA}{}^2\ P_{HCA}}\right] \int_0^{a_p} da_p$$

Then

$$\frac{aR}{aP} = \left(\frac{K_P}{K_S}\right)_{AVG} \left[\frac{N_{RA}\ N_{HCA}\ T_A}{N_{PA}{}^2\ P_{HCA}}\right] \qquad (12)$$

For two pyrolysis coils of different configurations and operating at a different combination of average residence times and hydrocarbon partial pressures, the pyrolysis selectivity is considered identical when:

$$\left(\frac{aR}{aP}\right)_{\text{Coil 1}} = \left(\frac{aR}{aP}\right)_{\text{Coil 2}}$$

and

$$(aR)_{\text{Coil 1}} = (aR)_{\text{Coil 2}}$$

Where the relative rates are the average for each coil.

Then:

$$\left[\left(\frac{(kp)}{(ks)}\right) \frac{N_{RA}\ N_{HCA}}{N_{PA}{}^2} \frac{T_A}{P_{HCA}}\right]_{\text{Coil 1, Avg.}}$$

$$= \left[\left(\frac{(kp)}{(ks)}\right) \frac{N_{RA}\ N_{HCA}}{N_{PA}{}^2} \frac{T_A}{P_{HCA}}\right]_{\text{Coil 2, Avg.}}$$

Where

- T_A is the average pyrolysis temperature of each coil

$$T_A = \frac{1}{ao} \int_{o}^{ao} Tg \; da$$

and Tg = Gas temperature along the length of the coil.

- P_{HCA} is the average hydrocarbon partial pressure of each coil.

$$P_{HCA} = \frac{1}{ao} \int_{o}^{ao} P_{HC} \; da$$

- $(kp/ks)_{Avg.}$ is the ratio of the specific velocity constants of each coil at the average temperature of each coil.

For two coils with identical pyrolysis selectivity,

$$\left(\frac{N_{RA}\, N_{HCA}}{N_{PA}^2}\right)_{\text{Coil 1}} = \left(\frac{N_{RA}\, N_{HCA}}{N_{PA}^2}\right)_{\text{Coil 2}}$$

i.e., if two coils produce essentially identical yields, the average moles of reactant, primary products, and total hydrocarbon should satisfy the above equality. Average quantities are defined as follows:

$$N_{RA} = \frac{1}{ao} \int_{o}^{ao} N_R \; da$$

$$N_{HCA} = \frac{1}{ao} \int_{o}^{ao} N_{HC} \; da$$

$$N_{PA} = \frac{1}{ao} \int_{o}^{ao} N_P \; da$$

If the definitions of the average parameters associated with the above derivation are correct, then the following relationship must also be valid:

$$\frac{\left(\frac{kp}{ks}\right)_{\text{Coil 1 Avg.}}}{\left(\frac{kp}{ks}\right)_{\text{Coil 2 Avg.}}} = \frac{P_{HCA1}\, T_{A2}}{T_{A1}\, P_{HCA2}} \qquad (14)$$

This can be confirmed by experimentation.

Using the Arrhenius equation, the ratio of reaction velocity constants can be expressed as follows:

$$\frac{\left(\frac{kp}{ks}\right) \text{ Coil 1 Avg.}}{\left(\frac{kp}{ks}\right) \text{ Coil 2 Avg.}} = e^{(Es - Ep)\left\{\frac{1}{T_{A1}} - \frac{1}{T_{A2}}\right\}\frac{1}{R}} \tag{15}$$

Where Es and Ep = Activation energies of primary and secondary reactions. In evaluating Equations (14) and (15), average values have been used in order to consider the overall tubular reactor length.

Based on the work published by Messrs. Hirato, Yoshioka, and Tanaka, (9) the most probable overall average values for (Es - Ep) have been estimated to about - 11000 Kcal/mol. This estimate has resulted from considering a variety of possible combinations of primary and secondary reactions which have been postulated to occur in thermal pyrolysis.

Equation (14) has been applied to the two pyrolysis coils used in the experimental work previously discussed. The calculated average residence times, gas temperatures and hydrocarbon partial pressures for Coils 1 and 2 are presented in Table II.

TABLE II

Average Parameters for Coils 1 and 2

Coil No.	1	2
θ_A, Secs	0.0893	0.1558
TA, ^{o}F	1468	1441.5
P_{HCA}, Psia	13.6	12.5

From Equation (15),

$$\frac{\left\{\frac{kp}{ks}\right\} A_2}{\left\{\frac{kp}{ks}\right\} A_1} = 0.9298$$

Then:

$$P_{HCA2} = \frac{T_{A2}}{T_{A1}} (P_{HCA1}) \frac{(Kp/Ks)_{A2}}{(Kp/Ks)_{A1}}$$

$$= \frac{1901.2\ (13.6)}{1927.7} (0.9298)$$

$$P_{HCA2} = 12.5 \text{ Psia} \qquad (16)$$

which is in agreement with the value shown in Table II and confirmed experimentally in Figure 3. This is a confirmation of the above discussed theoretical considerations.

Therefore, Equation (15) which is based on theoretical considerations, defines conditions for identical selectivity.

Shape of Pyrolysis Selectivity Lines

The application of Equation (15) to pyrolysis coils operating at average residence times in the millisecond region leads to the important conclusion that the constant pyrolysis selectivity lines are not straight. A typical pyrolysis selectivity chart extended into the millisecond region is schematically shown in Figure 9.

The above analysis of the factors affecting pyrolysis selectivity now permits the selection of a locus or conditions (i.e., a selectivity line) which will achieve a given pyrolysis selectivity and thus, a certain yield structure. Since this constant selectivity line covers a wide range of combinations of average residence times and average hydrocarbon partial pressures, the question remains "what other factors should be considered by the designer to select a point on this selectivity line which is optimum for large-scale commercial production of olefins?"

To answer this question, it is necessary to consider the effects of simultaneous heat, mass and momentum transfer in the pyrolysis coil.

SELECTION OF COIL CHARACTERISTICS

A pyrolysis coil of a given selectivity can be designed to lie at any point on the desired selectivity line. The selection of a coil operating at a lower residence time and a higher hydrocarbon partial pressure will result in a coil with a higher gas outlet temperature than a coil operating at a higher residence time and lower hydrocarbon partial pressure

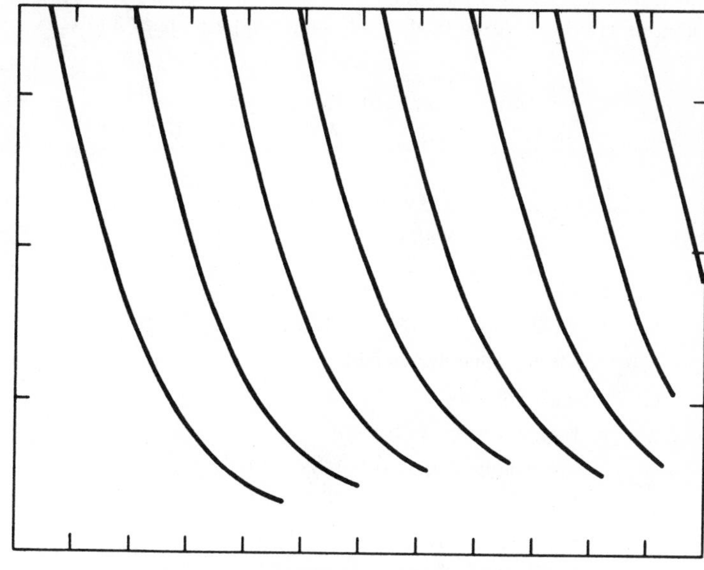

Figure 9. Schematic pyrolysis selectivity chart

on the same selectivity line. The implications of this higher outlet temperature are significant.

A commercially acceptable pyrolysis coil design is constrained by the limitations in available metallurgy. For any selected material, there exists a maximum operating temperature beyond which tube life is sharply reduced. This limitation in metal temperature must be considered in setting the coil design.

The coil with the higher gas outlet temperatures requires higher heat transfer coefficients and/or higher heat transfer surface-to-volume ratio in order to not exceed the tube metal temperature limitation. This is achieved by utilizing a coil with relatively small diameter outlet tubes. This design, however, results in other undesirable effects due to simultaneous heat, mass and momentum transfer.

In order to clearly discuss these effects, it is interesting to refer to the theory of the relationships between momentum, heat, and mass transfer.

TRANSPORT PHENOMENA

The transport phenomena which operate at a given point in a pyrolysis coil are related according to the following fluid analogies: (1)

$$Sc^{2/3}\ \frac{Km(MW)}{G} = \frac{hPr^{2/3}}{GC_p} = \frac{(F/S)\,gc}{M/A} = \frac{f}{2} = \frac{0.023}{(DG/\mu)^{0.2}} \tag{17}$$

$$(F/S) = A\Delta P/S = \frac{\pi D^2 \Delta P}{4\pi D L} = \frac{D\Delta P}{4L} \tag{18}$$

where:

Km = Molal mass transfer coefficient
MW = Fluid molecular weight
G = Mass velocity
h = Heat transfer coefficient
C_p = Specific heat at constant pressure
F/S = Drag force/unit surface area of pipe
gc = Universal gravitational constant
M = Momentum flow
A = Pipe cross sectional area
f = Fanning friction factor
D = pipe diameter
μ = Viscosity
Sc = Schmidt Number, $(\mu/\rho\mathscr{D})$
Pr = Pradtl number, $(Cp\mu/k)$
M/A = G^2/p
G = W/A
k = Thermal conductivity
$\mathscr{D}$ = Diffusivity
ΔP = Pressure drop
L = Axial pipe length
ρ = Mass density
W = Mass flow rate

Combining the above relationships, the following expressions are derived:

$$S_c^{2/3}\ Km(MW) = \frac{hPr^{2/3}}{C_p} = \frac{0.023W^{0.8}\ \mu^{0.2}}{D^{1.8}\,(\pi/4)^{0.8}} \tag{19}$$

and

$$\frac{\Delta P \rho gc}{4L} = \frac{0.023 W^{1.8} \mu^{0.2}}{D^{4.8} (\pi/4)^{1.8}} \tag{20}$$

From Equations (19) and (20), it is clear that the small diameter coil, while enhancing the heat transfer coefficient, also tends to increase the pressure drop and mass transfer coefficient.

The effect of increased pressure drop per unit length has been demonstrated in the pyrolysis selectivity experiment described above. This experiment showed that the shorter residence time coil, i.e. Coil 1, because of the higher pressure drop, and therefore higher average hydrocarbon partial pressure, achieved the same selectivity as the larger diameter, longer residence time, lower partial pressure coil.

The effect of an increased mass transfer coefficient is significant in terms of coil coking which limits run length due to increase in metal temperature and pressure drop.

COKING CHARACTERISTICS

Coking in a commercial pyrolysis coil is a highly complex process which has not been modelled in precise mathematical terms.[11, 12, 13] However, as is often the case, a simplified model utilizing certain empirically defined constants, has proven to adequately describe the coil coking phenomenon.[14]

The coil coking model postulates a two-step mechanism:

1. Mass transfer of coke precursors from the bulk of the gas to the walls of the tube.

2. The chemical reaction of coke precursors at the tube wall resulting in the formation and deposition of coke. This simplified two-step process can be described as follows:

$$Rm = Km\,(y - y_i) \tag{21}$$

$$Rr = Kr\,(y_i \frac{P}{RT}) \tag{22}$$

where;

Rm = molal rate of mass transfer
Rr = Rate of chemical reaction
Km = Mass transfer coefficient
Kr = Reaction velocity constant

y = mole fraction of coke precursor in bulk fluid
y_i = mole fraction of coke precursor at tube wall
P = Total pressure
R = Universal gas constant
T = Absolute temperature at tube wall

The chemical reaction is assumed to be first order for the purpose of this derivation. However, the conclusions are shown to be independent of this assumption.

At any given location at the tube wall:

$$Rm = Rr \tag{23}$$

Then:

$$Km(y - y_i) = Kr\,\frac{y_i P}{RT}$$

$$y_i = \frac{Kmy}{\frac{KrP}{RT} + Km}$$

$$\therefore\ R = Rm = Rr = Kmy\left(1 - \frac{Km}{\frac{KrP}{RT} + Km}\right) \tag{24}$$

Where R = Rate of coke formation

In a commercial pyrolysis coil, the temperature at the tube wall is considerably higher than the bulk temperature of the gas. At these high temperatures, studies have indicated that the reaction velocity constant, Kr, is much higher than the mass transfer coefficient, Km.

Since Kr $>>$ Km

$$\frac{Km}{\left[\frac{KrP}{RT} + Km\right]} \rightarrow 0 \qquad (25)\ (26)$$

and R = ykm

Therefore, the rate of coil coking is a mass transfer controlled process. From Equation (19),

$$R = \frac{\text{coke deposited}}{\text{Day}} = \frac{K^* \, W^{0.8}}{(D\text{-}2\Delta)^{1.8}} \qquad (27)$$

Where W = Total mass flow rate
D = Inside diameter of pyrolysis tube
Δ = Coke thickness
K* = Constant which is a function of feedstock, cracking selectivity, dilution steam ratio, cracking severity and other system properties.

Then, the run length of the pyrolysis coil, θRL can be predicted as follows:

$$\theta RL = \frac{\Delta \text{ max.}}{R} \qquad (28)$$

Where Δ max. is the coke thickness which coincides with the maximum allowable tube metal temperature or the maximum allowable pressure drop.

Equations (19) and (27) indicate that the rate of coke formation is higher for a small diameter tube to the same extent that the rate of heat transfer is greater for a small tube. Also the rate of coke formation in the small tube increases more rapidly during the length of a run because coke build-up in a small tube increases the mass transfer coefficient more rapidly than in a larger tube as indicated by the term $(D\text{-}2\Delta)^{1.8}$

The coil coking model has been tested in a number of plants. Table III presents these data. The agreement between the predicted and the actual run length is very good, which is a confirmation of the assumptions made in the development of the model.

SELECTIVITY LOSS DURING RUN

A further, extremely important characteristic of a small outlet tube is that the pyrolysis selectivity decrease during the length of a run will be greater than that of a large outlet tube. This results from the effect of the deposited coke on the tube cross sectional area. A given thickness of coke in a small tube increases the coil pressure drop more rapidly than in a large tube. As a result, the increase in average hydrocarbon partial pressure of a small tube during the run length will cause a more significant decrease in selectivity than in the larger tube coil. Figure 10 illustrates this effect. The two coils studied in the selectivity experi-

TABLE III
TEST OF THE PYROLYSIS COIL COKING MODEL

Olefins Plant	Feedstock**	Run Length, Days	
		Plant Data	Calculated
A	FRN	59	58
B	FRN	90	82
C	FRN	100	100
D	FRN	45	45
E	FRN	45	50
F*	LGO	83	82

*Gas Oil Prototype Unit in Japan
**FRN = Full Range naphtha
LGO = Light Atmospheric Gas Oil

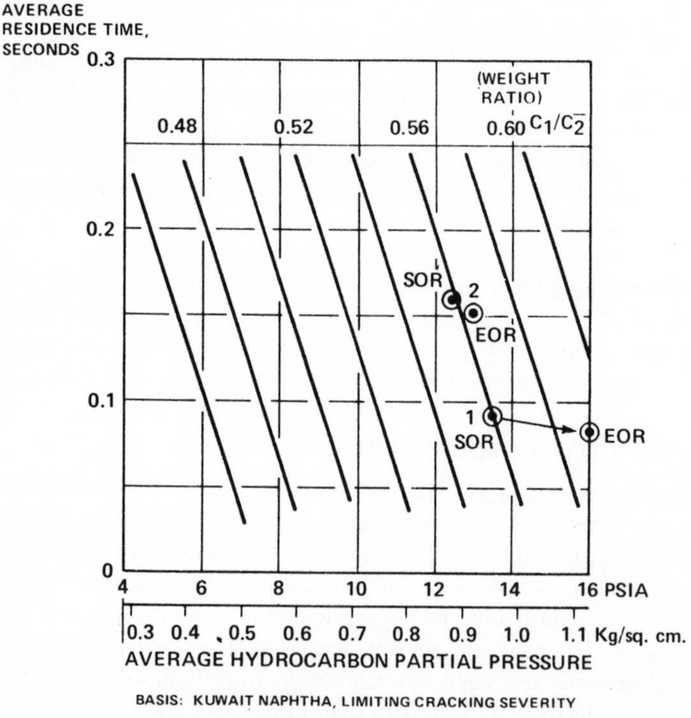

BASIS: KUWAIT NAPHTHA, LIMITING CRACKING SEVERITY

Figure 10. Pyrolysis selectivity chart

ment are compared in terms of end of run selectivity. The average selectivity of the larger outlet tube coil is greater than that of the shorter, small tube coil, despite the fact that the start of run selectivities are identical.

In summary, not only will a large tube coil coke at a lower rate, but also the effect of whatever coke does form is less significant in terms of continued pyrolysis selectivity.

OPTIMUM COIL DESIGN

Before discussing the characteristics of an optimum coil design, it is worthwhile to summarize the conclusions presented up to this point:

1. A given pyrolysis selectivity can be achieved by designing for an average residence time/average hydrocarbon partial pressure combination which can be anywhere on a locus of points which describe a line of constant selectivity.

2. The question facing the designer is where on this selectivity line is the point of optimum coil design. The choice is often between a short, small outlet diameter coil which emphasizes low residence time and a longer, large outlet diameter coil which emphasizes low hydrocarbon partial pressure.

3. The short coil, small diameter tube option has the disadvantage of a higher rate of coke formation and a greater loss of cracking selectivity as the run progresses.

In addition, the capacity of the large outlet tube coil is greater than that of the short coil and as such, the quench complexity and costs are reduced. The capacity of the large diameter coil compared to the small diameter coil studied in the selectivity experiment was 4.3 times greater for the same operating selectivity. This higher coil capacity and simpler overall system result in a pyrolysis reaction module of a much lower cost for a fixed production of olefins.

Having developed the arguments favoring a coil which achieves a given selectivity by emphasizing low hydrocarbon partial pressure, (longer, large outlet diameter), the remaining question is what are the specific characteristics of this coil type?

Any optimum design should achieve a given result at minimum cost. This minimum cost is achieved by minimizing total heat transfer surface

TABLE IV

COMPARISON OF PROCESS CHARACTERISTICS*

Coil No.	1	2 (SRT III Furnace)
Relative Hydrocarbon Feed Rate	1	4.31
Relative Avg. Residence Time	1	1.78
Relative Pressure Drop	1	0.81
Relative Avg. Hydrocarbon Partial Pressure	1	0.92
Relative Coil Coking Rate	1	0.75

* Design basis as described previously in this work.

and minimizing the complexity of the quench system. The large capacity, low cost quench system as used by Lummus is a by-product of the large outlet diameter, high capacity coil and has been discussed elsewhere.(15, 6)

To further illustrate these conclusions, the process characteristics of the Coil No. 1, small diameter, and Coil No. 2, SRT III heater, previously discussed are compared in Table IV and in Figure 10.

SUMMARY

The **average** hydrocarbon partial pressure and the **average** residence time are the factors which affect pyrolysis heater selectivity. The available data on pyrolysis selectivity have been successfully correlated as a function of these two parameters. The temperature and partial pressure profiles along the pyrolysis coil are considered in the definition of average residence time and average hydrocarbon partial pressure. The results of this work have shown that, while the average residence time and hydrocarbon partial pressure are both key factors in determining pyrolysis selectivity, the hydrocarbon partial pressure is somewhat more important than previously realized by investigators.

It has been experimentally shown that the pyrolysis selectivity of two coils of different geometry, tube diameters and temperature profiles can be identical provided their average residence times and hydrocarbon partial pressures fall on the same selectivity line. The same conclusions have also been drawn from kinetic considerations. This approach has been used to extend the pyrolysis selectivity lines into the millisecond region.

A mathematical model for calculating the rate of coking in the pyrolysis coil has been discussed. These models have been confirmed with performance data from commercial pyrolysis reactors.

The pyrolysis selectivity correlation and the coking models have been combined with momentum heat and mass transfer models to design pyrolysis coils. This application has led to the SRT III pyrolysis reactors which emphasizes low hydrocarbon partial pressure by employing a coil design with large diameter outlet tubes.

The approach of a high capacity, high selectivity and lower complexity reactor is viewed to best meet the needs of the olefins plants of the 1980's.

ABSTRACT

The hydraulic and kinetic factors affecting pyrolysis coil performance are defined. These factors are correlated to quantify their effects on pyrolysis yields and coking tendencies for a given feedstock. The correlation is supported by theoretical and experimental considerations.

Based on momentum, heat and mass transfer, the effect of coil diameter on module performance is discussed.

A logical result of the above considerations is the Lummus SRT® III pyrolysis module.

ACKNOWLEDGEMENT

The authors express their appreciation to Dr. K.W. Li of Lummus for his assistance in reviewing and offering invaluable advice on the contents of this paper.

REFERENCES

1. Williams, K.D. and H.G. Davis, "Mechanistic Studies of Ethane Pyrolysis at Low Pressures," American Chemical Society, Philadelphia Meeting, April, 1975.

2. Dorn, R.K. and M.J. Maddock, "Design Pyrolysis Heater for Maximum Profits," *Hydrocarbon Processing,* November, 1972.

3. Fernandez-Baujin, J.M. "Factors Affecting Pyrolysis Selectivity," Safety and Reliability of Large Single Train Ethylene Plants, Unpublished Report, New York, May, 1974.

4. Maddock, M.J., "SRT Heater Design and Engineering Characteristics", Safety and Reliability of Large Single Train Ethylene Plants, Third Ethylene Seminar-Lummus, Unpublished Report, New York; May, 1972.

5. Brooks, M.E. and J. Newman, "Gas Oil Cracking – The Problems That Had to be Solved," VIII World Petroleum Congress, Moscow, U.S.S.R., June, 1971.

6. Fernandez-Baujin, J.M., and A.J. Gambro, "Technology and Economics for Modern Olefins Plants," VI Interamerica Congress of Chemical Engineering, Caracas, Venezuela; July, 1975.

7. Chambers, L.E. and W.S. Potter, *Hydrocarbon Processing,* p. 121, January, 1974; p. 95, March, 1974; p. 99, August, 1974.

8. Smith, J.H., *Chemical Engineering Kinetics,* McGraw-Hill Book Company, New York, 1970.

9. Hirato, M., Yoshioka, S., and M. Tanaka, "Gas Oil Pyrolysis by Tubular Reactor and its Simulation Model of Reaction," *Hitachi Review* 20:8,326, 1971.

10. Bennett, C.O. and J.E. Myers, "Momentum, Heat and Mass Transfer," McGraw-Hill Book Company, New York, 1962.

11. Chen, J. and M.J. Maddock, "How Much Spare Heater for Ethylene Plants?" *Hydrocarbon Processing,* May, 1973.

12. DeBlick, J.L. and A.G. Goossens, "Optimize Olefin Cracking Coils," *Hydrocarbon Processing,* March, 1971, p. 76.

13. Mol, A., "How Various Parameters Affect Ethylene Cracker Run Lengths," *Hydrocarbon Processing,* July, 1974.

14. Maddock, M.J., "Coking on Pyrolysis Heaters," Private Communication, Lummus, November, 1971.

15. Chen, J. and W. Vogel, "Fouling of Transfer Line Exchangers in Ethylene Service," 74th National Meeting, A.I.Ch.E., New Orleans, La. March, 1973.

21

Pyrolysis of Naphtha and of Kerosene in the Kellogg Millisecond Furnace

HARRY P. LEFTIN, DAVID S. NEWSOME, and THOMAS J. WOLFF

Pullman Kellogg, Research & Engineering Development Laboratory, Houston, Tex.

JOSEPH C. YARZE

Pullman Kellogg, Northeast Operations Center, Hackensack, N.J.

Since net feedstock cost is the most significant item of olefins production cost, selection of cracking conditions conducive to high yields of the desired products is a very important feature of an economically successful olefins manufacturing venture. For any specific feedstock, residence time, temperature, degree of conversion and hydrocarbon partial pressure are the variables that influence the product distribution achieved in a steam-pyrolysis process. Of these, residence time and temperature are the most significant variables and the mechanism and kinetics of the thermal cracking of hydrocarbons indicate that optimum selectivity to olefins and maximum feedstock utilization should result from operations carried out at high temperatures and short contact times.

Advances in tube metallurgy have made it possible for furnace designers to reduce contact times for commercial pyrolysis plants. Thus over the past decade, design contact times have been reduced in several stages from over 2 seconds to 0.25 seconds, with each step providing an increase in ethylene yield and a decrease in relative tail gas production.

In the early stages of pyrolysis development, Kellogg designed a conventional pilot plant reactor (1) (0.25 to 1.5 seconds) in order to study the effect of temperature and contact time on product yields from the pyrolysis of pure hydrocarbons and complex mixtures, ranging from light naphthas to heavy vacuum gas oils. Yields and operating conditions obtained from this pilot plant have been used successfully to design many commercial olefins plants. The accuracy of the operating conditions and yields measured and predicted from this pilot plant have been confirmed by commercial experience. Kellogg's evaluation of the effects of contact time on pyrolysis yields was further stimulated by innovations in metallurgy, which would enable commercial furnace designs at higher temperatures and shorter contact times than those that could be examined in the existing pilot plant.

In order to ascertain whether pyrolysis in the low fractional second contact time range would afford ethylene and co-product yield advantages sufficient to justify still further development of a process, bench scale work was initiated in the Kellogg laboratory in 1965. These bench scale studies provided experimental evidence on the trends of product yields over a contact time range from 0.01 to 0.10 seconds at reaction temperatures from 1400°F (760°C) to over 2000°F (1093°C). Investigations were performed on a wide variety of gaseous (2) and liquid pure hydrocarbons, simple mixtures of pure hydrocarbons and on a variety of liquid feedstocks, including raffinates, naphthas and gas oils.

It was clearly evident from these bench scale studies that ethylene and butadiene yields could be increased substantially and gaseous and liquid fuel products decreased proportionately by operations at elevated temperatures in this short contact time range. It was also found that maximum olefins yield can be obtained within a narrow range of contact times well above 0.01 seconds, but below 0.10 seconds. With very high temperatures and contact times below about 0.01 second, greatly increased yields of acetylene, methylacetylene and allene were obtained.

Data from the bench scale studies clearly indicated sufficient incentive for further study. Consequently in 1968, Kellogg designed and constructed a pilot plant reactor system which could be used for cracking a variety of feedstock types in the optimum contact time range established by the bench scale work. In this design, the process variables were incorporated in a manner that assured that the data obtained would be comparable with eventual full scale yields and process conditions. The pilot plant will be described in detail in the present paper.

Data from an extensive series of tests carried out in this Millisecond Pilot Plant Reactor very closely confirmed all of the trends previously established by the bench scale studies. In addition, pilot plant and bench scale experiments with the same feedstock served to demonstrate the excellent correspondence between these two units as regards product yields and selectivity. Reasonably close correspondence was obtained for temperature and contact times and reliable correlation factors were developed to relate these variables for the two units.

The high level of confidence on the economic advantages of pyrolysis in this contact time range resulting from both bench scale and pilot plant studies led to in-house development work on a furnace design which would be needed to commercialize this process. About 1970, Kellogg and Idemitsu Petrochemical Company (IPC) agreed on a joint development effort to test these concepts on a full size installation, which would also serve to provide the additional ethylene capacity that IPC required at that time. The furnace was installed in the IPC No. 2 Ethylene Plant at Tokuyama, Japan(3). Although this Millisecond Furnace was described earlier, (4) some of the details of its construction and operation are recounted briefly in the present paper. This furnace, having an ethylene design capacity of 25,000 MT/yr, was operated with naphtha feedstock in 1972 and 1973. Subsequently, successful operations were also performed on kerosene, raffinate and gas oil feedstocks. Close correspondence of product yields and operating conditions were obtained between the commercial furnace and the laboratory reactors. These observations confirmed the development of the Millisecond Furnace and Pyrolysis process, which provides substantial improvements in yields and feedstock utilization, into a design that incorporates reaction temperatures of 1650-1700°F (899-927°C) and contact times of less than 0.10 seconds. The Millisecond Furnace and Pyrolysis process probably represents the last important improvement which can be made with respect to these critical operating variables since operations at shorter contact times and consequently higher temperatures unavoidably lead to the production of substantial quantities of acetylene. Within the contact time range of the Millisecond Furnace and Pyrolysis process, however, ethylene yields in excess of 32 wt % can easily be obtained in the pyrolysis of a typical wide range naphtha along with a concomitant tail gas/ethylene of well below 0.5 weight/weight.

Results and Discussion

The effects of decreasing contact time and increasing temperatures on product yields and feedstock conversions in the steam pyrolysis of a wide range naphtha and a kerosene were studied in the laboratory bench scale pyrolysis unit and in the Millisecond Pyrolysis Pilot Plant. The laboratory studies covered wide ranges of operating variables such as reactor outlet temperature [1400-1750°F (760 to 954°C)], contact

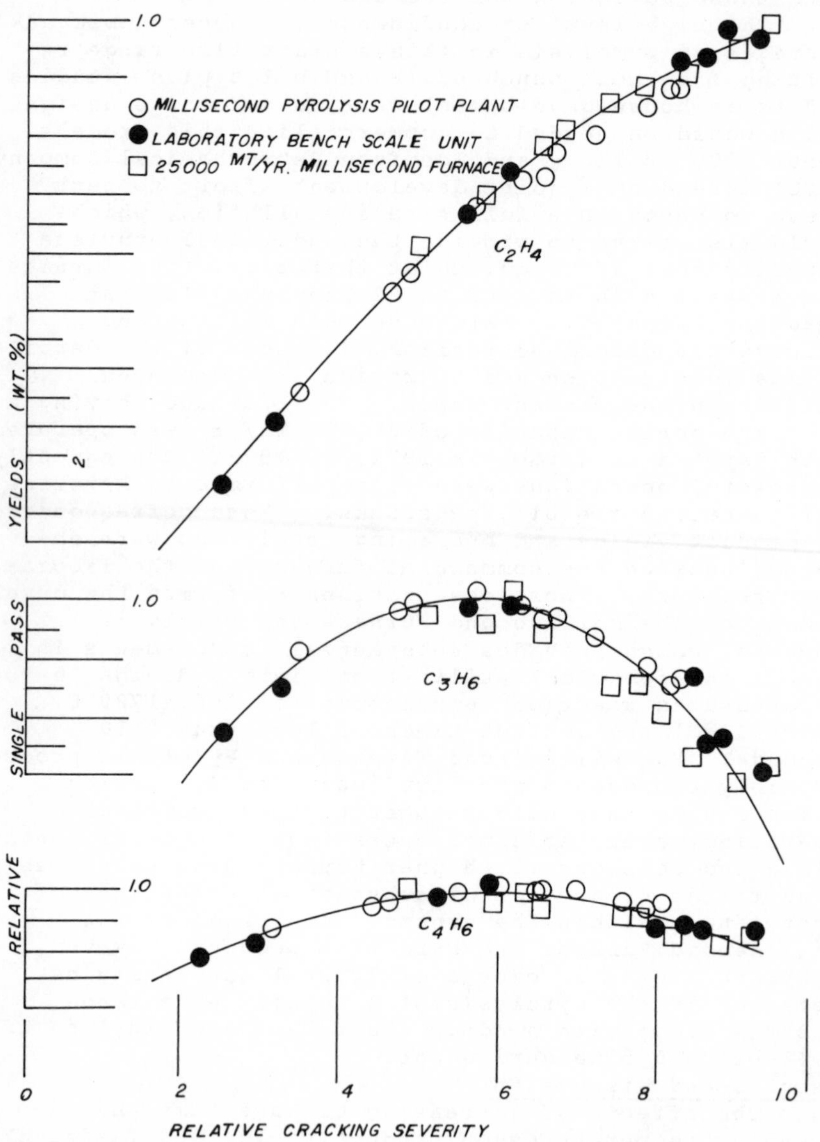

Figure 1. Naphtha pyrolysis yields vs. severity at millisecond furnace contact times

time (0.01 to 0.40 seconds), outlet pressure and steam to oil ratios. Simultaneously, runs using the same feedstocks were being performed at Tokuyama, Japan in the Idemitsu 25,000 metric ton per year Millisecond Furnace. The later measurements were over a more restricted range of variables, specifically, those conditions specified in the Kellogg Millisecond Furnace and Pyrolysis process. Table I summarizes the inspections and properties of the feedstocks employed in these studies. The naphtha is a wide range Kuwait naphtha and is fairly typical of the naphtha feeds encountered in many olefin units.

TABLE I
PYROLYSIS FEEDSTOCK INSPECTIONS

	NAPHTHA	KEROSENE
API°	71.0	46.0
Distillation, °F		
Vol%		
IBP	99	315
10	131	345
20	143	355
40	166	372
60	195	392
80	234	417
90	281	437
95	331	453
E.P.	385	476
Type Analysis, Vol%		
Paraffins	73.5	49.1
Cyclic Paraffin	21.2	28.2
Aromatics	5.3	18.2
Molecular Weight	93	157
H/C Atomic Ratio	2.16	1.93

Figure 1 shows the yields of principal products obtained from the steam pyrolysis of the naphtha feedstock as severity is varied, while contact times are maintained within 0.01 to 0.1 second. The open circular points are data obtained in the Kellogg Millisecond Pilot Plant, while the closed points are data obtained in the Millisecond Bench Scale Unit, and the square points are the data observed with the commercial Millisecond Furnace at Tokuyama (3). These data clearly establish that excellent correspondence exists between the experimental results from these units. The observed agreement is even more remarkable when one considers that the scale-up factors between the bench scale-pilot plant and the pilot plant-commercial furnace are of the order of 10^2 and 10^3, respectively, for an overall scale-up factor of 10^5.

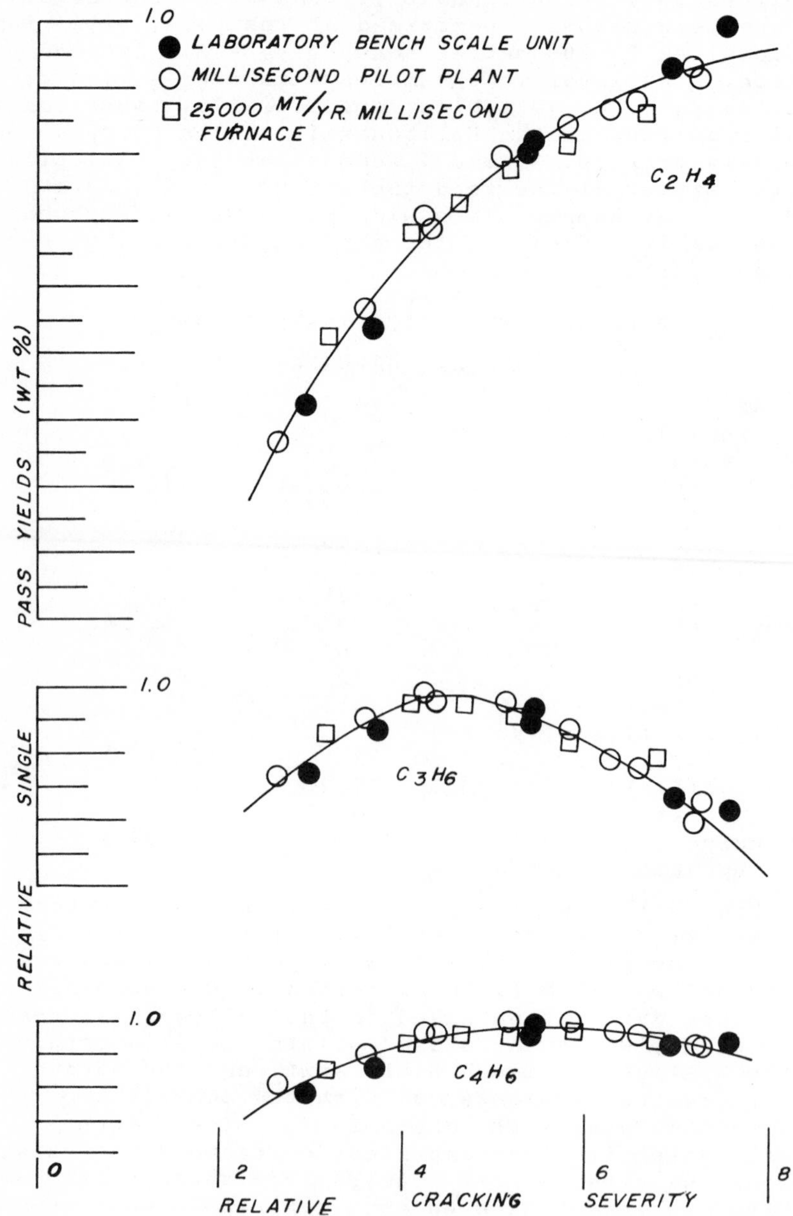

Figure 2. Kerosene pyrolysis yields vs. cracking severity at millisecond contact times

Comparable data for the pyrolysis of kerosene feedstocks are shown in Figure 2. Again the yields and selectivities of the products are in excellent agreement.

These observations provide strong evidence of the reliability of Kellogg's pyrolysis test facilities to provide data that can be used directly for plant design. As a result, it is now possible to determine rapidly, and in advance, the economic differences between candidate feedstocks using the bench scale unit; and to set the basis for the commercial plant design within the framework of the desired product slate flexibility utilizing the pilot plant unit.

A complete discussion of the economic advantages of the Millisecond Furnace and Pyrolysis process is beyond the scope of the present paper as this would involve considerations of utilities efficiencies as well as capital investment, in addition to yield and product slate values. It is, however, desirable to illustrate the order of magnitude of the yield and feedstock utilization advantages that acrue to Millisecond Furnace operations. Table II shows the yields and conversions obtained in pilot plant studies of a wide range naphtha carried out under conditions of conventional short residence time pyrolysis (0.35 seconds) and also under Millisecond Furnace contact time. It will be noted that at the shorter contact time of the Millisecond Furnace, substantial gain in single pass ethylene and butadiene yields are obtained. Similar yield advantages are observed over the entire range of operating severities. Propylene yield, on the other hand, appears to be but little effected by this change in contact time. Consequently, it can be concluded that the Millisecond Furnace provides higher ethylene and butadiene yields and reduced methane production at any severity level. Moreover, at any fixed ethylene capacity significantly less feedstock is required.

Experimental

A. Bench Scale. The experimental arrangement is shown in Figure 3. Liquid feedstock and water were separately metered from pressurized feedtanks into a preheater-vaporizer and finally into the pyrolysis reactor contained in an electrically heated furnace. Power input to the furnace transformer was controlled manually. Temperature profiles were measured with a calibrated Chromel-Alumel thermocouple manually driven along the entire length of the reactor. Pressures were measured with calibrated Bourdon gages sensitive to ±3 torr. On leaving the reaction zone, the gases were

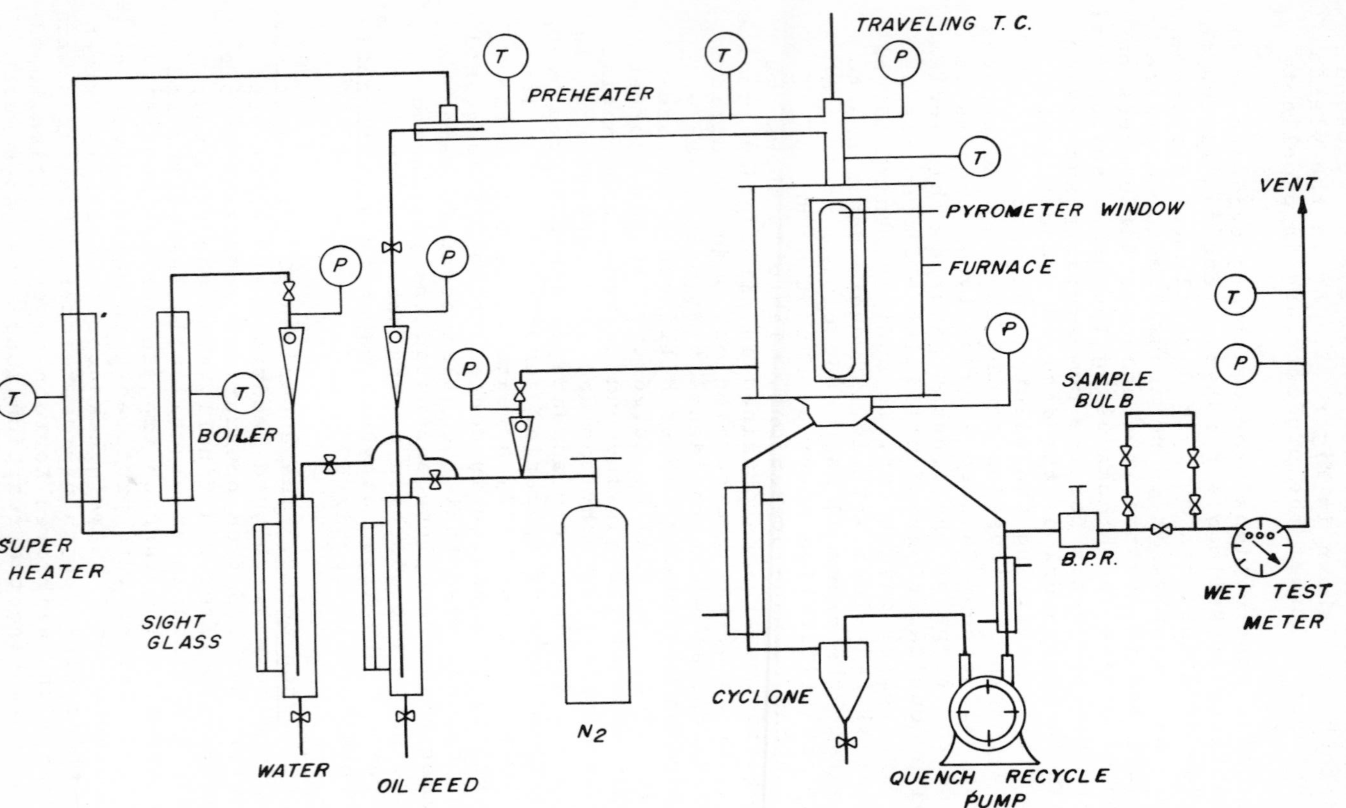

Figure 3. Bench scale apparatus used for millisecond pyrolysis studies

Table II
Pyrolysis of a Wide Range Naphtha
(Pilot Plant Data)

Furnace Type	Conventional	Millisecond
Contact Time, sec.	0.3	Less than 0.10
Severity	High	High
Tail Gas	17.6	16.7
C_2H_2	0.6	1.0
C_2H_4	28.2	31.0
C_2H_6	3.9	3.3
C_3H_4	0.6	1.3
C_3H_6	12.5	12.7
C_3H_8	0.5	0.2
C_4H_6	3.6	4.9
C_4H_8	2.7	2.8
C_4H_{10}	0.2	0.5
C_5^+	29.6	25.6

rapidly cooled by admixture with a recycled stream of cooled product gas using the method described by Happel and Kramer(5). In this way, a rapid direct quench was achieved without the usual complications that result from introducing large volumes of inert quench diluent. Prior to entering the recycle gas pump, the quenched products were further cooled against tap water in an indirect heat exchanger and the condensed water and liquid products were separated by means of a small cyclone separator. Both the heat exchanger and cyclone were integral parts of the quench gas recycle system. Net product gas passed through a sampling valve and a wet test meter and was then vented.

The reaction zone was an annulus between an outer tube of 1/8" NPS Sch 40 pipe and an inner tube of 3/16" OD which served as the thermocouple well. Both tubes were 347 stainless steel. The furnace comprised a water-cooled cylindrical shell equipped with FiberFrax (Union Carbide Corp.) insulation, a ceramic baffle and a split cylindrical graphite heating element that was concentric with the reactor. Stainless steel fittings were used to fix the reactor to the top of the furnace, while at the bottom it was free to move within a packing gland at the point where the reactor outlet entered the quench gas mixing chamber. The packing gland also served to effect a gas seal between the nitrogen-filled furnace shell and the reactor outlet. Nitrogen pressure in the furnace shell was maintained slightly above that of the reactor outlet so that a small amount of nitrogen leaked continuously into the product stream at the reactor outlet. Nitrogen content of the total gas sample was determined along with the product analysis by mass spectrometry and was subtracted from the gas yield in calculating the overall material balance for the runs.

The reactor walls were pretreated with a mixture of steam, hydrocarbon, and ethyl mercaptan prior to each series of runs. Wall poisoning was maintained by continuous addition of mercaptan throughout the course of each run. The observation that the maximum level of carbon oxides produced in any of the runs never exceeded 0.1 mol % of the total product gas indicated that the sulfiding procedure was effective in reducing complications due to catalytic reaction at the tube wall. Distilled water, used for the generation of reaction steam, was degassed and freed of carbon dioxide prior to being charged to the reservoir. Rotameters were used only to establish and maintain the instantaneous flow rates. Integral feed rates were determined periodically from readings of the calibrated sight glasses on the feed reservoirs.

Carbon deposits from previous runs were burned out and the reactor was sulfided for two hours at 600°C and one hour at 750°C prior to each series of runs. Furnace temperature was slowly raised to run conditions with steam and nitrogen flowing through the reactor. Final adjustment to the desired temperature was made with feed and water at the required flow rates. Furnace power was controlled to maintain a constant reactor temperature and the run was started only after this had remained constant for at least 45 minutes. After all initial readings had been made, the complete temperature profile was recorded (this required approximately 25 minutes). At the completion of the profile, the thermocouple was returned to the position of maximum temperature and a gas sample was removed for analysis. Complete temperature profiles were recorded continuously during the entire run period which normally ranged between one and three hours. Two gas samples were taken for duplicate analyses by mass spectrometry and one for gas chromatography. These were spaced between the first third and the final third of the run period. Usually the results of the duplicate mass spectrometric determinations fell within the established limits of this analytical method, and the averaged values were then used. However, occasionally the deviations were larger and in those cases the runs were repeated until good agreement was obtained. All calculations of yields and conversions are based on mass spectroscopic data and gas chromatography was used only for identification of the isomers with a particular mass number and to estimate the isomer distributions. Runs were carried out on both feedstocks at temperatures between 1400° and 1750°F (760° and 954°C) with dilution steam corresponding to 0.5 and 0.75 weight ratio on the naphtha and kerosene, respectively. All runs were isobaric at a total pressure of 22 psig. Material balances generally fell within 100±3% for all experiments.

B. Pilot Plant Reactor. Basically the pilot plant is capable of providing data that can be used with confidence in the design of full scale commercial units. To achieve these goals, a pilot reactor was developed, which included in its design such commercial furnace relationships as contact time, temperature profile, pressure drop and pressure profile, bulk-tube skin temperature relationship, turbulent flow regime, time to quench, etc. This was done so that it truly represented its commercial counterpart. In addition, a stringent set of operating ground rules were established upon which all pilot operations would

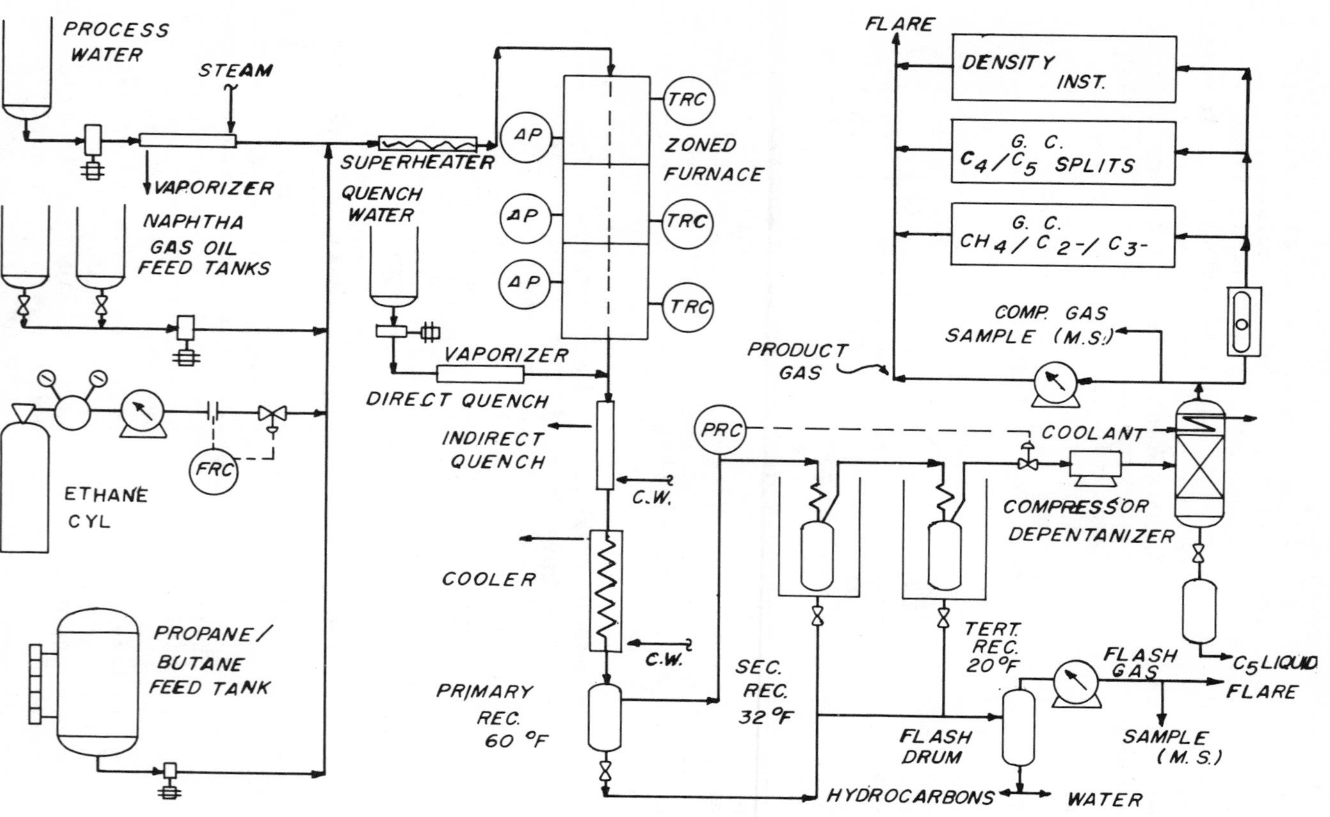

Figure 4. Pilot plant flow sheet

be based. Most important of these were: (1) material balances better than 100±1.5% must be obtained consistently; (2) good reproducibility must be obtained over long operating time periods; (3) yields must be determined with high degrees of accuracy; and (4) the relationship between commercial and pilot plant reactors must be known.

A simplified flowsheet of the pilot plant is shown in Figure 4. Process steam is constantly generated at a predetermined rate from weighed and metered water which is vaporized by steam.

Three systems for metering and measuring hydrocarbon feeds have been incorporated: one for gaseous feeds such as ethane, one for liquids under pressure such as propane or butane, and one for normally liquid feedstocks such as naphthas or gas oils. Each type feed can be fed independently or two or more can be fed simultaneously as would be required in co-cracking operations. Two feed tanks have been installed in order to permit side-by-side comparisons of two or more liquid feedstocks. It is a simple matter, therefore, to run each feedstock at identifical reaction conditions.

Liquid stocks are metered by dual-head Lapp reciprocating pumps, and hourly feed rates are determined by weight. The hydrocarbons are injected into superheated steam at 400°F (204°C), flashed, and mixed with the steam. The mixture is then superheated to about 800-1000°F (427°-538°C) as required before injection into the reactor.

The reactor section consists of a vertical electrically-heated furnace in which the reactor tube is placed. The furnace has separately controlled zones with appropriate temperature and heat input measurement facilities for each zone. Fluid temperature measurements are made along the length of the reactor with calibrated couples located in adiabatic zones and temperature profiles can be varied. Reactor pressures are continuously monitored and pressure drop (ΔP) and pressure profile across the reactor can be controlled. Since each reactor represents a specific commercial coil, multiple reactors have been designed to cover the commercial contact time range of interest. In these studies a reactor designed for operation between 0.01 and 0.10 seconds was employed.

Reactor effluent is quenched either by direct injection of water or steam, or by indirect heat exchange against water, depending on the particular requirements. These facilities not only permit simulation of commercial transfer line and quench technique, but also permit study of the effects of transfer line residence time on product yields. The quenched gas is further cooled to room temperature with cold water. The condensed heavy hydrocarbons and water are separated from the gas in the primary receiver. The resulting gas is successively cooled to 32°F (0°C) and 20°F (-6.7°C) and additional liquid hydrocarbons are separated in these secondary and tertiary receivers. All liquid products are combined and sent to a flash drum for depressuring which results in the separation of a flash gas which is measured and analyzed by mass spectrometry. Liquid products are withdrawn from this drum and the oil and water are separated and weighed.

At this point the product gas contains essentially all the C_4's and lighter and most of the C_5's. In order to obtain a gasoline product containing the C_5's, as is done commercially, the product gas is compressed and then fed to a debutanizer, where 95% of the C_5's and heavier are fractionated out of the gas. In order to avoid loss of the light ends and C_4's, the liquid product from the debutanizer is collected in a bomb and added to the other liquid products during final debutanization in the laboratory.

A product gas sample is collected throughout the duration of a run and this composite is analyzed by a mass spectrometer. In addition, a number of on-line analytical instruments have been provided for monitoring unit performance and product gas composition. A Beckman Model A-3 density instrument continuously records the product gas density which, together with the gas measurement and liquid hydrocarbon weight, enables a material balance to be developed within minutes after a run is terminated. Since density (composition) of the product gas is very sensitive to operating conditions, it is invaluable in relating the performance of the reactor and the steadiness of a run.

An on-line gas chromatograph is also used to determine the concentrations of the more important components such as H_2, CH_4, C_2H_4, and C_3H_6. These analyses, in conjunction with the product gas volume measurement, permit yields of the major components to be calculated within 30 minutes after completion of a run. More importantly, the on-line instrument permits the effect of reactor conditions on product yields to be screened rapidly, and enables the selection of reactor conditions required to achieve a particular product distribution.

Calibrated scales and gas meters were employed

throughout this work. Gas molecular weights were determined by direct measurement on a densitometer, and by calculations based on mass spectrometer analyses. A review of 61 consecutive runs showed an average material balance of 99.6% with a standard deviation of ±1.2%. A further check on the reliability of the material balance is obtained by individual C and H balances which are based on analytical C/H determinations. These comparisons are normally in good agreement.

Accuracy with which yields can be determined is largely a function of the analytical techniques employed. The analytical scheme is shown in Figure 5. Final yields of gaseous components are based on mass spectrometer analyses which have an accuracy of ±2% relative to the amount present for the major olefinic components.

All liquid products are combined, and after complete water separation, the hydrocarbon portion is debutanized. The resulting stabilized liquid is fractionated into two overhead cuts for analytical expediency. Yields of individual components through the C_9 aromatics, as well as fuel oil yields, can be determined quantitatively.

In certain situations, "screening" of operating conditions may be desired before "balance runs" are made. This is accomplished by altering the process variables and observing the effects of these alterations on the gas densitometer and the on-line product gas chromatograph, which continuously monitor the reactor effluent. It is simple, for example, to prescribe a reasonable ethylene yield and tail gas ratio and locate the best set of process variables to give these results. Maximizing yields or optimizing operating conditions, therefore, can readily be accomplished.

Hundreds of "balance runs" and innumerable screening runs have been made in the pilot plant on feeds ranging from ethane through heavy gas oil. The results of the most recent tests at high severity were employed in the design of the commercial Millisecond Furnace.

Commercial Millisecond Furnace

Working in close cooperation with Idemitsu Petrochemical Company, Kellogg designed a prototype of the Millisecond Furnace with a nominal ethylene capacity 25,000 metric tons a year. This furnace, shown in Figure 6, also served as an expansion to Idemitsu's No. 2 Ethylene Plant at Tokuyama, Japan.

The very short contact times required in the Millisecond Furnace are obtained through the use of

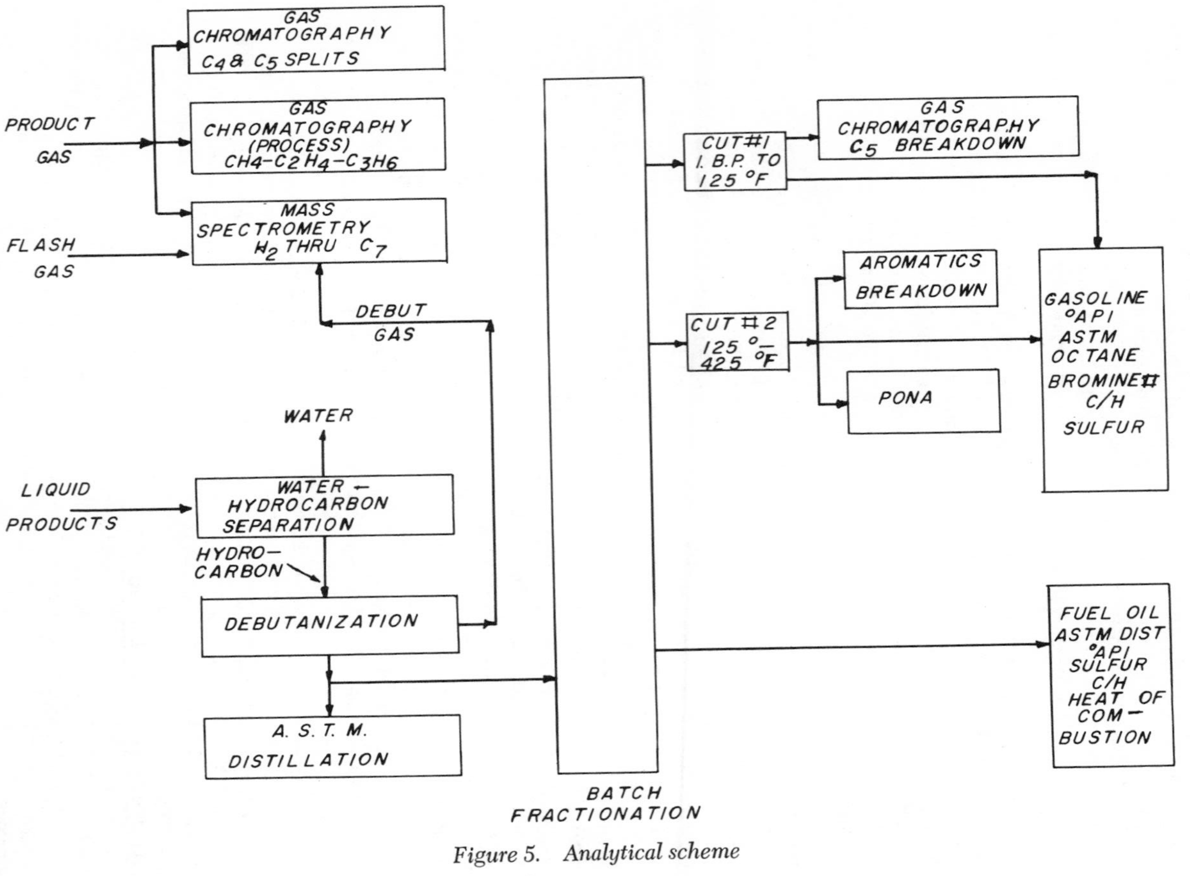

Figure 5. Analytical scheme

Figure 6. 25,000 MT/yr millisecond furnace

small diameter tubes. These pass straight through the radiant section and are fired from both sides by burners located in the furnace arch.

Preheated process feed plus dilution steam is distributed uniformly to the tubes through individual leads and is rapidly heated to reaction temperation. The olefins-rich gas produced then exits the radiant section floor and is immediately quenched to halt the reaction and fix the product composition.

Combustion flue gases travel downward parallel to the tubes and pass through tunnels at the radiant section floor into the convection section where a high level of heat recovery is obtained assuring maximum fuel efficiency and optimum operating economics.

In the course of commercial operations, extensive yield data were collected on various liquid feedstock operations over the full short residence time range and varying hydrocarbon partial pressure. Yields were determined in the field by a special sample conditioning and analysis system (6) operating on the furnace effluent. The results of these studies substantiated the laboratory and pilot plant findings and confirmed that ethylene yields with the Millisecond Furnace can be increased by 10-20 wt % relative to those obtained with conventional cracking. Similar increases were achieved in the yields of other valuable co-products while the yield of methane was significantly reduced in all cases.

Conclusions

Innovations and improvements in ethylene plant design has been, and continues to be, an area of major creative effort at Kellogg. This is reflected in the Company's more than 30 years of construction and operating experience. The Bench Scale and Pilot Plant units described in this paper constitute reliable research tools that provide the accurate laboratory data which serve as a basis for commercial furnace design. In addition, feedback from the many commercial units has provided the information vital to the development of Kellogg's numerous and significant improvements in the design and application of process equipment. The Millisecond Furnace is the latest advance and affords the olefins producer a number of desirable processing features, among which are: higher yields, lower utilities, greater feedstock flexibility and utilization, and high pressure steam regeneration.

Abstract

The effect of decreasing contact time and increasing severity on product yields and feedstock conversion in the steam pyrolysis of naphtha and of kerosene were studied in a laboratory bench scale pyrolysis unit. Operating conditions leading to maximum product values, thus defined, were subsequently verified in a pilot plant unit and also in a full sized commercial furnace. Experimental data and the design and operation of the pilot plant that provided the basic data for the development of the Millisecond Furnace and Pyrolysis process are described. The excellent agreement observed between data obtained in the pilot plant unit and those from the commercial furnace provides a high level of confidence in the use of the pilot plant unit for feedstock evaluation and yield predictions from the Millisecond Pyrolysis Furnace. By comparison with results obtainable under conventional pyrolysis conditions, the process is shown to afford substantial increases in olefins yields and greatly improved feedstock utilization.

Literature Cited

(1) Lambrix, J.R., Wallace, B.A., and Yarze, J.C., Proc. 7th World Petroleum Congress, Mexico City, Mexico, (1967), **5**, 219.

(2) Leftin, H.P. and Cortes, A., I&EC Process Design & Development, (1972), **11**, 613.

(3) Anon., Oil & Gas Journal, March 17, 1975.

(4) Ennis, B.P., Symposium on High Temperature Chemical Reaction Engineering, Brit. Institution of Chemical Engineers, Harrogate, England, June 16, 1975.

(5) Happel, J. and Kramer, L.C., Ind. & Eng. Chem., (1967) **59**, 39.

(6) Leitner, W., Lemieux, E.J., Severino, F.T. and Yarze, J.C., Proc. Div. of Refining, A.P.I., (1968), **45**, 917.

22

The Advanced Cracking Reactor: A Process for Cracking Hydrocarbon Liquids at Short Residence Times, High Temperatures, and Low Partial Pressures

HUBERT G. DAVIS and ROBERT G. KEISTER

Research and Development Department, Union Carbide Corp., South Charleston, W.Va. 25303

The potential advantages of operating at shorter times, higher temperatures and lower average partial pressures than is possible in conventional steam crackers have been recognized within Union Carbide for many years. Some time ago we developed, through small pilot plant scale, a process and apparatus for practical operation at such conditions. The process involved mixing of a stream of liquid hydrocarbons with a stream of highly superheated steam in such a way that extremely rapid vaporization occurs as well as extremely rapid mass and thermal mixing. For experimental purposes we produced our superheated steam by combustion of a fuel gas with oxygen. The burner used was scalable to plant size, but we recognized the possibility that another type of superheater, such as a regenerator, might be substituted. The technology was dubbed the "Advanced Cracking Reactor" and is usually referred to by the initials ACR. These will be used throughout the present report.

About the time of our initial success, we became aware of the development of a crude oil cracker by the Kureha Chemical Industry Company and realized that there must be many similarities. Further, after the presentation of a paper by Gomi (1) at the Moscow World Petroleum Conference, we learned that Kureha, with the Chiyoda Chemical Engineering and Construction Company, had developed both burner and regenerator techniques up to commercial size. After an exchange of information, we entered into a joint development agreement with the two Japanese Companies. Since the Kureha-Chiyoda process is operating dependably at a 100,000 ton/year scale, we have essentially chosen to use their cracker technology. Operating conditions are varied from the very high temperatures needed by Kureha for their desired product mix (1:1, C_2H_4/C_2H_2). Extensive data on a wide variety of feedstocks have been and are being obtained in pilot plants operated by both Kureha and Union Carbide. Graphs and observations reported here are based on results from both sources.

Besides operating our ACR pilot plant, we at Union Carbide are doing extensive research, development and engineering design work, primarily on separations and purifications. Here, because of our long experience in olefins production - we operate plants with about 2,000,000 metric tons of ethylene capacity - we have considerable expertise. A prototype ACR unit - actually a small commercial plant of about 25,000 tons/ year ethylene output - is now in the design phase. We expect to build the unit in one of our existing U.S. plants within the next 4 years.

Background - Chemistry of Steam Cracking

Union Carbide pioneered the development of products derived from ethylene and propylene in the 1920's and 30's. Our original cracking furnaces were designed and built with very limited knowledge of the chemistry of hydrocarbon pyrolysis. Information, especially quantitative information, on the effect of such variables as pressure, partial pressure of furnace gas, and the time/temperature relationship was almost non-existent. Steam was added to the cracking mixture early - but only because it was found that steam eliminated formation of carbon particles which were rapidly eroding out furnace tube bends.

Over the years - mainly since the 40's - we, and others in the industry have filled in these details of chemistry. Steam has been recognized as essentially an inert in the cracking operation. Therefore it can be used to modify partial pressures and residence times and thus influence product yields as well as coking rates. Hydrocarbon pyrolysis became "steam cracking". The interaction of components of mixed feedstocks - especially of paraffins with olefins - is understood and can sometimes be used to advantage. Effects of residence time - or more accurately of the particular combination of residence time and temperature required to give a desired depth of cracking - have been recognized and used.

However, we, and the rest of the industry, have found that the constraints of the conventional tube cracking furnace severely limit the yield improvements which can be made. Indeed, we are amused to find that our furnaces built in the 30's and 40's made ethylene with as high efficiency as the best modern furnaces. Thus, we see the conventional furnace development as

TABLE I

MEASURES OF CRACKING SEVERITY

"Average" Gas Temperature plus Residence Time

Equivalent Time, τ, at characteristic temperature (tube outlet temperature).

Severity Function, S, = the equivalent temperature at a standard time, usually one second

$$S = T_C^{\circ} \; t^{0.062}$$

% Decomposition of key feed component

butanes
pentanes

Percentage of (or yield of) key product

methane
ethylene
propylene
total C_3^- gas

Ratio of key products

Mol % CH_4/Mol % C_3H_6

Wt % C_2H_4/Wt % C_3H_6

aimed primarily at higher and higher productivity per furnace, with loss of efficiency to ethylene limited as much as possible.

To understand the chemical effects of the various process variables, we need a definition or two. First "severity", which is, subjectively, how hard we crack to get a desired product mix. A typical qualitative definition is the following (3): "Severity is the summary effect of the increments of residence times through a pyrolysis reactor at their corresponding temperatures." Making the concept quantitative is difficult. Various expedients are used to obtain measures of severity - some based on actual times and temperatures, others on internal analytical criteria. Some of these measures are shown in Table I.

Another term requiring definition is "partial pressure". Here we mean the partial pressure of the "furnace gas", normally taken to be everything except the mostly inert diluent, steam. Hence

$$pp = \frac{\text{Mols (F.G.)}}{\text{Mols (F.G.)} + \text{Mols } (H_2O)} \cdot P,$$

where P is the total absolute pressure. We use the partial pressure at the furnace tube outlet, pp_o, and an average partial pressure, $\overline{pp}$, which takes into account the ΔP across the furnace tubes. Our $\overline{pp}$ is usually an "equivalent time" - weighted average.

The major variables affecting product yields are severity and feed composition. If we hold all the other variables constant, we totally define a detailed product yield when we fix severity by any of the measures of Table I. Thus, we can draw standard yield curves, or interpolate between tables or yields in computer programs, to account for the actual values of the defined variables. The programming is simplified by the observation that gas product analyses (and, with certain reservations, liquid product analyses) tend to be the same for all straight-run petroleum feedstocks. The major feedstock variable affecting yields is the elemental hydrogen analysis (or the H/C ratio). The approximate sensitivity of yields of ethylene to the operational variables of conventional steam cracking is shown in Table II. Charts can be made showing effects of variables on each of the cracking products.

For the present discussion, we are primarily interested in two variables - partial pressure and residence time. We see that there is a considerable dependence of ethylene yield on partial pressure. The coefficient, $-\Delta Y/\Delta\overline{pp}$, is not linear as might be implied

TABLE II

FACTORS AFFECTING YIELD OF ETHYLENE

PARAMETER	Sensitivity of C_2H_4 Yield, Y, wt % of feed, to Change in Parameter
1) Feed Composition Variables	
a. % H (or H/C)	$\frac{\Delta Y}{\Delta(H)} \cong$ + 5 wt %/wt % H
b. Isoparaffins in feed (replacing n-Paraffins)	$\frac{\Delta Y}{\Delta(Iso)} \cong$ -0.18 wt %/wt % iso in feed
c. Cyclopentane homologs replacing cyclohexane homologs	$\frac{\Delta Y}{\Delta(CP)} \cong$ -0.1 wt %/wt % cyclopentanes in feed
2) Operational Variables	
a. Partial pressure* of "furnace gas"	
a-1. Outlet partial pressure PPo	$\frac{\Delta Y}{\Delta PPo} \cong$ -3.4 wt %/Atm.
a-2. Average partial pressure $\overline{pp}$	$\frac{\Delta Y}{\Delta \overline{pp}} \cong$ -3.0 wt %/Atm.
b. Residence time at constant severity	
b-1. Equivalent time,	$\frac{\Delta Y}{\Delta \log_{10} \tau} \cong$ -7 wt %
c. Other variables with qualitative effects: tube length and diameter, quench system, tube material (surface), sulfur, temperature and pressure gradients.	

* Determined by absolute pressure, steam dilution, pressure drop.

from the table. It actually increases somewhat as the pressure drops. As partial pressure is reduced, yields of acetylene, butadiene and hydrogen also increase, while yields of methane, ethane, heavy liquid products and (slightly) propylene fall. Similarly, decreasing the time at constant severity increases acetylene and hydrogen and decreases methane, ethane and heavy liquids. The effects of dropping residence time and partial pressure are remarkably similar and arise from the same basic causes. Acetylene and ethylene decompose by reactions which are kinetically of higher order than first and have relatively low energies of activation. Polymerization reactions leading to formation of heavy liquid products and disappearance of reactive diolefins, acetylenes, etc., are also of higher order. The cracking reactions which lead to the formation of ethylene and acetylene are basically first order reactions with high energies of activation. The influence of the equilibrium among ethane, ethylene, acetylene and hydrogen is also important and in the same direction.

Any discussion of the energies of activation of decomposition of the paraffin components of feeds is complicated by the complex mechanisms of these reactions. In every case, the decomposition reaction is highly product-inhibited, so that, at any reaction temperature, calculated first order rate constants fall off rapidly with time. The inhibition is so severe, that it is necessary to assume kinetic orders far above second to fit the decomposition-time curves (if one can fit them with any simple order at all). An example, from some work by K. D. Williamson (2) is shown in Figure 1. As a result, if we compare rates at different temperatures at constant residence time we calculate relatively low activation energies (of the order of 45,000 cal/Mol) (3). This is because at the higher conversions of the higher temperatures, product inhibition is greater than at the lower temperatures.

The temperature dependence found at constant residence time is illustrated in Figure 2. The apparent energy of activation corresponding to the slopes shown is 46 kcal/Mole - low for a first order hydrocarbon pyrolysis. All hydrocarbons shown, paraffin or olefin - even ethylene - behave about the same. The figure, therefore, offers little encouragement that one can much influence the product spectrum by changing temperature other than along prescribed curves of yield vs. severity.

If, however, we plot, as in Figure 3, integral first order rate constants measured at constant severity (i.e., constant decomposition of a key compound), a different result is obtained. The activation energies are in general higher. Ethylene, however, still has a low activation energy, probably because its decomposition is really of a considerably higher order than first. From plots such as these we conclude that we can influence the yield pattern toward ethylene by operating at higher temperature and shorter

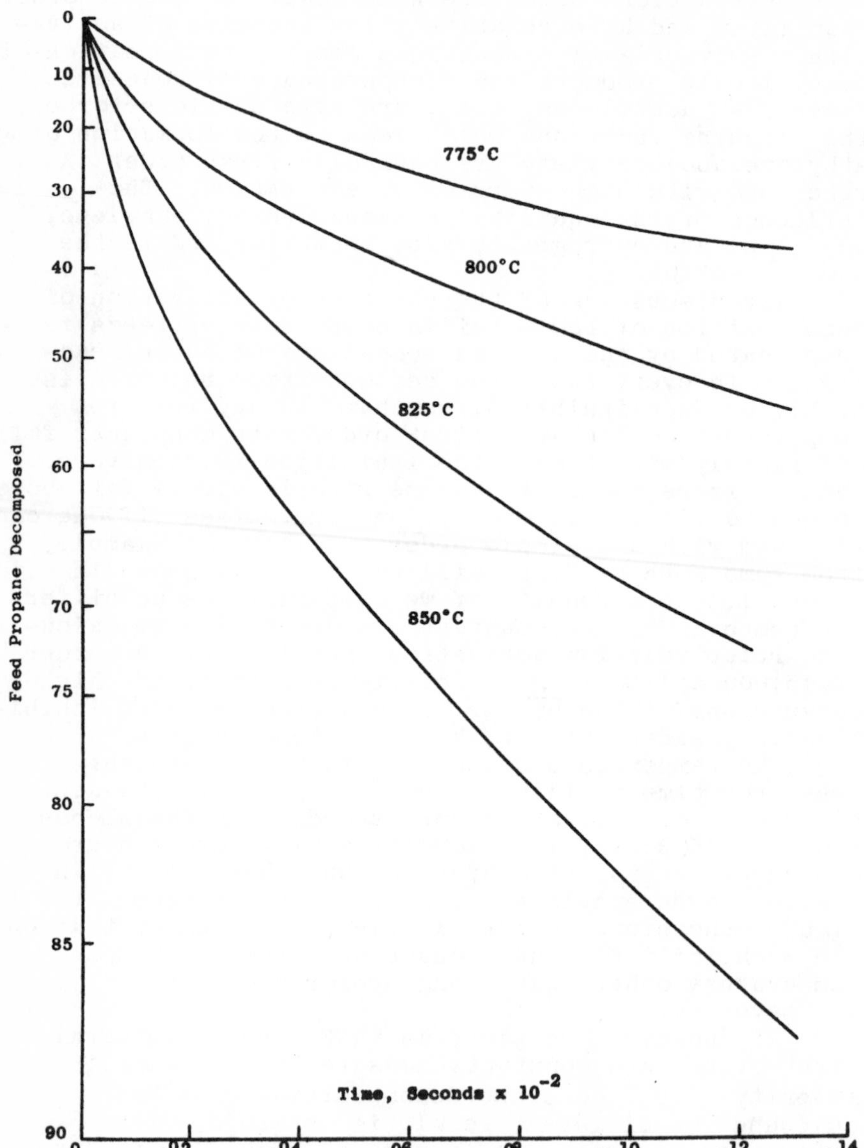

Figure 1. Decomposition as a function of time at constant temperature during pure propane pyrolysis

time. An apparent intermediate order of the ethylene decomposition reaction suggests a mixture of first and higher order reactions (shown schematically by the dotted lines of Figure 3). If we go high enough in temperature, the first order decomposition becomes important, and we stop increasing ethylene. However, in this region acetylene, which decomposes by a second order reaction of still lower activation energy (<40 kcals/mol) becomes a major product.

It is over-simplifying to attribute the total

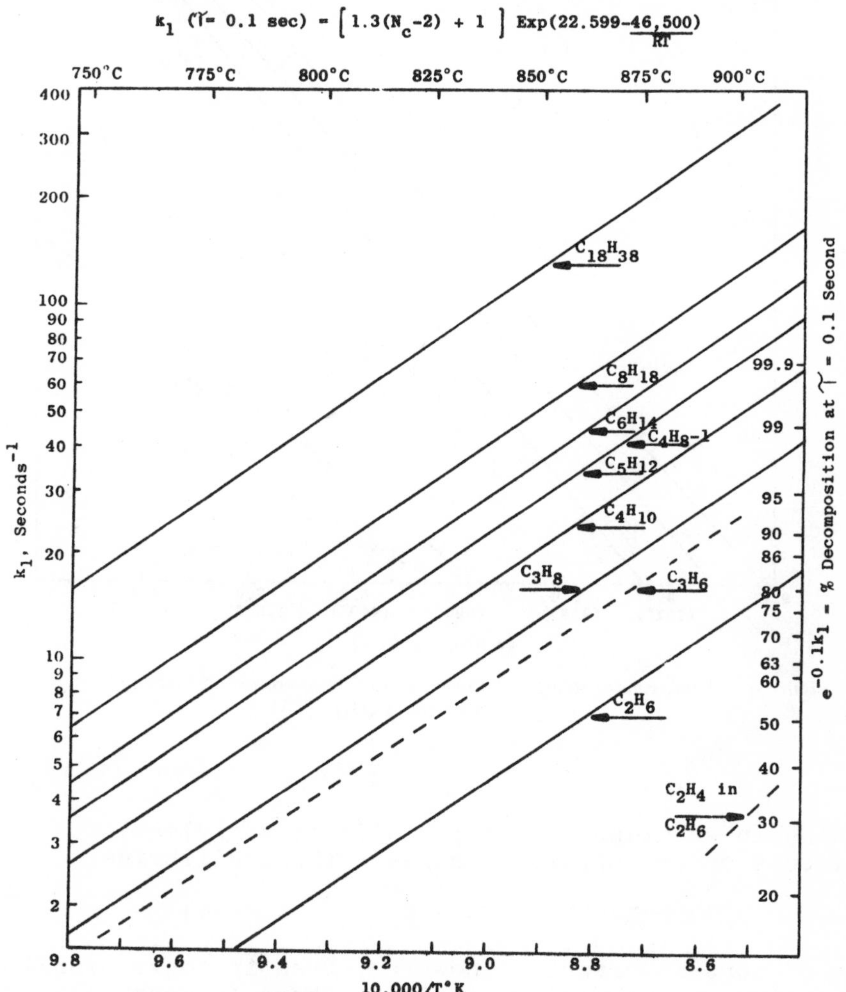

Figure 2. Dependence of rate of pyrolysis on temperature

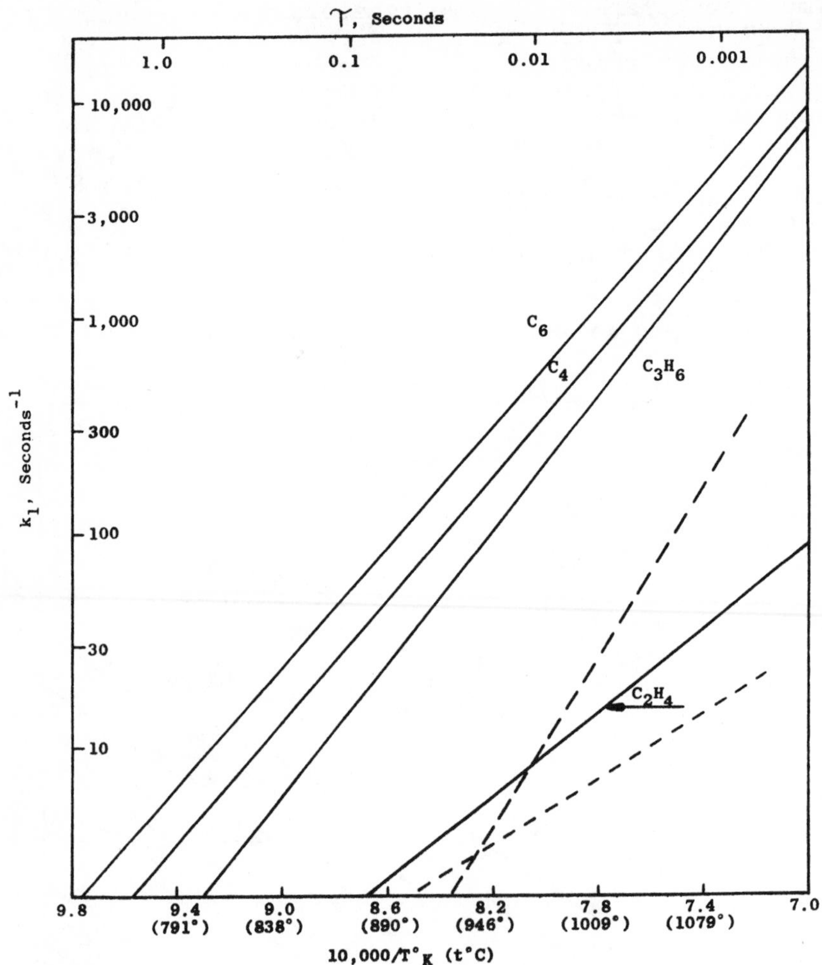

Figure 3. Approximate relation between k_1 *and temperature at constant severity (% decomposition* $C_4H_{10} = 95$*)*

gains in ethylene (and especially in acetylene) to effects on decomposition rates. The equilibrium

$$C_2H_6 \rightleftharpoons C_2H_4 + H_2 \qquad 1)$$

is of some importance. However, even in conventional cracking ethane yields are low. There is generally more ethane product than the equilibrium constant predicts (ethane though minor is a primary product, and part of the ethylene is formed from it). The equilibrium

$$C_2H_4 \rightleftharpoons C_2H_2 + H_2 \qquad 2)$$

is much more limiting. High yields of acetylene are obtained only at relatively high temperatures and (generally) low partial pressures. Other equilibria, involving, for example, butenes, butadiene and vinyl acetylene, affect yields of some byproducts.

Reduction of both partial pressure and residence time at constant severity decreases the ratio of H to C in the gas product without much affecting the H/C ratio of the liquid (C_5+) product. Simple stoichiometry therefore says we must increase the total gas yield at the expense of liquid. This is illustrated by Figure 4.

A kinetic model of the cracking process, covering both conventional cracking and the ACR range, must include all the features of the decomposition reactions we have mentioned, plus the equilibrium effects, plus the kinetic and equilibrium relations among the liquid products. It is obviously highly complex. A detailed model is not necessary, however, to point towards the

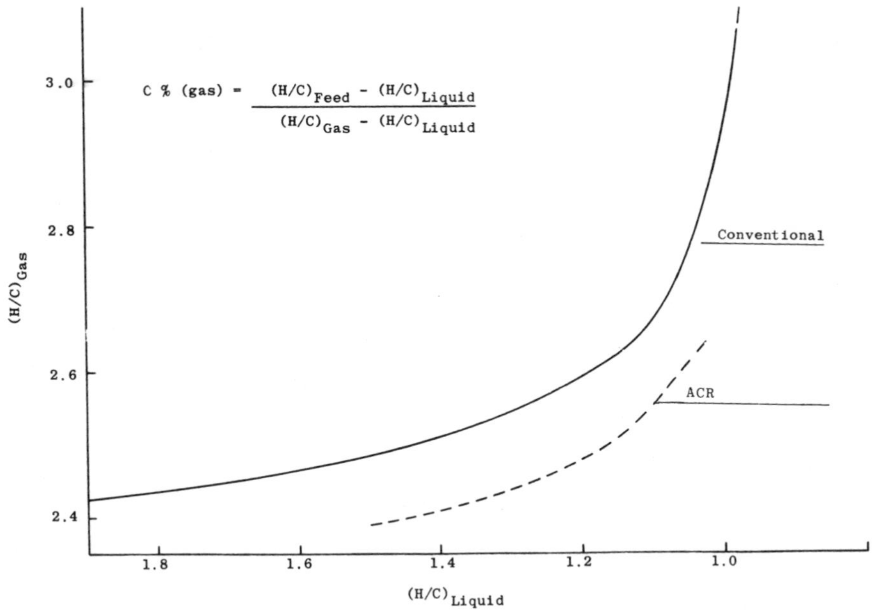

Figure 4. Typical relation between H/C atom ratios in C_4^- gas and C_5^+ liquid

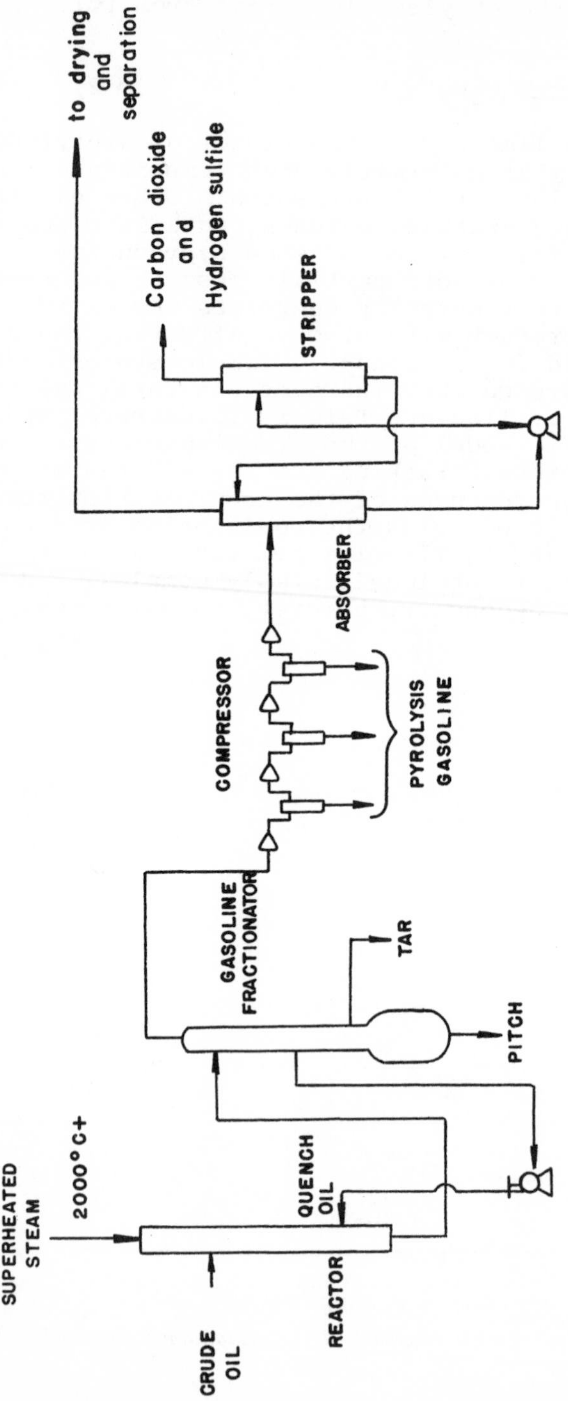

Figure 5. Schematic of crude oil cracking

need for a high-temperature, short-time regime for higher C_2 yields. It was on the basis of semi-quantitative arguments like those presented here that we started work on short-time, high-temperature processes many years ago.

Advanced Cracking Reactor, The ACR Process

Figure 5 shows a simplified schematic diagram of the process. Preheated hydrocarbon feedstock is injected as fine droplets into a high velocity stream of highly superheated steam (or steam plus other inerts). Conditions for extremely rapid mass mixing and thermal mixing, as well as almost instantaneous vaporization, are maintained. After a short residence time for pyrolysis, product gases are quenched to stop reaction and decomposition of desired products. Actual length of time needed for reaction will depend on the products desired, ranging from a very few milliseconds in the high acetylene regime needed by Kureha to perhaps 5 to 50 ms for product mixes more widely used in the petrochemical industry.

Figure 6 and 7 - taken from an earlier Kureha publication (2) - indicate the rates of appearance and disappearance of some of the major products in different time and temperature ranges. Greatest interest here is in the lower temperature range. We have converted the most important molar yields of Figure 7 into weight yields and show these as a function of time in Figure 8.

At the lower temperature level, we see that we have considerable leeway in the choice of product mix. If we fix, for example, the ethylene/propylene ratio, however, we fix severity and - as with conventional cracking - the other product yields from a given feedstock are now determined. Exceptions to this are minor dependencies - especially of acetylene - on the partial pressure variables.

It must be confessed that it is a little naive to talk of "temperature level". A heat source - steam - is mixed as rapidly as possible with a cracking stock which is heated at as close to zero time as we can achieve, reaching a maximum temperature which immediately starts to fall as pyrolysis proceeds. The maximum temperature is not easily measurable and the measurable temperatures after reaction are not always easily interpretable. What is generally known and controllable is the enthalpy added per unit mass of cracking stock, so that Figure 6 refers to a "higher enthalpy" regime and Figures 7 and 8 to a "lower enthalpy" regime.

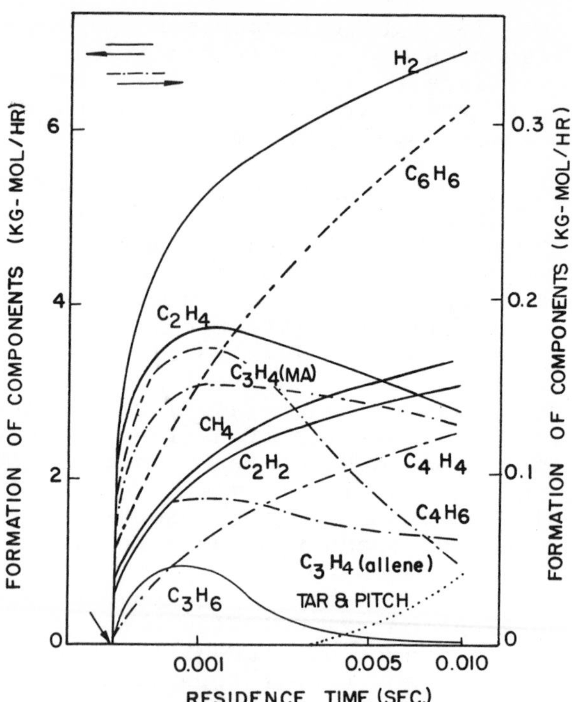

Figure 6. Formation of cracked gases (low ethylene to acetylene)

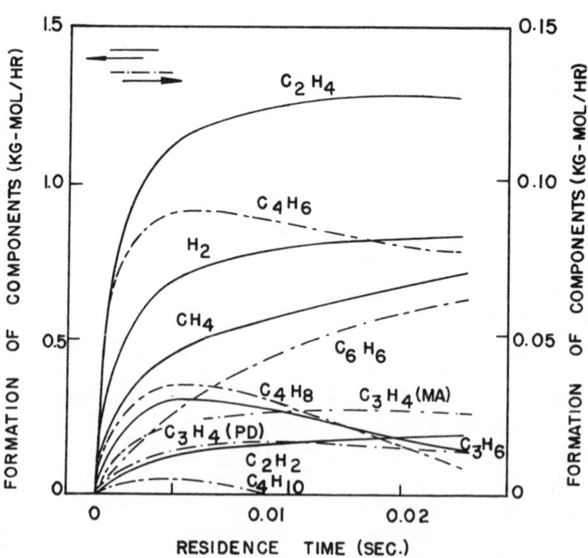

Figure 7. Formation of cracked gases (high ethylene to acetylene)

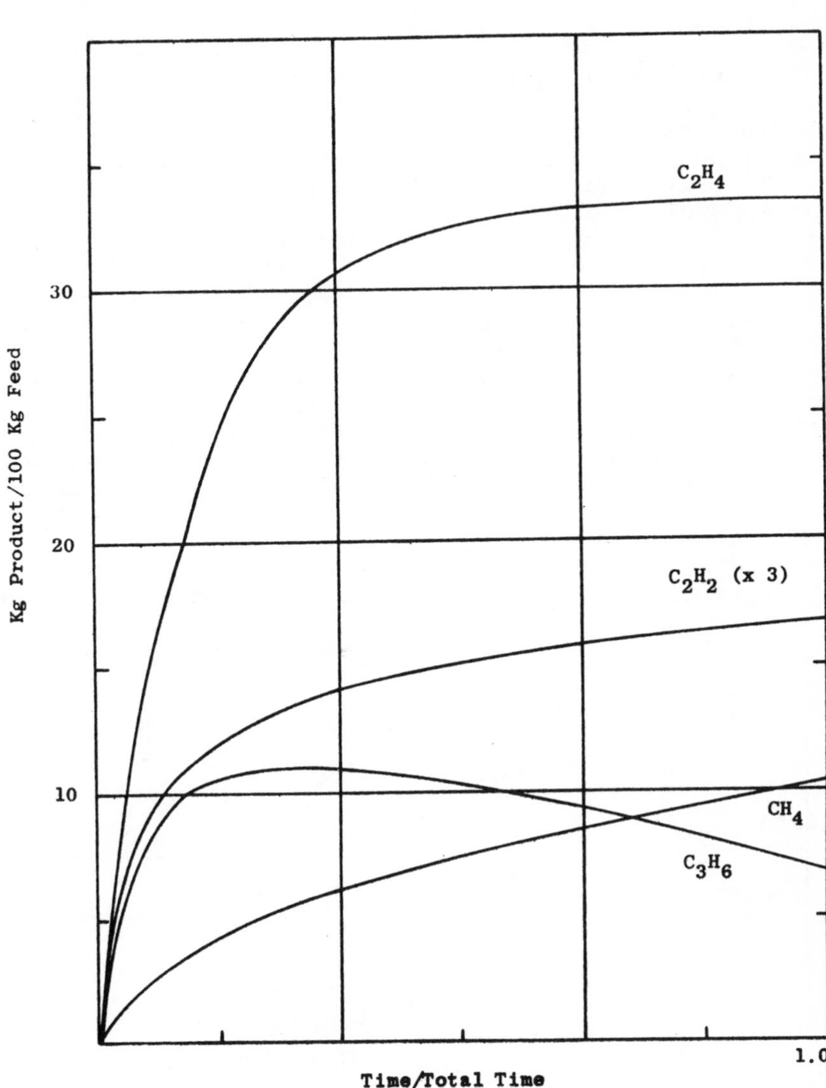

Figure 8. Rate of formation of major products at $CH_4/C_3H_6 = 4.0$ (yield vs. time/total residence time)

For each of the infinite number of enthalpy regimes, there is an optimum time corresponding to an economically optimum product mix. This optimum can be calculated either from actual rate data such as shown in Figures 6-8, or from a kinetic model of the process. It obviously will depend on the value of and demand for the various products. This ability to choose both enthalpy input and residence time gives us a degree of control not possible in conventional cracking furnaces.

Figures 9 and 10 show yields of major products as functions of severity, here measured by the CH_4/C_3H_6 molar ratio. Both time and temperature were varied to vary severity. For comparison we show yields by conventional cracking at about 1/1 by weight steam dilution. the yields shown were obtained by cracking a feedstock with a hydrogen/carbon atom ratio (H/C) of 1.89. This would correspond to a light mid-east crude oil, hydrogenated to reduce sulfur, or to a wide range distillate (naphtha through vacuum gas oil) from the same crude or to an atmospheric gas oil from one of these crudes. Since crude oils cannot be cracked in conventional plant furnaces, the data shown correspond to a high-quality atmospheric gas oil.

The ethylene yield curve is the most important, for it illustrates one of the two major reasons for our

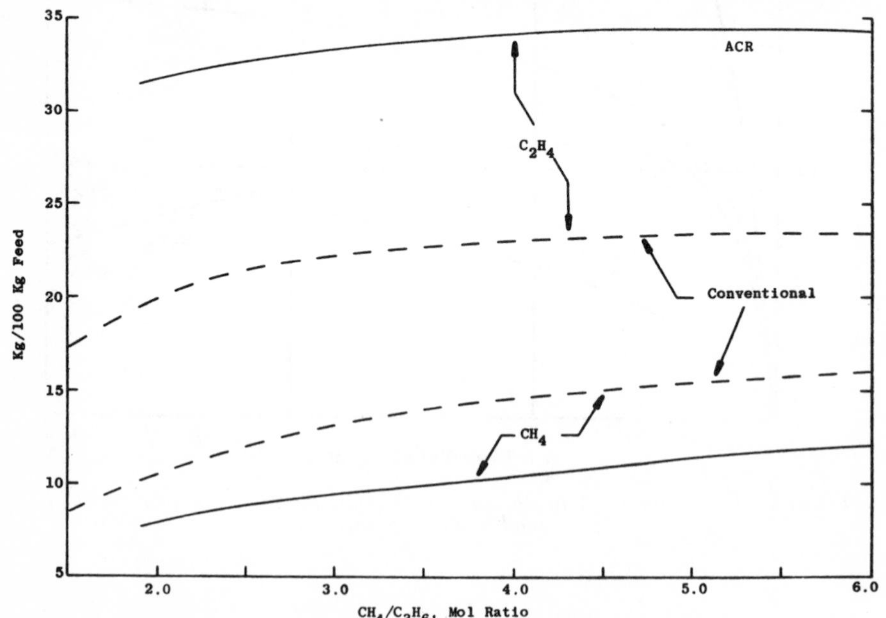

Figure 9. Single-pass yields, ethylene and methane

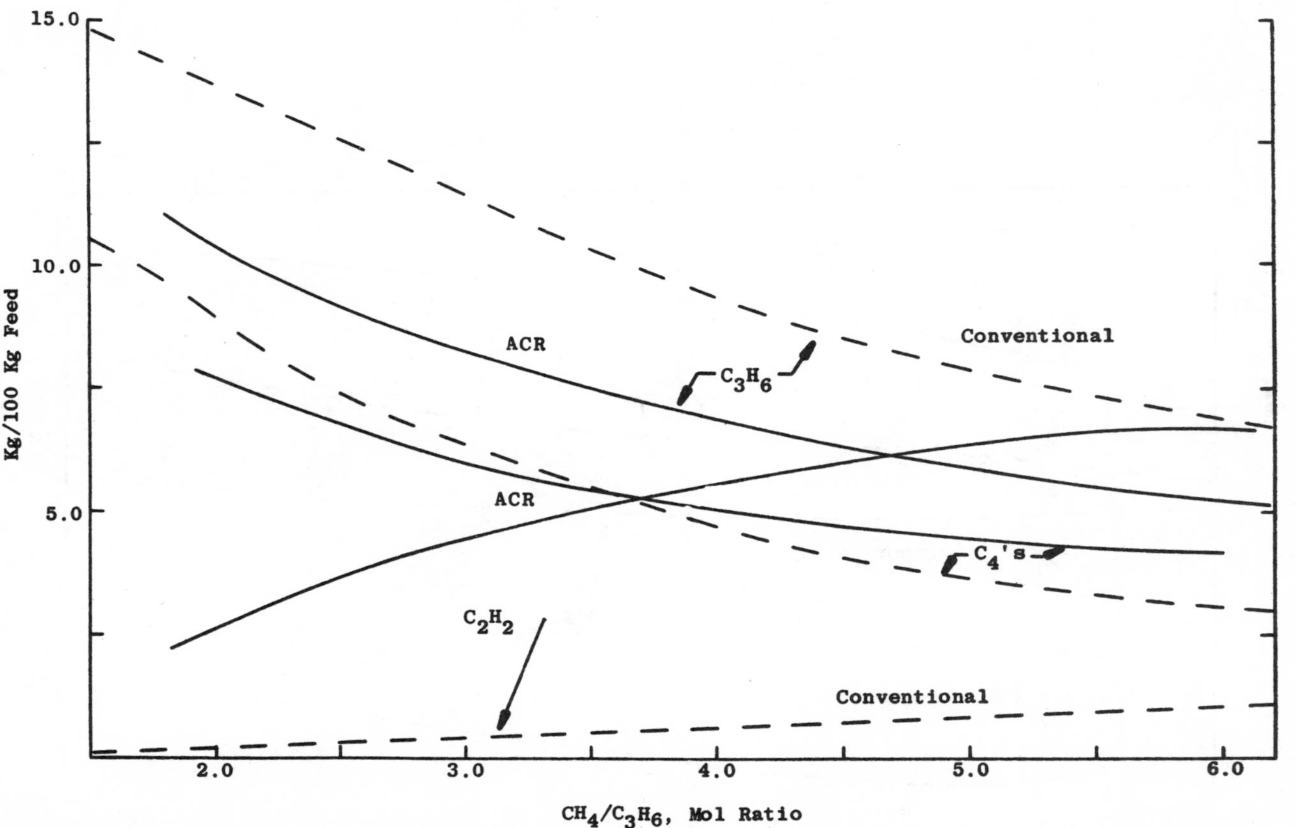

Figure 10. Propylene, acetylene, and C_4 yields

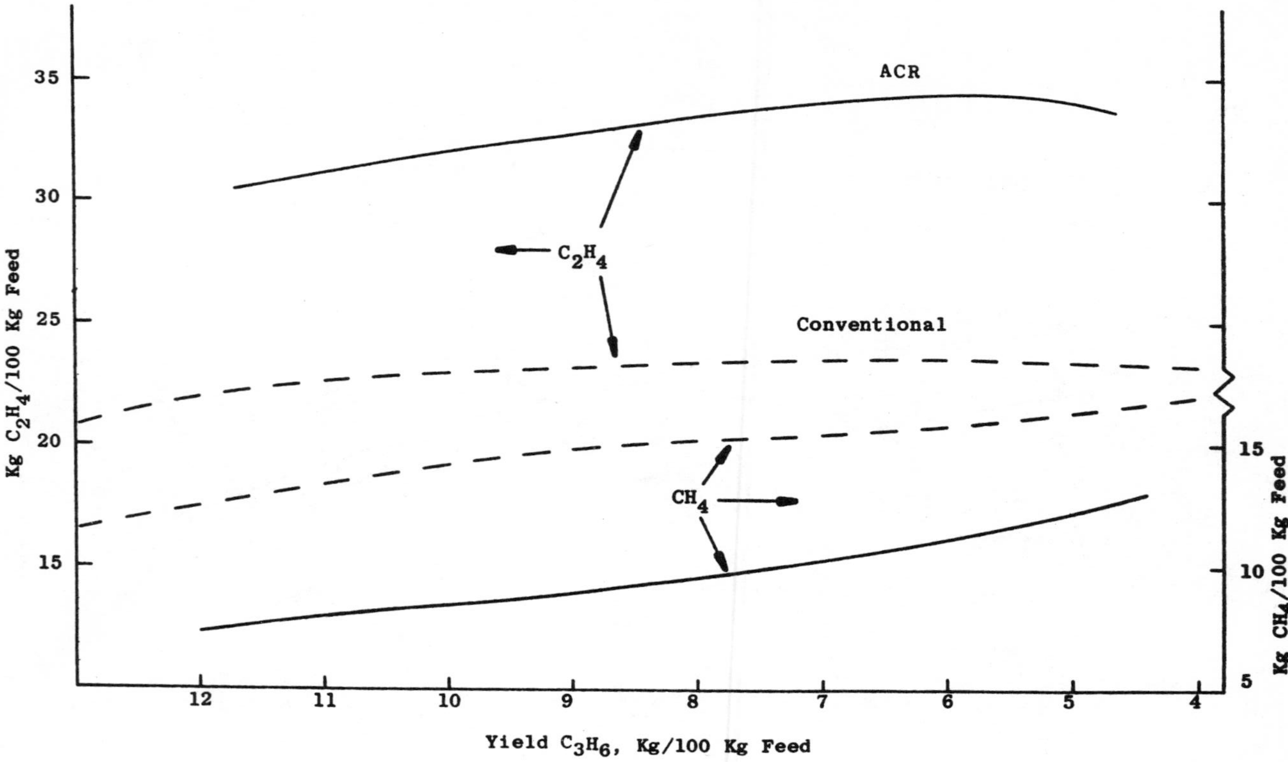

Figure 11. Ethylene and methane yield

great interest in the process (the other being the ability to crack crudes or other heavy feeds). ACR cracking, at constant CH_4/C_3H_6, gives some 10 kg/100 kg feed higher single-pass yield than does conventional cracking. This comes in part from reduced methane (see curve), ethane, and propylene (see curve). Surprisingly, little if any comes from reduced C_4's (see curve) and the C_4's are richer in butadiene than the C_4's of conventional cracking. When the ethylene yields are compared at constant propylene yield, most of the ethylene advantage remains (Figure 11), so that high ethylene yields are compatible with at least moderate propylene yields.

Acetylene becomes a much more important coproduct than it is in conventional cracking. It is indeed important enough to justify revival of acetylene-based processes now falling into disuse. We therefore tend to think in terms of ethylene plus acetylene yields. Figure 12 shows how this summary yield depends on feed-

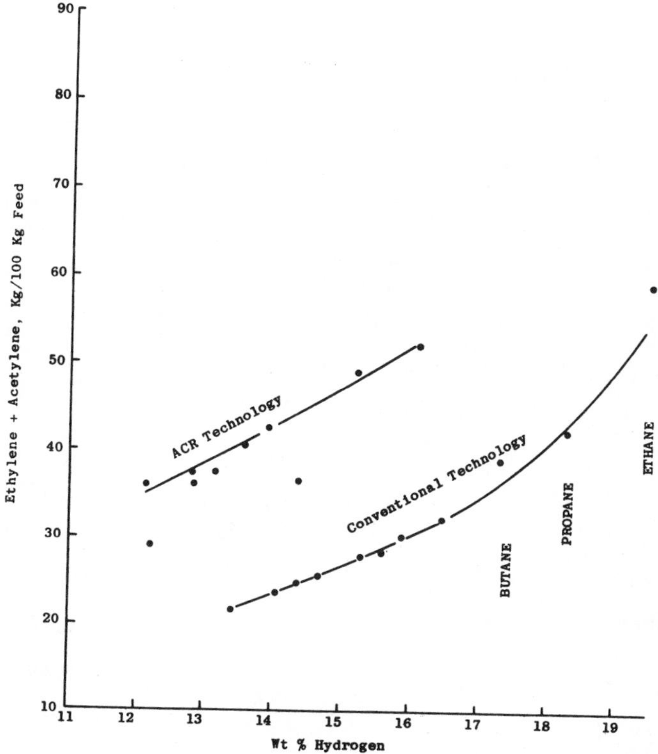

Figure 12. C_2 yield: comparison of ACR technology and conventional

TABLE III

CRUDE OIL VERSUS NAPHTHA CRACKING

RAW MATERIALS AND PRODUCTS

MM LB/YEAR

	Crude Oil Cracking	Naphtha Cracking
Raw Material		
Crude Oil	3050	(16,000)
Naphtha		3780
Products		
Ethylene	1000	1000
Acetylene	200	15
Propylene	135	590
Crude Butaidene	90	359
Crude BTX	435	990
Pitch	583	---
Fuel	600	840

TABLE IV

HOW CARBIDE PLANS TO CUT THE

COST OF ETHYLENE IN THE 1980's

(BUSINESS WEEK 24/8/74)

	Conventional Technology (Naphtha Feedstock) Cents/kgC_2H_4	ACR (Crude Oil Feedstock) Cents/kgC_2H_4
Raw Materials	53.2	26.8
Conversion and Overhead	13.0	17.6
Credit for Byproducts	(44.2)	(27.1)
Depreciation, Tax, Profit	12.1	14.5
Total	34.1¢	31.8¢

stock H/C ratio at constant severity. At all values of H/C, the difference (ACR-conventional) is strikingly large.

When a given quantity of feed is constrained to give more of some product, it must, of course, give less of something else. The increased C_2 yields are, as was pointed out, only partially compensated by drops in methane and propylene. As expected from the earlier discussion the total gas yields are raised by ACR cracking. Total liquid yields are reduced. Liquid yields and analyses are so dependent on feedstock and cracking severity that it is difficult to generalize. The "good" liquid products (benzene, toluene) are reduced along with the "bad" products (heavy fuel oils, dienes and other gum-formers). The overall quality of the liquid seems to be improved somewhat. For example, BTX (Benzene, Toluene, Xylenes) yield is reduced but the proportion of BTX in the C_6-C_8 fraction is increased and the proportion of B in BTX is increased. So there is less liquid coproduct, but it's a little better material.

Thus, the ACR has several attractions which warrant our continuing great interest in its development. Briefly these are flexibility in choice of feedstock, improved efficiency of ethylene plus acetylene formation and flexibility of choice of reaction time and temperature leading to flexibility of choice of product spectrum. These qualitative advantages are translated into a quantitative economic advantage. The great savings in raw material and the ability to optimize for a wide range of product mixes more than counterbalance a somewhat increased investment and operating cost and a reduced coproduct value. How this works out in one particular case is shown in Table III (4) and Table IV.

Literature Cited

(1) Araki, S., and Gomi, S., "Pyrolysis of Hydrocarbons Using Superheated Steam (2000°C) as Heat Carrier", World Petroleum Congress, Moscow, September, 1971.

(2) Williamson, K. D., Union Carbide Corporation, South Charleston, W. Va., unpublished work.

(3) Davis, H. G., and Farrell, T. J., Ind. and Eng. Chem. Process Design and Development, 12, 171-181 (1973).

(4) Wilkinson, L., and Gomi, S., "Crude Oil Cracking for Olefins", Presented at National Petroleum Refiners Association 72nd Annual Meeting, Miami, Florida, March 31-April 2, 1974.

23

Pyrolysis Gasoline/Gas Oil Hydrotreating

C. T. ADAMS and C. A. TREVINO

Shell Development Co., Westhollow Research Center, P.O. Box 1380, Houston, Tex. 77001

The use of heavier hydrocarbons as feeds to pyrolysis units for production of light olefins has increased significantly in recent years. Stocks such as atmospheric gas oils and even heavier fractions are key raw materials for several large-scale plants each with a capacity for producing over a billion pounds per year of ethylene. A highly significant feature of the use of the heavier feedstocks is the by-product yield structure. Table I based on data of Zdonik et al (1), presents yield data for cracking of naphtha, gas oil and vacuum distillate stocks. From these data, it is apparent that the user of the heavier feedstocks must be prepared to deal with substantial quantities of C_5 plus liquid pyrolyzate. The liquid pyrolyzate is rich in diolefins, styrenes, and aromatics and its by-product value potential is substantial. For economic reasons, one may not choose to recover all of these by-products. Table II itemizes some of the potential uses and process requirements for components of the liquid pyrolyzate. Since the pyrolysis liquid is rich in diolefins, styrenes, and other reactive species, it is highly unstable and tends to form polymers or gums when heated. Hydrogenation as a means of stabilizing the liquid is a key element in the downstream processing of the pyrolyzate and is of particular significance in optimization of by-product values.

Several processes have been described for hydrotreatment of pyrolysis liquids over the past twenty-five years (2, 3, 4, 5, 6). Most of the processes outlined in the literature have concentrated on hydrorefining of pyrolysis gasoline derived primarily from pyrolysis of naphtha or lighter stocks. The trend to cracking of heavier feeds has provided the incentive for development of new techniques for handling the increased volume of pyrolysis gasoline and gas oil. Further incentive has been provided by the fact that gasolines from gas oil pyrolysis have higher concentrations of olefins, diolefins and sulfur compounds than those derived from naphtha and lighter stock. (4)

TABLE I

TYPICAL ONCE-THROUGH YIELDS FROM PYROLYSIS OF HEAVY FEEDSTOCKS

(Range in liquid by-product yields from typical liquid feeds)

	Naphtha	Gas Oil	Vacuum Distillate
Yield Range, %wt			
C_5 - 204°C Gasoline	22-27	20-21	17-19
Pyrolysis	3-6	19-22	21-25

TABLE II

POTENTIAL USES OF PYROLYSIS BY-PRODUCT LIQUIDS

Component	Uses	Process Requirements
C_5 Cut	petrochemicals, e.g., isoprene, cyclopentadiene, cyclopentene	fractionation extraction selective hydrogenation
C_5 - 213°C Gasoline	high octane motor fuel blending	hydrogenation (selective)
C_6-C_8 Cut	benzene, toluene, xylenes source	hydrogenation (full olefin saturation) fractionation extraction
Pyrolysis Oil 213°C plus	fuel oil, fuel oil blending, hydrocracker feed, carbon black production	fractionation hydrotreatment

During the past several years Shell has carried out development studies on the hydrotreatment of pyrolysis liquids from gas oil cracking. Prime goals in the program were to minimize potential fouling problems and to optimize by-product values. This article is concerned with the hydrorefining of combined pyrolysis gasoline/gas oil streams and describes a process scheme which provides a high degree of flexibility in handling the liquid pyrolyzate.

Process Considerations

Process reliability, particularly as reflected in on-stream time, is a major consideration in selection of any process for hydrotreatment of pyrolysis by-products. The on-stream factor is of increased importance in the case of the increased volume and of the highly reactive nature of the liquid pyrolyzate. The reactivity of the pyrolyzate from gas oil cracking has two significant aspects. The first part concerns the increased concentrations of reactive species (diolefins, etc.), and the influence of these materials on the fouling tendency of the liquid. The tendency of pyrolysis liquids to form deposits and foul heat exchange equipment and catalysts quickly at elevated temperatures is well known and long has been a major concern as regards process reliability. Watkins (2), for example, has reported that exposure of the liquid to normal hydrodesulfurization conditions of 288-371°C could result in deposits amounting to 2%w of the feed charge. Such deposition rates would shut down a plant within 24 hours. Significant formation of polymeric deposits occur at much lower temperatures in the processing of pyrolysis liquids.

Table III presents data on coking tendencies of by-product liquids from gas oil pyrolysis. The data show that the pyrolysis gasoline is the most reactive of the components with significant formation of deposits at a temperature of 177°C. The gas oil, which contains fewer diolefins and alkenyl aromatics than the gasoline, is more stable and may be viewed as a diluent for the gasoline; i.e., the blend of pyrolysis gasoline and gas oil can be processed at a higher temperature than the gasoline alone. The pyrolysis gas oil is, however, a reactive liquid and can form significant deposits at relatively mild conditions. Therefore, for many applications, the gas oil requires hydrotreatment prior to further processing. The combined processing of the gasoline/gas oil thus appears to offer an advantage relative to the processing of the individual components.

A second aspect of the reactivity of the gas oil pyrolyzate is the impact it has on catalyst performance. The catalysts in many of the hydrorefining schemes developed for pyrolysis gasoline are particularly sensitive to control of the distillation end point and to the inclusion of substantial quantities of sulfur,

TABLE III

COKING TENDENCIES OF BY-PRODUCT LIQUIDS FROM GAS OIL PYROLYSIS

Stream	Temperature of Incipient Formation of Deposits*
Pyrolysis Gasoline	177°C
Pyrolysis Gas Oil (204-371°C Cut)	260°C
Pyrolysis Gasoline/Gas Oil (yield proportions)	204°C

*Deposits measured on stainless steel inserts exposed to the pyrolysis liquids under typical flow conditions of a hydrorefining unit.

nitrogen, and oxygen containing compounds that are found in gas oil pyrolyzate. (4)

These considerations led us to studies of the combined processing of pyrolysis gasoline and gas oil. Catalyst performance, in particular, was recognized as a critical factor. When it was determined that catalyst life of available commercial hydrotreating catalysts was too short to give the desired plant onstream time, catalyst development became an integral part of the effort.

Process Description

Figure 1 outlines a multiple bed hydrotreating scheme which employs temperature staging for processing of a combined pyrolysis gasoline/gas oil stream. For ease of description, the plant can be divided into two sections, hydrotreating and product separation. The basic elements in the hydrotreating sections are a trickle phase reactor train with two reaction stages, a third stage-vapor phase reactor (which processes only gasoline boiling range materials), quench and recycle oil systems, and a hydrogen system to provide make-up and recycle hydrogen.

The primary purpose of the first of the three stages (Stage 1) is conversion of all diolefins in the feed. The objective of the second stage is to saturate mono-olefins in the gasoline and sufficient unsaturates in the gas oil to make the oil thermally stable. The third stage is designed to desulfurize the gasoline portion, for example, to provide a sulfur-free concentrate for aromatics recovery.

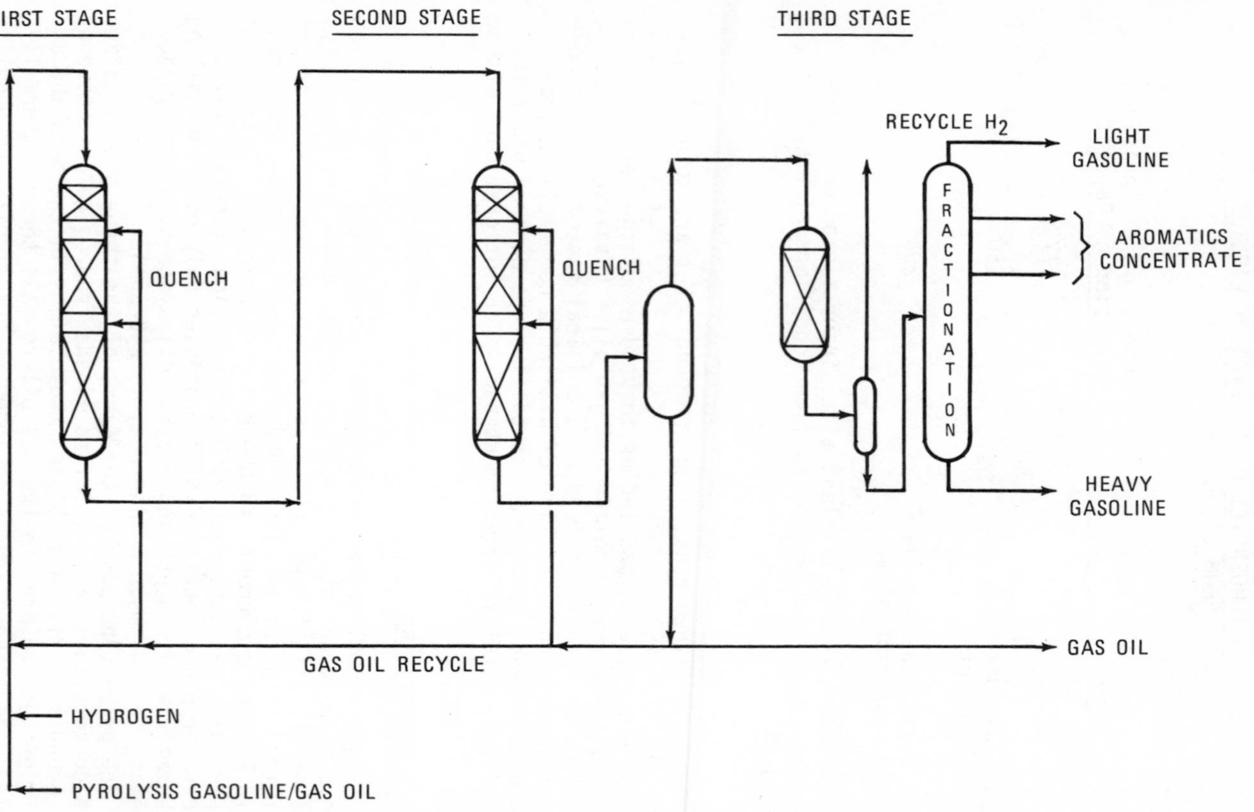

Figure 1. Flow schematic of three-stage hydrotreater for full saturation of olefins

Figure 2 outlines a hydrotreater of similar design as presented in Figure 1 but with the basic difference that selectively hydrotreated gasoline or its components are recovered after Stage 1 processing.

In both cases the pyrolysis gasoline/gas oil feed enters at the top of the first reactor. A recycle stream of hydrotreated gas oil is injected with the feed. The recycle streams serve as a reactant diluent, a heat sink to aid in reactor temperature control and as a solvent for polymer removal. Multiple catalyst beds are employed in the reactors to aid in temperature control. Quench oil is injected between the beds for reaction heat control. The first stage is operated at temperatures in the range of 107-177°C and at hydrogen partial pressures of the order of 48-68 atmospheres.

After diolefin conversion in Stage 1, the products and hydrogen are heated to 232°C prior to entering Stage 2. In Stage 2, which also employs multiple beds, olefins are saturated and substantial sulfur (about 70%) is removed from the gas oil. Reaction quench between beds is obtained by recycle of cooled hydrogenated gas oil.

The catalyst employed in the first two stages is a proprietary nickel oxide-molybdenum oxide/alumina formulation developed for this service.

Separation of the gasoline (or other fractions) and gas oil can be effected after either the first or second stage. Stage 3 processes either the full range gasoline from Stage 2 or alternatively an aromatics cut. The feed is heated and vaporized before entering Stage 3 at temperatures of the order of 316-343°C. Hydrogen pressure in the reactor is approximately 44 atmospheres. A lower pressure could be used as far as the process is concerned but the high pressure allows use of one compressor for all three stages. The third stage reactor has only one catalyst bed and quench is not required. Aromatics hydrogenation also is minimal in this stage. The catalyst for the third stage is a cobalt oxide-molybdenum oxide/alumina formulation.

Performance Tests

Product Properties. Properties of gasoline products from the first and second stage operations are compared with those of the feed in Table IV. The diene (measured by maleic anhydride value) and bromine number (a measure of mono-olefins) data show that essentially complete diolefin hydrogenation is obtained with substantial mono-olefin retention in the first stage processing. The quality data indicate the selectively hydrotreated gasoline is a valuable high octane blending component. The data presented for processing the gasoline/gas oil blend through both the first and second stages indicate that olefin saturation is complete. Gas-liquid chromatographic analyses of feed and product C_6-C_8 hydrocarbons showed little or no saturation of the

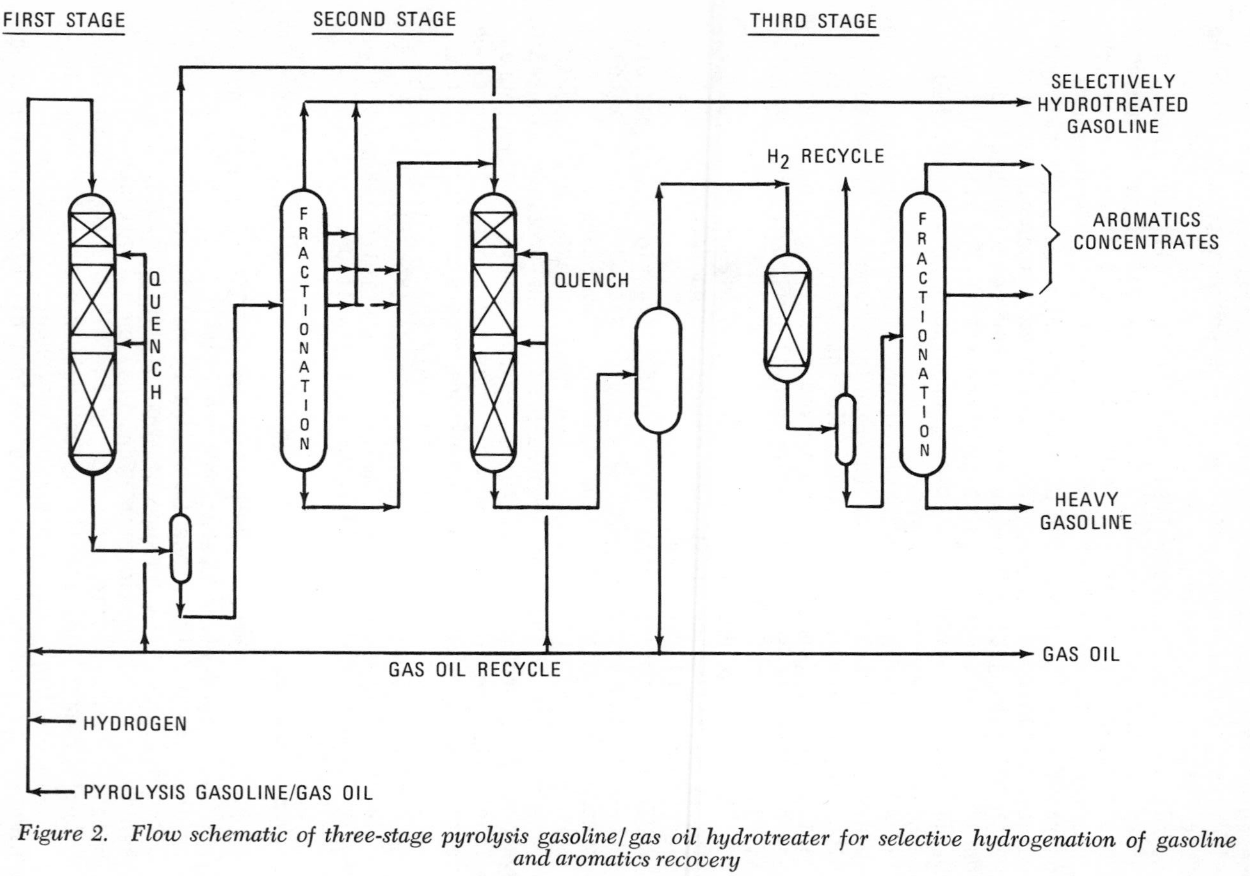

Figure 2. Flow schematic of three-stage pyrolysis gasoline/gas oil hydrotreater for selective hydrogenation of gasoline and aromatics recovery

TABLE IV

PROPERTIES OF HYDROTREATED GASOLINE FROM COMBINED PROCESSING OF PYROLYSIS GASOLINE/GAS OIL

	Gasoline Feed	Gasoline Product 1st Stage	Gasoline Product 2nd Stage
Maleic Anhydride Value	199	0-2	0
Bromine Number	106	40-50	0
Existent Gum, mg/100 ml	-	1	1
Sulfur Content, %wt	.01	.01	<.01
Research Octane No. (Unleaded)	-	99	96

aromatics in this boiling range.

Even though olefin saturation is essentially complete in Stage 2, C_6-C_8 aromatics cannot be obtained directly from this stream because of the sulfur content. Processing of the gasoline from Stage 2 at Stage 3 conditions, however, readily yields aromatics of the desired quality. The benzene fraction, for example, from such processing has a sulfur content of 0.3 ppm or less. The sulfur removal can be accomplished with high selectivity, i.e., very little or no aromatics saturation.

The gas oil derived from Stage 2 processing has been tested in a variety of services and found to be of comparable thermal stability to typical refinery stocks. The gas oil has been used in hydrocracker feed, as a blending component for fuel oil, and other services.

Influence of H_2S. The inclusion of hydrogen sulfide in the gas feed to the first stage has marked influence on catalyst performance and product properties. Table V compares the gasoline product from the processing of a combined gasoline/gas oil pyrolyzate in either the presence or absence of hydrogen sulfide. The inclusion of H_2S in the hydrogen feed results in markedly higher first stage temperature requirement to achieve the desired diolefin saturation. The most significant difference is in the sulfur content of the two products. The inclusion of H_2S in the H_2 feed results in a marked increase in sulfur in the gasoline product. Analysis of sulfur containing compounds show that most of the sulfur is present as mercaptans and disulfides. The reaction is believed to proceed sequentially with the first reaction being that of a diolefin with H_2S.

$$R\text{-}CH{=}CH\text{-}CH{=}CH_2 + H_2S \rightarrow R\text{-}CH{=}CH\text{-}CH_2\text{-}CH_2\text{-}SH$$

A disulfide is formed in a subsequent step through reaction of the mercaptan with the diolefin.

TABLE V

HYDROGEN SULFIDE IN RECYCLE GAS INCREASES PRODUCT SULFUR IN SELECTIVE HYDROTREATMENT OF PYROLYSIS GASOLINE

		Product Gasoline	
	Feed	0.5% vol H_2S in H_2 Feed	No H_2S in H_2 Feed
Average Reactor Temp., °C		166	121
Maleic Anhydride Value	191	< 1	< 5
Bromine Number	100	53	58
Total Sulfur, %wt	.03	.96	.02
Mercaptan Sulfur, ppm	40	3900	10

$$R\text{-}CH{=}CH\text{-}CH_2\text{-}CH_2\text{-}SH + CH_2{=}CH\text{-}CH{=}CHR \rightarrow (R\text{-}CH{=}CH\text{-}CH_2\text{-}CH_2)_2S$$

The substantial formation of sulfur compounds is detrimental to use of the gasoline in motor fuel blending. If the gasoline, with such mercaptans and disulfides present, is processed through the second stage, the sulfur is removed as hydrogen sulfide. Our proprietary catalyst can be operated either in the presence or absence of H_2S.

Catalyst Stability

Several pilot plant tests have been made with gas oil pyrolyzate and using Shell's proprietary catalyst in both Stages 1 and 2. Typical data from pilot plant processing of gasoline/gas oil pyrolyzate under relatively severe conditions (i.e., 1.5%v H_2S added to H_2 feed) are presented in Table VI. The results indicate good catalyst performance.

The hydrotreating scheme outlined in this paper has been used in Shell's commercial scale gas oil cracking plant at Houston for hydrogenation of the gasoline/gas oil pyrolyzate. The hydrotreater has been in operation for approximately 5 years on the initial catalyst charge. The Shell catalyst has been proven regenerable in both pilot and commercial scale operations.

Summation

This paper describes a process for hydrotreating of gasoline/gas oil pyrolyzate from gas oil cracking; this process has proved successful in commercial scale operation. The technology outlined overcomes some of the problems created by present day gas oil cracking, i.e., high volumes of liquid pyrolyzate which

TABLE VI

STABILITY OF Ni-Mo CATALYST IN PROCESSING COMBINED PYROLYSIS GASOLINE/GAS OIL

Feed: Maleic Anhydride Value 134

Bromine Number 90

Conditions: Pressure - 68 atm, 1.5% vol H_2S in H_2

Catalyst Age, Days	25	70
1st Stage		
Temperature, °C	149	149
Maleic Anhydride Value	7	7
2nd Stage		
Temperature, °C		270
Bromine Number	-	< 1

is of increased reactivity due to high concentrations of diolefins, styrenes, and other reactive species. The flexibility of the process is beneficial to realization of the full by-product value of the liquid pyrolyzate.

Abstract

The pyrolysis of gas oils for ethylene manufacture can yield four fractions:

Olefinic gases	C_4 and lighter
Pyrolysis gasoline	C_5 - 213°C
Pyrolysis gas oil	213-371°C
Pitch	371°C+

A versatile, multiple-bed hydrotreating process that employs temperature staging has been developed which permits hydrotreating of the combined gasoline/gas oil fractions. In the system the gasoline component can either be selectively hydrotreated to remove diolefins, yielding a high-octane gasoline blending component, or fully saturated for either aromatics recovery or control of gasoline sensitivity. The hydrotreated gas oil is sufficiently stable to be used as a hydrocracker feed, as cutter stock in No. 6 fuel, or directly as fuel.

Literature Cited

1. Zdonik, S. B., Bassler, E. J., and Hallee, L. P., Hydrocarbon Processing, (Feb. 1974), p. 73.
2. Watkins, C. H., "Advances in Hydrofining of By-Product Gasolines from Thermal Cracking in Ethylene/Propylene Manufacture", Elsevier Pub. Co., 7th World Petroleum Congress, Mexico City, April 2-9, 1967, pp. 207-215.
3. Sze, M. C. and Bauer, W. V., Chemical Engineering Progress (1969), 65 (2), p. 59.
4. Smith, J. I., Hurwitz, H., Hwa, F. C. S., and Yarze, J. C., Proceedings Division of Refining American Petroleum Institute (1969), 49, p. 473.
5. Espino, R. L., Boyurn, A. A., and Rylander, P. N., Proceedings Division of Refining, American Petroleum Institute (1970), 50, p. 98.
6. Derrien, M. L., Andrews, J. W., Bonnifay, P., and Leonard, J., Chemical Engineering Progress (1974), 70 (1), p. 74.

24

Kinetic-Mathematical Model for Naphtha Pyrolyzer Tubular Reactors

V. ILLES, O. SZALAI, and Z. CSERMELY

Hungarian Oil and Gas Research Institute, Budapest-Veszprem, Hungary

Most olefins of petrochemical interest are produced by thermal cracking of naphtha feed stock, yielding about 12-14 million metric tons/year ethylene[1]. A Hungarian olefin plant, completed in 1975, is also operated on a naphtha feedstock. Yields and relative amounts of the main products greatly depend on the qualities of the naphtha feedstock pyrolyzed and the parameters of the cracking operation[2,3]. A detailed study of the pyrolysis is, therefore, of great industrial significance.

Calculation of the ethylene yield of naphtha cracking using the severity function introduced by Linden et al.[4] is wide spread. Recently, Zdonik et al.[5] introduced the so called kinetic severity function (KSF) characterizing the cracking severity with the n-pentane conversion achieved within the experimental conditions, or, alternatively, with the term $\int K_5 dT$ calculated for the n-pentane from the first order kinetic equation. Several scientists including [6] and [7] have accepted this method. There are a few publications dealing with the statistical mathematical models, based on the results of experimental naphtha feedstock cracking operations[8,9].

This present paper presents the kinetic-mathematical model developed to describe the overall decomposition rate and yields of the naphtha feedstock cracking process. The novelty and practical advantage of the method developed lies in the fact that the kinetic constants and yield curves were determined from experiments carried out in pilot-plant scale tubular reactors operated under non-isothermal, non-isobaric conditions and the reactor results could readily be applied to simulate commercial scale cracking processes as well. During the cracking experiments, samples were withdrawn from several sample points located along the reactor. Temperature, as well as pressure were also monitored at these points[2,3].

Several investigators discuss the determination of the overall decomposition kinetics of hydrocarbons on the basis of non-isothermal experimental data. Kershenbaum and Martin[10], Kunzru et al.[11], and Leftin and Cortes[12], studied the pyrolysis pro-

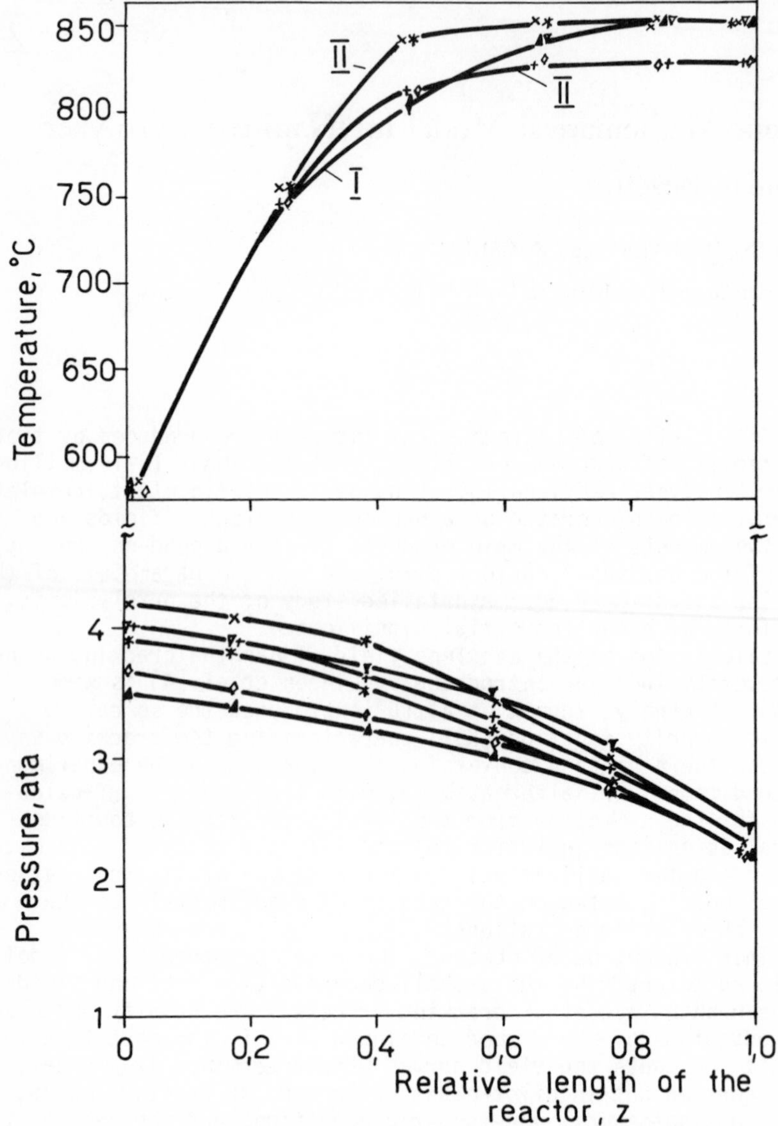

Figure 1. Naphtha cracking temperature and pressure as a function of relative length of the reactor

Expt. No.	*Symbol*	*Expt. No.*	*Symbol*
44	▲	*47*	◇
45	▽	*48*	×
46	+	*49*	*

cess of propane, n-nonane, and isobutene, respectively in bench scale flow reactors. Experiments were conducted with identical temperature profiles and feed-rates but different hydrocarbon partial pressures. Samples were taken only at the reactor outlet. According to our best knowledge, a descriptive method has never been published that is based on the true decomposition processes of naphtha feedstocks.

A kinetic-mathematical model is presented here that was developed based on experimental pyrolysis data obtained on straight run Romashkino (Soviet Union) crude naphtha cuts. Characteristics of this naphtha feedstock are summarized in Table 1.

Table 1

Characteristics and group analysis data of the naphtha feedstock used for the pyrolysis experiments

Characteristics		Group analysis	weight %
Density, 20°C:		n-paraffins	30.9
0.7224 g/cm^3		iso-paraffins	29.5
Average molecular weight:		naphthenes	28.1
100 g/mole		aromatics	11.5
Engler distillation:			
initial temp., ^{o}C	50.5	hydrogen	
10% dist. temp.	80.2	content	14.68
50% dist. temp.	110.5	sulphur	
90% dist. temp.	137.0	content	0.035
end point temp.	174.7		

Pyrolysis experiments were conducted in the pressure, temperature, and residence time range of commercial interest, under 50 weight percent steam in the feed stream. Figure 1 presents the temperature and pressure profiles kept throughout the experiments. The six experiment series were made using two different temperature-profile shapes (I. and II.) and 820°C and 850°C reactor exit temperatures, respectively.

Description of Overall Decomposition Rate of Naphtha Feedstocks

Determination of the rate equation and rate constants. Numerical characterization of the decomposition of naphtha feedstocks was made with the so called "degree of decomposition" term, introduced formerly by the authors to characterize the decomposition of hydrocarbon mixtures[13], defined for naphtha feedstocks as follows:

$$X = \sum_{j}^{K} y_j x_j \qquad (1)$$

where: y_j is the mole fraction of the decomposing components of the initial naphtha feedstock

x_j is the conversion of the decomposing components of the initial naphtha feedstock.

According to our experiences, aromatic constituents of the initial naphtha feedstock do not decompose in the early stages of the reaction (their quantity is practically unchanged); therefore, for the present, they can well be considered as inert components. That means that their mole fractions in Equation 1 are equal to zero. (Degree of decomposition calculated for aromatics-free naphtha feedstocks.) Consequently, the degree of decomposition of naphtha feedstocks lies in the range of 0 - 1, just as do the conversion values obtained while pyrolysing pure compounds.

Based on the measurements made at the inlet, exit, and intermediate test points of the tubular reactor, the following data are available to determine the kinetic rate constants:

a) Curves of conversion [X(z)] and gas expansion [E(X)].
b) Temperature of the gaseous material flowing in the reactor, measured at discrete points, [T(z)].
c) Pressure in the reactor, measured at discrete points, [P(z)].

The following differential equation[14] describes the overall decomposition rate of naphtha feedstocks (that is, the change of the degree of decomposition with the relative length of the reactor):

$$\frac{C^0}{\omega^0}\frac{dX}{dz} = K\left[\frac{C^0(1-X)}{E_e}\right]^n \qquad (2)$$

where X the degree of decomposition defined in Equation 1,

z the relative length of the reactor

ω^0 the apparent reaction time, calculated from the governing parameters at the inlet point of the reactor, seconds

C^o the overall concentration of the decomposing components of the naphtha feedstock at the inlet of the reactor, mole/liter

k the decomposition rate constant of the naphtha feedstock investigated, $(\text{mole/liter})^{1-n}/\text{sec.}$

n the overall kinetic order

E_e the effective expansion for the degree of decomposition X.

The effective expansion is the quantity representing the joint effects of the change of the moles caused by the chemical reaction (chemical expansion, E_R), and the change of the temperature and pressure (physical expansion, E_f):

$$E_e = E_R \cdot E_f = \frac{E(X) + s}{1 + s} \cdot \frac{T\,P^o}{T^o\,P} \tag{3}$$

where: E(X) the hydrocarbon expansion belonging to the degree of decomposition X

s the amount of the steam feed relative to that of the naphtha feedstock, mole/mole

T^o the temperature of the reaction mixture at the inlet of the reactor, ^oK

T the temperature of the reaction mixture in the reactor, ^oK

P^o the pressure of the reaction mixture at the inlet point of the reactor, ata

P the pressure of the reaction mixture in the reactor, ata.

The C^o, ω^o, T^o and P^o are constants derived from measurements. The $T = T(z)$, $P = P(z)$, and $E = E(X)$ functions are approximated by the following empirical functions:

The temperature profile:

$$T = a \log \frac{z + b}{b} + cz^3 + T^o$$

and

$$T + a[1 - \exp(-bz)] + T^o \tag{4}$$

respectively.

The pressure profile:

$$P = P^o \sqrt{1 - az^2 - bz} \tag{5}$$

The expansion:

$$E = 1 + aX + \frac{bX}{c-X} \tag{6}$$

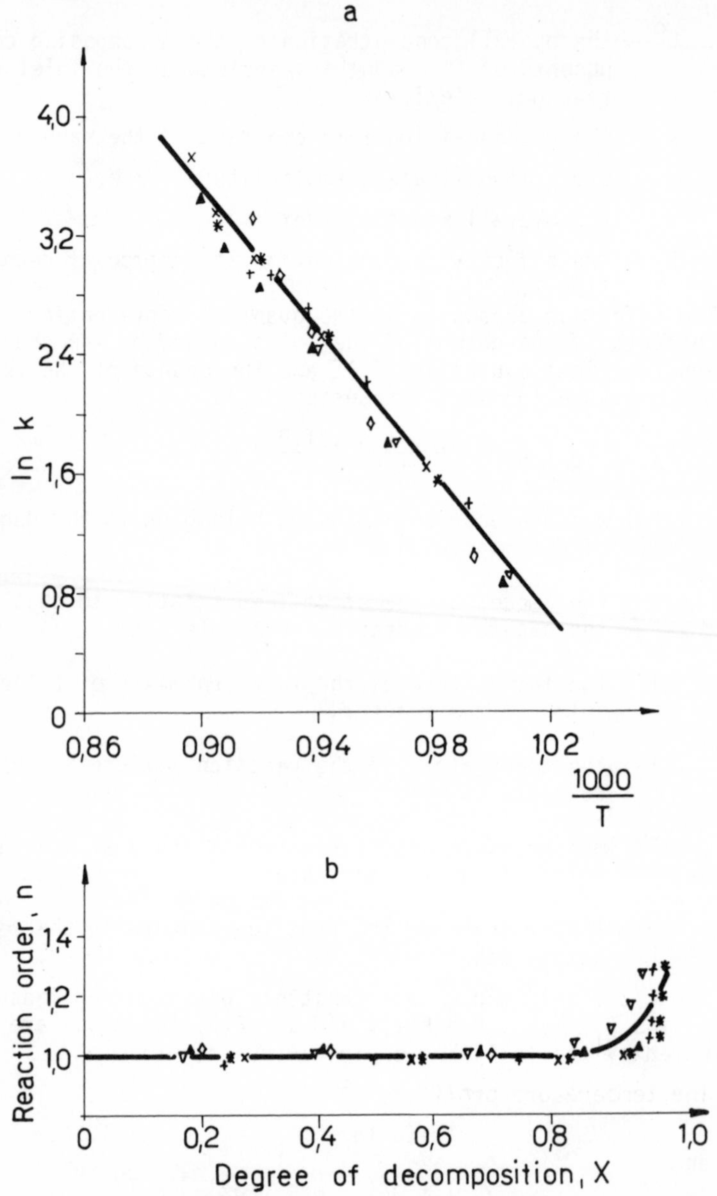

Figure 2. Dependence of the naphtha cracking decomposition reaction rate constant on temperature (a); dependence of the naphtha cracking overall kinetic order on conversion (b)

Expt. No.	*Symbol*	*Expt. No.*	*Symbol*
44	▲	*47*	◇
45	▽	*48*	×
46	+	*49*	*

The least square fitting method was applied to determine the constants of the above functions from the experimental data.

The differential equation (2) can not be solved analytically; therefore a numerical method, assuming values for the kinetic constants, had to be applied. The so called "differential method", extended for the case of non-isothermal, non-isobaric reactions, was applied to determine the $K(A, E_a)$ and n kinetic constants.

To do so, the differential equation (2) was rewritten in logarithmic form:

$$\ln\left[\frac{C^o}{\omega^o}\cdot\frac{dX}{dz}\right] = \ln k + n\ln\left[\frac{C^o(1-X)}{E_e}\right] \qquad (7)$$

To determine the $\frac{dX}{dz}$ values of Equation 7, the measured X(z) values were approximated with the tangent hyperbolic function:

$$X = c[\text{th } a(b-z) + \text{th } ab] \qquad (8)$$

Then, applying the n = 1 approximation, the ln k values corresponding to the different values of the X(z) curve (that is, to the different temperature settings), were determined from Equation 7. The ln k values thus obtained were plotted - according to the Arrhenius equation - against the reciprocal temperature values. As shown on Figure 2, a straight line can be fitted to the points of the six experiments within the $0.15 < X < 0.90$ range of the degree of decomposition values. The slope of that line yields the activation energy, E_a, while the intercept leads to the pre-exponential factor, A. (Figure 2/a does not contain the points above the $X > 0.9$ limit.) Substituting these values into Equation 2 and solving the differential equation over the range of the experimental parameters with an appropriate (e.g. Runge-Kutta) numerical method, the calculated X(z) curves are obtained. The calculated curve runs higher than the experimental values in the $X > 0.9$ range indicating that the decomposition rate - due to the inhibiting effect of the reaction products - is slower than that predicted by the first order reaction kinetics.

To increase the accuracy of the description, the overall kinetic order, n, was considered in that range as a variable depending on the degree of decomposition. Values of n corresponding to different conversion rates were determined from Equation 7 applying the E_a and A values obtained from Figure 2/a and are plotted on Figure 2/b.

From Figure 2 it can be concluded, that the overall decomposition rate of the naphtha feedstock investigated follows the first order kinetics over the $0.15 < X < 0.90$ degree of decomposition range. To achieve more accurate a description in the higher degree of decomposition range, an overall kinetic order not equal to unity and increasing with X, was applied.

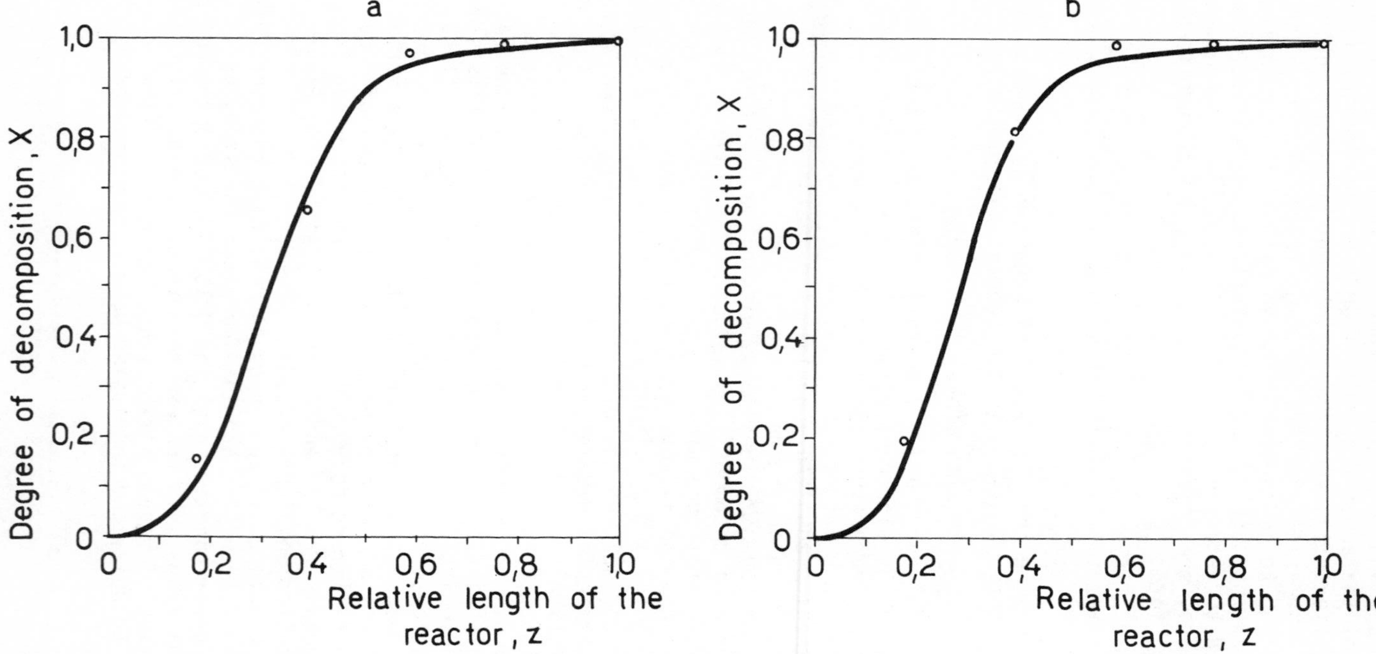

Figure 3. Change of the degree of decomposition with the relative length of the reactor in the naphtha cracking experiments No. 44 (a) and No. 49 (b), respectively. ——— full line: calculated curve.

Thus the kinetic constants of the naphtha feedstock investigated are as follows:

Preexponential factor: $A = 8.59 \cdot 10^{10}$
Activation energy: $E_a = 47{,}800$ cal/mole
Overall kinetic order: $n = 1 + 0.25\ X^{20}$.

To demonstrate the method, Figure 3 presents the measured and calculated X(z) values for the experiments No. 44 and No. 49. The good agreement obtained between the measured and calculated values proves that the method described is readily applicable to describe the overall decomposition rate and to determine the kinetic constants.

<u>Calculation of the overall kinetic severity function and the true residence time</u>. Differential equation 2 can advantageously be rewritten in the following form, within the integration limits proper [13, 14]:

$$\int_0^X \frac{dX}{(1-X)^n} = \int_0^z \left(\frac{C^o}{E_e}\right)^{n-1} k \frac{\omega^o\, dz}{E_e} \qquad (9)$$

The integral term on the right hand side describes the cracking severity of the naphtha feedstock (characterized by the overall decomposition rate constant $k(A, E_a)$ and the overall kinetic order n) achieved over the 0-z part of the reactor, under the given temperature and pressure profile and feed-in parameters, C^o, ω^o. Corresponding to that severity there is a decomposition of determined depth, described by the integral term on the left hand side.

The integral term on the right side of Equation 9 is called the overall kinetic severity function, BKSF:

$$\text{BKSF} = \int_0^z \left(\frac{C^o}{E_e}\right)^{n-1} k \frac{\omega^o\, dz}{E_e} \qquad (10/a)$$

or, considering that

$$\frac{\omega^o\, dz}{E_e} = dT$$

it reads as:

$$\text{BKSF} = \int_0^T \left(\frac{C^o}{E_e}\right)^{n-1} k\ dT \qquad (10/b)$$

If the degree of decomposition can accurately be measured with sufficiency over the entire conversion range, the value of the overall kinetical severity function can be calculated according to the left hand side of Equation 9. In the special case, where n = 1, the overall kinetic severity function becomes BKSF = -ln (1-X).

The purpose of introducing the overall kinetic severity function, BKSF, instead of the degree of decomposition, X, is, first of all, that due to the "scale expansion" effect, the yield curves can be more precisely calculated in the $X > 0.95$ range, which is of great significance in studying and describing commercial reactors.

The overall kinetic severity function (BKSF), just as the degree of decomposition, X, defined in Equation 1, is dimensionless, thereby allowing the kinetic model and the yield curves to be directly transferred to larger reactors (tube furnaces).

The true residence time, T in the kinetic model is calculated according to the following equation:

$$T = \int_0^T dT = \omega^0 \int_0^z \frac{dz}{E_e} \tag{11}$$

The kinetic model described, based on the numerical solution of the differential equation 2 can be readily applied to calculate the degree of decomposition (X), overall kinetic severity function (BKSF), and true residence time as a function of the relative length of the reactor on the bases of the feedstock parameters, temperature, and pressure profile.

Product Distribution of Plant-scale Naphtha Feedstock Cracking Operation, Description of Yield Curves

Change of the reaction products with the relative length of the reactor was determined from cracking experiments carried out in a pilot-plant scale system. (Sampling along the reactor made this determination feasible.) The number of the products was decreased to 25, grouping together those of similar molecular weight and structure. For convenience, yields were expressed in mole product/mole naphtha feed; using the respective molecular weights, conversion to weight per cent values becomes fairly simple.

Two basic variables have been defined in the previous section to quantify the degree of decomposition (severity of the pyrolysis). They are as follows:

a) degree of decomposition, X
b) overall kinetic severity function (BKSF).

Equations 9 and 10 relate these two variables and thus, having obtained any of them, the other can be calculated. That is to say that the two variables are interchangeable in describing the cracking process. However, considering practical points, the BKSF function has certain definite advantages.

To demonstrate the validity of our previous statement, Figures 4 and 5 present the ethylene, propylene, pentene and hexene

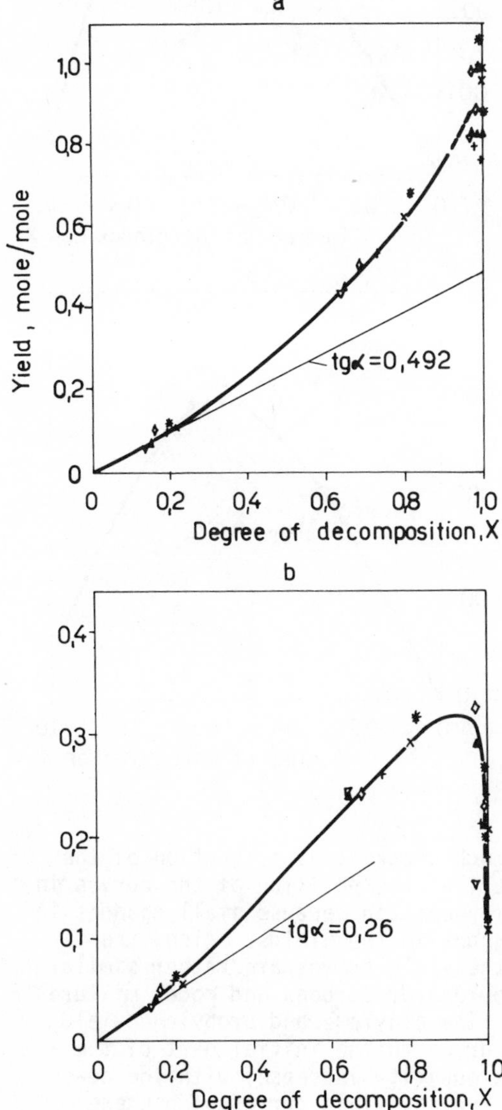

Figure 4. Change of ethylene yield (a) and propylene yield (b) of naphtha cracking with degree of decomposition

Expt. No.	*Symbol*	*Expt. No.*	*Symbol*
44	▲	*47*	◇
45	▽	*48*	×
46	+	*49*	*

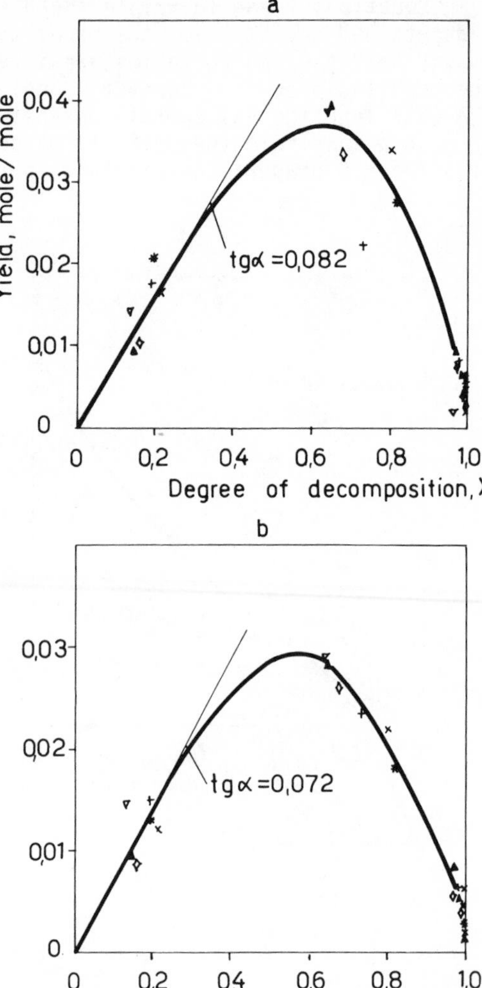

Figure 5. Change of the pentene yield (a) and hexene yield (b) of naphtha cracking with degree of decomposition

Expt. No.	*Symbol*	*Expt. No.*	*Symbol*
44	▲	*47*	◇
45	▽	*48*	×
46	+	*49*	*

yield curves of naphtha feedstock cracking as a function of the degree of decomposition. Note that the position of the curves in the X > 0.95 region is rather uncertain because small changes in X bring about considerable changes in the yields. Also, the shape and characteristics of the yield curves are rather similar to those obtained pyrolyzing pure hydrocarbons and model mixtures made up of them [15, 16, 17]. The ethylene and propylene yield curves are above the tangents drawn to the initial part of the curves, meaning their relative quantity increases with the degree of decomposition. Contrarywise, yield curves of pentene and hexene are below the initial tangent line, i.e. their relative quantity decreases with the degree of decomposition. Apart from ethylene, yield curves of the olefins produced follow maximum curves, their mode being shifted towards lower degrees of decomposition with increasing molecular weight. (Figures to be shown later, will reveal that even ethylene follows a maximum curve if yield is plotted against BKSF.)

Considering the yield curves of other products obtained (and due to lack of space not discussed here), it can be concluded, that increasing degree of decomposition brings about changes similar to those found cracking pure hydrocarbons or their model mixtures [15-17]. That is to say, that the main primary and secondary reactions are the same.

Because the yield versus degree of decomposition plots lack the required precision in the X > 0.95 range, detailed analysis of the experimental data, as well as the model applicable to calculate tube furnaces, where rather based **on** the use of the overall kinetic severity function (BKSF).

Figures 6 through 10 present the yield curves of the more

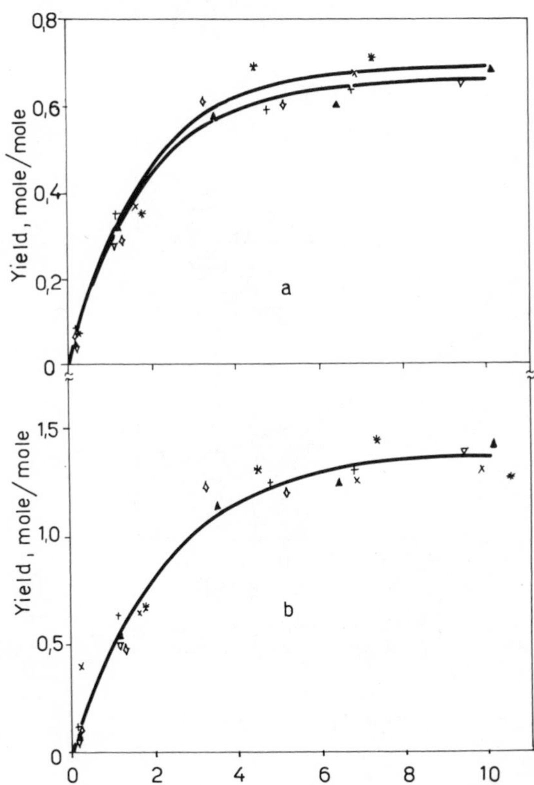

Figure 6. Change of hydrogen yield (a) and methane yield (b) of naphtha cracking with the overall kinetic severity function (BKSF)

Expt. No.	*Symbol*	*Expt. No.*	*Symbol*
44	▲	*47*	◇
45	▽	*48*	×
46	+	*49*	*

important products as a function of the BKSF.

Studying the curves, it can be concluded that apart from hydrogen and ethylene, the yield curves of the products are invariant to the exit temperature and the temperature profile within the range of the variables studied. Higher exit temperature and/or steeper temperature-profile (II), at a given degree of

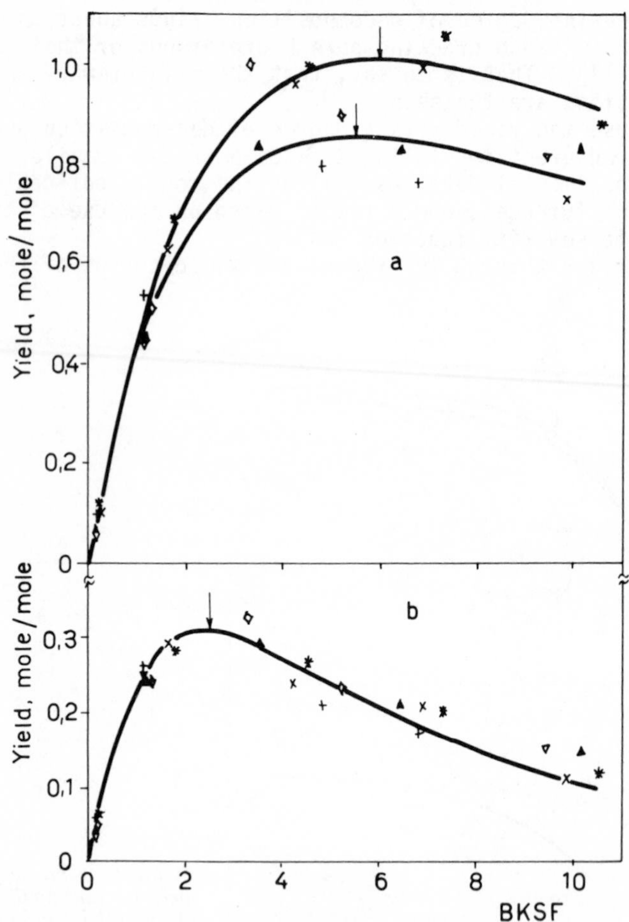

Figure 7. Change of ethylene yield (a) and propylene yield (b) of naphtha cracking with the overall kinetic severity function (BKSF)

Expt. No.	*Symbol*	*Expt. No.*	*Symbol*
44	▲	*47*	◇
45	▽	*48*	×
46	+	*49*	*

decomposition, result in higher hydrogen and ethylene yields. These findings agree perfectly with the mechanism of the decomposition [14-17].

Furthermore, it can be concluded, that product yields as functions of the BKSF follow only a few characteristic curve shapes, thus making their mathematical approximation greatly simplified.

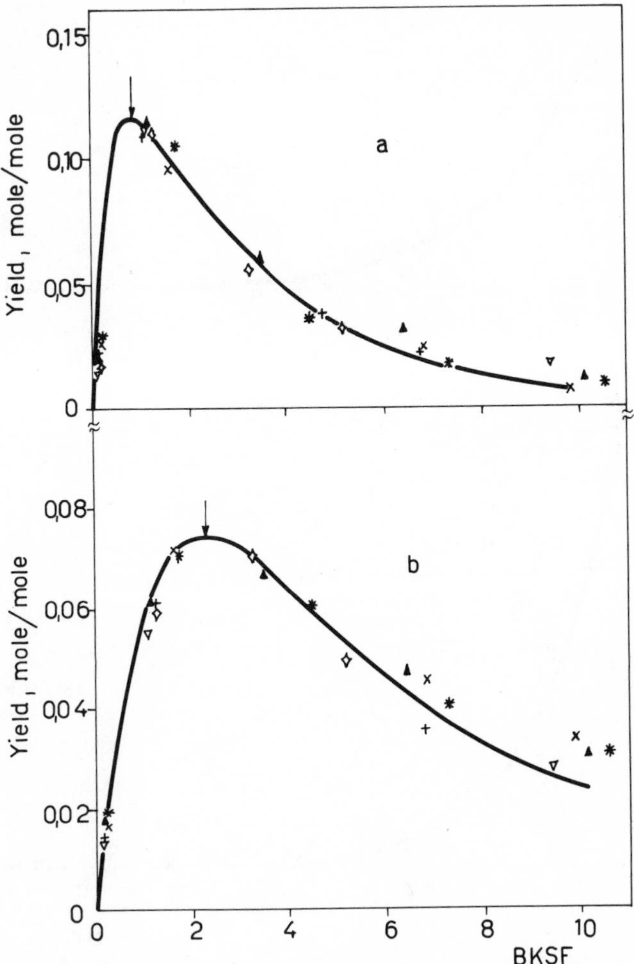

Figure 8. Change of butylene yield (a) and butadiene yield (b) of naphtha cracking with the overall kinetic severity function (BKSF)

Expt. No.	*Symbol*	*Expt. No.*	*Symbol*
44	▲	47	◇
45	▽	48	×
46	+	49	*

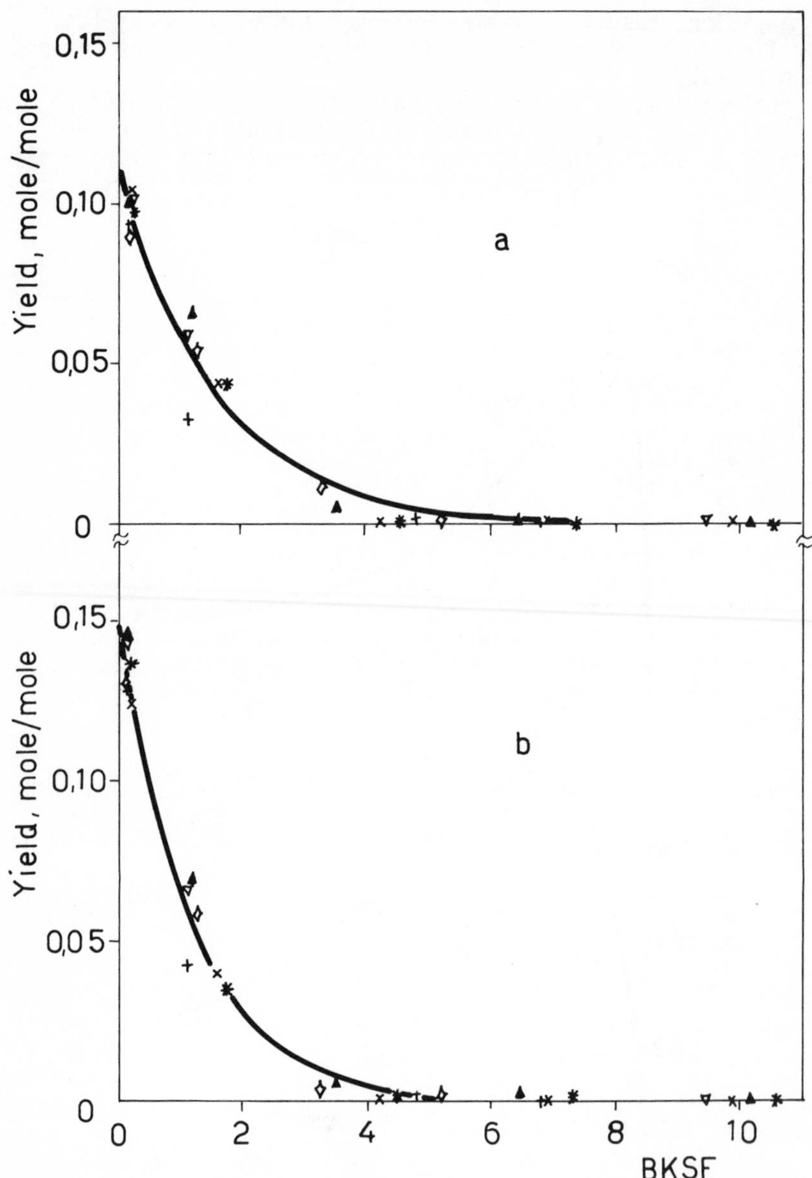

Figure 9. Change of pentane yield (a) and hexane yield (b) of naphtha cracking with the overall kinetic severity function

Expt. No.	*Symbol*	*Expt. No.*	*Symbol*
44	▲	*47*	◇
45	▽	*48*	×
46	+	*49*	*

The following functions were used to approximate the yield curves:

a) "Yield curves" of the decomposing components of the initial naphtha feedstock start from a definite initial value and exponentially approach either the zero or a given, nonzero values (such as for pentane and hexane in Figure 9). The general form of the function describing similar yield curves is:

$$H_i = a\,e^{-b\,BKSF} + c \tag{12}$$

b) Yield curves of the stable products of naphtha cracking start either from zero (hydrogen, methane, in Figure 6) or from a given initial value (benzene, toluene, in Figure 10) and exponentially approach a limiting value. The general form of the function describing similar yield curves is

$$H_i = a(1 - e^{-b\,BKSF}) + c \tag{13}$$

c) Yield curves of the so-called intermediates formed in the cracking process (olefins, diolefins, ethane, propane, in Figures 7 and 8) start from zero, pass a maximum value and continuously decrease from there on. The general form of the function describing similar yield curves is:

$$H_i = a(e^{-b\,BKSF} - e^{-c\,BKSF}) \tag{14}$$

d) Yield curves of a few products of secondary reactions (e.g acetylene, higher aromatics) follow an S shape, well approximated with the tangent hyperbolic function, already mentioned in connection with the conversion curves:

$$H_i = c[\mathrm{th}\ a(BKSF-b) + \mathrm{th}\ ab] \tag{15}$$

e) Yield curves of the alkylaromatics start from a determined initial value and passing a minimum value and slowly continue to rise (Figure 10).

They are approximated with the function:

$$H_i = a\,e^{-b\,BKSF} + c\,BKSF \tag{16}$$

Constants of the functions approximating the yield curves have been determined by the so-called least square method and occasionally, from so-called basic points. Yield curves drawn in full line, presented in Figures 6 through 10, were calculated from the above approximating functions.

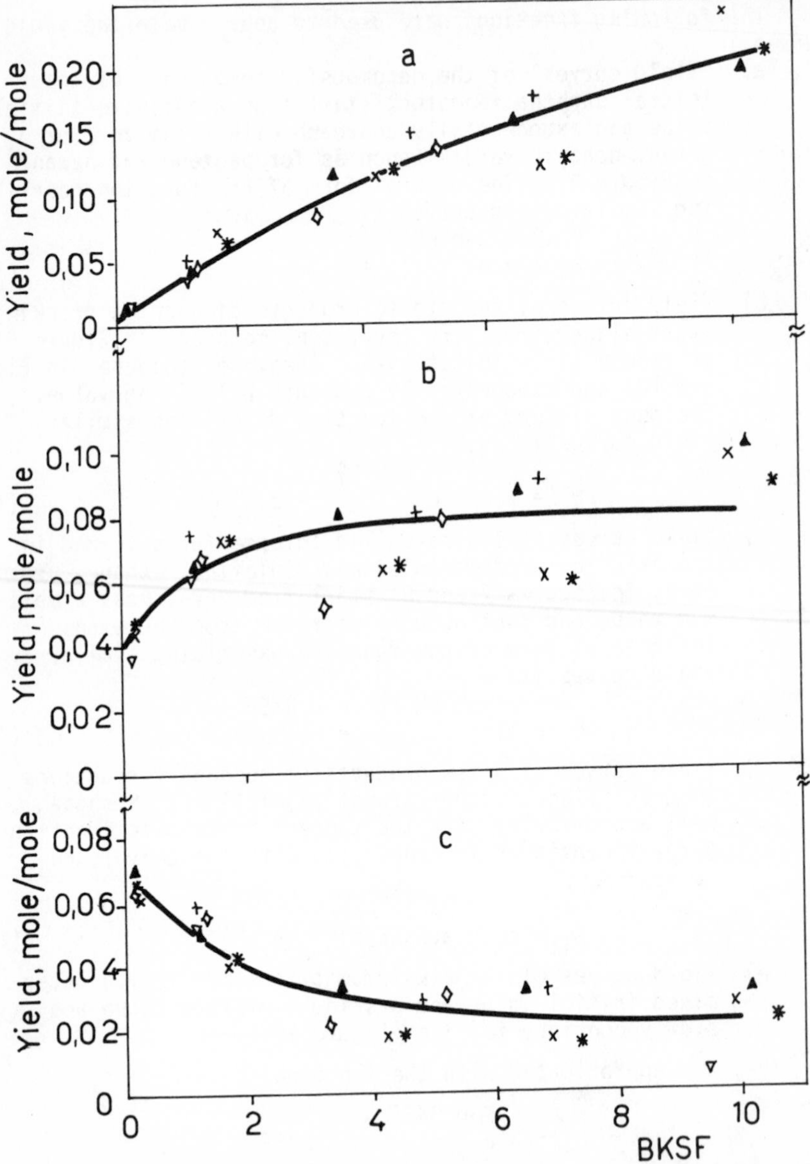

Figure 10. Change of benzene yield (a), toluene yield (b), and xylenes yield (c) of naphtha cracking with the overall kinetic severity function

Expt. No.	*Symbol*	*Expt. No.*	*Symbol*
44	▲	*47*	◇
45	▽	*48*	×
46	+	*49*	*

Kinetic-Mathematical Model of Naphtha Feedstock Cracking

Previous sections presented the so-called kinetic model of naphtha feedstock cracking, which allows the calculation of the degree of decomposition, X, the overall kinetic severity function, BKSF, and the true residence time, T, as functions of the relative length of the reactor, z.

Studying the product distribution, it was shown that the yield versus BKSF plots follow a few characteristic curves readily approximated by appropriately selected mathematical functions.

Coupling the kinetic model and the mathematical relationships describing the yield curves, the so-called kinetic-mathematical model is obtained that can readily be used to calculate the product distribution as well.

Figure 11 presents the measured yield values and the calculated yield values of the more important reaction products (drawn in full line) obtained from the kinetic-mathematical model applied for the naphtha feedstock cracking experiment No. 44. Plotted are also the curves of the degree of decomposition (X) and true residence time (T).

Studying the figures, it can be concluded that the measured and calculated data agree rather well, when either the degree of decomposition or the product yields are considered. Small random deviations can be attributed to the experimental (measurement) errors.

Furthermore, it can be established, that to obtain maximum ethylene yield in the naphtha feedstock cracking process, (and the process is advantageously interrupted at that point) the reaction mixture should have a true residence time between the 0.46 - 0.52 seconds limits, depending on the severity of cracking (exit temperature, temperature-profile).

The kinetic-mathematical model presented here can well be applied to determine cracking parameters, to obtain the required product yields of a given naphtha feedstock, or to calculate the product yields from the given cracking parameters. The model was completed by adding the hydrodynamic and heat transfer equations used to calculate the pressure drop and the amount of heat transferred in the reactor.

Abstract

Based on naphtha pyrolysis experiments conducted in a bench scale tubular reactor (suitable for the simulation of industrial tubular furnace operations and taking into account the changes of expansion, temperature and the pressure in the reactor), a kinetic model has been developed for the calculation of the degree of decomposition, the actual residence time, and the severity of cracking.

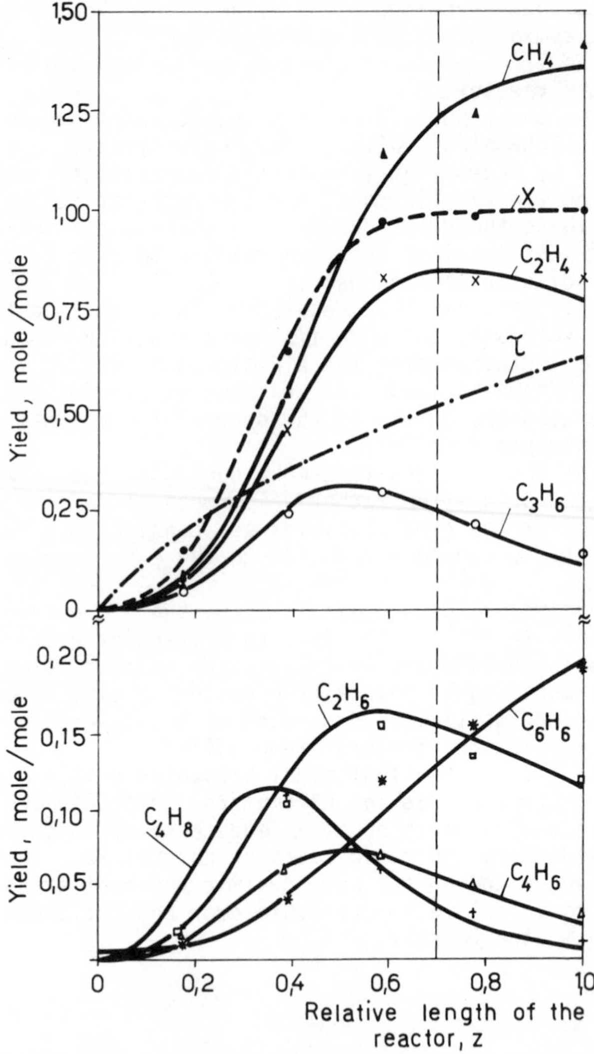

Figure 11. Change of measured and calculated yields of the main products, degree of decomposition, and true reaction time of naphtha cracking with the relative length of the reactor: experiment No. 44

It has been found, that with the exception of a few products (e.g. hydrogen and ethylene), the yields of reaction products are functions of the severity, and may be plotted with "single curves" independent of the shape of the temperature profile. The yield curves were approximated with different types of mathematical functions.

A computer program and a kinetic-mathematical model suitable for calculating product distribution have been developed by the combination of the kinetic relations describing overall decomposition rate, and mathematical relations representing yield curves.

The program can be operated when the temperatures of the reaction mixture or of the tubewall are known. Application of the model is illustrated with examples.

Literature Cited

[1] Lambrix, J. R., Morris, C. S.: Chem. Eng. Progr. 68(8), 24 (1972).

[2] Illes V., Rauschenberger J., Szepesy L.: Magy. Kem. Lapja. 26 (12), 624 (1971).

[3] Illes, V. Szepesy, L., Csikos, R., Rauschenberger, J.: Erdol u. Kohle. 25(10), 589 (1972).

[4] Linden, H. R., Reid, J. M.: Chem. Eng. Progr. 55, 71 (1959).

[5] Zdonik, S. B., Green, E. J., Halle, L. P. : The Oil and Gas J., July 10, 192 (1967).

[6] de Blieck, J. L., Goossens, A. G.: Hydrocarbon Proc., March, 76 (1971).

[7] Chambers, L. E., Potter, W. S.: Hydrocarbon Processing, January, 121 (1974).

[8] Gunschel, H., Nowak, S.:"Alkene 73" Bilaterale Colloq. Marianske Lazne, 6/19-21/73.

[9] Muhina, T. N., Babas, SZ.e., Guas, E. I., Tjatyenko, T. M., Szobolev, O. B.: Proidzvosztvo Nyizsih Olefinov. NIISZSZ,

[10] Kershenbaum, L. S., Martin, J. J.: A.I.Ch.E. Journal, 13 (1), 148 (1967).

[11] Kunzru, D., Shah, Y. T., Stuart, E. B.: Ind. Eng. Chem. Process Des. Develop., 11(4), 605 (1972).

[12] Leftin, H.P., Cortes, A.: Ind. Eng. Chem. Process Des. Develop. 11(4), 613 (1972).

[13] Illes, V.: Acta Chim. Acad. Sci. Hung. 72(2), 117 (1972).

[14] Illes, V., Horvath, A.: Hung. J. of Ind. Chem. 3(3), 369 (1975); ibid. 3(3), 391 (1975).

[15] Illes, V., Pleszkats, I., Szepesy, L.: Acta Chim. Acad. Sci. Hung. 79(3), 259 (1973).

[16] Illes, V., Welther, K., Szepesy, L.: Acta Chim. Acad. Sci. Hung. 80(1), 1 (1974).

[17] Illes, V., Pleszkats, I., Szepesy, L.: Acta Chim. Acad. Sci. Hung. 80(3), 267 (1974).

25

Mechanism and Kinetics of the Thermal Hydrocracking of Single Polyaromatic Compounds

JOHANNES M. L. PENNINGER* and HENDRIK W. SLOTBOOM

Laboratory for Chemical Technology, Eindhoven University of Technology, The Netherlands

The thermal interaction of polyaromatic hydrocarbons with hydrogen is of renewed industrial interest since the utilization of coal as a raw material for petrochemical intermediates is now being examined on a rapidly increasing scale. In this context the solvent refining of coal by heating a slurry of coal in a high boiling polyaromatic solvent with hydrogen under pressure has to be especially mentioned (1). A different line of approach is the hydrogenolytic extraction of coal, by contacting plain coal with hydrogen at elevated temperature and pressure in the absence of a solvent. Hydrogen degrades chemically the complex structure of coal and acts physically as a leaching agent for the removal of degradation products out of the coal matrix.

The present paper presents a survey of the reactions taking place when single polyaromatic components, present in coal tar, are contacted with hydrogen at elevated temperature and pressure. The experiments aiming at a revealing of the hydrocracking sequence of single components were carried out batchwise in a 316 SS autoclave with a volume of approximately 100 ml. The kinetic data were determined under flow conditions in a SS tubular reactor of 15 ml volume. The conversion of the parent compound was of the order of a few percent and was regarded as being only differential. By a strict temperature control the deviations were in a range of 1°C and there was no detectable temperature difference between the wall and the center of the autoclaves. The temperature of the experiments was chosen so that distinct stages in the hydrocracking sequence were observed; they ranged from 450 to 540°C. The pressures ranged up to 300 bar.

* Present correspondence address:
Akzo Chemie B. V. P. O. Box 247, Amersfoort, The Netherlands

Hydrocracking sequence of single polyaromatic compounds

Introductory experiments with naphthalene revealed that hydrogenation to tetralin occurred prior to the formation of a complex mixture of alkylated benzenes, which had already been observed by Spilker in early 1926 (2). The alkylated benzenes are thus formed by a cracking of the naphthenic ring of tetralin.

Using tetralin as a starting component a product composition with the reaction time, as is illustrated in Figure 1, was obtained. Analog simulation of the hydrocracking sequence, as presented in Figure 2, gave the full lines drawn in Figure 1. It appears that by cracking the aromatic C-naphthenic C bond the

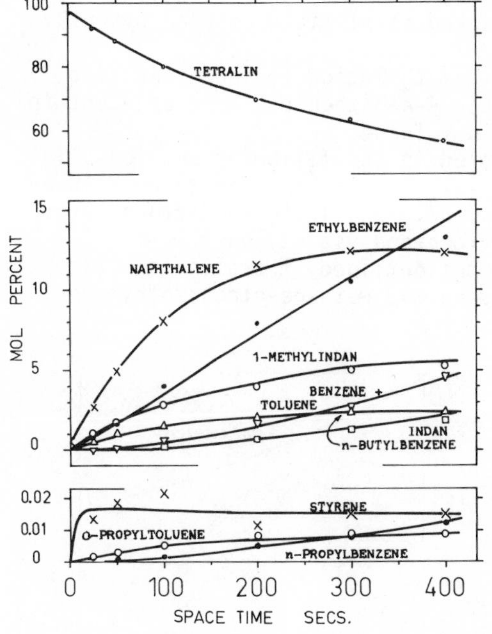

Figure 1. Hydrocracking of tetralin. Temp., 540°C; total press., 80 bar; molar ratio hydrogen–tetralin, 10; tubular flow reactor —— model (see Figure 2); ○×•○△▽□ experiment.

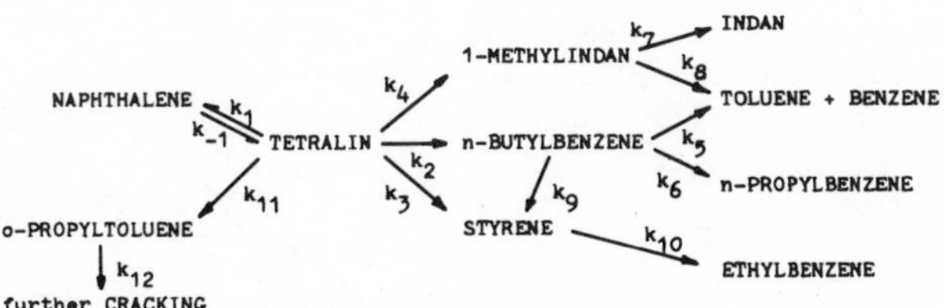

Figure 2. Hydrocracking sequence of tetralin. First-order rate constants (× 10^3) at 540°C: $k_1 = 1$, $k_{-1} = 6$, $k_2 = 0.2$, $k_3 = 0.4$, $k_4 = 0.4$, $k_5 = 1$, $k_6 = 0.2$, $k_7 = 1$, $k_8 = 2$, $k_9 = 5$, $k_{10} = 200$, $k_{11} = 0.0006$, $k_{12} = 4$.

following products are formed, namely n-butylbenzene, styrene and ethylene, 1-methylindan. By cracking of a naphthenic C-C bond o-propyltoluene results. Further, dehydrogenation of tetralin to naphthalene occurs. By consecutive cracking reactions the primary cracking products are further transformed. n-Butylbenzene undergoes side-chain cracking mainly by fission of ethylene and forms styrene; both are hydrogenated to ethane and ethylbenzene respectively. Second in significance is the formation of toluene from n-butylbenzene by propylene fission. These reactions were confirmed from separate experiments with n-butylbenzene as the starting compound (Figure 3)(3). This further revealed that formation of n-propylbenzene, by fission of a methyl group, is comparatively a very slow reaction.

The hydrogenation of styrene, that results directly from tetralin and via n-butylbenzene, is a very fast reaction; the resulting ethylbenzene is cracked mainly into toluene by methylene split-off.

1-Methylindan undergoes demethylation to indan and ring opening, mainly at the aromatic C-naphthenic C bond adjacent to the methyl group, resulting in toluene.

Indan, as will be discussed in the following sections, is transformed mainly to toluene-ethylene-ethane and to much lower concentrations of n-propylbenzene. The latter undergoes methylene split-off and forms ethylbenzene via styrene.

So, via the cracking routes outlined, tetralin is ultimately transformed into benzene-toluene and methane-ethane/ethylene-

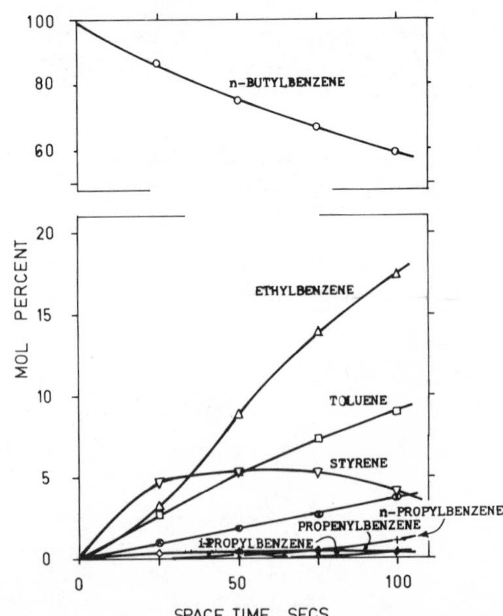

Figure 3. Hydrocracking of n-butylbenzene. Temp., 500°C; total press., 80 bar; molar ratio hydrogen–n-Bubz, 10; tubular flow reactor.

propane/propylene in which the C_2 fraction is dominant.

Phenanthrene, which reacted much slower than tetralin, gave a product distribution as is illustrated by Figure 4 (4). Ring hydrogenation resulted mainly in 1, 2, 3, 4-tetrahydrophenanthrene and by naphthenic ring cracking naphthalene and its 2-ethyl- and 2-methyl derivates were formed in concentrations higher than methyl-ethyl naphthalene and dimethylnaphthalene. Second in significance was ring hydrogenation to 9, 10- dihydrophenanthrene; cracking products derived therefrom were 2-ethyl biphenyl, 2-methyl biphenyl and biphenyl and, in lower concentrations, also diphenylmethane.

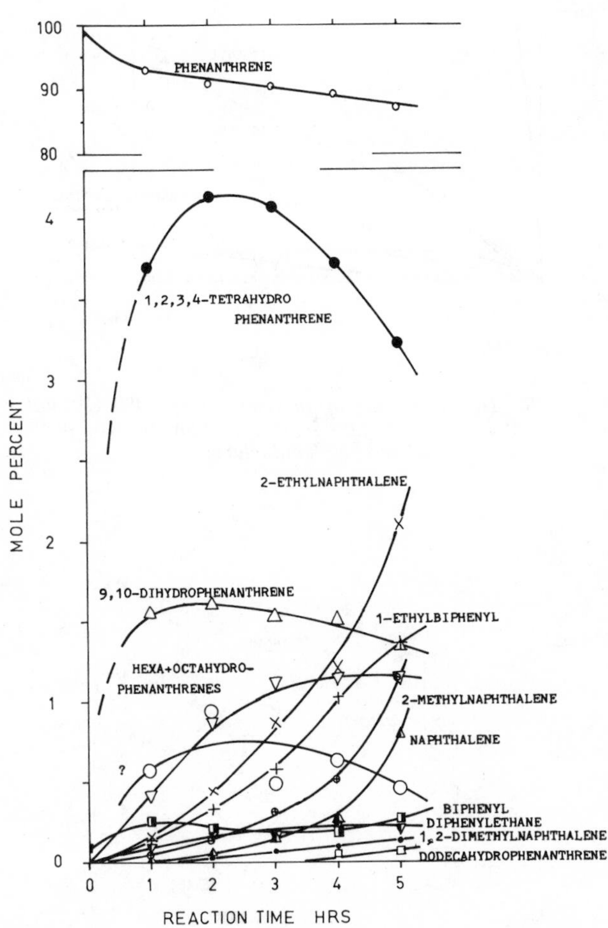

Figure 4. Hydrocracking of phenanthrene. Temp., 495°C; total press., 80 bar; molar ratio hydrogen–phen., 4; batch autoclave.

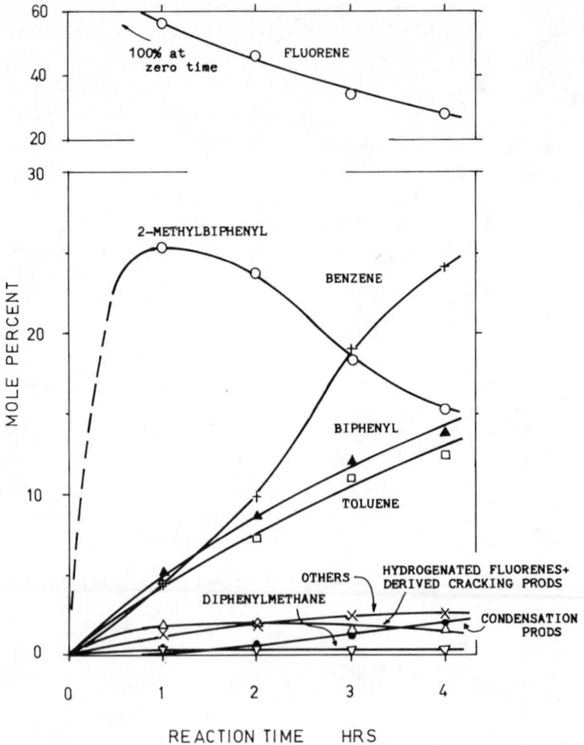

Figure 5. Hydrocracking of fluorene. Temp., 480°C; total press., 140 bar; molar ratio hydrogen–fluorene, 6.6; well-stirred batch autoclave.

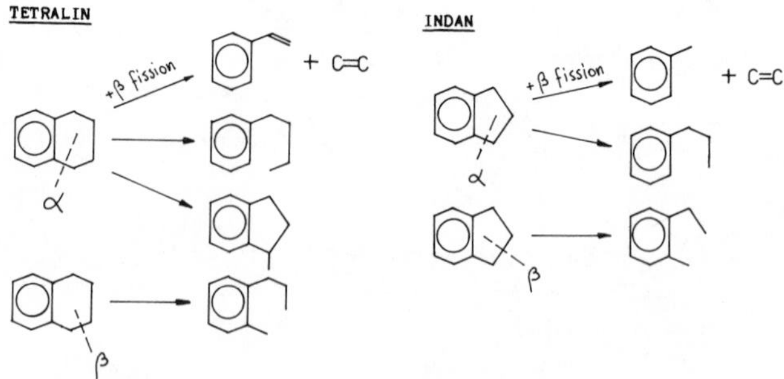

Figure 6. α ring opening and β ring opening of indan and tetralin.

With polyaromatics possessing a methylenic group hydrodemethylation was preferred to ring hydrogenation. Thus, fluorene, under intensive mixing with hydrogen in a well-stirred autoclave* has, as main products 2-methyl biphenyl and biphenyl; diphenylmethane was found only in very low concentrations (5). Hexahydrofluorene, formed by ring hydrogenation, and the cracking products derived therefrom viz. alkylindans, were formed in concentrations much lower than the biphenyls (Figure 5). Although structurally similar to 9, 10-dihydrophenanthrene, fluorene exhibited a much lower reactivity to cracking of the phenylphenyl bond.

2-Methylnaphthalene gave naphthalene in concentrations much higher than the two ring hydrogenation products 2-methyltetralin and 6-methyltetralin. By successive hydrocracking the methyltetralins gave quite different cracking products. 6-Methyltetralin underwent hydrodemethylation to tetralin preferentially to ring cracking; 2-methyltetralin formed products typical for ring cracking namely toluene and n-propylbenzene (6).

It can therefore be conclusively stated that non-alkylated polyaromatic hydrocarbons upon hydrocracking are first hydrogenated, so that a part of the aromatic skeleton is transformed into a naphthenic system. This napthenic part is cracked until a (poly) aromatic structure remains, which again by hydrogenation and cracking proceeds further. Polyaromatic compounds possessing methylenic groups undergo hydrocracking at the methylene-phenyl bond preferentially to ring hydrogenation.

Thus, by the sequence of successive hydrogenation cracking steps a polyaromatic hydrocarbon disintegrates into low molecular weight hydrocarbons as methane, ethane/ethylene, propane/propylene and aromatics as mainly benzene and toluene.

The kinetics and mechanism of the cracking of naphthenic systems

The characteristic reaction in the hydrocracking sequence of polyaromatics viz. the cracking of a naphthenic system attached to aromatic rings, was studied in further detail with indan as a model substance. The "primary" cracking products, formed at low conversion levels of indan, indicate the occurrence of two different ring opening reactions (Figure 6). Firstly, the α ring opening resulting in n-propylbenzene and toluene-ethylene; secondly the β ring opening resulting in o-ethyltoluene and o-xylene successively. From the ratio of these cracking products it was clear that the α ring opening is a much faster reaction than the β ring opening. The formation rates of the individual cracking products were correlated with the rate equations as given in Table I. The same table includes the rate equations found for the hydrocracking of tetralin. This illustrates that identical

* Without mechanical mixing considerable coking resulted.

Table I Thermal hydrocracking of indan and tetralin; experimental rate equations for the α and β ring opening reactions.
Temperature range : 460-540 °C
Partial pressure of hydrogen: ~ 73 bar.

Type of cracking reaction	Indan hydrocracking*	Tetralin hydrocracking**
α ring opening	$r_T = k_T[\text{indan}]^{0.5}[\text{hydrogen}]$ $r_{Pr} = k_{Pr}[\text{indan}][\text{hydrogen}]^{1.5}$	$r_{St} = k_{St}[\text{tetralin}]^{0.5}[\text{hydrogen}]$ $r_{Bu} = k_{Bu}[\text{tetralin}][\text{hydrogen}]^{1.5}$
β ring opening	$r_{o\text{-}Et} = k_{o\text{-}Et}[\text{indan}]$	$r_{o\text{-}Pr} = k_{o\text{-}Pr}[\text{tetralin}]$

* T = Toluene, Pr = n-Propylbenzene, o-Et = o-Ethyltoluene

** St = Styrene, Bu = n-Butylbenzene, o-Pr = o-Propyltoluene

reactions obey identical rate equations.

The kinetic similarity between indan and tetralin was found also in the temperature dependence of the rate constants, as is illustrated in Figure 7. The relatively low values of the Arrhenius parameters may indicate a radical chain mechanism for the α ring opening; the Arrhenius parameters of the β ring opening are significantly higher and suggest a non-chain mechanism with the thermal dissociation of a naphthenic C-C bond as first and rate determining step. This leads to the mechanism proposed for the α ring opening and β ring opening in the thermal hydrocracking of indan, as are delineated in Figure 8.

Assuming a steady state with respect to the concentration of intermediate radicals and atoms, theoretical rate equations were derived and those were found to be identical with the experimental equations (see Table I and Table II-A) for the relatively low hydrogen partial pressures of up to $\sim$ 73 bar. In further testing the applicability of this mechanism under more extended experimental conditions, it was found that the kinetics of the toluene formation was different at a hydrogen partial pressure of 230 bar. At the latter hydrogen pressure a rate equation of

$$r_T = k_T'' \text{ (indan)}^{0.75} \text{ (hydrogen)}^{0.75}$$

was found better in agreement with the experimental data. Also the Arrhenius parameters were changed, e.g. the activation energy was lowered by 4 to 5 kcal/mole. As is shown in Figure 9 said increase of the hydrogen pressure had no influence on the kinetics of the n-propylbenzene formation. This effect of the hydrogen pressure is explained with the assumption that the steady-state concentration of the benzyl radical at higher hydrogen pressures is controlled by the reactions "3" and "5" (Table II-B) rather than "3" and "-3" (Table II-A). In the former case a theoretical rate equation of

$$r_T = k_T'' \text{ (indan) (hydrogen)}^{0.5}$$

will result with an energy of activation of 8 kcal/mole less than in the latter case (7). So, the experimental data found at a hydrogen pressure of 230 bar probably indicate a transient situation between the conditions A and B of Table II.

The initiation and termination reactions

Molecular dissociation of hydrogen and recombination of hydrogen atoms, as Figure 8 presents, do not provide a completely satisfactory answer to the question of chain initiation and termination. Although on this basis theoretical rate equations are readily derived which agree with experimental reaction orders, many other questions are still possible. A thermal dissociation of molecular hydrogen as an initiation reaction would give an overall energy of activation for n-propylbenzene and toluene of

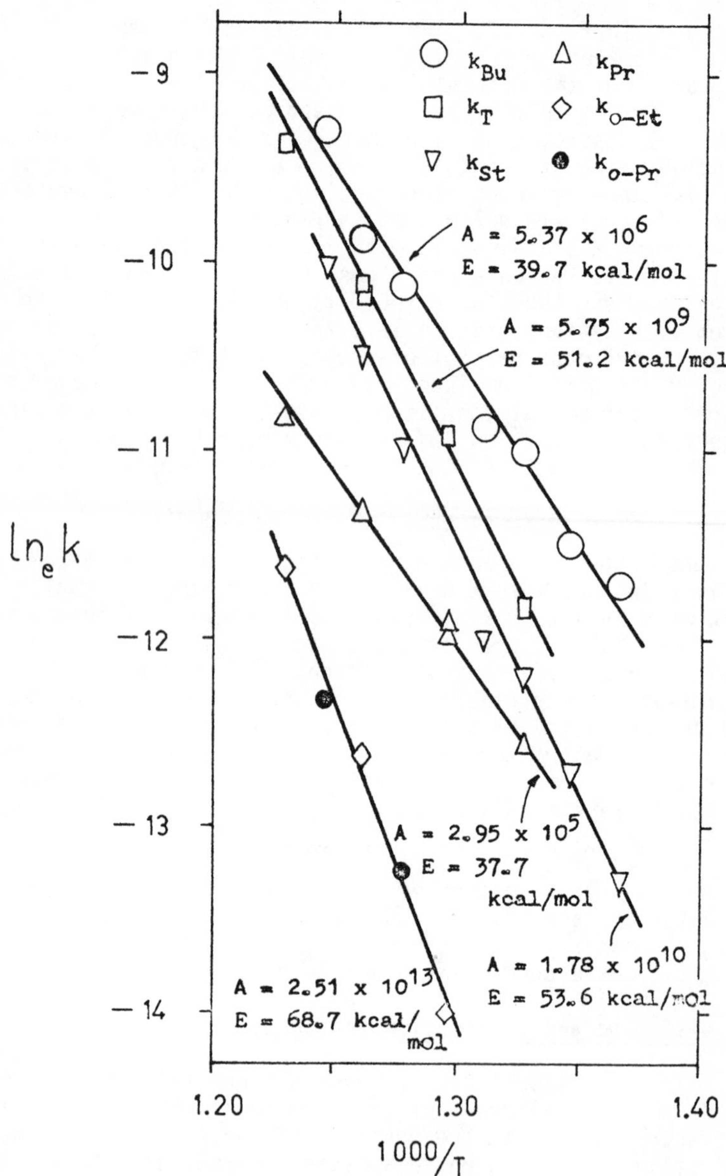

Figure 7. Arrhenius plot of primary hydrocracking reactions of indan and tetralin. Individual k *values were calculated from rate data according to the rate equations listed in Table I.*

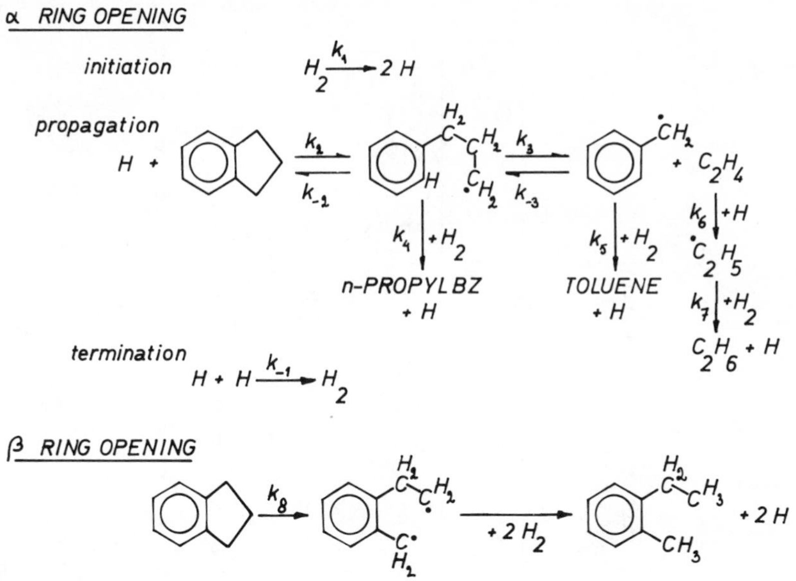

α RING OPENING

initiation $H_2 \xrightarrow{k_1} 2H$

propagation H + indan $\underset{k_{-2}}{\overset{k_2}{\rightleftharpoons}}$ radical $\underset{k_{-3}}{\overset{k_3}{\rightleftharpoons}}$ $\dot{C}H_2$ radical + C_2H_4

k_4 $+H_2$ → n-PROPYLBZ + H

k_5 $+H_2$ → TOLUENE + H

C_2H_4 $\xrightarrow{k_6,\ +H}$ $\dot{C}_2H_5$ $\xrightarrow{k_7,\ +H_2}$ $C_2H_6 + H$

termination $H + H \xrightarrow{k_{-1}} H_2$

β RING OPENING

indan $\xrightarrow{k_8}$ diradical $\xrightarrow{+2H_2}$ o-ethyltoluene + 2 H

Figure 8. *Mechanism of the α ring opening and β ring opening of indan*

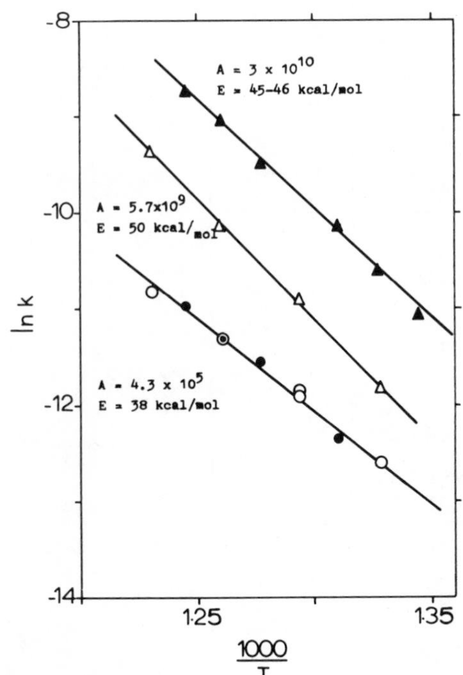

Figure 9. *Arrhenius parameters of the α ring opening products of indan; effect of the hydrogen partial pressure.* △ ▲ *toluene;* ○ ● n-propylbenzene. *Empty symbols: hydrogen pressure ~ 73 bar; total press. 80 bar; full symbols: hydrogen pressure ~ 230 bar, total press. 250 bar.*

Table II Thermal hydrocracking of indan; theoretical rate equations according to the α ring opening chain reaction.

	Ratio of rate constants of intermediate reactions	Rate equations of toluene and n-propylbenzene
A	$k_{-2} \gg k_3 + k_4 [H_2]$ $k_{-3} [C_2H_4] \gg k_5 [H_2]$	$r_T = (2K_1\ K_2\ K_3\ k_5\ k_6)^{0.5} [\text{indan}]^{0.5} [\text{hydrogen}]$ $r_{Pr} = (2K_1)^{0.5} K_2\ k_4 [\text{indan}] [\text{hydrogen}]^{1.5}$
B	$k_{-2} \gg k_3 + k_4 [H_2]$ $k_{-3} = 0$	$r_T = (2K_1)^{0.5} K_2\ k_3 [\text{indan}] [\text{hydrogen}]^{0.5}$ $r_{Pr} = (2K_1)^{0.5} K_2\ k_4 [\text{indan}] [\text{hydrogen}]^{1.5}$

approximately 12 kcal/mole higher than the observed values. From an energy point of view the thermal dissociation of a naphthenic C-C bond forming a biradical would be more likely as an initiation reaction. However, any of the other combinations of initiation-termination reactions examined, except for the dissociation and recombination of hydrogen, gave theoretical rate equations with reaction orders different from the experimental orders (7). Furthermore, the dehydrogenation of indan to indene and the reverse reaction, which take place under all experimental conditions simultaneously with the hydrocracking, as a source of initiating hydrogen atoms, does not agree with the experimental rate equations.

Many observations suggest that the problem of initiation/termination is strongly related to the nature of the reactor wall. The following observations made with the SS reactor are relevant in this respect (8). Firstly, there was a correlation between the conversion/cracking selectivity of indan and the ageing time of the reactor; secondly, the temperature level of the previous experiment affected the indan conversion in the early stages of the present experiment (so called "temperature history effects"); thirdly, the conversion and selectivity was not affected by an increase of the surface-to- volume ratio of the aged reactor; and fourthly, the gradual increase of the indan conversion with ageing time starting from almost zero conversion in a gold plated reactor. These observations strongly support the hypothesis that the α ring opening chain reaction proceeds only in the presence of an "active" reactor wall. Apparently, both the initiation and termination reactions are wall reactions, viz. the dissociation of molecular hydrogen into atoms and the recombination of same occur at the reactor wall. An unaged SS reactor has naturally occurring active sites, e.g. the nickel and chromium crystallites; gold is not active in hydrocarbon reactions. After ageing, a carboneous deposit is formed on the reactor wall. It was demonstrated elsewhere that coke deposits indeed have an enhancing effect on the thermal cracking of hydrocarbons (9) and that this can be attributed to free radicals present in the coke (10). Thus, by ageing, the original activity of the SS reactor wall is gradually replaced by the activity of coke deposits, as occurs also in the gold plated reactor. Ultimately, the reactor wall activity is determined by the deposits; experimentally it was confirmed that the conversion of indan and the cracking selectivity was almost independent of the original reactor wall material after the ageing period.

Deutero cracking of indan

In order to elucidate more details of the reactions occurring under hydrocracking conditions, the cracking of indan was carried out in deuterium (11). Although a final evaluation of the data is not complete at present, the general picture is much more complicated now, than originally appeared from the hydrocracking experiments. Some main observations are a slower overall cracking rate of indan; a comparatively fast rate of hydrogen-deuterium exchange of indan, faster than the cracking rate. In addition, the primary cracking products (n-propylbenzene and toluene) are deuterated up to only approximately 20% at low cracking conversion levels of indan; in such cases there is apparently major hydrogen transfer from indan to the intermediate radicals of the cracking sequence. The complications arise from the considerable isotopic effects of hydrogen and deuterium; a close analysis of the date will show whether cracking of indan in deuterium is the right model reaction to disclose the desired elementary details of the hydrocracking reaction. At present, the deuterium experiments have produced only astonishment that the simple mechanism as presented in this paper can describe such a complex of reactions as seem to occur in the thermal hydrocracking of indan and other hydro-aromatic components.

Literature Cited

(1) (a) Baldwin, R. M., Golden, J. O. Gary, J. H. Bain, R. L. and Long, R. J., Chem. Eng. Prog., (1975) 71, 70: (b) **Schmid,** B. K., ibid., (1975) 71, 75; (c) Anderson, R. P., ibid., 71, 72.

(2) Spilker, F., Z. F. Angew. Chem., (1926) 39, 1138.

(3) Slotboom, H. W. and Penninger, J.M.L., Erdol u. Kohle, (1974) 27, 410.

(4) Penninger, J. M. L. and Slotboom, H. W., ibid.,(1973) 26,445.

(5) Oltay, E., Penninger, J. M. L. and Konter, W.A.N., J. Appl. Chem. Biotechn., (1973) 23, 573.

(6) Oltay, E., Penninger, J. M. L. and Koopman, P. G. J., Chimia, (1973) 27, 318.

(7) Van Boven, M., Roskam, G. J. and Penninger, J. M. L., Intern. J. Chem. Kinet.,(1975) VII, 351.

(8) Slotboom, H. W. and Penninger, J. M. L., Ind. Eng.Chem. Proc. Des. Develop., (1974) 13, 296.

(9) Cramers, C. A., Diss. Eindhoven - The Netherlands (1967).

(10) Holbrook, R. A., Walker, R. W. and Watson, W. R., J. Chem. Soc. B., (1968) 1089.

(11) Publication in preparation, Professor G. Dijkstra and K. Versluis of Utrecht State University are acknowledge for analyzing the deuterated fractions by GLC-MS array.

26

Thermal Decarbonylation of Quinones in the Presence of Hydrogen

TOMOYA SAKAI and MASAYUKI HATTORI

Department of Chemical Reaction Engineering, Nagoya City University, Tanabedori, Mizuhoku, Nagoya, 467 Japan

It will be an ultimate problem for hydrocracking of coal or heavy petroleum residues to transform effectively polycondensed aromatics to lower hydrocarbons. Two simplified model compounds of polycondensed aromatics were adopted, i.e., anthracene (ANT) which is linear and phenanthrene (PHN) which is angular cata-condensed aromatics. Processes for degradation of these molecules to lower hydrocarbons, preferably to monocyclic aromatics, have been explored.

It is well known (1,2,3,4,5) that, during the usual catalytic hydrogenation processes at around 450°C and with a hydrogen pressure of ca. 100 atm, ANT and PHN are hydrogenated on the side ring to produce 1,2,3,4-tetrahydroanthracene and 1,2,3,4-tetrahydrophenanthrene, respectively; subsequent C-C bond cleavage produces mainly naphthalene homologues.

To achieve the cleavage at the center ring, the chemical activation at the 9,10- position of both ANT and PHN is necessary. In this connection, the reductive activation, i.e., to use 9,10-dihydroanthracene and 9,10-dihydrophenanthrene as tentative starting materials of cracking, was not an appropriate way of activation. Thermal cracking of these compounds resulted in the formations of ANT and PHN, respectively. Moreover, they proved to be intermediate compounds of the above-mentioned catalytic hydrogenation of ANT and PHN to tetrahydroanthracene and tetrahydrophenanthrene. On the contrary, the oxidative activation, i.e., to use 9,10-anthraquinone (ANQ) or 9,10-phenanthrenequinone (PHQ) as starting materials for cracking, proved to be an effective way to achieve the cleavage of tri-condensed aromatics at their center rings.

Decarbonylation of ANQ and PHQ proceeded at 500 - 600°C in the presence of hydrogen to form initially fluorenone (FLR), followed by successive hydrogenolyses of FLR to biphenyl and CO, finally of biphenyl to benzene. Progress of similar reactions were examined for 1,4- and 1,2-naphthoquinone (NPQ) and p-benzoquinone (BNQ) (6, 7).

There have been few works on the hydrocracking of quinones or related reactions. Davies and Ennis (8) reported that zinc dust distillation of polycyclic o-quinones related to 9,10-PHQ produced fluorenone derivatives in moderate yields at 300 - 320°C. They found their result to be useful in the recognition of o-quinones which reacts with o-phenylenediamine with difficulty. Brown and Solly (9) noted the formation of FLR, ANQ and PHQ in their study on the pyrolysis of indanetrione at 500 - 800°C and 0.2 - 0.4 mm Hg. They stated however in the same paper that the ready loss of CO from ANQ on pyrolysis is not to be expected, nor has it been observed. Brown and Butcher (10) reported also the smooth decarbonylation and recyclyzation of 3,3,6,8-tetramethyltetralin-1,2-dione to yield indanone structure at 650°C and 0.1 mm Hg. Liquid-phase cleavage of ANQ was reported by Hausigk (11) and Davies and Hodge (12). They used excess potassium t-butoxide in toluene/ether mixed solution or butoxide in 1,2-dimethoxyethane at 85 - 150 °C to afford mixtures of benzoic and/or phthalic acids in high yields.

Experimental

Commercially available ANQ, 1,4- and 1,2-NPQ and p-BNQ were used with or without purifications (13,14,15). PHQ was synthesized according to the method reported earlier (16). Purities of feed materials were confirmed to be more than 99.9% for ANQ and PHQ and 80 - 90 % for 1,4- and 1,2-NPQ and p-BNQ by gas chromatographic analyses. Hydrogen from a cylinder was purified by passing it through heated copper gauze and then a silica gel column.

Experiment was conducted by use of a transparent quartz ampule of 20 mm i.d. and 130 mm length as a reaction vessel. Quartz was used to attain an improved temperature profile throughout the reaction period by sudden heating at the start and quenching at the end of the experiment. The amount of quinone employed in an ampule was 70 - 100 μ moles, and the hydrogen pressure was 400 - 1000 mm Hg at room temperatures (700 - 1000 μ moles). Six sealed quartz ampules were subjected to the reaction at one time by inserting them to a six-holed stainless steel block furnace maintained at a designated temperature in the range of 500 - 600°C. After fixed reaction periods varying between 2.5 and 160 minutes, each ampule was taken out of the furnace one by one to quench immediately in a water bath.

The reaction products were gaseous and solid. They were treated with acetone in the absence of light to obtain an acetone solution of products plus also a gas phase product. Gaseous products were analyzed by gas chromatographs furnished with molecular sieve 5A and silica gel columns. A chromatographic unit furnished with a squalane and silicone GE-SE30 columns was employed for the acetone solutions. In order to measure the absolute amounts of products in these solutions, pyrene or o-terphenyl were used as standards.

Results and discussion

Typical experimental results are summarized in Tables I, II and III for ANQ, PHQ and FLR, 1,4- and 1,2-NPQ and p-BNQ, respectively. The indicated temperature corresponds to that of the furnace at the withdrawal of each ampule from the furnace. Conversion (%) was calculated based on the amount of unreacted quinones remained in a quenched ampule. Mass balance (%) was defined to be a ratio of the mass of total compounds recovered which were measured from gas chromatographic peaks to the mass of feed quinone. Accordingly, whenever any polymerized compounds of high molecular weight were formed, there was a shortage in mass balance (%).

Decarbonylation of ANQ and PHQ. As listed in Table I, main products in the reaction of ANQ and PHQ were common in both cases, i.e., CO, benzene, biphenyl and FLR; and minor products were fluorene, CO_2 and ANT in the case of ANQ and fluorene, CO_2 and PHN in the case of PHQ. A small amount of high molecular weight substances was also formed in both cases but in smaller amounts in the case of ANQ. By use of helium instead of hydrogen, FLR was the sole product other than CO in the case of PHQ; however, in the case of ANQ, the reaction proceeded to a slight extent beyond FLR yielding a small amount of biphenyl and benzene. Reaction of FLR in the presence of hydrogen proceeded in an expected way yielding CO, biphenyl, benzene and fluorene. By replacing hydrogen with helium, the rate of decarbonylation was decreased very significantly as expected.

Figures 1 and 2 illustrate the change of product distribution with reaction period for the decarbonylation of ANQ at about 570°C and that of PHQ at about 560°C, respectively. There is an apparent induction period for both compounds. The batch system apparatus adopted in these experiments was responsible for this induction period since 2 - 3 minutes were needed to heat up the inserted ampule to a temperature within 95% of the temperature in °C of that of the furnace block. It was concluded that no real induction period exists in these reactions.

Based on Figures 1 and 2, typical consecutive reaction schemes 1 and 2 occur during the decarbonylation reactions of ANQ and PHQ. Such a conclusion seems justified even though the reac-

$$\xrightarrow{-CO} \quad \xrightarrow[+H_2]{-CO} \quad \xrightarrow[+H_2]{} \; 2 \qquad (1)$$

$$\xrightarrow{-CO} \quad \xrightarrow[+H_2]{-CO} \quad \xrightarrow[+H_2]{} \; 2 \qquad (2)$$

Table I Typical Data for Hydrodecarbonylation

	ANQ				
Temperature, °C	568	567	614	620	600*)
Reaction period, min	10	100	7.5	80.0	60.0
H_2/ANQ, PHQ or FLR, molar ratio	10.9	10.1	11.3	12.4	11.8
Conversion, %	24.2	93.9	47.2	100.0	71.1
Mole of obtained compounds per 100 moles of feed					
CO	4.0	114.4	42.8	178.4	89.7
Benzene	0.0	6.7	0.0	84.8	1.5
Biphenyl	0.0	47.4	12.4	34.3	10.9
Fluorene	0.0	16.7	2.4	5.6	0.2
FLR	6.2	11.5	25.9	0.0	21.3
ANT or PHN	0.1	7.2	0.9	1.8	0.0
ANQ or PHQ	75.8	6.1	52.8	0.0	28.9
Total mole	86.1	210.0	137.2	304.9	152.5
Mass balance, %	81.7	88.3	92.9	91.3	68.3

*) In the presence of He instead of H_2.

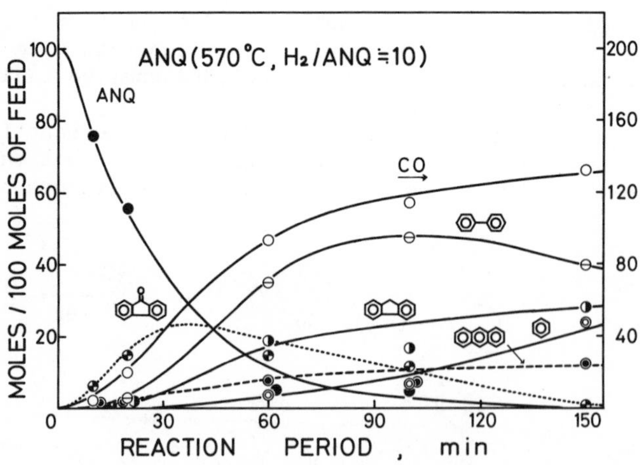

Figure 1. Time course of reaction of ANQ at 570°C

of ANQ, PHQ and FLR

PHQ					FLR	
560	556	583	588	558*)	600	616*)
5.0	15.0	7.5	120	10.0	15.0	7.5
10.3	11.0	11.1	11.4	10.2	12.1	13.2
52.6	93.9	100.0	100.0	74.2	61.8	20.0
37.8	78.7	84.1	163.5	64.3	49.5	9.7
0.0	0.0	0.0	65.9	0.0	4.6	0.0
0.8	4.6	8.2	30.7	0.0	48.0	4.2
3.4	4.3	4.5	4.7	1.1	3.2	0.5
30.1	65.1	78.5	0.0	63.2	38.2	80.0
0.0	0.0	0.0	2.3	0.0	-	-
47.4	6.1	0.0	0.0	25.8	-	-
119.5	158.8	175.3	267.1	154.4	143.5	94.4
81.9	80.0	88.9	75.2	90.0	92.0	85.5

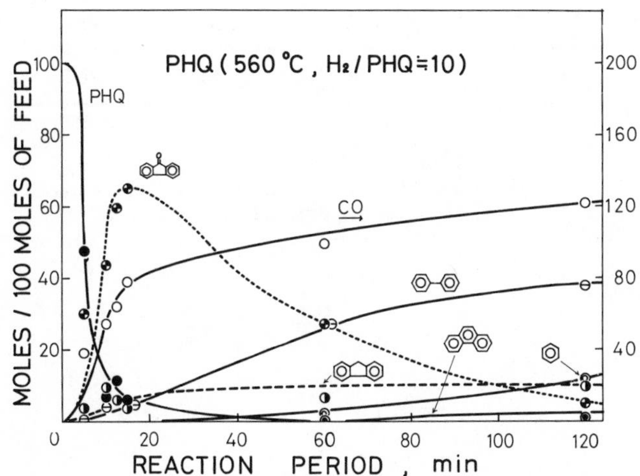

Figure 2. Time course of reaction of PHQ at 560°C

Table II Typical Data for Hydrodecarbonylation

	1,4-NPQ		
Temperature, °C	581	582	582
Reaction period, min.	5.0	15.0	30.0
H_2/NPQ, molar ratio	19.5	20.4	19.4
Conversion, %	51.3	100.0	100.0
Mole of obtained compounds per 100 moles of feed			
CO	54.4	155.0	167.0
CO_2	2.7	7.2	5.1
Methane	0.0	16.1	36.3
Ethane	0.0	6.6	15.6
Benzene	0.7	11.7	24.9
Toluene	0.4	8.8	16.2
Ethylbenzene	1.1	18.8	7.1
Styrene	7.3	2.8	0.4
Indene	0.0	0.4	0.9
Peak No.5*)	0.0	2.7	2.8
Naphthalene	0.4	1.4	3.6
1,4- or 1,2-NPQ	48.7	0.0	0.0
Total mole	115.7	231.5	279.9
Mass balance, %	65.1	61.0	70.1

*) Unidentified substance.

**) Sealed ampule was treated at 200°C for 10 min. before putting into the furnace maintained at 580°C.

of 1,4- and 1,2-NPQ

1,2-NPQ			
577	572	578	579**)
5.0	15.0	30.0	15.0
21.2	22.0	20.8	20.7
100.0	100.0	100.0	100.0
43.1	51.0	48.2	28.2
16.8	16.4	17.4	18.9
0.9	2.6	5.2	2.0
0.3	1.2	2.1	0.5
0.7	3.0	4.2	2.0
0.6	1.3	2.2	0.6
2.3	5.3	2.7	1.4
2.1	0.0	0.0	0.0
0.0	0.0	0.0	0.0
0.0	0.0	0.0	0.0
0.0	0.9	2.0	0.0
0.0	0.0	0.0	0.0
66.8	81.7	84.0	53.6
16.1	20.6	21.1	12.9

Table III Typical Data for Hydrodecarbonylation of p-BNQ

Temperature, °C	585	589	591
Reaction period, min.	2.5	7.5	30.0
H_2/BNQ molar ratio	9.9	9.6	10.0
Conversion, %	57.6	100.0	100.0
Mole of obtained compounds per 100 moles of feed			
CO	26.9	89.8	108.1
CO_2	2.6	7.4	7.5
Methane	0.0	2.7	17.0
Ethane	0.0	1.8	13.4
Ethylene	0.0	1.1	0.9
Acetylene	0.0	0.5	4.6
1-Butene	0.0	0.5	0.3
t-2-Butene	0.0	0.8	0.6
c-2-Butene	0.0	0.6	0.5
Butadiene	12.0	0.3	0.3
Benzene	0.5	3.0	5.8
Toluene	0.0	0.6	1.4
Ethylbenzene	0.0	0.8	0.2
Styrene	0.3	0.2	0.0
BNQ	42.4	0.0	0.0
Total mole	84.7	110.1	160.6
Mass balance, %	57.0	32.3	45.0

tion temperature was 30 - 40°C lower for the case of PHQ than that of ANQ to attain the corresponding conversions.

<u>1,4- and 1,2 NPQ and p-BNQ</u>. In the cases of 1,4- and 1,2-NPQ and p-BNQ, as listed in Tables II and III, respectively, the shortages in mass balances were serious and at the same time fairly large amounts of CO_2 were formed.

For the case of 1,4-NPQ, however, the decarbonylation is seen to be still a main reaction step based upon the mass balances obtained as well as the change in the product distribution with time as illustrated in Figure 3. Although indenone was not detected as one of the products in the decarbonylation of 1,4-NPQ, the formation of styrene, ethylbenzene, and CO during the initial stage of the reaction enabled us to propose the following reaction scheme 3.

$$\text{1,4-NPQ} \xrightarrow{-CO} [\text{indenone}] \xrightarrow[+H_2]{-CO} \text{styrene} \xrightarrow{+H_2} \text{ethylbenzene} \begin{cases} \xrightarrow{+H_2} \text{toluene}, \text{benzene}, CH_4 \\ \xrightarrow{+H_2} \text{benzene}, C_2H_6, CH_4 \end{cases} \quad (3)$$

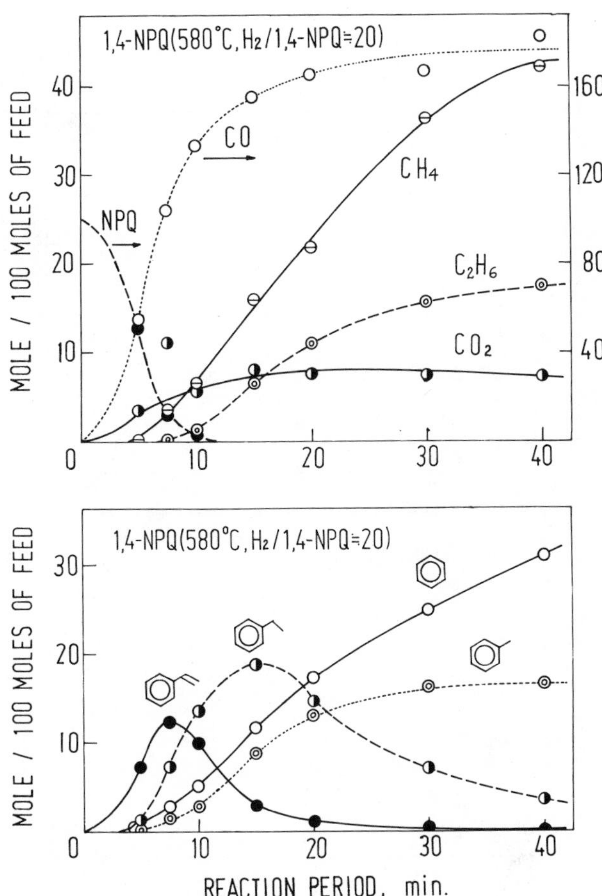

Figure 3. Time course of reaction of 1,4-NPQ at 580°C

Similar schemes 4 and 5 can be postulated for the decarbonylations of 1,2-NPQ and p-BNQ, see Figures 4 and 5, respectively; the shortages in mass balances were relatively large in both cases and are probably caused by the polymerization of the feed quinones.

$$\xrightarrow{-CO} \quad \xrightarrow[+H_2]{-CO} \quad \xrightarrow{+H_2} \quad \xrightarrow{+H_2} \quad C_2H_6,\ CH_4 \qquad (4)$$

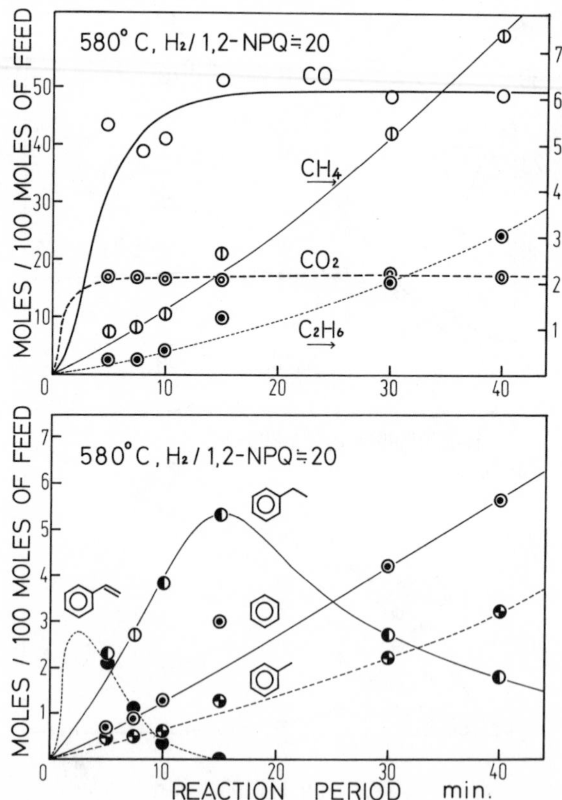

Figure 4. Time course of reaction of 1,2-NPQ at 580°C

$$\xrightarrow{-CO} \xrightarrow[+H_2]{-CO} \quad \xrightarrow{+H_2} C_4H_8\text{'s}, C_2H_6, C_2H_4, CH_4 \qquad (5)$$

With high speed liquid chromatography, several unidentified substances with molecular weights higher than those of the feed quinones were detected in the product streams. Smaller percentages in mass balance in the result of experiment listed in the last column of Table II than the other corresponding experiments revealed that polymerization of 1,2-NPQ proceeded even at 200°C in the presence of hydrogen and that the polymerization depressed the

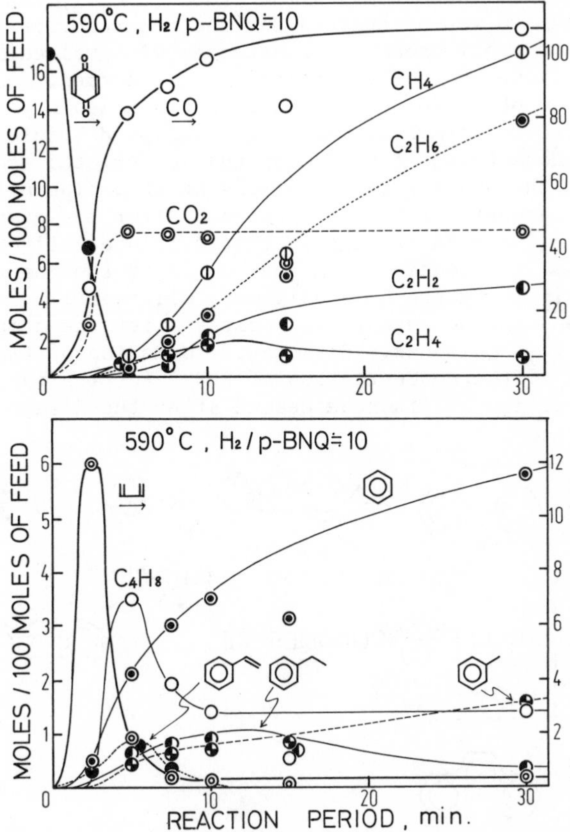

Figure 5. Time course of reaction of p-*BNQ at 590°C*

decarbonylation considerably. It is not clearly understood currently why the shortages in mass balance were large especially in the cases of 1,2-NPQ and p-BNQ. The reactive double bonds in the structures of these compounds may play some important role for the formation of high molecular weight substances.

Formation of carbon dioxide was observed during the initial stage of the reaction. It equilibrated rapidly after the consumption of feed quinones maintaining a constant amount throughout the reaction period. The amount of carbon dioxide formation was in the order 1,2-NPQ > p-BNQ > 1,4-NPQ > PHQ > ANQ. No carbon dioxide was detected in the decarbonylation of FLR to biphenyl and benzene. Information obtained above suggests that carbon dioxide is formed solely from quinones and that it is produced in a unimolecular way, i.e., not through intermolecular reaction between two carbonyl groups.

Kinetics and Mechanism on the Decarbonylation of ANQ and PHQ. Based on the facts summarized below from a) to d), the overall reaction scheme 6 was postulated for the decarbonylations of ANQ and PHQ. a) A small amount of substances with molecular weights higher than those of the feed quinones were formed. Although identification of these substances is not available yet, their formation were confirmed by means of high speed liquid chromatography. b) More CO was formed than that calculated stoichiometrically from the amounts of formations of FLR, biphenyl and benzene. Small amounts of CO were produced after the disappearance of quinones. These facts suggest that a quinone oligomer, Oligomer I, gradually released CO in the course of the reaction converting itself to oxygen-lean oligomer, Oligomer II. c) Anthracene and phenanthrene were formed mostly after the complete consumption of feed quinones. d) Rate of formation of fluorene was proportional to the concentration of FLR. For the pyrolysis of FLR, the formation of fluorene ceased after the disappearance of FLR.

$$\text{ANQ, PHQ} \xrightarrow[-CO]{k_{1A},\,k_{1P}} \text{FLR} \xrightarrow[-CO,\ +H_2]{k_2} \text{biphenyl} \xrightarrow[+H_2]{k_3} 2\ \text{benzene}$$

$$\text{FLR} \xrightarrow[+2H_2,\ -H_2O]{k_4} \text{fluorene}$$

$$\text{ANQ, PHQ} \xrightarrow{k_{5A},\,k_{5P}} \text{OLIGOMER (I)} \xrightarrow[-CO]{k_{7A},\,k_{7P}} \text{OLIGOMER (II)}$$

$$\text{OLIGOMER (I)} \xrightarrow[+2H_2,\ -2H_2O]{k_{6A},\,k_{6P}} \text{anthracene, phenanthrene} \quad (6)$$

In order to grasp the overall reaction kinetically, the first-order reaction kinetics were assumed for all steps of the scheme.

Notations k_{1A}, k_{5A}, k_{6A}, k_{7A} and k_{1P}, k_{5P}, k_{6P}, k_{7P} mean the first-order rate constants of indicated steps for ANQ and PHQ, respectively. Assumption of first-order kinetics is reasonable at least for the main reaction course, i.e., steps 1A, 1P, 2 and 3 because the molar ratio of H_2 to quinone was more than 10. Estimation of k values was carried out by means of RUNGE-KUTTA method at four temperature levels for the reactions of ANQ and PHQ as well as for those of FLR.

Figures 6 and 7 illustrated the results of applied kinetics for the reactions of ANQ at 600°C and FLR at 570°C, respectively. Deviations between the experimental plots and the calculated curves for fluorene and benzene in Figure 6 and FLR and CO in Figure 7 were left unaltered for the reason that the optimization of k values was carried out on all data obtained at different temperatures and by different feed stocks. If any deficiency exists in the reaction scheme 6, it will be the oligomerization of FLR. Deviation of calculated curve for fluorene in Figure 7 suggests that FLR also participates in the oligomerization reaction with or without quinones. In general, there are reasonably good fits between the experimental data and the calculated curves. The k values obtained were normalized to 540°C and 600°C, and they are listed in Table IV accompanied with kinetic parameters for all reaction steps in the scheme 6. For step 6 or 7, k_{6A} and k_{6P} or k_{7A} and k_{7P} were indistinguishable from the obtained data, so that both were combined to k_6 or k_7, respectively. Arrhenius plots for k_{1A}, k_{1P}, k_2 and k_3 are given in Figure 8.

From Table IV, it is obvious that PHQ decomposes more easily than ANQ, FLR or biphenyl, that the activation energies for decarbonylation reactions are nearly equal or a little smaller than that of the hydrodecomposition of biphenyl, and that the decarbonylation reactions are much more important than polymerization at higher temperatures; it is inadequate however to assume first-order kinetics for polymerization step k_{5A} or k_{5P}.

Mukai et al. (17) reported that in the thermal decarbonylation of tropone the three membered ring compound might be a possible reaction intermediate.

$$\text{tropone (=O)} \longrightarrow [\text{bicyclic intermediate (=O)}] \longrightarrow \text{benzene} + CO \quad (7)$$

They said that the formation of such an intermediate is thermally allowed from the orbital symmetry rule (18). Since the same symmetrical situation is expected for the structure of ANQ, a prediction of the thermal decarbonylation of ANQ to proceed via such an intermediate may be possible in accordance with the symmetry rule. However, the same reasoning is difficult in the case of PHQ in which the decarbonylation proceeds more rapidly than ANQ.

Preferable thinking is to postulate homolytic cleavage at the C-C bond next to the oxygen, as exemplified below in the case of PHQ.

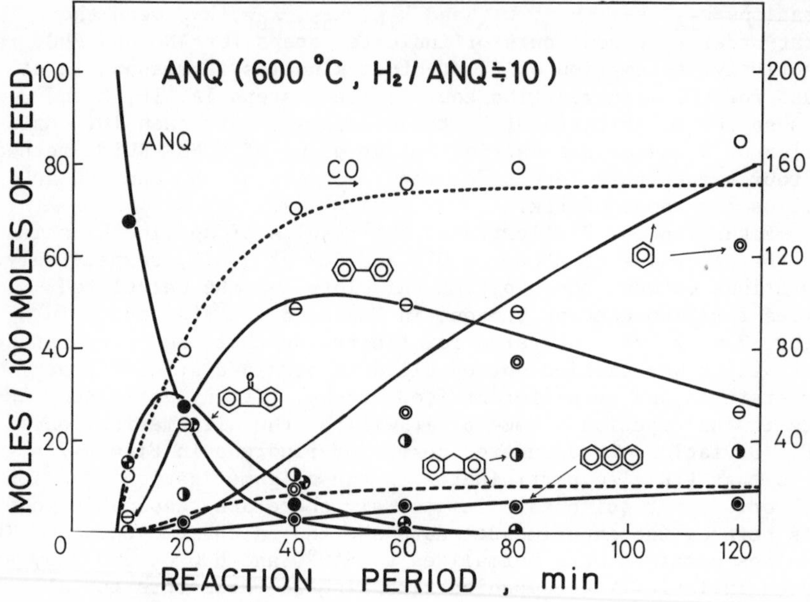

Figure 6. Applied kinetics for reaction of ANQ at 600°C based on reaction scheme 6

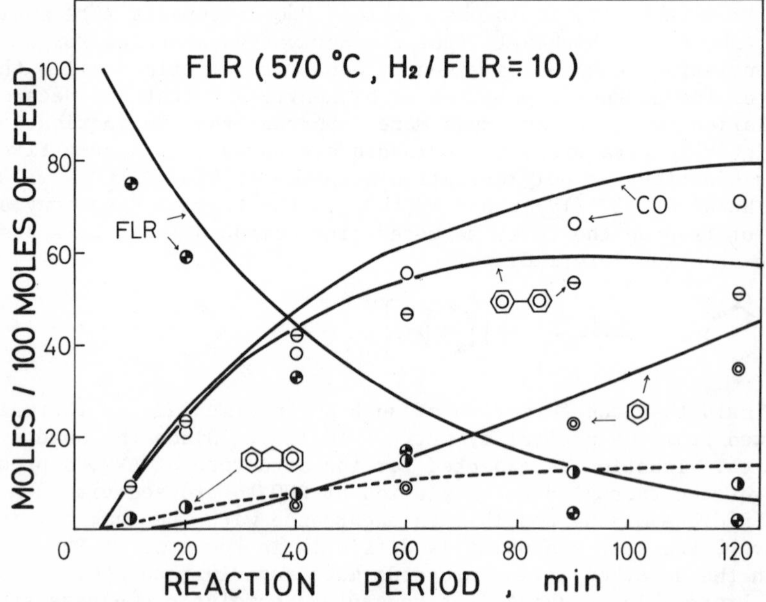

Figure 7. Applied kinetics for reaction of FLR at 570°C based on reaction scheme 6

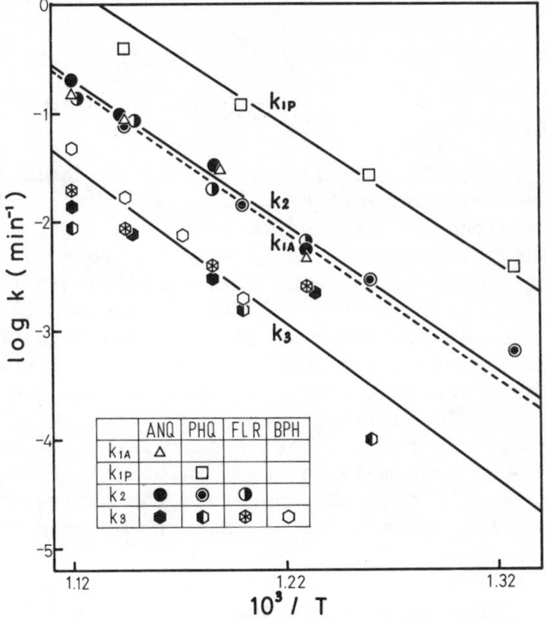

Figure 8. Arrhenius plots for k_{1A}, k_{1P}, k_2, *and* k_3

Table IV. Rate Constant and Kinetic Parameter for Each Step in Reaction Scheme (6)

	Rate const. (min^{-1}) 540°C	600°C	E (kcal/mole)	logA
k_{1A}	0.0047	0.090	62.0	14.5
k_{1P}	0.052	0.40	55.8	13.8
k_2	0.0055	0.09	61.4	14.4
k_3	0.0016	0.009	64.4	14.4
k_4	0.0010	0.013	60.0	14.9
k_{5A}	0.0044	0.023	42.6	9.1
k_{5P}	0.023	0.10	39.5	9.0
k_6	0.0009	0.007	61.7	13.6
k_7	0.0008	0.003	59.0	12.3

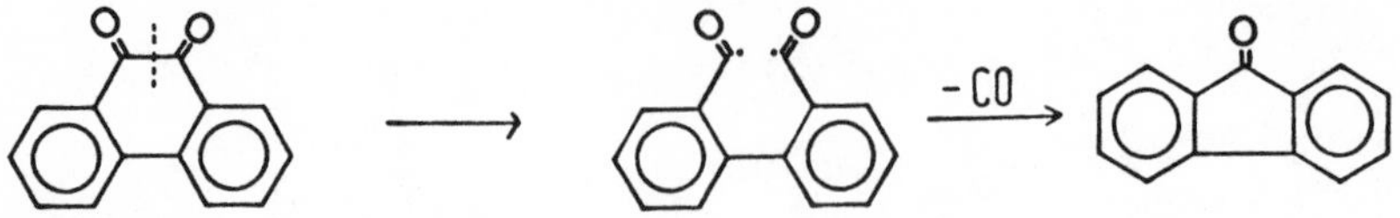

(8)

Brown et al. (19) took a similar standpoint in their study on pyrolysis of phthalic anhydride at 700 - 1100°C and 0.1 - 10 mm Hg to yield biphenylene and triphenylene and other related polycarbonyl compounds, although they suggested the possibility that the intermediate is zwitterion rather than diradical in their case of pyrolysis of phthalic anhydride (20,21). Reaction 8 is supported by a smaller π-bond population at the C-C bond between 9 and 10 carbons than that between 10 and 11 carbons in the molecular diagram of PHQ, which was calculated to be 0.254 and 0.443, respectively, by the simple Hückel program (22). Moreover, the experimental evidences that the activation energies of decarbonylations are between 56 - 62 kcal/mole in their values, that PHQ decarbonylates more easily than ANQ, and that the activation energy of the reaction of FLR is as large as that of ANQ encourage the homolysis mechanism.

When hydrogen was replaced with helium in reactions of PHQ, ANQ and FLR, a remarkable change took place in the reaction profile. The results are illustrated in Figures 9 and 10 for PHQ and ANQ respectively. In the case of PHQ, nearly the same molar amount of FLR and CO were formed as two main products of the reaction and the rate of FLR formation was as large as that in the case of hydrogen dilution. This was the expected result in line with the first step in the scheme of the reaction. In the case of ANQ, however, the rate of formation of FLR was remarkably decreased, i.e., there was observed explicitly a long induction period. Moreover, oligomerization of ANQ took place and small amounts of biphenyl and benzene were formed successively. Pyrolysis of FLR in helium proceeded in a similar way as that of ANQ.

Namely, in the case of ANQ and FLR, the action of hydrogen on the substrate is necessary for the progress of decarbonylation even at the first step of the reaction.

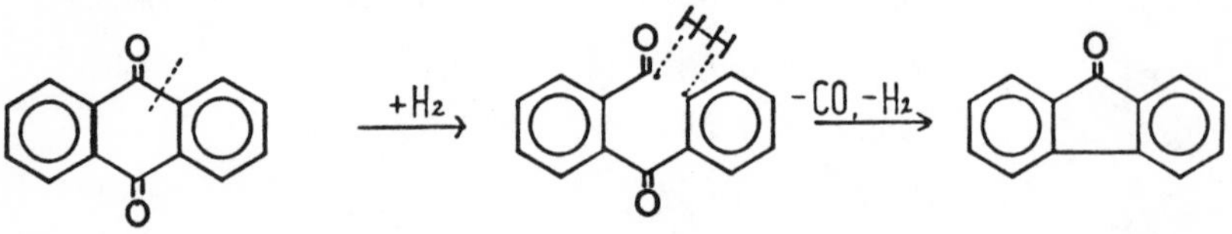

(9)

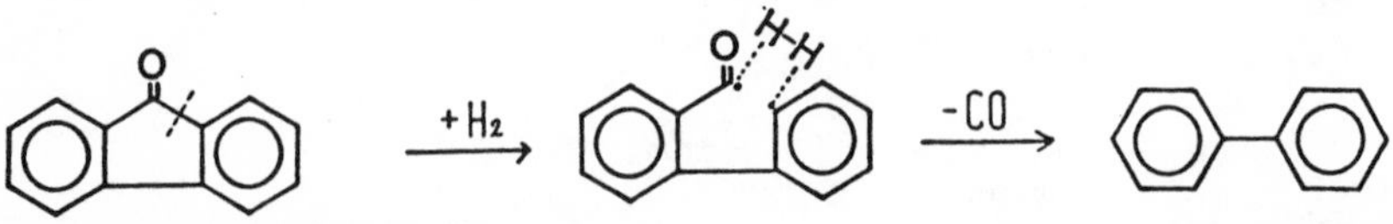

(10)

The facts that no CO_2 is formed in the hydrodecarbonylation of FLR

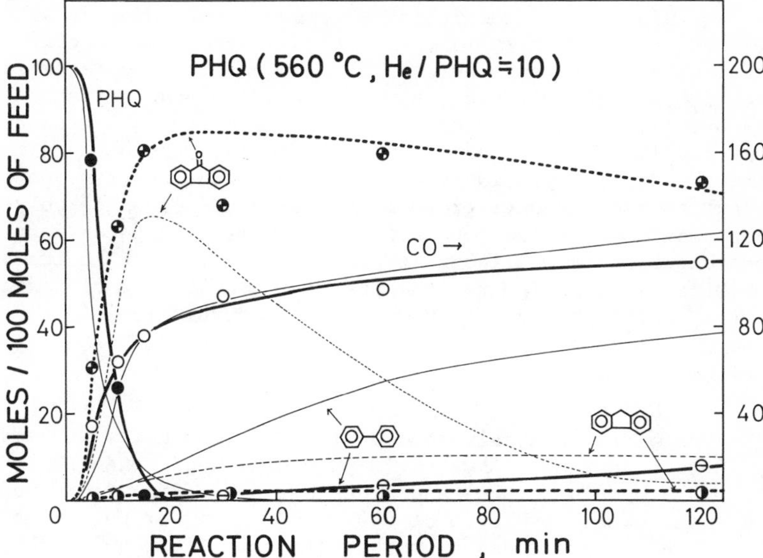

Figure 9. Decarbonylation of PHQ in He (thick line) instead of H_2 (thin line) at 560°C

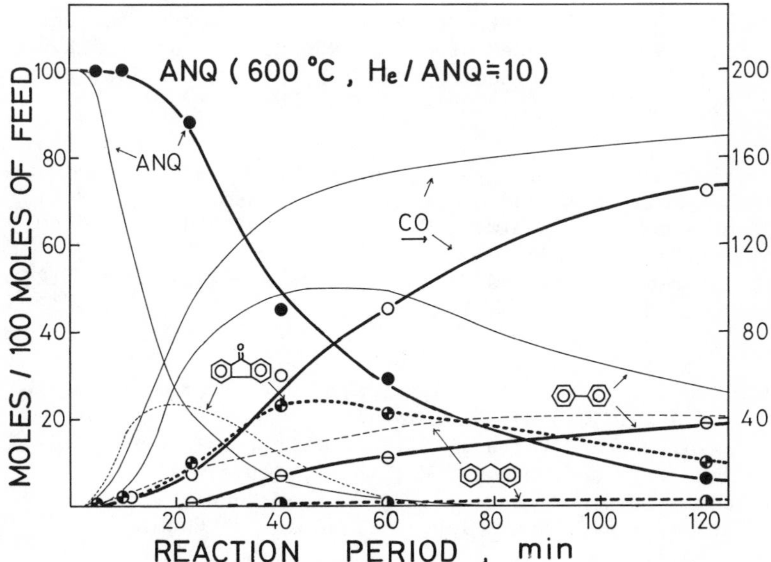

Figure 10. Decarbonylation of ANQ in He (thick line) instead of H_2 (thin line) at 600°C

and a negligibly small amount in the case of ANQ coincide with the participation of hydrogen in the initial step of the reaction.

Structual difference between PHQ and ANQ and the resemblance between ANQ and FLR will be the main reason for the different reaction profile of PHQ as compared to the other two. In the cases of 1,4- and 1,2-NPQ and p-BNQ, an analogous discussion is difficult on the decarbonylation mechanism due to the reactive double bond next to the carbonyl group in these compounds which polymerized easily giving serious shortages in mass balances in the present experiments. Although the action of hydrogen in reactions 9 and 10 is still ambiguous, the postulated homolytic cleavage will be a most probable mechanism in the decarbonylation of quinones.

Abstract

Non-catalytic thermal decarbonylation of quinones proceeded for 9,10-anthraquinone (ANQ), 9,10-phenanthrenequinone (PHQ), 1,4- and 1,2-naphthoquinone(NPQ) and p-benzoquinone (BNQ) in the presence of atmospheric hydrogen at 500 - 600°C. ANQ or PHQ decarbonylated in the presence of hydrogen to form fluorenone (FLR) at the first step, followed by successive hydrodecarbonylation and hydrocracking to form biphenyl and benzene. In the case of 1,4-NPQ, the reaction did not proceed in any obvious stoichiometric relation. Only 60 - 70 % of 1,4-NPQ decarbonylated and was hydrogenated yielding styrene, ethylbenzene, toluene, benzene and lower hydrocarbons; the remainder polymerized or partly polycondensed. Similar, or less selective results were obtained for 1,2-NPQ and p-BNQ. Analyses of kinetic treatments are reported for the consecutive reaction schemes of ANQ or PHQ as well as of FLR. The kinetic parameters for decarbonylation, polymerization, polycondensation reactions of quinones are discussed.

Literature Cited

(1) Campbell, K.N. and Campbell, B.K., Chem. Rev., (1942) 31, 77.
(2) Friedman, S., Kaufman, M.L., Wender, I., J. Org. Chem., (1971) 36, 694.
(3) Galbreath, R.B. and Van Driesen, R.P., Panel Disc. 12, Proc. 8th WPC (Moscow), 1971.
(4) Horne, W.A. and McAfee, J., "Adv. Petrol. Chem. & Refining, Vol.III", pp.215-218, 1960.
(5) Qader, S.A. and Hill, G.R., Sym. Div. Petrol. Chem. ACS (Chicago), A63, 1970.
(6) Sakai, T. and Hattori, M., Chem. Letter (Japan), (1974) 617.
(7) Sakai, T. and Hattori, M., ibid., (1975) 1150.
(8) Davies, W. and Ennis, B.C., J. Chem. Soc., (1960) 1488.
(9) Brown, R.F.C. and Solly, R.K., Aust. J. Chem., (1966) 19, 1045.
(10) Brown, R.F.C. and Butcher, M., ibid., (1970) 23, 1907.
(11) Hausigk, D., Tetrahedron Lett., (1970) (No.28) 2447.

(12) Davies, D.G. and Hodge P., J. Chem. Soc. (C), (1971) 3158.
(13) Braude, E.A. and Fawcett, J.S., Org. Syn., (1963) Coll. Vol. IV, 698.
(14) Fieser, L.F., ibid., (1943) Coll. Vol.II, 430.
(15) Underwood, Jr., H.W. and Walsh, W.L., ibid (1943) Coll. Vol. II, 553.
(16) Wendland, R. and Lalonde, J., ibid., (1963) Vol.IV, 757.
(17) Mukai, T., Nakazawa, T. , Shishido, T., Tetrahedron Lett., (1967) (No.26) 2465.
(18) Woodward, R.B. and Hoffmann, R., J. Am. Chem. Soc., (1965) 87, 395, 2046.
(19) Brown, R.F.C., Gardner, D.V., McOmie, J.F.W., Solly, R.K., Aust. J. Chem., (1967) 20, 139.
(20) Brown, R.F.C. and Butcher, M., ibid., (1969) 22, 1457.
(21) Brown, R.F.C., Gream, G.E., Petero, D.E., Solly, R.K., ibid., (1968) 21, 2223.
(22) Kuboyama, A., Bull. Chem. Soc. Japan, (1959) 32, 1226.

27

Fuels and Chemicals by Pyrolysis

JAMES R. LONGANBACH and FRED BAUER

Occidental Research Corp., 1855 Carrion Rd., LaVerne, Calif. 91750

The Occidental Research Company, a subsidiary of Occidental Petroleum Corporation, is developing a highly versatile flash pyrolysis process to produce gases, liquids and chars from a variety of organic materials. Flash pyrolysis has the potential of producing maximum yields of gases and liquids from coal and organic solid wastes such as municipal refuse, tree bark, cow manure, rice hulls and grass straw using simple process equipment. The main features of the process are near ambient pressure, no requirement for added chemicals, low capital investment, high feed throughput, flexibility of feedstock, variability of temperature, and minimum feed pretreatment.

Process Description

A diagram of the process is shown in Figure 1. The organic material is heated by contact with hot recycle char made in the process and carried in a gas stream through a reactor where pyrolysis occurs at very short residence times and high heat-up rates. This method maximizes the volatile yield and protects the products from further cracking. The char-tar-gas mixture is separated in a cyclone, and the char is heated by partial combustion with air in a second entrained bed, separated from the combustion gases and recycled. Excess char is removed and can constitute a sizeable by-product depending on the starting material used. The product gas and tar are separated and are then available for processing.

This process maximizes the yields of high value products, both gaseous and liquid and produces a char which can be used as a fuel. The detailed nature of these products will be discussed along with the conditions used and yields obtained in the various modifications of the basic process.

Coal Gasification

Two coal processing modifications are being developed. In the first, coal gasification, subbituminous coal has been used. An analysis of a typical subbituminous coal is shown in Table I.

TABLE I

SUBBITUMINOUS COAL
USED IN COAL GASIFICATION

Analysis	Weight % As Fed	Weight % Dry Basis
Moisture	15.1	0.0
Ash	10.3	12.1
Volatile Matter	31.1	31.1
Fixed Carbon	43.5	56.8
C	55.3	65.1
H	5.3	4.3
N	0.7	0.8
S	1.3	1.5
O	27.1	16.2
Heating Value, Btu/lb	9268	10913
Relative Ignition Temperature, °F	830	

The preferred temperature for coal gasification has been 1600°-1700°F. The residence time is kept as short as possible to minimize secondary cracking of the gas. The amount of hydrogen and carbon monoxide which can be stripped from the coal increases with temperature but the maximum gasification temperature is determined by the char heating step. The char must be at least 100°F hotter than the desired reactor temperature to provide driving force for heat transfer. The softening point for char ash from typical American subbituminous coals is approximately 2000°F. Typical yields at 1700°F are 42% char, 9% tar, 37% gas and 12% water.

A typical analysis of gas produced at 1600°F is shown in Table II.

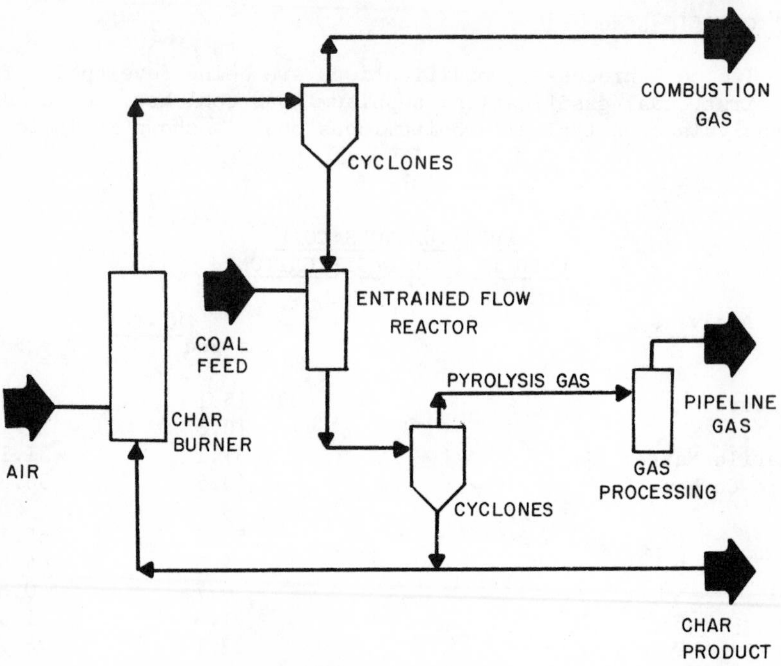

Figure 1. Occidental flash pyrolysis process schematic

C_2H_5 OH

H H H N

$C_{23}H_{19}NO$

MOLECULAR WEIGHT	325 g/mole
AROMATIC CARBON	83 %
ALIPHATIC CARBON	17 %
AROMATIC HYDROGEN	53 %
ALIPHATIC HYDROGEN	42 %
PHENOLIC HYDROGEN	5 %
AROMATIC SULFUR	
AROMATIC NITROGEN	
PHENOLIC OXYGEN	

Figure 2. Typical tar molecule from coal gasification

TABLE II

GAS ANALYSIS FROM COAL GASIFICATION AT 1600°F

Gas	Mole %[(a)]
H_2	26.8
CO	30.0
CO_2	8.5
CH_4	22.4
C_2H_2	0.7
C_2H_4	8.9
C_2H_6	1.0
C_3H_6	1.2
C_4+	0.3
Molecular Weight	20.0 g/Mole
Heat of Combustion (b)	626 Btu/SCF

(a) Dry, N_2-Free, H_2S-Free

(b) Calculated, 25°C, HHV

The ethylene yield is 8.9%. Optimization of the ethylene yield for possible separate sale is one of the goals of the coal conversion program. The gas is medium Btu gas, 620 Btu/SCF, and may be processed to obtain high Btu pipeline gas using existing technology.

Tar is produced as a by-product during coal gasification. The tar yield is ∿5% at 1700°F and increases to >15% at 1200°F. A structure of a typical tar molecule made at 1650°F is shown in Figure 2. The typical tar molecule contains 3-4 condensed aromatic rings with aliphatic side chains. The carbon content is 83% aromatic and 17% aliphatic. The hydrogen distribution is 53% aromatic, 42% aliphatic and 5% phenolic. The tar also contains aromatic nitrogen and sulfur and phenolic oxygen. Typical tar properties are shown in Table III. Because of the high heat of combustion (15,666 Btu/lb) and low sulfur content (0.7%), the tar could be used directly as a fuel. The high aromatics content makes the tar attractive for use as a petrochemical feedstock. The tar might also be cracked further to yield additional

TABLE III

TAR FROM COAL GASIFICATION AT 1650°F

Analysis	Weight %
Moisture	0.0
Ash	0.0
C	86.5
H	5.7
N	1.0
S	0.7
O	6.0
Density, (at 100°C)	1.14g/ml
Viscosity, (at 100°C)	33 cps
Heating Value	15,666 Btu/lb
Molecular Weight, (Number Average)	315g/mole

Boiling Range

Percent Distilled	Temperature °C
IBP	100
13 %	100-200
20.5%	200-300
30 %	300-400
10.5%	400-600
Residue (16%)	>600

Distillate, % by Volume (100-180°C Fraction)

Acids	20.4
Bases	0.7
Neutral Oil	78.9

gas. However, the tar yield and the amount of potential gas available by further cracking decrease as the coal gasification temperature is increased. Options available for use of the tar are presently being studied.

The char from gasification of subbituminous coal is an ideal pulverized fuel. A typical analysis is shown in Table IV for a char which has been devolatilized at 1600°F but has not been partially burned to process heat, and for char which has been partially burned at 1900°F.

TABLE IV

CHAR FROM GASIFICATION OF SUBBITUMINOUS COAL

Analysis	Weight % Devolatilized 1600°F	Weight % Partially Combusted 1900°F
Moisture	0.0	0.0
Ash	19.5	21.6
Volatile Matter	17.2	6.7
Fixed Carbon	63.3	71.7
C	69.1	73.5
H	2.5	0.8
N	0.9	1.0
S	1.2	0.6
O	6.7	2.5
Heating Value, Btu/lb	11,111	10,994
Relative Ignition Temp., °F	740	870
Surface Area, m^2/g	123	312
Pore Volume, ml/g	0.12	0.17

To provide the process heat about 12% of the char is burned. The heating value of the char changes only slightly, the surface area is increased and the sulfur content is decreased by partial combustion.

The relative ignition temperatures of the chars were determined by dispersing them in oxygen in a hot tube at short residence time to simulate ignition of pulverized fuel. The relative ignition temperature of the unoxidized char is lower than the parent subbituminous coal (830°F). The partially oxidized char ignites at a temperature slightly higher but comparable to the ignition temperature of the coal. Full scale combustion tests have been made on these chars by an outside contractor.

Most Western subbituminous coals will yield a char with a sufficiently low level of sulfur to meet emission standards as a pulverized fuel so no effort has been made to desulfurize these chars.

Coal-to-Liquids

The second process utilizing coal is the conversion of coal-to-liquids. The coal to liquids process was originally developed to convert hvc bituminous coal from the Western Kentucky Hamilton Mines of Island Creek Coal Co., also a subsidiary of Occidental Petroleum Corporation, into liquid fuels. A typical analysis of the coal is shown in Table V.

TABLE V

HAMILTON BITUMINOUS hvc COAL

Analysis	Weight %	
	As Fed	Dry Basis
Moisture	2.6	0.0
Ash	9.7	10.0
Volatile Matter	37.6	37.6
Fixed Carbon	50.2	52.4
C	70.7	72.6
H	5.2	5.0
N	1.5	1.5
S	2.7	2.8
O	10.3	8.1
Heating Value, Btu/lb	12,754	13,082
Relative Ignition Temp., °F	940	

Fischer Assay, (As Fed)	
Char	73.4
Tar	14.5
Water	8.2
Light Oil	0.0
Gas	3.9

The optimum yield of liquid product is obtained in the coal-to-liquids process at approximately 1075°F. A typical product distribution for coal-to-liquids processing using this coal is 56% char, 35% tar, 7% gas, and 2% water. The residence time for this process is also kept as short as possible which maximizes the yield and prevents further cracking of the liquid product. The effect of flash pyrolysis on the liquid yield is shown by the fact that the tar and light oil yield for this coal from slow, fixed bed heating in the Fischer Assay is 14.5% while the liquid yield from flash pyrolysis is ∿35%. The liquid product from hvc bituminous coal has the analysis shown in Table VI.

TABLE VI

TAR FROM THE COAL-TO-LIQUIDS PROCESS AT 1000°F

Analysis	Weight %
Moisture	0.0
Ash	0.0
C	80.3
H	7.0
N	1.4
S	2.1
O	9.2
Density (at 100°C)	1.04g/ml
Viscosity (at 100°C)	127 cps
Heating Value	15342 Btu/lb
Boiling Range	
Percent Distilled	Temperature, °C
IB P	75
20%	75-200
25	200-350
22.5	350-520
Residue (32.5%)	>520
Distillate, % by Volume (75-350°C Fraction)	
Acids	25%
Bases	--
Neutral Oils	75%

There is a large amount of oxygenated material which is primarily tar acids. Work will be done to try to isolate these as a separate product fraction.

The primary intent of the process is to hydrogenate the tar liquid to produce a low sulfur fuel oil or possibly a synthetic crude oil.

Conversion of the tar to chemicals by a second stage pyrolysis of the tar-gas stream before tar condensation has also been investigated. The goal of this work is an economically viable yield of BTX or low molecular weight gases such as ethylene or methane.

Typical char analyses from the coal-to-liquids process are shown in Table VII.

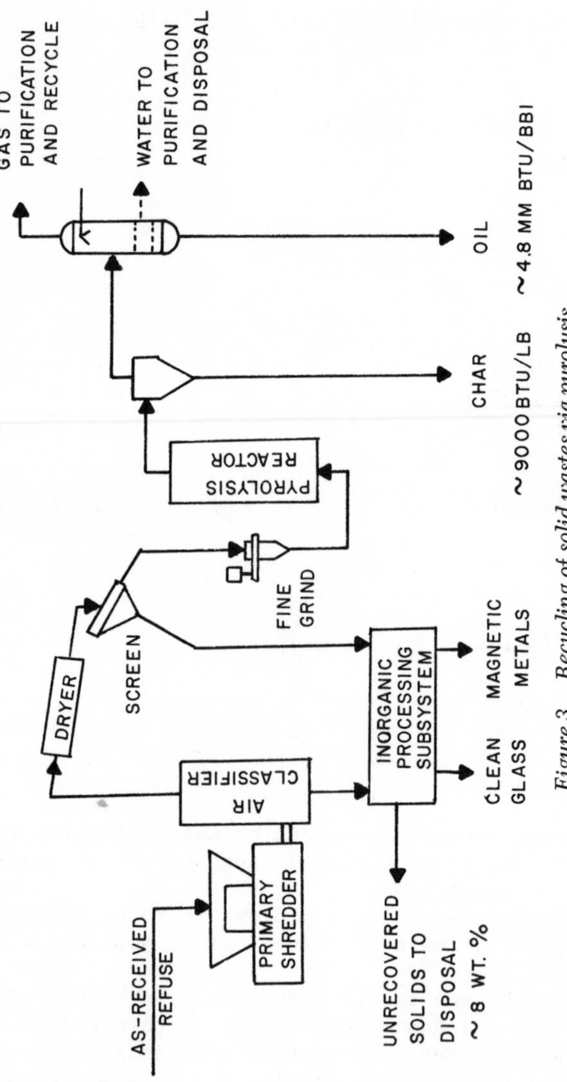

Figure 3. Recycling of solid wastes via pyrolysis

TABLE VII

CHAR FROM THE COAL-TO-LIQUID PROCESS USING BITUMINOUS hvc COAL

Analysis	Weight %		
	Devolatilized	Partially Combusted	
	1000°F	1335°F	1600°F
Moisture	0.0	0.0	0.0
Ash	20.2	22.8	25.2
Volatile Matter	14.1	14.2	8.7
Fixed Carbon	65.7	63.0	66.1
C	68.4	70.7	68.8
H	2.3	2.2	1.5
N	1.6	1.6	1.5
S	2.4	1.9	1.9
O	5.1	0.8	1.0
Heating Value, Btu/lb	10,942	11,198	10,901
Relative Ignition Temperature, °F	970	1,030	1,050
Surface Area, m^2/g	29.1	124	269
Pore Vol., ml/g	0.03	0.07	0.13

Char made from flash devolatilization of the coal is shown along with chars partially burned to produce process heat at 1335°F and 1600°F. The char combustion temperature is expected to fall within this range. Again, partial combustion results in a char with lower sulfur, higher surface area, adequate volatile matter for use as a pulverized fuel and very similar heating value to the devolatilized char. The relative ignition temperature is increased by partial combustion.

Char produced from bituminous hvc coal contains too much sulfur to be used as pulverized fuel without sulfur removal procedures. Sulfur removal from the char seems preferable to stack gas cleanup. The desulfurization of char will not be discussed in this paper but a promising process has been developed.

Solid Waste Liquefaction

The ORC flash pyrolysis process has been used with minor modifications to process municipal and industrial solid wastes. The emphasis is on liquid yield. More preparation of the feed is involved than for coal gasificaltion or coal liquefaction. A process schematic is shown in Figure 3. In addition to grinding and drying, an air classification process and screening are used

TABLE VIII

FLASH PYROLYSIS OF SOLID WASTE FEEDSTOCKS

	ANIMAL WASTE	RICE HULLS	FIR BARK	GRASS STRAW	MUNICIPAL SOLID WASTE
Feed Analysis					
C	39.3	39.4	48.3	45.0	44.2
H	4.7	5.5	5.3	6.0	5.7
N	2.3	0.5	0.2	0.5	0.7
S	0.6	0.2	0.0	0.5	0.2
O	28.1	36.1	34.3	42.0	42.3
Cl	1.7	0.2	0.2	0.4	0.2
Ash	23.3	18.2	11.7	5.7	6.7
Char Analysis					
C	34.5	36.0	49.9	51.0	48.8
H	2.2	2.6	4.0	3.7	3.3
N	1.9	0.4	0.1	0.5	1.1
S	0.9	0.1	0.1	0.8	0.4
O	7.9	11.5	24.3	19.2	12.8
Cl	3.7	0.2	0.2	0.5	0.3
Ash	48.8	49.2	21.4	24.3	33.3
Btu/lb	5449	6100	8260	8300	9000
Yield, %	48.1	35.9	41.6	23.2	20.0
Gas Analysis					
H_2	6.1	23.3	6.3	9.6	11.7
CO	21.9	28.5	14.7	53.9	34.9
CO_2	55.9	40.8	64.7	26.9	35.4
CH_4	6.1	3.5	10.5	3.8	5.7
C_2H_4	0.5	0.8	1.0	0.8	4.0
C_2+	1.8	0.8	1.2	2.8	2.4
H_2S	7.8	2.4	1.7	2.2	0.4
Btu/SCF	226	215	222	349	350-500
Yield, %	10.8	12.1	8.2	5.2	30.0
Oil Analysis					
C	64.8	62.4	60.5	58.6	57.0
H	6.9	5.8	6.0	5.6	7.7
N	7.0	1.4	0.5	1.3	1.1
S	0.2	0.1	0.1	0.1	0.2
O	19.8	29.4	30.7	33.9	33.6
Cl	0.2	0.3	0.2	0.1	0.2
Ash	1.1	0.6	2.1	0.5	0.2
Btu/lb	11800	10400	10300	9400	10500
Yield, %	20.0	44.2	28.7	35.7	40.0

to obtain an organic feed which is relatively free from metals, glass, and other disposable solids.

As in coal pyrolysis, the incoming feed is heated by contact with recycled char and pyrolyzed to form liquids, gas and char. The char is separated by cyclones, liquid is separated with a quench fluid and the gas is recycled or purified. The residence time in the reactor again is very short. The temperature of pyrolysis found effective is 850°F to 1050°F. Quench fluids are chosen for their immiscibility, low vapor pressure and rapid phase disengagement. A desirable fluidity in the product oil may be achieved by varying the moisture content. The moisture content of the pyrolytic oil is controlled by the quench oil temperature. A moisture content of 14-18% gives a suitable viscosity for handling municipal solid waste oil.

Several pyrolysis experiments have been carried out with each type of feed. The data in Table VIII are from typical experiments but are not necessarily optimum with regard to liquid yield and fuel value. Animal waste, rice hulls, fir bark, grass straw and municipal solid waste have been pyrolyzed under nearly identical conditions.

The feedstock from municipal solid waste contains only 6.7% ash, 0.2% chlorine, 0.2% sulfur and 0.7% nitrogen. It is a highly oxygenated material with a C/O ratio of 1.39. The other feedstocks are also highly oxygenated but they contain more undesirable components, particularly animal waste, which has 2.3% nitrogen, 1.7% chlorine and 23.3% ash.

Yields of char vary from 20% for municipal solid waste to 48% for animal waste. Heating values range up to 9000 Btu/lb for char from municipal solid waste, similar to the value for as-received subbituminous coal. All of the chars contain large amounts of ash varying from 21% for fir bark char to 49% for chars from rice hulls and animal waste.

The gases produced have low heating value, from 200 to 500 Btu/SCF, because they contain from 27 to 65% CO_2. They contain from 1.7 to 7.8% H_2S except for municipal solid waste gas which contains 0.4% H_2S.

The oil analyses shown in Table VIII are given on a moisture-free basis. Oils as condensed contain 6-20% moisture, however. The pyrolytic oils derived from the solid waste feeds are fundamentally similar to pyrolyzed cellulose. Variation in plastic content, vegetable waxes, lignins and ash composition causes variation in yield, fuel value, and viscosity. A spectrum of molecular weights can be seen in Figure 4, ranging from 100 to 10,000 for bark oil. A similar distribution is anticipated for other cellulose-derived pyrolysis oils. Functional groups in the product oils include hydroxyl, carboxyl, carbonyl and ether due to the high oxygen content of the feed materials. Solvents for the pyrolytic oil are dilute aqueous caustic, dimethyl formamide and dimethyl sulfoxide. Although the pyrolytic

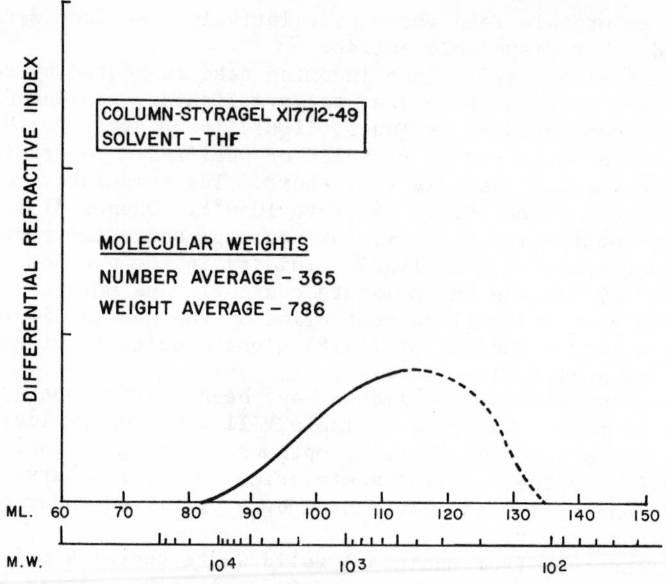

Figure 4. Molecular weight distribution of bark pyrolytic oil

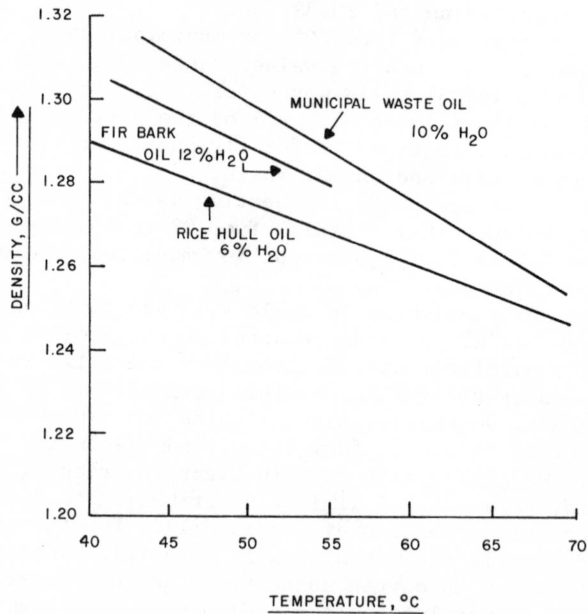

Figure 5. Fluid density vs. temperature

oil is miscible with limited amounts of water, only 35% is water soluble.

Pyrolysis oils from cellulosic feedstocks have a pugent order. Their density is 30 to 50% greater than fuel oils. Their density of municipal waste oil, fir bark oil and rice hull oil versus temperature are shown in Figure 5. It can be seen that rice hull oil is inherently less dense than either fir bark oil or municipal waste oil. The dependence of density on moisture content was determined at room temperature for fir bark oil and municipal waste oil and is shown in Figure 6.

Their viscosity is somewhat higher than most No. 6 fuel oils. The viscosity of municipal solid waste oil is shown in Figure 7. A family of curves at five different moisture content values from 8-17% H_2O is plotted as a function of reciprocal absolute temperature. Viscosity is also a function of thermal degradation since reheating the oil for extended periods above 180°F causes

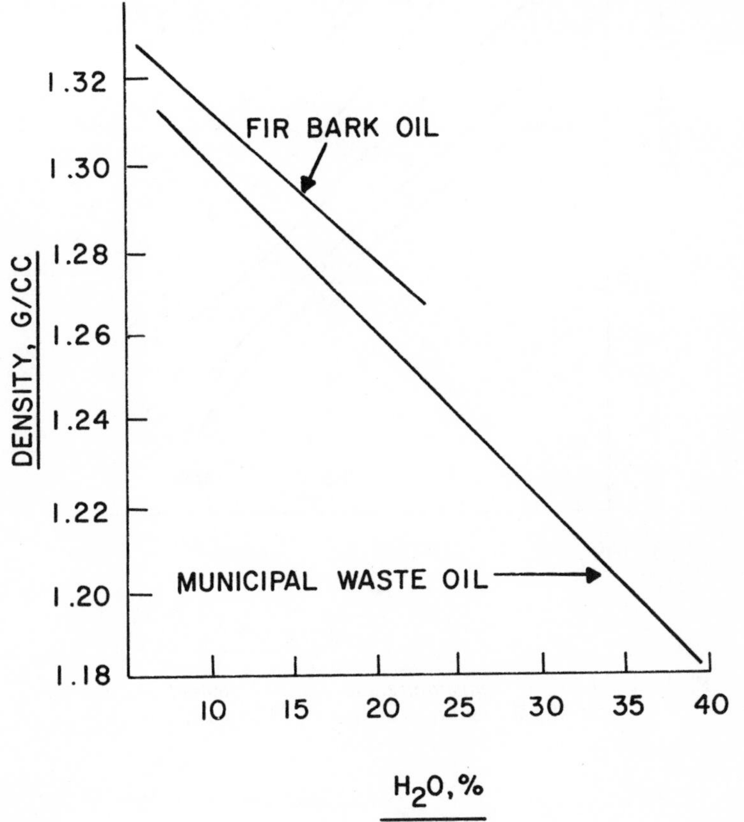

Figure 6. Fluid density vs. moisture (27°C)

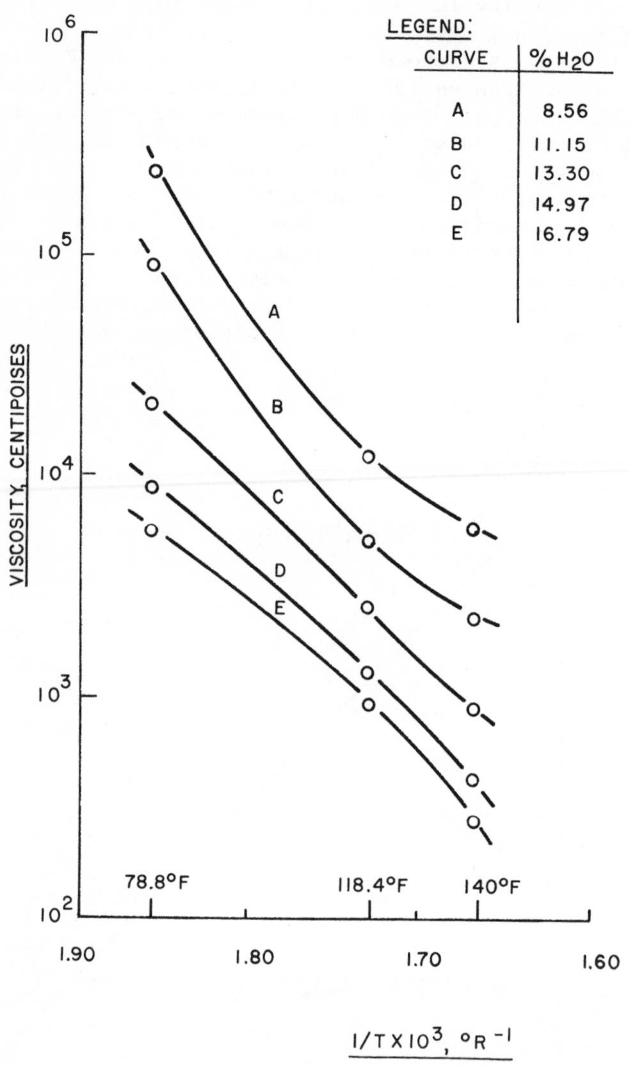

Figure 7. Effect of temperature and moisture content on viscosity of M.S.W. pyrolytic oil (run 61-73, drum 473)

some polymerization. In general, the fluids are very non-Newtonian, showing a hysteresis especially at lower temperatures and a shear dependent viscosity.

The acidity of pyrolytic fluids is due primarily to carboxylic acids. Corrosion of pretreated 304 stainless steel at pyrolytic oil pumping temperatures, 130-140°F, has been measured at 0.05 to 0.10 mils per year. Corrosion rates are substantially reduced by emulsificalation of municipal solid waste oil with No. 6 fuel oil.

Summary

A variety of feedstocks including bituminous and subbituminous coal and industrial and municipal solid waste have been pyrolyzed to give char, liquid fuels, and gas. The char may be used for process heat or pulverized boiler fuel after desulfurization if required. The liquids are a potential replacement for fuel oil as well as sources of chemicals or gas with further cracking. Gases are variable in Btu value and may be carrier gas, a source of additional process heat, low Btu gas or high value gas convertible to pipeline quality. The process has a short residence time and therefore high conversion rate of feedstock. Pressures are near ambient with no external chemical costs involved. Flash pyrolysis offers a flexible and economical means of converting large quantities of low cost solid materials into a solution for our expanding energy requirements.

28

Effect of Heating Rate on the Pyrolysis of Oil Shale

CHARLES ARNOLD, JR.

Polymer Science and Engineering, Div. 5811, Sandia Laboratories, Albuquerque, N.M. 87115

Oil shale is a finely-textured, laminated marlstone that contains a waxy material called kerogen. Kerogen is insoluble in organic solvents but at elevated temperatures decomposes to yield an oil which is suitable for use as a refinery feedstock. The two principle routes used for the recovery of oil from shale are 1) above ground retorting of the mined shale, and 2) in-situ retorting. To recover oil from oil shale by the in-situ technique excavation and fracturing are required to create a sufficient distribution of porosity for the injection of retorting gases. Heat for retorting is then provided by partial combustion of the oil shale with air. Work on the development of an in-situ process for recovery of oil from shale has been in progress for several years at the Laramie Energy Research Center (Bureau of Mines) and by several oil companies. Some of the potential advantages of in-situ processing are 1) reduced costs attributable to elimination of the mining and waste disposal requirements, and 3) amenability to recovery of oil from deep beds that are inaccessible to conventional mining methods. The main obstacles to development of a viable in-situ process are 1) suitable fracturing techniques have not yet been developed, and 2) scarcity of data on the most efficient way of retorting of oil shale.

In-situ retorting of oil shale involves simultaneous movement of a combustion and pyrolysis zone through the bed which results in dynamic retorting conditions and for this reason, the effect of heating rate on the pyrolysis of oil shale should be studied. Although the kinetics of oil shale pyrolysis has been extensively investigated, (1-6) no data are available on the effect of heating rate on the pyrolysis of oil shale. Work, however, is under way at the Bureau of Mines to study the effect of heating rate on in-situ oil shale retorting using a controlled-state retort.(7) This work is being carried out on a relatively large scale using a 13 foot long retort and crushed shale with a particle size range of 1/8-1/2 inches. The study initiated here at Sandia to assess the effect of heating rate on the decomposition of oil shale was done on a much smaller scale with finely divided samples of oil shale using the techniques of dynamic thermogravimetry and thermal chromatography. Specifically, data will be obtained on the effect of heating rate on oil yield, boiling range of the oil and char yield. In addition, two of the kinetic parameters, namely the activation energy (E_a) and reaction order (n), will be determined as a function of conversion level.

Experimental

Materials. The origin of the three shale samples used in this study was the Piceance Creek Basin in Rio Blanco County, Colorado. These shales, which were obtained through the courtesy of the Bureau of Mines at Laramie, had Fischer Assays of 10.1, 25.6 and 49.4 gallons per ton (gpt.) and were pulverized to a size range of 2-150 microns. Thermogravimetric analysis (TGA) and thermal chromatography experiments were carried out on these samples as received.

Thermogravimetric Analysis. The TGA runs were carried out with a Perkin Elmer Thermogravimetric Analyzer (Model TGS-I). Temperature control was achieved by use of a Perkin Elmer Programmer (Model UU-1) and the readout (temperature and weight loss) was recorded on a Servo/Riter II two pen recorder. In a typical run, 7 mg of the sample was weighed in the platinum boat of the thermobalance and the system flushed out thoroughly with helium. After adjustment of the helium flow rate to 28 cm^3/min, the temperature was programmed from ambient temperature to about 700°C at the following rates; 0.625, 20, 80 and 160°C/min. The removal of trace quantities of oxygen and water from the helium carrier gas was accomplished by installation of Oxisorb (Supelco Co.) and molecular sieves into the carrier gas line.

The total weight of organic matter present in the oil shale, i.e., kerogen plus bitumen, and the residual char weight for pyrolysis were needed to define the extent of reaction and to calculate the order of the reaction. The total weight of organic material corresponds to the total weight loss brought about by combustion of the kerogen and bitumen at conditions under which there would be no weight loss due to the decomposition of the inorganic carbonates present. It was found by thermal chromatography (see below) that the rate of decomposition of inorganic carbonates was negligibly slow in the heating rate range of 4-40°C/min up to about 600°C. In practice, the shale was pyrolyzed in helium to 480°C at a programmed heating rate of 20°C/min; then air was introduced and the remaining organic matter and char were burned off isothermally at 480°C. The residual char weight for each shale sample was determined by pyrolyzing the shale in helium at a programmed temperature rise rate of 20°C/min to 480°C and then holding at this temperature until no further weight loss was observed.

The char yield was found to vary with heating rate, thus the determination of the conversion level was somewhat arbitrary. The reaction rates, $1/w_o\ dw/dt$, were determined by graphically constructing tangents to the TGA curves. Here w_o is the original sample weight, w is the remaining weight of the sample at any time and t is time. The kinetic constants were obtained by application of linear regression analysis to the thermogravimetric curves. This method of dynamic thermogravimetry avoids the major disadvantage encountered with isothermal kinetics, namely that decomposition of the sample often occurs as the temperature is being raised to the desired temperature. Kinetic parameters derived from multiple heating rate TGA experiments are generally considered more reliable than those determined from a single TGA trace; the latter method involves the determination of degradation rates from steep, rapidly changing slopes.(8)

Thermal Chromatography. The effect of heating rate on oil yields, composition and rate of formation was carried out with a commercial thermal chromatograph (Model MP-3, Chromalytics, Inc.). This instrument contains the following elements: 1) a temperature programmable pyrolyzer, 2) an evolved gas analyzer with both flame ionization (F.I.D) and thermal conductivity (T.C.) detectors, 3) a product collection trap, and 4) a gas chromatograph. In applying the technique of thermal chromatography to pyrolysis of oil shale, 25 mg of the shale was heated in a flowing helium stream at four different heating rates (4, 20, 28 and 40°C/min) from 50°C to ~625°C and the evolved pyrolysis products were monitored with both the F.I.D. and T.C. detectors. Only the organic gases and liquid are recorded by the F.I.D. detector. From the evolved gas analysis (EGA) record of the F.I.D. detector, data on the relative yield and rates of formation of oil plus organic gases can be obtained. The magnitude of the signal response is proportional to the rate of formation, and the area under the curve, after normalization with respect to heating rate, is proportional to the yield. The pyrolysis products are collected downstream from the T.C. detector in a trap consisting of a short column of chromatographic materials (Poropak Q and SE 30 on chromosorb). Upon heating the trap to 250°C in a stream of helium, the products are backflushed into a gas chromatograph (G.C.). Gas chromatograph simulated distillation (9) was used to determine the boiling range of the products.

Kinetic Analysis. The activation energy (E_a) and reaction order (n) for the pyrolysis of oil shale were derived from the dynamic thermogravimetric data using the differential kinetic analysis of Friedman.(10) In analogy to homogeneous reaction kinetics, the following assumptions were made in this kinetic treatment (11): 1) $dw/dt = f(w)\ K(T)$, where w = fractional remaining weight, t = time, and T = absolute temperature; 2) $f(w) = w^n$, where n = the reaction order; 3) $k(T) = Ae^{-Ea/RT}$, i.e., the Arrhenius relationship is assumed; and 4) f(w) is independent of k(T). The particle size of the oil shale samples used in this study was small (<150 microns); implicit in the analogy to homogeneous kinetics was the assumption that the particle size distribution below 150 microns did not affect the kinetics.

Results and Discussion

TGA Studies. The effect of heating rates on the TGA curves for the pyrolysis of 25.6 gpt oil shale are shown in Figure 1. Similar curves were obtained for the 10.1 and 49.4 gpt shales. The normal displacement of the TGA curves in the direction of higher temperatures with increased heating rate was observed. Because the TGA curves did not flatten out to a constant fractional weight but continued to lose weight at a steady but slow rate, determination of the completion of the reaction was difficult to assess by TGA alone. It was established by evolved gas analysis, using the F.I.D. detector of the thermal chromatograph, that the pyrolysis reaction was >98% complete at 570°C at the 20°C/min heating rate. It is also considered noteworthy that at the lowest heating rate (0.625°C/min) there was no residual char. The possibility that oxidative or water gas type reactions occurred is unlikely since Oxisorb and molecular sieves had been installed in the carrier gas line to remove trace quantities of oxygen and water. It is postulated that at slow heating with small particles metastable semi-chars are formed which continue to decompose forming volatile products. Volatilization of semicokes to form high molecular weight oils has been proposed by Snyder, et. al. (12)

The differential thermogravimetric (DTG) curves obtained for the 25.6 gpt shale are shown in Figure 2. Similar curves, not shown, were obtained for the 10.1 and 49.4 gpt shales. The absence of clearly defined multiple-maxima in these DTG plots suggests either that decomposition is occurring in a single step or that a number of competitive reactions are occurring simultaneously. Since it is known that the pyrolysis of oil shale is a rather complex process, the interpretation involving competitive reactions is most likely to be the correct one. The rate of volatilization increased with increasing heating rate as would be expected on the basis of kinetic theory, (13) providing that the rate of char formation does not increase dramatically at the higher heating rate. Although the amount of char produced was affected by the heating rate, no dramatic increase in the rate of char formation was observed.

In applying the kinetic analysis of Friedman to the TGA data it was found that the activation energy increased with increasing conversion (Figure 3). Thus, for the 25.6 gpt shale, the activation energy increased from 28 kcal/mole to 65 kcal/mole over the conversion range of 15-92%. The average activation energy for this shale was 46 kcal/mole. Although similar results were obtained with the 94.4 gpt shale (E_a range 28-60 kcal/mole, average

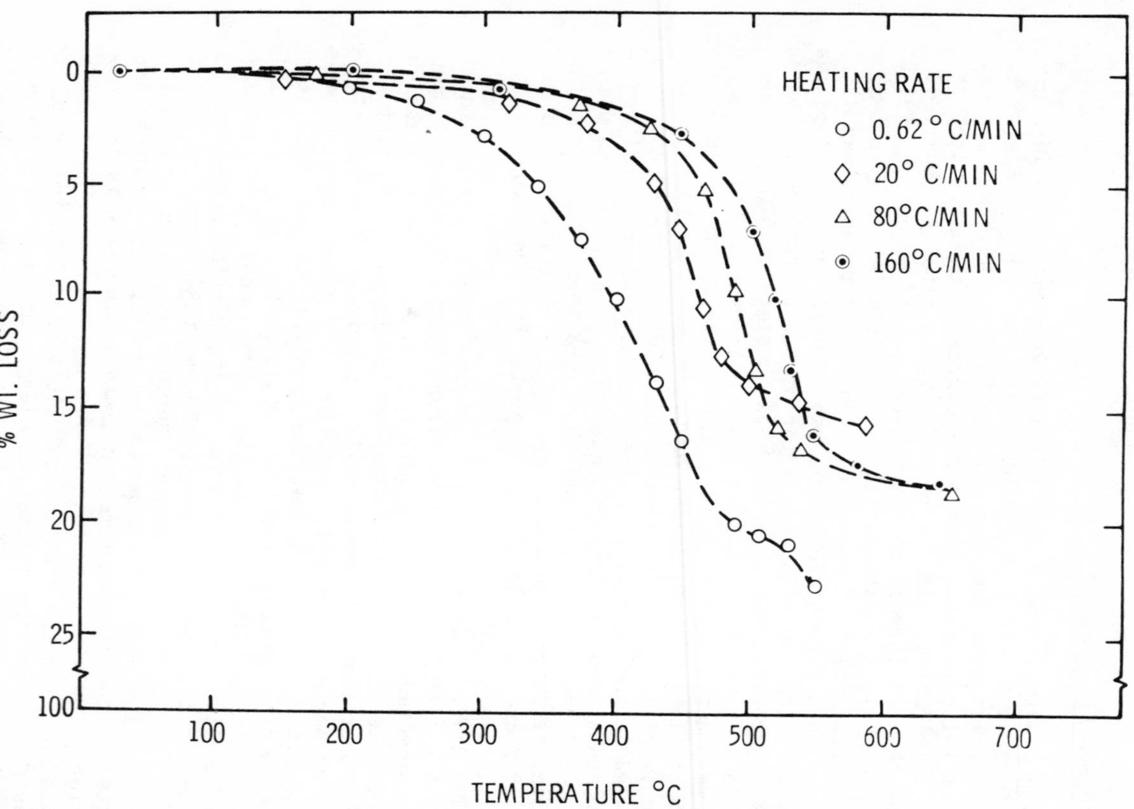

Figure 1. TGA plots of the decomposition of 25.6 gpt oil shale at different heating rates

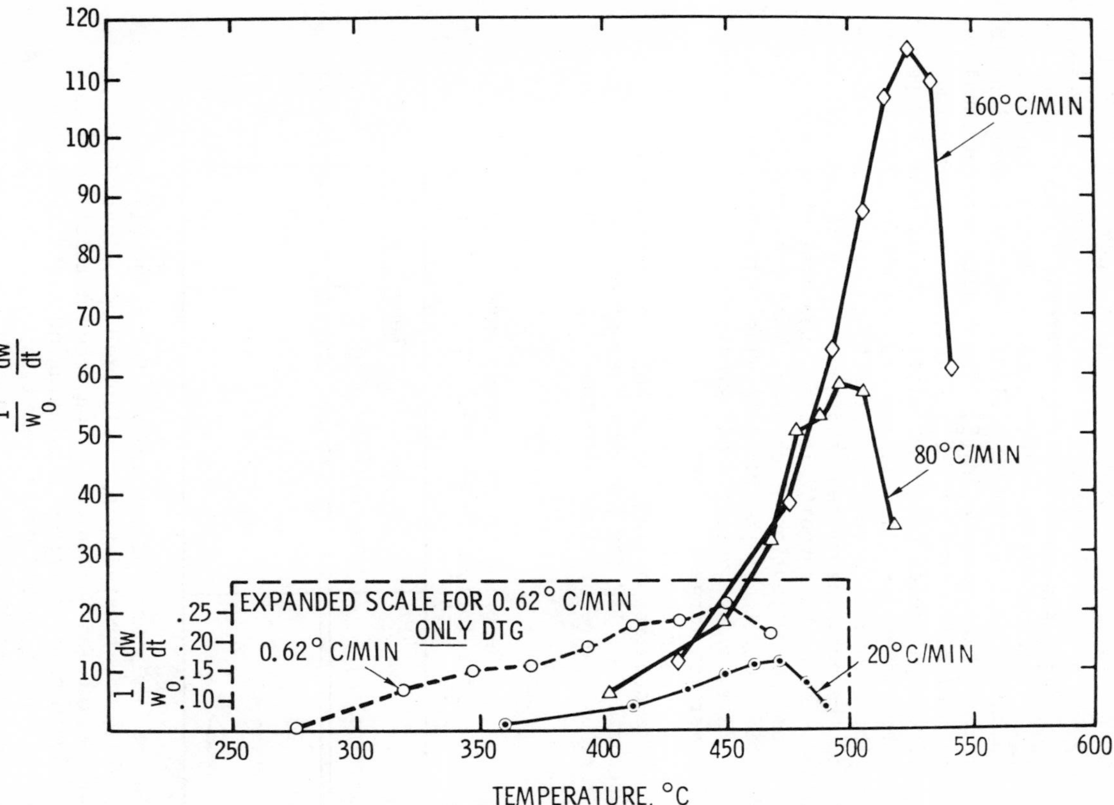

Figure 2. DTG plots of the decomposition of 25.6 gpt oil shale at different heating rates

E_a of 45 kcal/mole), the range of activation energies for the 10.1 gpt shale was narrower (25-49 kcal/mole) and the average activation energy was less (~36 kcal/mole). Activation energies that have been reported in the literature vary from 19 kcal/mole to 62 kcal/mole over rather wide conversion ranges.(1,6) Most of the previously published activation energies, however, were in the range of 40-48 kcal/mole without reference to conversion level.(2,3) Weitkamp and Guberlet found, as we did, that the activation energy varied from 22 kcal/mole at low conversions to 60 kcal/mole at high conversion.(5) These data were rationalized on the basis of diffusional effects; i.e., at low conversions the individual particles of shale, albeit, small, have low porosity and the process is diffusion controlled whereas at high conversions the particles have high porosity caused by the volatilization of the organics present and the process is not diffusion controlled. When the porosity is high, the activation energy obtained becomes a reflection of chemical bond breaking processes.

Kinetic analysis of the TGA data also indicated that the reaction order, n, was variable and greater than unity over the entire conversion range as shown by the data in Table I. Thus, the reaction order for the 25.6 gpt shale decreased from 5.6 to 1.2 over the conversion range up to 46% then remained essentially constant at 1.2 for the remainder of the decomposition. Similar results were obtained with the 10.1 and 49.4 gpt shales.

TABLE I

Reaction Orders Versus Conversion for 10.1, 25.6 and 49.4 gpt Oil Shales

	Reaction Order, n		
Conversion, %	Shale Assay → 10.1	25.6	49.4
20	5.6	5.6	4.7
30	1.6	2.6	2.9
~46-92	1.2	1.2	1.2

In the early work of Hubbard and Robinson, (1) the pyrolysis of oil shale was treated as a first order reaction. Recently, Faucett, George and Carpenter (4) proposed a model based on a thermal decomposition system consisting of two first order reactions and three second order reactions which provide a better fit to the Hubbard and Robinson data. Our results tend to confirm the existence of higher order effects and are generally in accord with the model proposed by Faucett, et. al. It is considered noteworthy that the highest reaction orders that we observed were associated with decomposition at the lower conversion levels where the activation energies were found to be low.

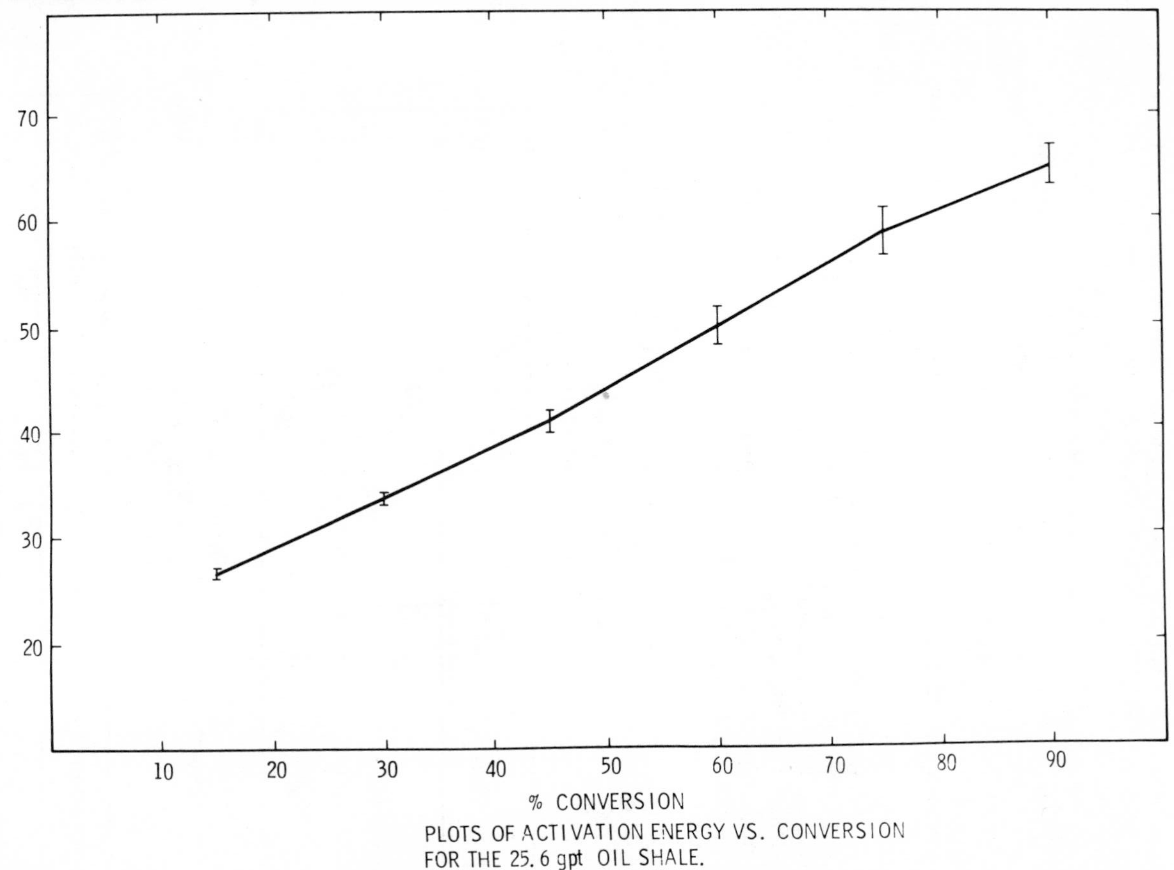

Figure 3. Plots of activation energy vs. conversion for the 25.6 gpt oil shale

The opposing trends of these two parameters suggest the possibility that autocatalysis is occurring during the early stages of the pyrolysis. Autocatalytic effects of bitumen on the decomposition of kerogen were previously proposed by Allred.(2) Although weight dependence orders greater than one are somewhat unusual for the decomposition of an organic polymer, a fifth order dependence was recently reported for a glass-supported phenolic.(10) Composite materials of this type resemble oil shale in that both consist of an organic matrix bound to an inorganic substrate.

Thermal Chromatography Studies. The effect of heating rate on the relative yield of hydrocarbons (oil and gas) produced as measured by the EGA module of thermal chromatograph using a F.I.D. detector is shown in Table II. For this range of heating rates it is evident that increasing the heating rate effected a decrease in yield of oil plus organic gas from oil shale. One possible explanation for these results is that at the more rapid heating rate, higher temperatures are attained in a shorter time span, and as a consequence, polymerization and char forming reactions are favored leading to lower oil yields.

TABLE II

Effect of Heating Rate on Yield of Oil and Organic Gas

Heating Rate (°C/min)	Relative Areas Under Thermogram Curves (Arbitrary Units)
4	6939 ± 341
20	6388 ± 2.6
28	6123 ± 173
40	5378 ± 158

TABLE III

Effect of Heating Rate on Boiling Range of Shale Oil

Boiling Range (°C) at 760 mm Hg	Percent Hydrocarbon in Boiling Range of 75-400°C	
	4°C/min	40°C/min
75-200	23	19
200-300	30	29
300-400	47	52

These results are consistent with the G.C. simulated boiling range data shown in Table III which indicate that a higher percentage of the higher boiling fraction was obtained at the higher heating rates.

Mechanism of Pyrolysis. Allred proposed the following mechanism for the pyrolysis of oil shale.(2)

$$\text{Kerogen} \xrightarrow{k_1} \begin{matrix}\text{Gas}\\ \text{Bitumen}\\ \text{Carbon Residue}\end{matrix} \xrightarrow{k_2} \text{Oil and Gas}$$

Allred also was the first to point out that the concentration of bitumen reached a maximum at the point where the decomposition of kerogen was essentially complete. At this stage of the pyrolysis about one half of the total oil and gas remained to be produced. Therefore, at conversion levels >50%, the rate of volatilization was attributable to the rate of decomposition of bitumen only. Furthermore, from the shape of the curves showing the rate of disappearance of kerogen, rates of appearance of oil, gas and char and rate of appearance and disappearance of bitumen, Allred concluded that in the early stages of the pyrolysis, the decomposition was more complex than the two consecutive first order schemes proposed by Hubbard and Robinson (1) and that an autocatalytic mechanism was probably involved. Our observation of low activation energy coupled with high reaction order in the early stages of the pyrolysis of oil shale tend to support this hypothesis.

Summary and Conclusions

The effect of increasing the heating rate from 4°C/min to 40°C/min on the pyrolysis of oil shale was to decrease the yield of oil and gaseous organic products by 23%. The higher heating rates also tended to favor the formation of higher boiling components in the shale oil obtained.

Under the conditions which prevailed in these small scale runs, namely small particle size and flowing streams of helium, it is postulated that metastable residues were formed which underwent complete volatilization (at slow heating rates). The extent to which these metastable residues were formed was a function of the heating rate. Maximal amounts of the residue were favored by the intermediate heating rate of 20°C/min.

The activation energies for all three shales studied increased by a factor of 2.0 to 2.5 over the entire temperature and conversion range. High reaction orders were observed in the early stages of the decomposition when kerogen was breaking down at the same time that bitumen and oil plus gas were being formed. The reaction order decreased with conversion to a value somewhat greater than unity then remained constant during the second half of the pyrolysis. These opposing trends of increasing activation energy and decreasing reaction order during the early stages of the the pyrolysis tend to confirm the hypothesis that autocatalytic effects were predominant at this stage of the decomposition.

Acknowledgment

The author wishes to express his gratitude to H. B. Jensen of the Bureau of Mines, Laramie, who supplied the oil shale samples and characterization data. The author also wishes to acknowledge L. K. Borgman for her experimental contributions to this work.

Abstract

Studies were made on the effect of heating rate on the yield of oil plus organic gases and the char yield during the pyrolysis of lean, medium and rich oil shale samples whose origin was the Piceance Creek Basin in Colorado. Increasing the heating rate in the range of 4 to 40°C/min effected a decrease in the yield of oil and gaseous organic products by 23% and promoted the formation of higher boiling constituents in the shale oil produced. In addition, activation energies and reaction orders for these shales were determined as a function of conversion using the kinetic analysis of Friedman. The results obtained tend to confirm the hypothesis that the early stages of the decomposition involve autocatalysis.

Literature Cited

1. Hubbard, A. B. and Robinson, W. E., "A Thermal Decomposition Study of Colorado Oil Shale," U.S. Bureau of Mines Rep. of Inv. 4744 (1950).
2. Allred, V. D., Chem. Eng. Prog., (1966) 62, 55.
3. Hill, G. R. and Dougan, P., Quart. Colo. School Mines (1967) 62, 75.
4. Faucett, D. W., George, J. H. and Carpenter, H. C., "Second Order Effects in the Kinetics of Oil Shale Pyrolysis," Bureau of Mines Rep. of Inv. Rl 7889 (1974).
5. Weitkamp, A. W. and Gutberlet, L. C., Ind. Eng. Chem., Process Design and Development, (1970) 9, 386.
6. Cummins, J. J. and Robinson, W. E., "Thermal Degradation of Green River Kerogen at 150 to 350°C. Rate of Product Formation," U.S. Bureau of Mines Rep. of Inv. 7620 (1972).
7. Jensen, H. B., unpublished work.
8. Anderson, H. C., Chapter 3 in Techniques and Methods of Polymer Evaluation, Eds. P. E. Slade and Lloyd T. Jenkins (Marcel Dekker, New York, 1966).

9. ASTM-D-2887-72T, "Tentative Method of Test for Boiling Range Distribution of Petroleum Fractions by Gas Chromatography" (1967).
10. Friedman, H. L., J. Polym. Sci., (1967) C6, 183.
11. Flynn, J. H. and Wall, L. A., J. Res. N.B.S., (1966) 70A 487.
12. Snyder, P. W., Jr., Timmins, T. H. and Johnson, W. F., "An Evaluation of In-Situ Oil Shale Retorting," paper presented at a Workshop on In-Situ Oil Shale Retorting sponsored by the National Science Foundation, University of California, San Diego, September 3-7, 1974.
13. Reich, L. and Stivala, S., Elements of Polymer Degradation, p. 101 (McGraw Hill Book Co., New York, 1971).

INDEX

G

H

I

K

L

M

N

O

P

Q

R

S